Laser Spectroscopy

Laser Spectroscopy

Edited by
Richard G. Brewer
IBM Research Laboratory
San Jose, California

and
Aram Mooradian
Lincoln Laboratory
Massachusetts Institute of Technology
Lexington, Massachusetts

PLENUM PRESS · NEW YORK AND LONDON

Library of Congress Cataloging in Publication Data

International Conference on Laser Spectroscopy, Vail, Colo., 1973.
  Laser spectroscopy; [proceedings]

  Sponsored by International Business Machines Corporation and others.
  Includes bibliographical references.
    1. Laser spectroscopy—Congresses. I. Brewer, Richard G., ed. II. Mooradian, Aram, ed.
III. International Business Machines Corporation. IV. Title.
QC454.L3157   1973   543'.085   74-12090
  ISBN 0-306-30802-9

*Proceedings of an International Conference on Laser Spectroscopy
held in Vail, Colorado, June 25-29, 1973*

Sponsors
International Business Machines Corporation
Massachusetts Institute of Technology
Air Force Office of Scientific Research
National Science Foundation
Spectra-Physics, Incorporated
Coherent Radiation, Incorporated
Chromatix, Incorporated

© 1974 Plenum Press, New York
A Division of Plenum Publishing Corporation
227 West 17th Street, New York, N.Y. 10011

United Kingdom edition published by Plenum Press, London
A Division of Plenum Publishing Company, Ltd.
4a Lower John Street, London, W1R 3PD, England

Printed in the United States of America

Preface

The Laser Spectroscopy Conference held at Vail, Colorado, June 25-29, 1973 was in certain ways the first meeting of its kind. Various quantum electronics conferences in the past have covered nonlinear optics, coherence theory, lasers and masers, breakdown, light scattering and so on. However, at Vail only two major themes were developed - tunable laser sources and the use of lasers in spectroscopic measurements, especially those involving high precision. Even so, Laser Spectroscopy covers a broad range of topics, making possible entirely new investigations and in older ones orders of magnitude improvement in resolution.

The conference was interdisciplinary and international in character with scientists representing Japan, Italy, West Germany, Canada, Israel, France, England, and the United States. Of the 150 participants, the majority were physicists and electrical engineers in quantum electronics and the remainder, physical chemists and astrophysicists. We regret, because of space limitations, about 100 requests to attend had to be refused.

Vail was selected because of its alpine location at 8000 feet elevation. Its atmosphere is free of distractions and conducive to relaxation and to an interchange of ideas. Mid to late June is particularly pleasant; the rivers are full, the higher ridges are still covered with snow, and the summer tourists are few. A Sunday reception in the early evening with cocktails and hors d'oeuvres at a pool side got the conference underway. Tuesday provided a gondola ride up to Mid-Vail at 10,000 feet, breaking the routine with a magnificent panoramic view and a cookout lunch on the terrace. As the program neared its conclusion, a banquet dinner on Thursday featured Edward Teller speaking on the important subject of "Lasers and Isotopes".

Financial support from our sponsors was generous and timely for which we express our gratitude and appreciation in helping to

further the new field of Laser Spectroscopy. Our congratulations
go to K. Nill of the MIT Lincoln Laboratory for his expert handling
of the local arrangements and to the staff at the Vail Lodge for
their cooperation and excellent choice of California wine.

Richard G. Brewer
Aram Mooradian

Contents

ASTROPHYSICS AND GEOPHYSICS

CHAIRMAN'S OPENING REMARKS

B. P. Stoicheff

University of Toronto

In keeping with our initial intoxication at the high altitude of Vail, and with the enjoyable reception last evening, this morning our opening scientific session features four illustrious spectroscopists: Herzberg, Javan, Shimoda and Cohen-Tannoudji. This is an international cast with each member having made important contributions to spectroscopy, and in particular, high resolution spectroscopy, even before the laser made its appearance and certainly since that time.

It seems appropriate in introducing our first speaker, Gerhard Herzberg, that I say a few words about his work which has spanned 46 very active years, and a large part of the optical spectrum, from the

Fig. 1.   Balmer series of H from H_α to the ionization limit.

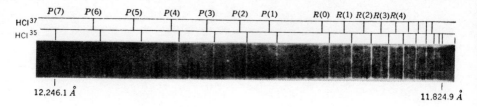

Fig. 2. Rotational structure in the 3-0 band of HCl at 1.18 μ.

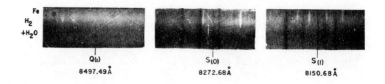

Fig. 3. Several lines of the 3-0 band in the quadrupole spectrum of H_2.

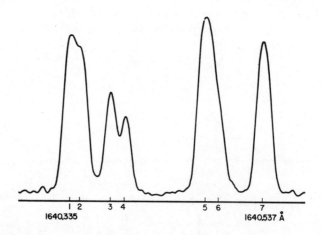

Fig. 4. Fine structure of the H_e^+ line at 1640 Å showing the Lamb shift. The line pairs 1, 2 and 3, 4 would be single lines if there were no Lamb shift.

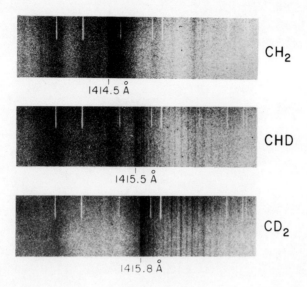

1414.5 Å

1415.5 Å

1415.8 Å

Fig. 5.   The 1415 Å band of CH_2, CHD and CD_2 under high resolution.

infrared through the visible region to the vacuum ultraviolet, in fact, the spectral region which we shall be discussing here with reference to the recent advances in laser spectroscopy.

His first work was on the hydrogen Balmer series (Fig. 1) from H_α to the ionization limit and the continuum beyond. Molecular spectroscopy became an early interest and included electronic spectra in the visible and ultraviolet regions, and when suitable photographic plates became available, he began work in vibrational spectra in the photographic infrared (Fig. 2). His interests in astrophysics led to his suggestion that molecular hydrogen in planetary atmospheres and in the interstellar medium might be detected through its quadrupole spectrum, and he was the first to observe this spectrum in the laboratory using a long-path cell of H_2 (Fig. 3). He pushed his research to the vacuum U. V., and amongst many other investigations, he observed the H_e^+ line at 1640 Å (the analogue of the H_α line) at resolution sufficient to resolve the structure due to the Lamb shift, as shown, in Fig. 4. Finally, I want to remind you of his work on free radicals by showing you the spectrum of CH_2 and its deuterated isotopes in the region of 1400 Å (Fig. 5).

It is a pleasure to call on Dr. Herzberg for his "Introductory Remarks. "

SPECTROSCOPY I

INTRODUCTORY REMARKS

G. Herzberg

National Research Coucil

Canada

I am greatly honoured by the invitation to present an introductory paper to this conference on Laser Spectroscopy.

The last fifty years have seen a number of extremely important, almost revolutionary, developments in spectroscopy.

(1) The discovery of the Raman effect in 1928 came at a time when the study of infrared spectra was still extremely difficult. When it was recognized that in some way the Raman spectrum corresponds to an infrared spectrum shifted into the conveniently accessible visible and ultraviolet region the optimists thought that the determination of vibrational frequencies of molecules would no longer require the difficult study of infrared spectra. Theoretical developments soon showed that the selection rules for the Raman spectrum are different from those for the infrared spectrum so that infrared and Raman spectra supplement one another rather than replace one another in the study of molecular vibrations, particularly in symmetric molecules. Yet, until about ten years ago books were published that dealt either exclusively with Raman spectra or exclusively with infrared spectra. The authors of these books through their intensive preoccupation with one of these two fields had apparently lost sight of the fact that both methods were part of one and the same field from different angles, namely, the field of the study of molecular structure. At least in retrospect, it appears that they were missing something important by leaving out of consideration one of the two methods of studying molecular vibrations.

(2) A similar situation arose in the development of microwave spectroscopy. As most of you know, except for the early work of Cleeton and Williams on NH_3, microwave spectroscopy started rather suddenly immediately after the last war through the work of Bleaney and Penrose,

Wilson, Gordy, Townes and many others. A microwave spectrum is, of course, nothing but a far infrared spectrum, that is, in most cases a rotation spectrum or, in a few cases, an inversion spectrum or Λ-type or K-type doubling spectrum. While far infrared spectra at the time were extremely difficult to investigate, the microwave method, once developed, proved to be very simple and had, compared to the infrared spectra, an almost unbelievable resolution, so that the rotational constants of many molecules whose spectra could not be resolved in the near or far infrared could be determined with very high precision.

Again, some authors became so exclusively attached to microwave spectroscopy that they had a tendency to ignore the results on molecular rotations obtained by other more conventional methods. Books on microwave spectroscopy were written in some of which infrared spectra were completely ignored. In some of the early papers on microwave spectra, molecular constants were derived on the basis of single microwave lines. In a few instances the constants so determined turned out to be wrong because of faulty calibration of the single line used. These mistakes could have been avoided if the particular authors had known, or taken more seriously the results from more conventional spectroscopic determinations.

Not very long ago an extensive and excellent table of rotational constants and geometrical molecular data was published which was based entirely on microwave spectra. If you want to know the rotational constants and geometrical parameters of CH_4, C_2H_6, C_2H_4, C_2H_2 and many other molecules with zero dipole moment you cannot find them in these tables even though these constants are well known from near infrared investigations.

(3) The invention of the photo- and electron multipliers and other detectors (PbS and similar cells), combined with an incredible improvement in amplifiers, has led to another revolution in spectroscopy. By means of these detectors it is now possible to record directly and automatically infrared, visible and ultraviolet laboratory spectra as well as astronomical spectra, while in previous years such recordings were almost always based on the recording of the blackening of photographic plates.

The new detectors have also opened up entirely new fields such as the study of electron scattering, of electron-loss spectra, photoelectron spectra and similar subjects. While all of these new fields are closely related to conventional spectroscopy they have certainly opened up new vistas and advanced the subject immeasurably.

Many of those who have used these new and modern methods sometimes seem to forget, in their enthusiasm, that the photographic plate is still a very useful method of recording spectra and has certain qualities that the modern methods do not have, at least not without undue complication.

(4) A somewhat similar situation is apt to arise in the new and revolutionary subject of laser spectroscopy to which this meeting is devoted. I take it, however, from the fact that I was invited to address this meeting (since I have done no laser work whatsoever) that the organizers were aware of the dangers of over-enthusiasm and of the merits of more orthodox and conventional methods of spectroscopy. Naturally as a spectroscopist I cannot help but be impressed by the extraordinary resolution that can be achieved by laser spectroscopy, but I am also very much aware of the many fields of spectroscopy where such high resolution is premature and where first a study of conventional methods is needed.

The four revolutionary developments which I have mentioned (and others which I have not mentioned) have, of course, meant an enormous expansion of the field of spectroscopy. It is no longer possible for one and the same person to know even approximately what is going on in all the various areas of spectroscopy. As a consequence, it has happened more than once that certain developments in one field are unknown to authors working in another field. In some instances this situation has led to minor anomalies in the notation and the vocabulary used for the description of various spectroscopic phenomena. As an example I would like to refer to the type of phenomenon first discovered by Auger in the X-ray field and later discovered in many other parts of physics under different names. It is, in my opinion a very useful and revealing insight to realize that the predissociation of molecules, the autoionization of atoms and molecules, the Breit-Wigner resonances in nuclear physics, the resonances in elementary particle physics, are all phenomena of the same type which might be described by the general name Auger process, that is, a process in which a continuous range of energy levels interacts with one or more discrete energy levels. I find it unfortunate that this generality of the phenomenon is somewhat lost in the presentation of the various fields of physics. Even within spectroscopy, soon after the discovery of the phenomenon of predissociation, a similar phenomenon in atomic spectroscopy corresponding to ionization was discovered but it was not called preionization but autoionization. I find it equally unfortunate that the inversion of this autoionization is now often referred to as dielectronic recombination when inverse preionization, or inverse autoionization, or inverse Auger process would be a much more easily understandable term. This is particularly so since there are many instances where the so-called dielectronic recombination involves more than two electrons but, of course, only one that actually recombines with the ion under consideration.

The reason why I am making these comments is solely to emphasize the need for those who work enthusiastically in a new field to reflect once in a while on the unity of physics (or science in general) and make sure that the new names and notations that they introduce are not in conflict with established notation in another field.

EXTENSION OF MICROWAVE DETECTION AND FREQUENCY MEASURING TECHNIQUES INTO THE OPTICAL REGION

A. Javan and A. Sanchez[*]

Department of Physics, Massachusetts Institute of

Technology, Cambridge, Massachusetts

ABSTRACT

Precise frequency measurements in the far-infrared and infrared have made possible highly accurate spectroscopic observations in these regions. This article gives a review of the general properties of the high-speed diode element used in the frequency measurements. Other future applications are discussed in the concluding section.

INTRODUCTION

Conventional spectroscopy in the optical and infrared region is based on precision wavelength measurements which utilize a variety of dispersive methods to compare different wavelengths and accurately determine their ratios. In the microwave region, on the other hand, precision measurements are done by measuring frequencies. In these, two different microwave frequencies, or a microwave and a radio frequency, are compared via frequency mixing in a nonlinear element.

The practical limits in the accuracies of the optical wavelengths measuring methods are set by the dispersive instruments, whose resolving widths are generally broad and considerably exceed the frequency spread of a stable monochromatic laser oscillator. In contrast, the ultimate accuracy of the microwave frequency measuring methods are determined by the fundamental limitation aris-

* Present address, Air Force Cambridge Research Laboratory
   L. G. Hanscom Field, Bedford, Massachusetts

ing from the purity and the frequency stability of the microwave
radiation. Because of this, the relative accuracy obtainable in
a precision microwave frequency measurement is in general appreciably higher than the relative accuracy of a precise wavelength
measurement in the optical region.

The existing wavelength measuring methods have been more than
adequate in a wide range of spectroscopic observations where the
widths of the optical spectral lines are broad and appreciably
exceed the resolving widths of the high-resolution wavelength measuring devices. In these cases, accurate molecular or atomic parameters are obtained by means of observations in which a large number of transitions belonging to the same atomic or molecular species
are studied over a broad range of the spectrum.

With the availability of stable monochromatic laser radiation,
it has become possible to make high-resolution spectroscopic observations in which optical resonances appear at much reduced
linewidths, considerably below the resolution limits of even the
best optical wavelength measuring instruments.[1] Early observations
followed the advent of gas lasers. These resulted in the discovery
of a number of nonlinear radiative processes in which the interaction of monochromatic radiation with an atomic resonance occurs
selectively over a narrow region of the atomic velocity distribution, resulting in narrow resonances which are completely free
from Doppler broadening.[1] Other approaches, such as the molecular
beam method,[2] have also enabled application of monochromatic laser
radiation to the observation of narrow resonances whose widths are
considerably less than their classical Doppler breadths. Beyond
these, in the far-infrared and the long-wavelength region of the
infrared, the doppler breadth of a molecular transition is inherently sufficiently narrow to allow the frequency of a monochromatic
source to be tuned to the center of the transition with an accuracy
much beyond the capabilities of a high resolution dispersive instrument.[3] The exploitations of these methods in high resolution
spectroscopic observations are subjects of our "laser spectroscopy",
the topic under review in this Conference.

In order to use the potentials of laser spectroscopy in highly
accurate observations, with an ultimate accuracy beyond the capabilities offered by classical wavelength measurements, the frequency
measuring methods previously used in the microwave region, are introduced into the optical regions. This radically different approach
to the art of measurements in optics became possible with the aid
of a new optical element which responds to an infrared frequency
in much the same way as a microwave diode responds to microwave
radiation.[4-9]

In the microwave region, the technology of microwave rectifier-
frequency mixer diodes has been the basis for all types of micro-

wave receivers and spectrometers. With the extension of this technology into the optical region, a host of new possibilities have
now become available. These include a number of possibilities
besides the frequency measuring applications, some of which do not
necessarily have an analogue in the microwave region. (See below)

In September of 1971, a review paper was presented (by A.
Javan) at the Esfahan Symposium on the topic of modern methods in
precision spectroscopy. This review included a summary of the
developments at M.I.T. in nonlinear laser spectroscopy and also
in the area of extending the microwave frequency measuring methods
into the optical regions. Since then, further advances have been
made at M.I.T. in both of these areas. In this paper, however,
rather than reviewing these recent results[10-13] (which are due to appear in several publications elsewhere), an account will be
given of the quantum mechanical mechanism occurring in the high
speed diode element which has enabled extension of the microwave
frequency measuring methods into the infrared region. This will
be done along with a description of the general properties of the
diode element. The concluding section will describe several novel
possibilities and future applications.

GENERAL PROPERTIES OF THE INFRARED METAL-METAL OXIDE-METAL ELECTRON TUNNELING DIODE

In the high speed diode element, alternating electric currents
flowing at infrared frequencies give rise to its general properties
as an infrared frequency mixer and rectifier. In addition to the
application in precise frequency measurements, the diode has been
used to parametrically generate infrared radiation at the mixture[14]
frequencies of two applied fields. The nonlinear current-voltage
characteristics of the element arises from a quantum mechanical
electron tunneling process occurring at its metal-metal oxide-
metal junction.

The diode element consists of a thin metallic wire antenna
mechanically contacted at its pointed tip to a polished metal post.
The thin wire antenna is a few microns in diameter and several
infrared wavelengths long. The antenna serves the purpose of coupling the diode to the radiation fields. The pointed tip of the
wire antenna has a dimension below 1000 Å (as verified by electron
microscope photographs of the tip.) Electron tunneling occurs
across a potential barrier formed by thin layers of oxides existing on the surfaces of the two metals in the small area of the
junction between the pointed tip and the metal base. The size of
this area, along with the thickness and the effective dielectric
constant of the oxide layers, determine the capacitive loading,
C, of the element across its tunneling junction. As is shown
below, in the elements studied the thickness of the barrier poten-

tial is below about 10Å. From this it follows that $C \stackrel{<}{\sim} 10^{-4}$ pico-
fard. Inspection shows that the diode's series resistance, R,
is essentially determined by the thin antenna's radiation resistance. From these general considerations, the RC time constant of
the diode which limits its high-frequency response, can be approximately estimated to be $RC \stackrel{<}{=} 10^{-14}$ sec.

Quantum mechanical electron tunneling across a potential barrier formed by a thin dielectric in between two metals, has been
the subject of extensive studies,[15] with some of the theoretical
considerations dating back to the early days of the quantum mechanics. Until now, the experimental studies in this field have
been mainly confined to the observations made in the dc and the
low frequency regions, in which the I-V characteristics of different types of metal-oxide-metal junctions are explored under varying conditions. These junctions, however, differ in important
ways from ours. The contact areas in the previously studied junctions are macroscopic in dimensions, having generally an area of
about 1 mm x 1 mm. Because of this, these are low speed junctions.
In our elements, the required high speed performance dictates a
microscopic size for the contact area, from which several important consequences follow. Considering that the junction resistance
is inversely proportional to its area, we note that the barrier
thickness in our junctions will have to be thinner than those
studied previously; otherwise, the junction resistance will be ex-
cessively large. (A reduction in the barrier thickness exponentially lowers the electron tunneling impedance, see below.) From
this it follows that at a given voltage drop across the oxide layer,
the corresponding electric field is larger in our element. Moreover, at a given bias field, the current density in our junctions
can become very large. This can be of particular importance,
since we believe that the charges within the dielectric layer
arising from the tunneling electron's wave functions inside the
classically forbidden region of the barrier potential can contribute
cooperatively to the shape of the potential barrier (in a way
similar to the role played by the space charge in a vacuum diode.)
The charge neutrality in this case is satisfied by the image charge
distributions in the two metals, which also cooperatively contribute to the potential barrier. As the junction is biased by an
external field, the resultant current (which arises from an imbal-
ance caused by a change in the wave functions of the oppositely
traveling electrons inside the barrier) will be accompanied by a
change in the overall charge distributions. This will in turn
result in a modification of the potential barrier.

The nonlinearities of the I-V characteristic give rise to
rectification and frequency mixing in the diode. The various orders
of frequency mixings are dictated by the derivatives of the I-V

characteristic curve, as can be readily seen by considering the
Taylor expansion of I versus V around a dc bias voltage, $V=V_0$.
For instance, when a radiation field is coupled to the diode, a
dc voltage appears across it due to rectification of current flowing through the diode.   At a low signal level, this dominantly
originates from the second order term in the Taylor expansion of
the I-V curve.   This term is proportional to the second order
derivative of I versus V. Hence, a plot of the resultant voltage
measured versus a simultaneously applied dc bias voltage gives a
curve identical to a second-order derivative of the I-V characteristic curve. At the same time, the amplitude of the current component flowing through the diode at the second harmonic of the frequency of the applied field is also due to the same second order
term, and hence, it will show a dependence on bias voltage which is
again identical to the second-derivative of the I-V characteristic.
In general, the current amplitude due to an nth order frequency
mixing, measured versus a dc field, follows the nth order derivative curve of the I-V characteristic. (This is true for low level
applied ac fields, otherwise, the n + 2, n + 4,... can also contribute to the nth order frequency mixing.)

     Fig. 1 gives the observed dc bias field dependence of the
amplitude of an infrared frequency current flowing through the
diode at the sum (or difference) frequency of a simultaneously applied 10.6-μ radiation field and a microwave field. The synthesis
is due to a second-order frequency mixing, (as in the case of
second-harmonic generation or the rectification of an ac current
resulting in its conversion to dc current.) This synthesis corresponds to currents flowing through the diode at the 1st microwave
side-bands on either side of the 10.6-μ frequency. The current
amplitude for either side-band will be proportional to I" (V),
the second-order derivative of I versus the bias field V. The
measurements are made by detecting the infrared radiation emitted
from the diode at the 1st microwave side-bands on either sides of
the frequency of the 10.6-μ radiation. (See Ref. 14) The detection is done by means of a superheterodyne method. In this case,
the detected signal is proportional to the electric field amplitude
of the side-band radiation and, hence, proportional to the corresponding diode current responsible for the emitted radiation. Note
that this method of detection is phase-sensitive and thus is capable
of recognizing a 180° phase-shift which can occur as the polarity
of the applied dc bias field is reversed. Accordingly, the measured signal is proportional to the magnitude as well as the algebraic sign of I" (V).

     The above observation represents measurements directly per-
formed in the infrared. A measurement of the rectified dc voltage
developed across the same diode when subjected to either the 10.6-μ
radiation or the microwave radiation alone also shows a dependence

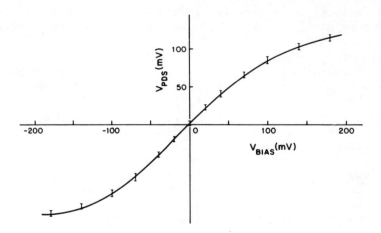

Fig. 1. Bias voltage dependence of the amplitude of an infrared frequency current, synthesized in the diode as a result of a second order mixing.

on the bias voltage identical to that of Fig. 1. These observations show the intricate relationship of all these processes, and the fact that they are all manifestations of ac currents flowing through the diode at the frequencies of the applied radiation. The non-linear I-V characteristic distorts the current waveform, resulting in the rectification and the frequency mixing.

Fig. 2 shows another frequency mixing example observed at an infrared frequency. This figure gives the bias field dependence of an infrared frequency current flowing through the diode aris-ing from a third order frequency mixing process. The diode is simultaneously subject to 10.6-μ and microwave radiation. In this case, the emitted radiation at the second microwave side-band on either sides of the frequency of the 10.6-μ radiation is detected in an external detector. (See Ref. 14). As in the previous case, the detection is done by means of a superheterodyne method, which gives a signal proportional to the amplitude of the detected side-band radiation, and hence, as before, the infrared frequency diode current responsible for the radiation. Fig. 2 gives the dependence of the detected signal on a dc bias field applied to the diode. This curve represents the third-order derivative of the I-V charac-teristic curve. In fact, it also agrees with the bias field depen-dence of a signal observed across the same diode due to a third order frequency mixing process in which both of the applied fields are in the microwave regions (see below).

Measurements of the 3rd-order derivative given in Fig. 2 is performed on a different diode than that used in determining the 2nd-order derivative curve given in Fig. 1. Note that the tunneling contacts in both diodes correspond to I-V characteristic curves which are somewhat asymmetrical with respect to the origin. However, the asymmetry is more pronounced in the Fig. 2 diode than that of Fig. 1.

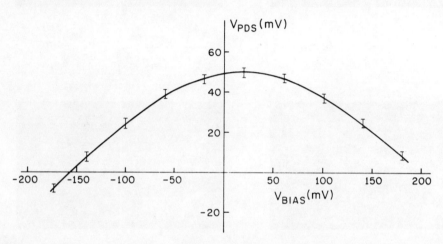

Fig. 2. Bias voltage dependence of the amplitude of an infrared frequency current, synthesized in the diode as a result of a third order mixing.

The similarity of the microwave frequency mixing curves and those obtained at the infrared frequencies show that in the elements studied there are no observable dispersive effects. (There are a number of dispersive possibilities which can be expected to become observable under appropriate conditions. These include effects when $h\nu$ becomes comparable to the barrier height.)

Observation of a high-order frequency mixing signal versus a bias-field is a convenient method of obtaining high-order derivatives of the I-V characteristic curve. Because of the broadband characteristics, the measurements can be more conveniently performed at a microwave or rf frequency. In addition to information on barrier thickness and height, comparison of these high order derivatives with theory can give additional useful information, particularly in regard to the shape of the potential barrier. The latter is a topic which has not been given sufficient attention in the previous studies.

Fig. 3 gives the observed derivatives up to the 6th order of the I-V characteristics. These have been obtained in the same

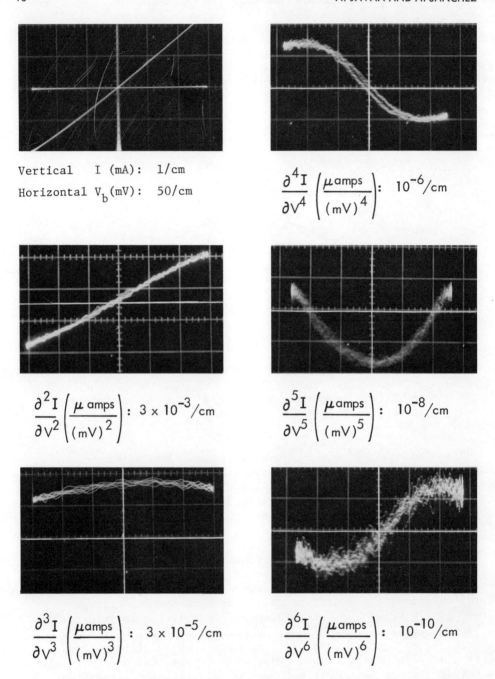

Vertical   I (mA):   1/cm
Horizontal V_b(mV):   50/cm

$$\frac{\partial^4 I}{\partial V^4}\left(\frac{\mu \text{amps}}{(\text{mV})^4}\right):\ 10^{-6}/\text{cm}$$

$$\frac{\partial^2 I}{\partial V^2}\left(\frac{\mu \text{amps}}{(\text{mV})^2}\right):\ 3 \times 10^{-3}/\text{cm}$$

$$\frac{\partial^5 I}{\partial V^5}\left(\frac{\mu \text{amps}}{(\text{mV})^5}\right):\ 10^{-8}/\text{cm}$$

$$\frac{\partial^3 I}{\partial V^3}\left(\frac{\mu \text{amps}}{(\text{mV})^3}\right):\ 3 \times 10^{-5}/\text{cm}$$

$$\frac{\partial^6 I}{\partial V^6}\left(\frac{\mu \text{amps}}{(\text{mV})^6}\right):\ 10^{-10}/\text{cm}$$

Fig. 3.   Experimental I-V characteristic of the MOM diode and its different Nth order derivatives (up the sixth order).

diode element by means of a frequency mixing procedure performed
in the microwave region: The element is subjected to the outputs
of two S-band microwave klystrons (tunable from 2 to 4 GHz) at
frequencies ν_1 and ν_2. The two microwave frequencies are tuned
to obtain a 40 MHz beat signal due to a frequency mixing of a
type in which $m_1\nu_1 + m_2\nu_2 = \nu_{IF}$, where m_1 and m_2 are positive or
negative integers and ν_{IF} is the IF frequency. The order of fre-
quency mixing is given by $|m_1| + |m_2|$. The 40 MHz beat note is
observed by feeding the signal across the diode to a fixed frequency narrow-band IF. By tuning the ν_1 and ν_2 frequencies, signals due to different orders of frequency mixing can be observed
in the same diode and studied in sufficient detail over a period
of several hours, during which time the diode characteristic remains unchanged. The studies are made by observing the dependence
of the 40 MHz beat notes versus a varying dc bias field applied to
the diode. A phase sensitive detector is used to obtain a signal
proportional to the amplitude of the 40 MHz beat note (rather than
its square obtainable with a square law detector.) This is done
in order to obtain information on the sign as well as the amplitude of the observed beat note versus the dc bias field. In the
phase sensitive system, the reference signal is obtained by simultaneously performing the same frequency mixing experiment in a
separate frequency-mixer consisting of an ordinary silicon diode.
The resultant 40 MHz IF signal provides the reference used in a
40 MHz lock-in amplifier whose signal channel is subjected to the
40 MHz beat note across the high speed tunneling element.[16]

A CLASSIFICATION OF THE OBSERVED CHARACTERISTICS

As is expected, due to the local inhomogeneity of the metal
base and the non-uniformity of its surface, the exact shape of
the I-V characteristic is found to vary appreciably as the process
of mechanically contacting the wire antenna to the metal post is
repeated. In fact, for the same diode element, several types of
I-V characteristics corresponding to different types of barrier
shapes and heights can be obtained by repeatedly breaking the
mechanical contact and attempting new ones. A disadvantage of this
is that, once a contact is burned-out or purposely disconnected,
it requires several trials in order to reproduce a contact with
approximately the same I-V characteristic curve. Because of this,
accurate comparisons of different properties of the diode must be
done by means of observations performed on the same contact and
before a change occurs in its I-V characteristic. Breaking the
contact and repeating it a number of times, however, gives an op-
portunity to conveniently produce junctions corresponding to a
varied set of tunneling barriers. This in turn makes it possible
to study different types of barriers whose properties are varied
and represent different manifestation of the tunneling effect. In
general, however, mechanical instability of a junction is trouble-

some and careful measurements require both delicate manipulation
and patience. Also, one must guard against spurious signals not
related to the electron tunneling process, which can appear across
the element and interfere with the measurements (See Ref. 1).

The different types of I-V characteristics observable in the
element can be classified as follows. One type consists of anti-
symmetric I-V curves for which the current versus voltage closely
satisfies the relationship $I(V) = -I(-V)$. In this case, rectification of an applied radiation field can occur only in the presence
of a dc bias field since the even order terms in the I-V expansion
are all zero at V=0. For the same two metals, a different type
of contact can be obtained with a non-symmetrical I-V characteristic around $V_0=0$. In this case, rectification of an applied
radiation field can occur in the absence of a bias voltage. The
polarity of the dc rectified voltage, however, is dependent on a
specific contact and can reverse its sign as the contact is broken
and repeated over again. This can be understood on the basis of
the relative values of the effective work functions of the two
metals in different contacts.

THEORETICAL DISCUSSIONS

In the previous studies of the large area tunneling junctions,
there have been a number of theoretical analyses to explain the observed I-V characteristics.[15] In general, the observed I-V curves
are found to be in good agreement with estimates based on reasonable
models for the barrier potentials.

The following discussion is intended to give a preliminary
account of the application of the existing theory to our explanation of the general features of our observed I-V characteristics
and their high-order derivatives.

The theory considers an individual electron with an energy E_x
(measured relative to the Fermi energy E_F) moving inside the metal
along a direction, x, normal to the plane of the metal-oxide-metal
junction. When the electron reaches the junction, it sees a potential barrier higher than its energy E_x; according to quantum mechanics, there is a probability that the electron will tunnel through
the barrier. Let us consider this process in the presence of an
external field causing a potential difference V to appear between
the two metals. From the W.K.B. approximation, the tunneling
probability can be written as:

$$P(E_x, V) = \exp\left\{-Q(E_x, V)\right\}$$

where
$$Q(E_x, V) = 2\frac{\sqrt{2m}}{\hbar} \int_{x_1}^{x_2} \sqrt{\phi(x,V) - E_x}\, dx$$

$\phi(x,V)$ is the barrier potential in the presence of the applied voltage, V. It can be written as

$$\phi(x,V) = \phi_o(x) - \frac{x-x_1}{x_2-x_1}\, eV$$

In this equation, $\phi_o(x)$ is the barrier potential for V=0 and x_1 and x_2 are the classical turning points which satisfy $\phi(x_1,V) = \phi(x_2,V) = E_x$. The thickness of the oxide layer is $L = x_2-x_1$, at the Fermi level ($E_x=0$), and no bias voltage (V=0).

The net current density can be expressed as the integral

$$J(V) = A \int_{-E_F}^{\infty} dE_x\, P(E_x,V)\, F(E_x,V)$$

where
$$A = \frac{4\pi me}{h^3}\, kT$$

and
$$F(E_x,V) = \ell n \frac{1 + \exp(-E_x/kT)}{1 + \exp(-(E_x+eV)/kT)}$$

the function $F(E_x,V)$ accounts for the Fermi-Dirac electron energy electron energy distribution. Inspection shows that the integrand in J peaks near $E_x=0$, (the Fermi level). In the previous treatments, approximate analytic expressions are obtained by expanding the integrand around $E_x=0$. The resultant expressions for different barrier heights and thicknesses provide good fits to the experimentally measured J-V characteristics. However, these analytic approximations are not applicable to our junctions where the barrier thickness is considerably less than those of the barriers studied previously. In a thin barrier, electrons with energies appreciably removed from $E_x=0$ can contribute sizably to the tunneling current.

The integral in equation (1), along with the high-order derivatives of J versus V, can be numerically calculated without resorting to an analytic approximation. This can be done with the aid of a digital computer. Here we give a brief summary of the results for several cases in which Q can be expressed analytically. This considerably simplifies the computer calculations. Specific-

ally the calculations are done for trapezoidal, rectangular, parabolic, and triangular barriers. The parabolic barrier is emphasized
below. In this case with

$$\phi_0(x) = 4\phi_m \frac{x}{L} \left(1 - \frac{x}{L}\right) \quad , \quad \phi_m = \phi_0\left(\frac{L}{2}\right)$$

the computer's expression for Q is:

$$Q(E_x, V) = \frac{\pi}{4} \; \alpha \; L \; \sqrt{\phi_m} \; \left((1 - \frac{V}{4\phi_m})^2 - \frac{E_x}{\phi_m}\right)$$

For a symmetrical barrier, the J-V curve as well as the curves
corresponding to the even-order derivatives of J versus V go
through the origin and have an inflection at that point (i.e. the
Taylor expansion of I versus V will contain only odd powers of V).
Let us now consider the behavior of J"(v), the second-derivation
of J versus V, for a symmetrical barrier. In this case, in the
region close to the origin, J" appears as a nearly straight line,
with a slope which varies at increasing values of V. Depending
on the shape of the barrier, its thickness and its height, the
slope can either increase or decrease when increasing V. An increasing slope versus V corresponds to a positive value for $J_o^{(5)}$
the 5th derivative of J versus V at V=0, while a negative value for
$J_o^{(5)}$ gives a decreasing slope. We have inspected this behavior
over wide ranges of values for barrier heights and thicknesses.
Interestingly, we find that for a rectangular barrier, $J_o^{(5)}$ appears
to remain positive for reasonable values of barrier thicknesses
and heights, corresponding to an increasing slope for J" versus
V. The experimental results show, however, that in our junctions
the observed second-derivative curves for symmetrical barriers
have a decreasing slope at increasing values of V.[19] In a parabolic
barrier, on the other hand, it is possible to find a range of values for the potential heights and thicknesses corresponding to decreasing values of slope versus V. Fig. 4 gives a typical computer
calculation. In this figure, in order to emphasize the nonlinear
features of J" versus V, a straight line has been subtracted from
J" as indicated. This is done so that each curve appears tangent
to the V axis at V=0. Note that the curves below the V axis have
decreasing slopes at increasing V   (i.e. $J_o^{(5)} < 0$).

The above example is mentioned in order to emphasize that
high order derivatives of the I-V characteristics can be used to
obtain some information on the shape of a potential barrier. From
the theory, it is possible to obtain a series of curves for J(V)
as well as J"(V) at different values of ϕ_m and L. These curves,

as noted above, are nearly straight lines for small voltages.
From the slope of J versus V, a value for the tunneling conductance
per unit area, Y, can be found at varying ϕ_m and L. We then note
that for given values of ϕ_m and L, the ratio of $J''(V)/J(V)$ is
constant at small V's. (Both J'' and J are nearly straight lines,
as noted above.) We define this parameter as $r(\phi_m,L)=J''(V)/J(V)$.

From the experiments, it is possible to measure the parameters
Y and r in a high speed symmetric junction. This can be done, e.g.,
at a radio frequency where the rf voltage across the junction can
be directly measured. The parameter Y can be measured by applying
a small dc bias voltage (.01 to 0.1 volts) to the diode and measuring the bias current density. The current density can be obtained from the measured current along with a knowledge of the
junction area A. From electron microscope photographs of the tip
of wire antenna, A is found to be close to $0.01(m\mu)^2$. The parameter r can be measured in the same contact by applying a radio
frequency field and measuring the rectified rf current versus the
dc bias current. This is done at a fixed (and known) value of
the rf voltage amplitude. (Note that r is proportional to v_{rf}^{-2}
and is independent of area).

The resistance of the junction varies in different contacts.
In a typical low impedance contact of 100 ohms, (corresponding to
$Y \simeq 1\text{mho}/(m\mu)^2$), the measured value of r is found to vary between
$2-4$ $(\text{volts})^{-2}$.

For a parabolic barrier, a class of curves for r and Y can
be found by varying ϕ_m and L. The results show that $Y=1$ $\text{mho}/(\mu m)^2$
and r = 3 volts corresponds to $\phi_m = 0.7$ ev and L = 6Å. A further
result is that these values are not a sensitive function of the
barrier shape (as is the sign of the $J^{(5)}$, see above). In fact
a rectangular barrier having an effective height the same as the
parabolic barrier with $\phi_m = 0.7$ volts and L = 6Å, will also give
the same values for Y and r. (For definition of effective height
see Ref. 17). As noted above, however, a rectangular barrier
predicts the incorrect sign for the curvature of J'' versus V.

In the face of mechanical instability of the junction, and
the difficulties in obtaining reproducible results, the above
estimates must be considered to give approximate values for the
barrier parameters. These results will yet have to be supported
by other independent observations, particularly at cryogenic
temperatures.

FUTURE POSSIBILITIES

The observed properties of the high-speed diode element arise
from interaction of the tunneling electrons with the thin dielec--

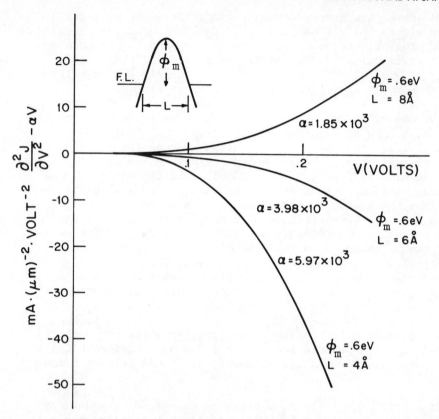

Fig. 4. Theoretical second derivative with respect to V of the J-V characteristic. The shape of the barrier is approximated by a parabola. A straight line, αV, has been subtracted to emphasize the nonlinear features of J".

tric barrier occupying a microscopic volume where the pointed tip of the thin wire antenna touches the metal base. Accordingly, we can generalize our thinking and consider this as a method for making observations on the physical properties of matter occupying a microscopic volume in the region of the point-contact. For instance, this approach can be used as a "micro-probe" for studying surface physics in metals. From previous observations made in large-area metal-to-metal tunneling junctions, it is known that resonant scattering of electrons off the local infrared or far-infrared vibrational modes in the thin dielectric barrier can contribute to the detailed features of the I-V characteristic curve.[18] These have been observed at cryogenic temperatures and appear as changes in the tunneling impedence whose onsets occur at fairly well-defined values of the bias voltages. These voltages correspond to the quantized energies of the vibrational modes

associated with the specific molecular bonds in the dielectric
barrier. Our high-speed tunneling junctions will permit these
types of observations to be made not only in the dc and the low
frequency regions, but also at the far-infrared or the infrared
frequencies where the quantized local modes of a barrier can
interact resonantly, or in their near-resonant regions, with the
ac tunneling current and the applied fields. These will result
in dispersive effects, causing phase-shifts and thus introducing
inductive or capacitive components in the tunneling impedence.
These effects will be nonlinear-functions of the applied fields.

A high speed metal-dielectric-metal tunneling junction can
be made via vacuum-deposition of thin films on a substrate. This
requires application of methods of microelectronics to achieve the
small area junction. These methods will also permit the junction
to be integrated with thin film antenna structures for coupling
to the radiation field. The deposited antenna length can be
designed to be resonant at the wavelength of an applied radiation
field. In addition, with microelectronics, it is possible to
simultaneously deposit a large number of these types of junctions
in a small area of a substrate. The various junctions along with
their resonant antenna structures can be placed in a geometrical
pattern to form a phased-array configuration.

At the time of the Vail Symposium, promising results were
obtained in a series of experiments designed to show the feasibility of achieving a high-speed tunneling junction by means of
thin-film vacuum deposition combined with microelectronics. These
experiments have been completed by the time of final preparation
of this manuscript for publication in the proceedings of the Vail
Symposium. The results are described in a manuscript due to appear
in publication.[12] The reader is referred to this reference for
the details of the experiments and the results.

With the possibility of depositing a large number of high-
speed diode elements as noted above, several novel applications
appear to be within the range of practical possibilities. These
include the possibility of applying a two-dimensional phased-array
configuration to obtain in real time a hologram of an image formed
on the array. Consider, for example, the image to be formed by
laser irradiation of an object. The image can be heterodyned at
each junction with radiation from a local oscillator uniformly
applied to all the junctions in the array. The local oscillator
frequency, if desired, can be shifted by a fixed frequency to
obtain a heterodyne signal at a convenient IF. The IF signals
across the various junctions will have information on both the
amplitudes and the phases of the image at the locations of each
junction. These real-time signals can be stored; their Fourier
transform, in turn, will reproduce the image holographically.
It should also be noted that this proposed method for holographic

imaging in real time, can be applied utilizing any two dimensional
array of detector elements whose speeds are sufficiently fast
to produce heterodyne signals at the IF frequency. A holographic
image in real time is of considerable importance in a number of
practical applications, both in the infrared as well as in the
visible region. Beside these, superheterodyne imaging offers
the advantage of enhanced detection sensitivity. There are a
number of variations in this imaging scheme. For instance, the
local oscillator used to obtain phase and amplitude information
can be chosen to have a frequency widely different from that of
the laser which illuminates the object. (In this case, the two
laser frequencies can be phased-locked, using a separate frequency
multiplier chain, as discussed in Ref. 1.) Another possibility
is to use an illuminator laser consisting of two closely spaced
frequencies, ν_1 and ν_2. With appropriate frequency components
in the reference local oscillator radiation, it is possible to
extract phase information (and a corresponding amplitude), at
the difference frequency $|\nu_1-\nu_2|$. This will enable holographic
imaging at a longer effective wavelength corresponding to interference between the two propagating waves at frequencies ν_1 and
ν_2, giving rise to spatial beats.

It is known theoretically that, under special conditions, a
tunneling junction can be obtained in a multi-barrier structure
(the super-lattice), in which the corresponding I-V characteristic[20]
will show regions of negative differential conductivity. A high-
speed junction of this type, if achieved, can be integrated with
a high frequency L-C tank configuration, to obtain a high-frequency
negative impedence oscillator. Inspection shows that, with the
available methods of microelectronics, an L-C tank circuit consisting of, e.g., a simple pair of lines, or a more complex configuration, can readily be made whose resonances lie in the far-
IR or the IR regions. The major challenge, however, is achievement of the negative impedence junction.

It is also known theoretically that, instead of a diode, a
triode configuration[21] can be envisioned with a control-grid for
controlling the current flow in the element. A high-speed junction
of this type will have interesting possibilities in novel high
frequency device applications.

The above examples are cited in areas other than frequency
measurements and spectroscopy. For a review of these and the
related topics, see Ref. 1.

REFERENCES

1. For a review see: A. Javan, Fundamental and Applied Laser
    Physics, Proceedings of the Esfahan Symposium, edited by
    M. S. Feld, A. Javan, and N. A. Kurnit, Wiley-Interscience,
    1973, Page 295.

2. See S. Ezekiel, this volume.

3. See A. Mooradian, this volume.

4. L. O. Hocker, D. R. Sokoloff, V. Daneu, A. Szoke, and A. Javan,
    Appl. Phys. Letters 12, 401 (1968).

5. V. Daneu, D. R. Sokoloff, A. Sanchez, and A. Javan, Appl.
    Phys. Letters 15, 398 (1969).

6. D. R. Sokoloff, A. Sanchez, R. M. Osgood, and A. Javan, Appl.
    Phys. Letters 17, 257 (1970).

7. K. M. Evenson, G. W. Day, J. S. Wells, and L. O. Mullen,
    Appl. Phys. Letters 20, 133 (1972).

8. K. M. Evenson, this volume.

9. A. Javan, Ref. 1.

10. A. T. Mattick, A. Sanchez, N. A. Kurnit, and A. Javan, Appl.
    Phys. Letters 23, 675 (1973).

11. F. Keilmann, R. L. Sheffield, M. S. Feld, and A. Javan, Appl.
    Phys. Letters 23, 612 (1973).

12. J. G. Small, G. M. Elchinger, A. Javan, A. Sanchez, F. J.
    Bachner, and D. L. Smythe, to be published in March 15 issue
    of Appl. Phys. Letters.

13. M. Kelly, J. Thomas, J. Monchalin, N. A. Kurnit, and A. Javan,
    to be published.

14. A. Sanchez, S. K. Singh, and A. Javan, Appl. Phys. Letters,
    21, 240 (1972).

15. For review of recent works see for instance, Tunneling
    Phenomena in Solids, edited by E. Burstein and S. L. Lunquist
    (Plenum Press, New York, 1969).

16. For detail see, A. Sanchez, Ph. D. Thesis, M.I.T., (1973), unpublished.

17. J. G. Simmons, J. Appl. Phys. <u>34</u>, 1793 (1963).

18. J. Lambe and R. C. Jaklevic, Phys. Rev. <u>165</u>, 821 (1968).

19. See also S. M. Farris, T. K. Guftanson, and J. C. Weisner, IEEE J. Quantum Electronics <u>QE-9</u>, 737 (1973).

20. See e.g. R. Tsu and L. Esaki, Appl. Phys. Letters <u>22</u>, 562 (1973).

21. See e.g. R. H. Davis and H. H. Horack, J. Appl. Phys. <u>34</u>, 864 (1963).

INFRARED-MICROWAVE DOUBLE RESONANCE

Koichi Shimoda

Department of Physics, University of Tokyo, Tokyo 113
and
Institute of Physical and Chemical Research, Wako 351

1. INTRODUCTION

Since the double resonance effect in molecules was first observed in 1955 with an ammonia-beam maser at 24 GHz, it has been
investigated as a tool for microwave and radiofrequency spectroscopy.   The recent development of infrared lasers has made the infrared-microwave double resonance experiment feasible by using the
vibration-rotation transitions in molecules.   Previous works on
double resonance were reviewed[1] in 1971, and they are not shown in
this paper.

The method of infrared-microwave double resonance is useful
for[1,2]
1)   assigning vibration-rotation transitions,
2)   observation of weak transitions,
3)   studies of relaxation processes and collision-induced tran-
     sitions,
4)   microwave spectroscopy of vibrationally excited molecules,
5)   modulation and detection as well as generation and amplifi-
     cation of infrared and far-infrared, and
6)   measurement of the infrared or microwave field intensity.

The infrared transitions are much stronger than the microwave
transitions in general.   The infrared laser output can easily be
concentrated to saturate the transition when the frequency is tuned
in.   Therefore, the infrared-microwave double resonance signal is
more easily observed than the microwave-microwave double resonance
signal.   Furthermore, the double resonance under strong infrared
and strong microwave radiations showing higher-order effects can be

observed with moderate powers of the order of milliwatt.

2.   HIGHER-ORDER EFFECTS IN DOUBLE RESONANCE

A semiclassial theory of coherent interactions may be employed
to describe most of the double resonance effects.[3,4]   The two-fre-
quency perturbation field at the position of a molecule is written
as

$$E(t) = E \cos(\omega t + \theta) + E_m \cos(\omega_m t + \theta_m)$$

where the subscript m stands for microwave.

Consider a three-level system as shown in Fig. 1, where tran-
sitions $2 \leftrightarrow 1$ and $3 \leftrightarrow 1$ are allowed at microwave and infrared
frequencies respectively.   The relaxation processes are phenomenol-
ogically taken into consideration.   The different rates between
longitudinal, transversal, and cross relaxations in conjunction with
Doppler effect of molecular velocities are considered in the theory.

The probability of induced transition $3 \leftrightarrow 1$ is given by

$$S = \frac{x^2}{2} \cdot \frac{\gamma}{\gamma^2 + (\omega - \omega_{31} - kv)^2} \tag{1}$$

where $k = \omega/c$, $x = |\mu_{31}|E/\hbar$, γ is the transverse relaxation rate,
and v is the component of molecular velocity along the direction
of the infrared radiation.   By neglecting the Doppler effect for
microwave transition $2 \leftrightarrow 1$, the microwave transition probability
is given by

$$S_m = \frac{x_m^2}{2} \cdot \frac{\gamma_m}{\gamma_m^2 + (\omega_m - \omega_{21})^2} \tag{2}$$

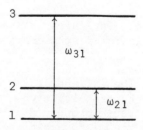

Fig. 1 A three-level system.

where $x_m = |\mu_{21}|E_m/\hbar$, and γ_m is the transverse relaxation rate for the microwave transition.

The rate equations for populations n_i of molecules having the velocity component v are written as

$$\frac{d}{dt} n_1 = (n_3 - n_1)S + (n_2 - n_1)S_m - \frac{n_1 - n_1^0}{\tau_1} \tag{3}$$

$$\frac{d}{dt} n_2 = (n_1 - n_2)S_m - \frac{n_2 - n_2^0}{\tau_2} \tag{4}$$

$$\frac{d}{dt} n_3 = (n_1 - n_3)S - \frac{n_3 - n_3^0}{\tau_3} \tag{5}$$

where τ_i is the lifetime of level i. On the assumption of Maxwell velocity distribution in the absence of radiation, the unperturbed value of n_i is written as

$$n_i^0 = \frac{N_i^0}{\sqrt{\pi}\, u} \exp(- v^2/u^2) \tag{6}$$

where $N_i^0 = \int_{-\infty}^{\infty} n_i^0 dv$ is the unperturbed population of level i, and u is the most probable molecular velocity.

The steady-state solution of eqs. (3) - (5) is obtained and integrated over the Maxwell velocity distribution at the Doppler limit for $ku \gg \gamma$ to calculate the absorbed power. The change in infrared absorption at $\omega \simeq \omega_{31}$ caused by the microwave power at $\omega_m = \omega_{21}$ is written for $N_3 \ll N_1$ in the form

$$\Delta P = \frac{\sqrt{\pi}\, N_1^0\, \hbar\omega}{2ku} \cdot \frac{\tau_1 x_m^2}{2\gamma_m + (\tau_1 + \tau_2)\, x_m^2} \cdot \frac{x^2}{\sqrt{1 + bx^2}}$$

$$\times \left[- \delta + \frac{\tau_1}{2\gamma} \cdot \frac{x^2}{1 + ax^2 + \sqrt{(1 + ax^2)(1 + bx^2)}} \right] \tag{7}$$

where $\delta = (N_1^0 - N_2^0)/N_1^0 \simeq \hbar\omega_m/k_B T$, $a = (\tau_1 + \tau_3)/2\gamma$, and

$$b = a - \frac{\tau_1}{2\gamma} \cdot \frac{\tau_1 x_m^2}{2\gamma_m + (\tau_1 + \tau_2)\, x_m^2} \quad .$$

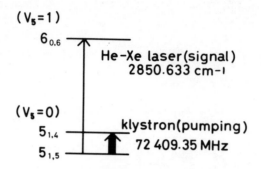

Fig. 2 Three levels in H_2CO.

When the laser field is weak, the second term in the square
bracket in eq. (7) may be neglected and the double resonance signal
ΔP is proportional to $- \delta \cdot x^2 x_m^2 / \{2\gamma_m + (\tau_1 + \tau_2) x_m^2\}$.   When
the infrared transition is saturated, on the other hand, so that
ax^2 and bx^2 can not be small, the second term in the square bracket
dominates the first term δ.   The rate-equation approximation of
the higher-order effects in double resonance is fully represented
by eq. (7).

Because the quantitative measurement of the infrared field intensity was rather difficult, the higher-order effect might well be
studied in the pressure dependence of double resonance in gas[3].

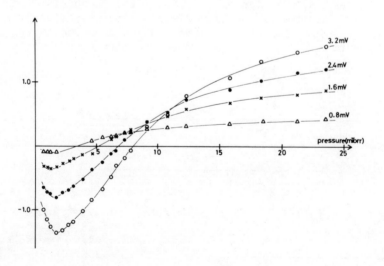

Fig. 3 Observed pressure dependence of double resonance in H_2CO.

The pressure dependence was studied with three levels of formaldehyde, H_2CO, as shown in Fig. 2.   A multi-mode Zeeman-tuned 3.5 µm
He-Xe laser of 60 cm length was employed to saturate the infrared
transition, while a frequency modulated Oki 70V11A klystron was employed to saturate the microwave transition.   The transmitted laser
power through the absorption cell of 1 m length was phase-sensitively detected.[3,4]

The observed pressure dependence of the infrared-microwave double resonance in H_2CO is shown in Fig. 3.   The nonlinear pressure
dependence seen in Fig. 3 at low pressure must be the result of
higher-order effects.   The theoretical pressure dependence is calculated by using parameters as $\gamma_m/2\pi = 10p$ in kHz, $\gamma/2\pi = (36 + 10p)$
in kHz, and $\tau_1 = \tau_2 = \tau_3 = \gamma^{-1}$, where p is the gas pressure in millitorr, and a value of 36 kHz corresponds to the molecular transit
time across the laser beam of 2 mm diameter.

Theoretical pressure dependence for four different values of
x^2 is shown in Fig. 4.   The agreement between theoretical and experimental results is very good.

If the matrix element of the dipole moment μ_{31} averaged over
the degenerate levels is known, the effective value of $x = |\mu_{31}|E/\hbar$
which is found by comparing Figs. 3 and 4 will give a value of the
infrared field intensity E.   The saturation parameter for the microwave transition can be determined from the pressure dependence
at higher pressures.   Figure 4 is calculated for the best fit value

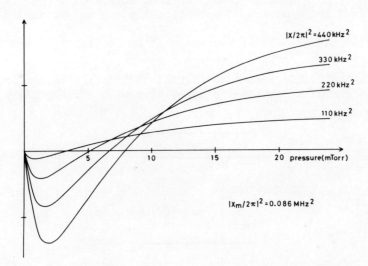

Fig. 4   Calculated pressure dependence from eq. (7).

of $x_m/2\pi$ = 290 kHz, which corresponds to about 1 mW through the waveguide of 4.78 × 2.39 mm^2 cross-section.   The average infrared field intensity may be evaluated by using the value of dipole moment estimated from the absorption constant.   The average value of E corresponding to $x/2\pi = \sqrt{440}$ kHz is 1.4 V/cm which gives a power of 90 μW through a circular cross-section of 1 mm radius.[4]   These values are subject to change if difference in saturation for degenerate levels is taken into consideration.

Although the experimental result of Fig. 3 is fitted very well with the theoretical result on the assumption of $\tau_1 = \tau_2 = \tau_3 = \gamma^{-1}$, it does not necessarily verify the alove assumption.   In fact, those relaxation constants can be determined separately, if the double resonance signal is quantitatively measured as functions of the laser power, the microwave power, and the gas pressure.

It may be mentioned here that these higher-order effects of saturation are easily observed with a power of less than 1 mW at low gas pressure.

3. ROTATIONAL TRANSITIONS IN A VIBRATIONALLY EXCITED STATE

When the vibrational frequency of a molecule is higher than 2000 cm^{-1}, the population of the vibrationally excited state is very small.   Laser pumping to enhance the population of the excited state to such an extent that allows detection of microwave absorption in the excited state is not promising.   The double resonance method, on the other hand, is found to become a powerful tool for microwave spectroscopy of vibrationally excited molecules.

Consider a three-level system as shown in Fig. 5, where levels 2 and 3 are in the excited state.   Since $N_3{}^0 \simeq 0$ and $N_2{}^0 \simeq 0$ in this case, the double resonance signal, that is the change in infrared absorption caused by the resonant microwave corresponding to

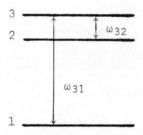

Fig. 5 A three-level system for microwave spectroscopy in the vibrationally excited state.

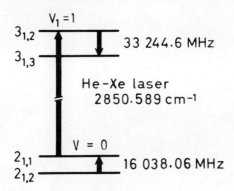

Fig. 6   Four levels in HDCO for infrared-microwave double resonance.

the 3 ↔ 2 transition, is expressed by omitting δ in eq. (7) with subscripts 1 and 3 interchanged. Then only the higher-order effects, the lowest term being $x^4 x_m^2$, appear in this case. The magnitude of the signal is expected to be almost as large as that of double resonance in the three-level system of Fig. 1 at low pressure. This type of double resonance method was employed by Shimizu[5] for the observation of radiofrequency transitions between Stark components in the ν_2 excited state of phosphine, PH_3.

The method of microwave spectroscopy of vibrationally excited molecules by the double resonance technique can conveniently be investigated in four levels of formaldehyde d_1, HDCO, as shown in Fig. 6, in which double resonance with the microwave transition either in the ground or excited state is observed.[6]

The output frequency of a multi-mode 3.5 μm He-Xe laser is Zeeman shifted by -1.33 GHz from the center to tune it in the $3_{12}(\nu_1 = 1) \leftarrow 2_{11}(\nu = 0)$ transition. The laser beam passes through the gas of HDCO in a K-band waveguide cell of 1 m length and enters an InSb detector. The increase in saturated absorption of the 3_{12} $(\nu_1 = 1) \leftarrow 2_{11}(\nu = 0)$ infrared transition, when either the $2_{11} \leftarrow 2_{12}$ $(\nu = 0)$ or the $3_{12} \rightarrow 3_{13}(\nu_1 = 1)$ transition is stimulated by microwaves, is observed.

The frequency of the $2_{11} \leftarrow 2_{12}$ transition in the ground state was known to be 16.03806 GHz. The pressure dependence of the double resonance signal with the 3.5 μm laser and the 16 GHz Varian X-12 klystron was observed as shown in Fig. 7. Because the common level for the two transitions is the second among three levels in this case, the two terms in the square bracket of eq. (7) have the same sign. It is seen in Fig. 7 that the contribution of the first term δ in eq. (7) dominates that of the second term at a pressure

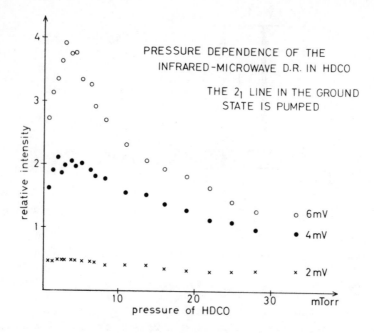

Fig. 7 Observed pressure dependence of double resonance in HDCO
with the 16 GHz transition in the ground state.

higher than about 30 millitorr.

The search for the $3_{12} \rightarrow 3_{13}$ microwave transition in the excited state in the range between 31.5 and 33.5 GHz with the Oki
35V11 klystron resulted in the observation of double resonance signal at 33.2446 GHz. The observed frequency was higher than that
of the $3_{12} \rightarrow 3_{13}$ transition in the ground state at 32.0726 GHz.
The pressure dependence of the double resonance signal was observed
as shown in Fig. 8.

The signal in Fig. 8 is seen to disappear at high gas pressure
while at low pressure it is almost as strong as that in Fig. 7 of
double resonance with the microwave transition in the ground state.
It is concluded, therefore, that the microwave transition in the
vibrationally excited state of negligible population can well be
observed with the double resonance technique. Because the homogeneous width for the infrared transition at a pressure of a few millitorr is less than 100 kHz while the Doppler width is about 100MHz,
the population of excited molecules under pumping with the laser is
of the order of 10^{-4} of the ground state population. Thus microwave power absorption by the laser-pumped molecules at such a low

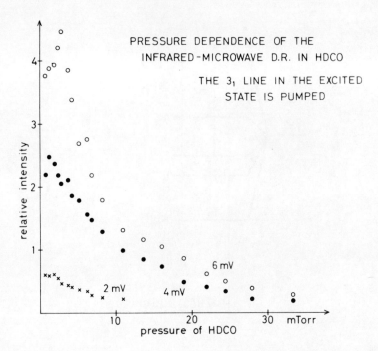

PRESSURE DEPENDENCE OF THE
INFRARED-MICROWAVE D.R. IN HDCO

THE 3_1 LINE IN THE EXCITED
STATE IS PUMPED

Fig. 8 Observed pressure dependence of double resonance in HDCO with the 32 GHz transition in the excited state.

pressure would be too small to detect.   The double resonance technique is much more sensitive.

4.   TRANSITIONS WITH VERY SMALL DIPOLE MOMENT

Very weak microwave transitions having small matrix elements of the molecular dipole moment can be observed with the infrared-microwave double resonance method.   Microwave spectroscopy of methane by using the 3.39 μm He-Ne laser is investigated for example. The CH_4 molecule is tetrahedral, having no dipole moment in the first approximation.   Thus no microwave transition of CH_4 is allowed. Because of the vibration-rotation coupling, however, it has a small electric dipole moment of 0.02 debye in the ν_3 state of vibration. Therefore, weak transitions may occur between F_1 and F_2 sublevels of the fine structure due to Coriolis interaction.[7]   The corresponding microwave absorption is many orders of magnitude smaller than the minimum detectable absorption of an ordinary microwave spectrometer.

One of the components in the P(7) branch of the ν_3 band of CH_4

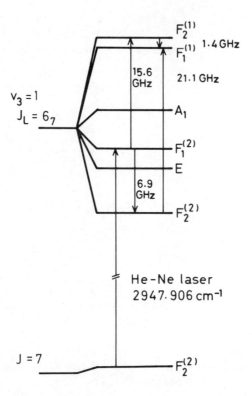

Fig. 9 Energy levels for the P(7) branch of the ν_3 band of CH_4 showing sublevels in the excited state.

is within the Doppler tuning range of the 3.39 μm He-Ne laser. The relevant energy levels for the P(7) branch of the ν_3 band are shown in Fig. 9.   The experimental apparatus for microwave spectroscopy of CH_4 by using double resonance is very similar to that was used for double resonance in H_2CO and HDCO.   Somewhat larger power, however, must be used in the double resonance experiment of CH_4. The resulting change in transmitted laser power through the cell is phase-sensitively detected with a large signal-to-noise ratio.

The frequency of the transition between $F_1^{(2)}$ and $F_2^{(2)}$ sublevels in the vibrationally excited state ($\nu_3 = 1$, $J_L = 6_7$) was observed by Curl and Oka[8] to be 6895.3 ± 0.3 MHz, followed by a more accurate measurement by Takami and Shimoda[9] of 6895.204 ± 0.010 MHz. The former workers empolyed watts of microwave power, while the latter observed the double resonance signal with a microwave power of some ten milliwatts through a waveguide cell fed by a 2K26 klystron.

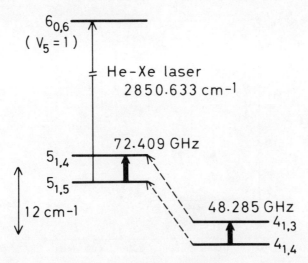

Fig. 10 Energy levels in H$_2$CO for the study of collision-induced transitions.

The microwave transition between $F_1^{(2)}$ and $F_2^{(1)}$ sublevels in the vibrationally excited state was also observed[9] at 15601.846 ± 0.010 MHz. The half-width at half maximum of the signal with respect to the microwave frequency was measured[9] to be about 100 kHz at a pressure of 4 millitorr. Similar transitions in the ground state have been observed by Curl, Oka, and Smith[10] at 423.02 ± 0.10 MHz and at about 1200 MHz respectively.

5. COLLISION-INDUCED TRANSITIONS

Collision-induced transitions between rotational levels have been extensively studied by Oka and his collaborators[11] by using the method of microwave double and triple resonance. Similar studies can be performed with infrared-microwave double resonance and triple resonance technique, which is more sensitive and reveals more detailed informations of the collision-induced transitions.

A preliminary work[12] on H$_2$CO by using the Zeeman-tuned He-Xe laser is described in the following. The relevant energy levels are shown in Fig. 10. The double resonance signal described in section 2 is observed with the 3.5 μm line corresponding to the 6_{06} ($v_5 = 1$) ← 5_{15}($v = 0$) transition, as the 5_1 line corresponding to the 5_{14} ← 5_{15}($v = 0$) transition is saturated. If the 4_1 line corresponding to the 4_{13} ← 4_{14}($v = 0$) transition at 48.285 GHz is saturated in addition, some of the population changes of the 4_{13} and

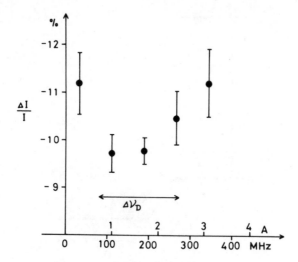

Fig. 11   The change signal $\Delta I/I$ as a function of the laser frequency. $2\Delta\nu_D$ is the full Doppler width of the absorption line of H_2CO.

4_{14} levels are transferred by collisions to the 5_{14} and 5_{15} levels. Since the probability of collisional transfer is different between the different pair of levels, the infrared-microwave double reso- nance signal in the three levels is modulated by saturation of the 4_1 line.

The double resonance signal in three levels in Fig. 10 is ob- served to decrease with saturation of the 4_1 line.   This is ex- plained by the dipole selection rule in the preferred transition induced by collisions.   The probability of collision-induced tran- sition is a complicated function of the molecular velocity.   Exper- imentally, the velocity dependence of collision-induced transitions can be observed with the technique of infrared-microwave triple re- sonance.

When a frequency-stabilized single-mode laser is detuned from the center of the Doppler-broadened 3.5 μm line of H_2CO, the double resonance effect will probe a group of molecules having the corres- ponding value of velocity along the direction of the laser beam. The effect of triple resonance with saturation of the 4_1 line can thus show the magnitude of collision-induced transitions between the 4_1 and 5_1 levels for a group of molecules in a narrow range of the velocity.

The observed change in the double resonance signal, ΔI, de- tected by the infrared absorption with saturation of the 4_1 line is shown in Fig. 11.   Here I is the double resonance signal with

3.5 μm and 72.4 GHz radiations, while I-ΔI is the triple resonance
signal with 3.5 μm, 72.4 GHz, and 48.3 GHz radiations.   Because the
infrared laser is probing the molecules in the 5_1 level, the observed value of ΔI/I is the rate of collision-induced $5_1 \leftarrow 4_1$ transitions relative to the total rate of relaxation of the 5_1 level.

Although the experimental scatter in Fig. 11 is rather large
as shown by error bars which are three times the standard deviations,
the observed result is in favor of an interpretation that the faster
molecule has the larger probability of collision-induced transition.

Saturation of the 3_1 line of H_2CO at 28.9 GHz is also observed
to slightly decrease the double resonance signal with 3.5 μm and
72.4 GHz radiations.   An average value of ΔI/I with saturation of
the 3_1 line at 28.9 GHz is about one percent.

Some refinement of the technique of triple resonance will enable one to study collisional processes in further detail.   If the
sensitivity and stability of the triple resonance method are in-
proved, experiments with space resolution as well as velocity resolution may be performed for example.

6.   MODULATION DOUBLING OF THE INVERTED LAMB DIP

Because of the coherent interaction effect, the double resonance signal in a non-degenerate three-level system must be a doublet.   It is well known in radiofrequency and microwave spectroscopy, and called modulation doubling.   At optical frequencies,
however, the line is broadened by the Doppler effect of the order
of 100 MHz.   Thus small modulation doubling in the double resonance
signal can not be observed.

When the absorption cell is in the standing-wave field of the
laser, saturated absorption by the gas exhibits an inverted Lamb
dip.   Its width is the homogeneous width with saturation broadening, which is much narrower than the Doppler width.

Splitting of the inverted Lamb dip under resonant modulation
was observed by Takami and Shimoda[4] on the 3.5 μm line of H_2CO, and
by Freund and Oka[13] on the two-photon (of 10.8 μm and 1.3 cm) transition in $^{15}NH_3$.   Oscilloscope traces of the inverted Lamb dip of
H_2CO at 3.5 μm are shown in Fig. 12.   At a pressure of 2 millitorr
the full width in Fig. 12 (a) is about 0.4 MHz and the splitting due
to resonant microwave is evident in (c).   In the case of J > 1
there are Stark components with different values of M, which exhibit
different magnitudes of modulation splitting at resonance.   Because
the observed doubling is smeared by the effect of inhomogeneous
field distribution, the M components have not yet been resolved.[4,13]

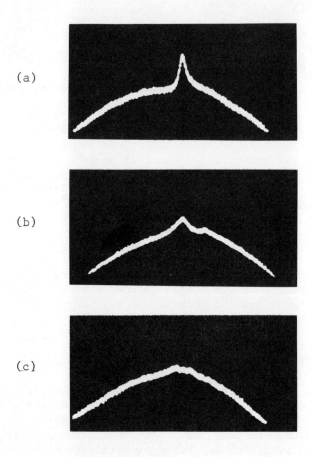

Fig. 12   Shapes of the inverted Lamb dip of H_2CO in double resonance.   The microwave is more than 20 MHz off-resonance in (a),
about 1 MHz off-resonance in (b), and at resonance in (c).

7.   TRANSIENT EFFECTS

In the main part of this paper, double resonance effects under
a steady-state condition are discussed.   It is briefly mentioned
here that the double resonance technique is particularly useful for
observation of transient effects.   One of the two radiations for
double resonance is easily separated from the other at the detector
to observe the transient signal, while the amplitude or frequency
of the other radiation is pulsed.   The observed temporal variation
of the double resonance signal immediately shows up a relaxation
effect.

Under proper conditions, the coherent interaction between molecules and radiations will manifest such effects as free induction decay, optical nutation, and photo-ehoes. These effects are originally known in magnetic resonance, and it is rather recent that their optical analogues in a (degenerate) two-level system of molecules have been investigated[14,15] by using lasers.

Those coherent transients in a three-level system were observed by Levy et al[16] by using the three levels in $^{14}NH_3$ used by Shimizu and Oka[17] for their steady-state double resonance experiment. The observation of transients by the double resonance technique can be performed in a wider variety of methods. Transient signals of the 3.5 μm absorption in the three-level system of H_2CO (Fig. 2), when the microwave frequency is suddenly shifted, are being observed by Takami and Amano.

Investigations of this sort will reveal much informations on dephasing relaxation and lifetimes of the involved levels. Power dependence and velocity dependence of these relaxation constants can also be studied.

ACKNOWLEDGMENT

In conclusion, the author expresses his gratitude to Dr. M. Takami, Dr. K. Uehara, and in particular Dr. T. Oka and his collaborators for helpful discussions and informations.

REFERENCES

1. K. Shimoda and T. Shimizu, in Progress in Quantum Electronics, eds. J. H. Sanders and S. Stenholm (Pergamon, Oxford 1972) vol. 2, pp. 43-139 and references therein.

2. K. Shimoda, IEEE J. Quantum Electron. 8, 603 (1972).

3. K. Shimoda and M. Takami, Optics Commun. 4, 388 (1972).

4. M. Takami and K. Shimoda, Japan. J. appl. Phys. 11, 1648 (1972)

5. F. Shimizu, Chem. Phys. Letters, 17, 620 (1972).

6. M. Takami and K. Shimoda, Japan J. appl. Phys. 12, 603 (1973).

7. K. Uehara, K. Sakurai, and K. Shimoda, J. Phys. Soc. Japan, 26, 1018 (1969).

8. R. F. Curl, and T. Oka, J. chem. Phys. to be published.

9. M. Takami, K. Uehara, and K. Shimoda, Japan. J. appl. Phys. 12 (1973), to be published.

10. R. F. Curl, T. Oka, and D. S. Smith, Chem. Phys. Letters, to be published.

11. P. W. Daly and T. Oka, J. chem. Phys. 53, 3272 (1970); A. R. Fabris and T. Oka, ibid. 56, 3168 (1972) and references therein.

12. M. Takami and K. Shimoda, Japan. J. appl. Phys. 12, (1973), to be published.

13. S. M. Freund and T. Oka, Appl. Phys. Letters, 21, 60 (1972).

14. R. G. Brewer and R. L. Shoemaker, Phys. Rev. Letters, 27, 631 (1971).

15. R. L. Shoemaker and R. G. Brewer, Phys. Rev. Letters, 28, 1430 (1972).

16. J. M. Levy, J. H.-S. Wang, S. G. Kukolich, and J. I. Steinfeld, Phys. Rev. Letters, 29, 395 (1972).

17. T. Shimizu and T. Oka, J. chem Phys. 53, 2536 (1970): T. Shimizu and T. Oka, Phys. Rev. A2, 1177 (1970).

NON-LINEAR EFFECTS AND COHERENCE PHENOMENA IN OPTICAL PUMPING

C. COHEN-TANNOUDJI

Ecole Normale Supérieure, Paris

The shape of resonances observable on atoms and corresponding to the resonant absorption of one optical or RF photon is modified when the intensity of the electromagnetic wave increases : the resonance is shifted and broadened; the variation of its amplitude is no more described by a simple power law. Furthermore, new higher order resonances appear, corresponding to the absorption of several photons by the same atom.

Observation of these non-linear effects becomes easier now, even in the optical range, as an increasing number of high intensity sources is available, and this explains the renewal of interest in the theoretical study of these effects.

I think however that some misleading assertions have recently been published on this subject. I would like in this paper to discuss some of these assertions.

Although the problem of atoms interacting with laser beams seems very different from the one of spins submitted to intense RF fields, the determination of the resonance characteristics is actually very similar in both cases : we have to evaluate the transition probability between 2 energy levels of a system under the influence of a sinusoidal perturbation. The finite number of energy levels in the case of spins simplifies the calculations. This explains why I have chosen to illustrate the discussion by considering some magnetic resonances observed in optical pumping experiments, in particular the "coherence resonances" which result from the preparation of atoms in a coherent superposition of Zeeman sublevels. These resonances turn out to be, in some cases, a very convenient tool for the study of higher order non linear effects.

In the first part of this paper, I will consider the problem of higher order terms in the radiative shift of a resonance; I will then discuss the possibility of getting non perturbative expressions for the amplitude of multiphoton processes.

I. HIGHER ORDER TERMS IN THE RADIATIVE SHIFT OF A RESONANCE

The radiative shift of a resonance is, at the lowest order, proportional to the <u>intensity</u> of the electromagnetic wave which gives rise to this resonance. For example, the Bloch-Siegert shift of the ordinary magnetic resonance induced by a linear RF field $\vec{B}_1 \cos\omega t$ perpendicular to the static field $\vec{B}_0$ is proportional to $B_1{}^2$ ([1]).

Chang and Stehle ([2]) have recently presented a quantum electro-dynamics (Q.E.D.) calculation of the higher order terms of the Bloch-Siegert shift (terms in $B_1{}^4$, $B_1{}^6$...). Their result, which exhibits a kind of oscillatory behaviour, disagrees with those of several other theoretical approaches, using a classical description of the RF field and called, for that reason, "semi-classical" ([3]) ([4])([5]). Chang and Stehle's conclusion is that semi-classical theories are not the appropriate limit of Q.E.D. at high intensities.

It seems very strange that pure quantum effects could appear in the RF domain where the number of quanta is extremely large. The same remark applies to the optical domain : the more intense is the optical wave, the less expected are the pure quantum effects. This does not mean that semi-classical theories are the only ones to be considered in these cases. We have recalculated ([6]) independently the higher order terms of the Bloch-Siegert shift using a quantum description of the RF field and the formalism of the "dressed atom" theory ([7])([8]). Our results are in complete agreement with those of "semi-classical" theories. We have also performed an experimental check ([9]) of the theory by measuring with great precision the position of a coherence resonance observable in transverse optical pumping experiments. The higher order terms in the radiative shift of this resonance, obtained from our calculation, fit very well the experimental results.

The misleading result of Chang and Stehle could give the impression that, in the RF domain, fully quantum mechanical calculations are more difficult to work out than semi-classical ones. This is not our opinion. We think on the contrary that, for high order effects, a correct quantum description of the field leads to calculations which are simpler as they proceed from time independent perturbation treatments.

To illustrate this last point, let's now briefly analyse our method of calculating the higher order radiative shifts.

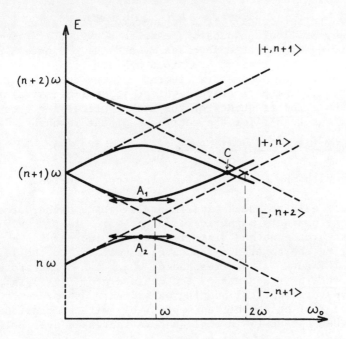

Fig. 1 : Energy levels of the system
"spin + RF photons"

The dotted lines of figure 1 represent the unperturbed energy
levels of the total system "spin + RF photons", drawn as a function
of ω_0 which is the Larmor pulsation in the static field B_0. The
$|+, n>$ state for example corresponds to the spin in the $|+>$ state
in the presence of n photons; the unperturbed energy of this state
is $(\omega_0/2) + n\omega$ where ω is the RF angular frequency. The coupling V
between the spin and the field perturbs these energy levels and
leads to the energy diagram $E(\omega_0)$ represented by the full lines of
fig. 1. The various magnetic resonances observable on the spin correspond to the "anticrossings" and to the crossings on this diagram.

For example, the first anticrossing which arises from the re-
pulsion between the levels originating from $|a> = |+, n>$ and
$|b> = |-, n+1>$ is associated to the ordinary magnetic resonance
(transition from $|->$ to $|+>$ by absorption of one RF quantum).
It can be shown that the center of the resonance is given by the
abscissa of the extremum A_1 (or A_2) of the curve $E(\omega_0)$. The difference between this abscissa and ω (abscissa of the crossing formed
by $|a>$ and $|b>$) is nothing but the radiative shift and comes
from the coupling induced by V between $|a>$ and $|b>$ and the other
unperturbed levels.

As shown in reference ([6]), a degenerate perturbation treatment using a Wigner-Brillouin expansion leads to a system of 2 implicit equations giving exactly the abscissa ω_0 and the ordinate E of extremum A_1 . Solving these 2 equations by iteration, one gets ω_0 to any desired order. The matrix elements of the coupling V can be easily related to the parameter ω_1 which, in semi-classical theories, represents the frequency of the Larmor precession around the RF field B_1. Our result for the centre ω_0 of the resonance is

$$\omega_0 = \omega - \frac{1}{\omega} \left(\frac{\omega_1}{4} \right)^2 - \frac{5}{4\omega^3} \left(\frac{\omega_1}{4} \right)^4 - \frac{61}{32\omega^5} \left(\frac{\omega_1}{4} \right)^6 + \ldots \tag{1}$$

in agreement with the result of semi-classical theories.

It is not easy to check experimentally these higher order effects on the ordinary magnetic resonance which is not only shifted, but also broadened and distorted when ω_1 increases. We preferred studying the coherence resonance associated to the level crossing C of figure 1. This resonance, which can be interpreted as a level crossing resonance of the dressed atom ([7])([8]),is observable in transverse optical pumping experiments and presents the great advantage of being shifted appreciably without considerable broadening when the RF field amplitude is increased over a large range.

Calculations similar to the ones mentioned above give the abscissa of C, i.e. the centre of the coherence resonance at any desired order. The result up to sixth order is :

$$\omega_0 = 2\omega - \frac{1}{6} \frac{\omega_1^2}{\omega} - \frac{7}{54\omega^3} \left(\frac{\omega_1}{2} \right)^4 - \frac{103}{2430\omega^5} \left(\frac{\omega_1}{2} \right)^6 + \ldots \tag{2}$$

The experimental results obtained on ^{87}Rb are compared to these theoretical predictions on figure 2.

The straight line (noted a) corresponds to the lowest order expression for the shift $2\omega - (\omega_1^2/6\omega)$. Curve b corresponds to expression (2). One sees that the experimental precision is sufficient to test the higher order terms (in ω_1^4 and ω_1^6). When ω_1/ω reaches a value (2.40) corresponding to the first zero of the J_0 Bessel function, it is possible to show that the coherence resonance vanishes at the position $\omega_0 = 0$ ([8]). Expression (2) is no more valid. A perturbative treatment of the Zeeman coupling of the spin with the static field is however possible and gives the theoretical curve c.

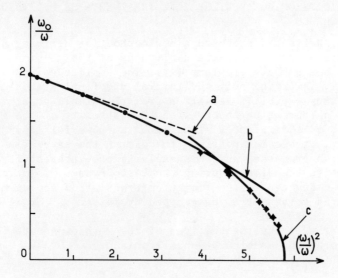

Fig. 2 : Position of the coherence resonance
         as a function of the RF intensity
         [measured by $(\omega_1/\omega)^2$]

    This good agreement between theory and experiment gives
confidence in the theoretical approach described above. Other experimental tests have been performed which check in detail the
whole variation with ω_0 of the energy levels represented in fig. 1
([10]). Let's also mention that the theoretical approach based on
continued fractions is very well adapted to a computer calculation
of the higher order effects ([5]).

    This shows that, for spins, higher order non-linear effects
are well understood and can be calculated with a great precision.
This is due mainly to the finite number of spin states. In the
case of atoms, the greater number of atomic states complicates the
perturbation treatment (in particular, the summation over intermediate states). Although some progress has been done in this direction for hydrogen atoms ([11]), it would be very important, for the
interpretation of experimental results, to derive new non-perturba-
tive methods. This leads us to the second part of this paper.

II. VALIDITY OF NON-PERTURBATIVE TREATMENTS OF MULTIPHOTON PROCESSES

In 1970, Reiss has proposed a new method for deriving non perturbative expressions for the amplitudes of the multiphoton processes in bound state problems ([12]). The amplitudes appear in a closed form, involving only the initial and final states of the transition; they exhibit a deviation from the simple power law predicted by perturbation theory when the intensity of the electromagnetic wave is sufficiently high. With the powerful laser sources now available, it becomes possible to observe the saturation of the multiphoton processes and this explains the interest for new non perturbative theories : during the last 2 years, a lot of papers have been devoted to the "momentum-translation approximation" (later referred to as m-t-a) developped by Reiss.

Let us briefly outline the idea of the method. An atomic system described by the unperturbed hamiltonian

$$H_0 = \frac{\vec{p}^2}{2m} + V(r) \tag{3}$$

starts at time t_1 from the state $|\phi_i(t_1)> = |\phi_i> e^{-iE_i t_1}$. One switches on adiabatically the potential vector $\vec{A}$ of an incident electromagnetic wave. A time t_2 later, $\vec{A}$ has been switched off. The state vector $|\psi(t_2)>$ of the system is given by the following exact integral equation :

$$|\psi(t_2)> = |\phi_i(t_2)> - i \int_{t_1}^{t_2} e^{-iH_0(t_2-\tau)} H'(\tau) |\psi(\tau)> d\tau \tag{4}$$

where

$$H' = -\frac{e}{m} \vec{A}.\ \vec{p} + \frac{e^2}{2m} \vec{A}^2 \tag{5}$$

describes the coupling between the atom and the wave. The transition amplitude from $|\phi_i(t_1)>$ to another field-free state $|\phi_f(t_2)> = |\phi_f> e^{-iE_f t_2}$ is finally given by :

$$U_{fi} = < \phi_f(t_2)\ |\psi(t_2)> \tag{6}$$

The main problem is to find a non perturbative expression for the state $|\psi(\tau)>$ appearing in the integral of (4).

The idea of m.t.a is to perform a unitary transformation $T(\tau)$ chosen in such a way that the transformed state

$$|\bar{\psi}(\tau)> = T(\tau)\ |\psi(\tau)> \tag{7}$$

obeys a Schrödinger equation in which the coupling between the atom and the field

$$H_I = T H' T^+ + i \frac{\partial}{\partial \tau} T \tag{8}$$

is sufficiently small to be neglected. The zeroth order solution of this equation

$$|\overline{\psi}^{(0)}(\tau) > = |\phi_i > e^{-iE_i\tau} \tag{9}$$

is then transformed back to give the approximate expression

$$|\psi(\tau) > \sim T^+(\tau) \; |\overline{\psi}^{(0)}(\tau) > = T^+(\tau) \; |\phi_i > e^{-iE_i\tau} \tag{10}$$

to be substituded for $|\psi(\tau) >$ in the integral of (4).

The unitary transformation chosen by Reiss is

$$T = \exp(-ie\vec{A}.\vec{r}) \tag{11}$$

and leads (in the dipole approximation) to the well known coupling

$$H_I = -e\vec{E}.\vec{r}$$

between the electric field of the incident wave and the atomic electric dipole e$\vec{r}$. It may be shown that for bound state problems, the ratio H'/H_I is of the order of ω_0/ω, where ω_0 is a typical atomic frequency and ω the frequency of the electromagnetic wave. If we are interested in multiphoton processes involving a large number N of photons, ω_0/ω which is of the order of N is large, and H_I much smaller than H'. Even if H' cannot be treated as a perturbation with respect to H_0, this can occur for H_I and is considered by Reiss as the justification for taking the zeroth order approximation for $|\overline{\psi}(\tau) >$.

Finally, the transition amplitude is approximated by

$$U_{fi} = \delta_{fi} - i \int_{t_1}^{t_2} e^{i(E_f-E_i)\tau} <\phi_f|H'(\tau) \exp\left[-ie\vec{A}(\tau).\vec{r}\right]|\phi_i> d\tau \tag{12}$$

and leads, when $|\phi_f > \neq |\phi_i >$ and when a sinusoïdal time dependence is assumed for $\vec{A}(\tau)$, to the following expression for the transition probability per unit time :

$$W_{fi} = 2\pi \sum_N |T_{fi}^{(N)}|^2 \; \delta(E_f - E_i - N\omega) \tag{13}$$

where

$$T_{fi}^{(N)} = i^N(E_i-E_f) < \phi_f|J_N(ea\vec{\epsilon}.\vec{r}) \; |\phi_i > \tag{14}$$

(J_N is the Nth order Bessel function, a and $\vec{\epsilon}$ the amplitude and polarization of the incident wave $\vec{A}$).

We think that the approximations made in m-t-a are not correct. Several objections concerning the validity of this method are explicitly described in reference ([14]). We will just summarize here the conclusions of this discussion.

(i) m–t–a does not predict at all any kind of radiative shift of the type described in the first part of this paper (the δ function of 13 gives $N\omega = E_f - E_i$). In addition, enhancements effects, which can occur when an intermediate state is quasi resonant for a p-quanta process with p < N, do not appear in the final closed expression of the transition amplitude.

(ii) When an exact solution is known (for a particular choice of the atomic potential $V(r)$), the results disagree with those given by m–t–a.

(iii) The unitary transformation performed by Reiss is in fact a well known gauge transformation ([13]) which, as quantum mechanics is gauge invariant, does not change the physical content of the theory. Treating in the new gauge H_I to zeroth order amounts to incorporate no interaction processes at all in the wave function (10) which approximates $|\psi(\tau) >$ in the integral of (4) (Note for example that the electronic density associated to this approximate wave function coincides with the unperturbed one) . When (10) is used, the only term which represents the coupling in (4) is $H'(\tau)$ and one does not understand how this procedure can describe correctly processes involving more than one photon.

(iv) One can try to improve the precision by approximating $|\psi(\tau) >$ by

$$T^{+}(\tau) \left(\; |\overline{\psi}^{(0)}(\tau) > + \; |\overline{\psi}^{(1)}(\tau) > \right) \tag{15}$$

which is better than (10) ($|\overline{\psi}^{(1)}(\tau) >$ is the first order contribution in H_I in the perturbation expansion of $|\overline{\psi}(\tau) >$) . One finds that the $|\overline{\psi}^{(1)} >$ term of (15) leads to expressions which have the same order of magnitude as the ones obtained by keeping only $|\overline{\psi}^{(0)}(\tau) >$. m–t–a is therefore non consistent as one neglects terms comparable to the ones which are kept.

The total expression obtained by using (15) is correct only to first order and, at this order, coincides with the results of ordinary perturbation theory. More generally, we have shown that, for a coherent calculation up to N^{th} order (this is the minimum required for interpreting a N-photon process), one has to use for $|\psi(\tau) >$ an expression at least as precise as

$$T^{+}(\tau) \sum_{p=0}^{N} |\overline{\psi}^{(N)}(\tau) > \tag{16}$$

(where $|\overline{\psi}^{(N)}(\tau) >$ is the Nth order contribution in H_I to $|\overline{\psi}(\tau) >$).

All these considerations show that, for bound states problems, it seems impossible to bypass perturbation theory by performing a unitary transformation and making some crude approximations in the new representation.

One can now ask about the possibility of applying non perturbative methods to bound state problems. It seems that the coupling
between a bound system and the electromagnetic field can be treated
to all orders at least in 2 simple cases :

a) situations where the "rotating wave approximation" can be
used, and which generalize the simple problem of a spin 1/2 interacting with a rotating RF field perpendicular to the static field.
This is interesting for the study of the one photon resonances and,
more precisely, for an exact treatment of the saturation effects
appearing as a result of successive <u>resonant</u> interactions of the
same atom with the electromagnetic field;

b) situations where the sinusoïdal interaction hamiltonian is
diagonal in the basis of unperturbed states (generalization of the
problem of a spin 1/2 interacting with a linearly polarized RF
field parallel to the static field).

We will say a few words about this last situation which is
less known than the first one, and which has been investigated in
some optical pumping experiments.

The diagonal character of the perturbation, which, on the one
hand, permits an exact solution of the equations of motion, exclu-
des, on the other hand, any resonant exchange of energy between the
atom and the field : the absorption (or induced emission) of a photon is not accompanied by a jump of the atom from one level to another, and the total energy cannot be conserved. However, virtual
processes involving one or several photons can occur, but they
require, in order to be detected, that we use a second probing
electromagnetic wave. So, we are faced with a new problem where the
atom interacts with 2 waves (1) and (2) : we can treat to all orders the coupling with wave (1) which can be as intense as we want,
but we must restrict to low intensities for the probing wave (2)
and use perturbation theory for describing its effect.

As a first example of such a situation, let us mention the
transverse optical pumping of atoms interacting with a RF field
$\vec{B}_1\cos\omega t$ parallel to the static field $\vec{B}_0$ (the wave (1) is the RF
field, while the optical pumping beam may be considered as playing
the role of wave (2)) . The resonances detected on the light absorbed or reemitted by atoms and which appear when the Larmor frequen-
cy ω_0 in $\vec{B}_0$ is equal to $n\omega$ (n = 0, 1, 2 ...), are associated with
the level crossings appearing in the energy diagram of the total
system spin + RF photons [7][8] (level crossing resonances of the
dressed atom). These coherence resonances can also be interpreted
as resulting from the interference between 2 different scattering
amplitudes [15]; each of these amplitudes, which starts from the
same initial state and ends at the same final state, involve the
absorption (or emission) of one optical photon and of an arbitrary

number of RF photons. The variation of the intensity of the resonances as a function of the RF field amplitude has been calculated
exactly and the saturation behaviour has been experimentally
checked in great detail (^{16}). This shows, as in the first part of
this paper, the interest of coherence phenomena associated to
transverse optical pumping experiments for the study of higher order
non linear effects.

As a second example, I will mention the study by Haroche of
the transitions between the 2 hyperfine levels of an alkali atom
and corresponding to the absorption of one microwave photon and
several radiofrequency photons (^{8}). The coupling with the microwave
(wave 2) is treated to first order; the coupling with the RF field
(wave 1) is treated to all orders (this coupling is diagonal in
zero static field, and the effect of an applied weak static field
is evaluated by perturbation theory). It has been possible to check
with a great accuracy all the theoretical predictions concerning
the radiative shifts and the saturation behaviour of these multi-
photon resonances for arbitrary values of the RF power.

This leads us to the following remark. The calculations of
non linear effects observed in the 2 experiments described above
are in fact equivalent to a partial resummation of the perturbation
series : one sums the contributions of all processes involving only
one photon of the wave 2 but an arbitrary number of photons of the
wave 1. One might then ask whether it would be possible, in some
other cases, to pick out by some physical arguments a certain class
of processes which are the most important for the phenomenon being
studied, and to sum up all the corresponding contributions. There
is perhaps in this direction a possibility for getting the non
perturbative expressions which are urgently needed at the present
time.

REFERENCES

(^{1}) F. Bloch, A. Siegert - Phys. Rev. 57, 522 (1940)
(^{2}) C.S. Chang, P. Stehle - Phys. Rev. A, 4, 641 (1971)
                                    Phys. Rev. A, 5, 1087 (1972)
(^{3}) J.H. Shirley - Phys. Rev. 138 B, 979 (1965)
(^{4}) D.T. Pegg, G.W. Series - J. Phys. B. Atom. Molec. Phys. 3,
             L33 (1970)
    D.T. Pegg, G.W. Series - Proc. Roy. Soc. 332, 281 (1973)
    D.T. Pegg - J. Phys. B. Atom. Molec. Phys. 6, 246 (1973)
(^{5}) S. Stenholm - J. Phys. B. Atom. Molec. Phys. 5, 876 and 890
             (1972)

(^{6}) C. Cohen-Tannoudji, J. Dupont-Roc, C. Fabre - Submitted to
            J. Phys. B. Atom. Molec. Phys. (1973)
(^{7}) C. Cohen-Tannoudji - Cargèse Lectures in Physics, Vol. $\underline{2}$,
            Ed. M. Lévy (New York), p. 347-93 (1968)
(^{8}) S. Haroche - Ann. Phys. Paris, $\underline{6}$, 189 (1971)
(^{9}) C. Cohen-Tannoudji, J. Dupont-Roc, C. Fabre - Submitted to
            J. Phys. B. Atom. Molec. Phys. (1973)
(^{10}) C. Landré, C. Cohen-Tannoudji, J. Dupont-Roc, S. Haroche -
            C.R. Acad. Sci. Paris, $\underline{270\ B}$, 73 (1970)
(^{11}) Y. Gontier, M. Trahin - Phys. Rev. $\underline{172}$, 83 (1968)
(^{12}) H.R. Reiss - Phys. Rev. A $\underline{1}$, 803 (1970)
       H.R. Reiss - Phys. Rev. D $\underline{4}$, 3533 (1971)
(^{13}) M. Goeppert-Mayer - Ann. Physik, $\underline{9}$, 273 (1931)
       J. Fiutack - Canad. J. of Phys. $\underline{41}$, 12 (1963)
(^{14}) C. Cohen-Tannoudji, J. Dupont-Roc, C. Fabre, G. Grynberg -
            Submitted to Phys. Rev. (1973)
(^{15}) C. Cohen-Tannoudji, S. Haroche - J. Phys. (Paris), $\underline{30}$, 125
            (1969)
(^{16}) N. Polonsky, C. Cohen-Tannoudji - C.R. Acad. Sci. Paris, $\underline{260}$,
            5231 (1965)

TUNABLE LASERS I

GENERATION OF ULTRAVIOLET AND VACUUM ULTRAVIOLET RADIATION*

S. E. Harris, J. F. Young, A. H. Kung, D. M. Bloom, and
G. C. Bjorklund

Microwave Laboratory
Stanford University
Stanford, California 94305

ABSTRACT

The paper describes the use of nonlinear optical techniques
for the generation of coherent radiation at ultraviolet, vacuum
ultraviolet, and soft x-ray wavelengths. Mixtures of metal vapors
and inert gases, and other mixed gas systems, allow generation to
regions of the spectrum where nonlinear optical crystals are opaque;
and also allow generation at high incident power and energy densities. Progress is reported on programs aimed at efficient conversion from 1.064μ to 3547 Å, and from 3547 Å to 1182 Å. To date,
the shortest wavelength generated by this technique is 887 Å. Theoretical considerations indicate that generation into the soft x-ray
region should be possible. Using 1182 Å radiation, a holographic
grating with a fringe spacing of 836 Å has been constructed and examined on an electron microscope.

I.   INTRODUCTION

The difficulty of obtaining laser oscillation increases sharply
as shorter wavelengths are approached. This is due to rapidly decreasing spontaneous lifetimes; to generally increasing optical loss

---

*The work reported in this paper was jointly supported by the
Office of Naval Research, the U.S. Army Research Office, the Air
Force Cambridge Research Laboratories, the Atomic Energy Commission,
and the National Aeronautics and Space Administration.

(both single-photon and multi-photon); and to more practical difficulties such as the fabrication of sufficiently flat high-reflecting
surfaces, and the difficulty of generating a sufficiently hot and
intense pumping source.

The difficulties with nonlinear optical techniques are not as
severe. Although nonlinear optical susceptibilities decrease with
decreasing wavelength, this is in the most part offset by the ability to utilize much higher incident pumping power densities, and by
the inverse square wavelength dependence of conversion efficiency
at constant nonlinear optical susceptibility and incident power
density.

To date, the most intense source of ultraviolet radiation is
obtained by frequency quadrupoling Nd:glass or Nd:YAG lasers at
1.064μ to yield 2660 Å.[1] Broadly tunable ultraviolet radiation is
best obtained by frequency doubling and summing of visible frequency dye lasers with nonlinear optical crystals. Using these
techniques, spectral coverage of 2610 Å to 3150 Å,[2] and tunable
peak powers of about 1 MW have been obtained.[3] The shortest wavelength thus far generated with nonlinear optical crystals is 2128 Å,[4]
which is about the limit of sufficient birefringence of known nonlinear crystals.

In this paper we will discuss nonlinear optical techniques
which make use of phase-matched mixtures of metal vapors and inert
gases,[5,6] or in certain cases of only inert gases.[7] These systems
have the advantage of transparency in regions of the spectrum where
nonlinear crystals are opaque. When properly chosen they are phase-
matchable, and have high nonlinear optical susceptibilities. Their
high ionization potential allows optical power densities of 10^{10} or
10^{11} watts/cm^2 to be employed.

In the following sections of this paper we first describe progress on an experiment aimed at efficiently generating 3547 Å radiation from incident 1.064μ radiation. Generation of vacuum ultraviolet radiation at 1773 Å, 1520 Å, 1182 Å, and 887 Å is described.
Theoretical considerations with regard to the further use of this
technique for generation of soft x-ray radiation are discussed.
Finally, we report the construction of a vacuum ultraviolet holographic grating with a fringe spacing of 836 Å.

II. CONVERSION FROM 1.064μ TO 3547 Å

A schematic of the basic configuration for third harmonic generation in a phase-matched mixture of a metal vapor and inert gas
is shown in Fig. 1. The metal vapor is negatively dispersive, i.e.,
its refractive index at the fundamental wavelength of 1.064μ exceeds

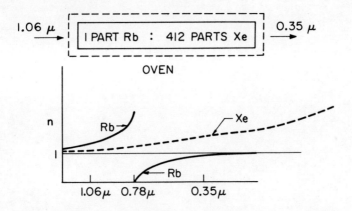

FIG. 1--Schematic of third harmonic generation in phase-
matched metal vapor inert gas systems.

its refractive index at the third harmonic wavelength of 0.3547μ.
The inert gas is positively dispersive, and at an appropriate ratio
of metal vapor to inert gas exactly compensates for the negative
dispersion of the metal vapor. For a mixture of Rb and Xe, this
ratio is 1 part Rb to 412 parts Xe.[6,8]

The choice of a metal vapor as the nonlinear ingredient is indicated not only by its negative dispersion, but also by its large
third-order nonlinearity. The large nonlinearity results from near
coincidences of the fundamental frequency, twice the fundamental,
and three times the fundamental, with the resonance lines of the
atomic specie. Figure 2 shows the relation of the incident beam
and its harmonics to the energy level structure of Rb. The calculated nonlinear susceptibility, $\chi^{(3)}$, for Rb is shown as a
function of incident optical wavelength in Fig. 3.[8]

Once phase-matching is achieved, the conversion efficiency of
a nonlinear process of this type increases as the square of the cell
length, metal vapor pressure, and incident power density. Figure 4
shows experimental conversion efficiency as a function of incident
1.064μ power. This experiment utilized 5 Torr of Na, phase-matched
with 806 Torr of Xe, and had an active optical path length of 40 cm.
A maximum conversion efficiency of 1% was obtained at an incident
power of about 2×10^8 watts. The power density was about 10^{11}
watts/cm^2 over the active 40 cm zone.

Theoretical considerations[8] indicate that the most important
limitation on the obtainable conversion efficiency will be the

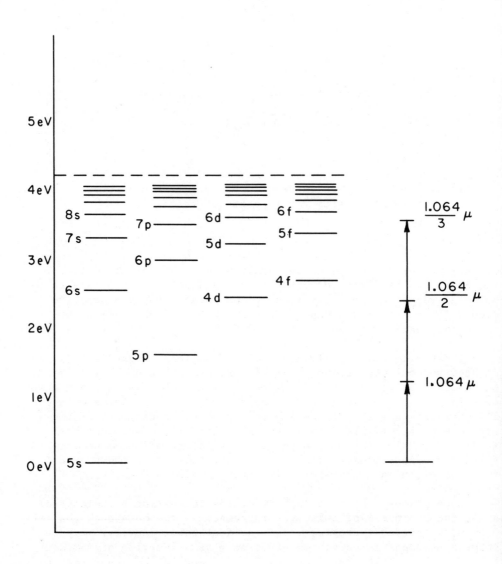

FIG. 2--Energy levels of Rb relative to the harmonics of 1.064µ
        radiation.

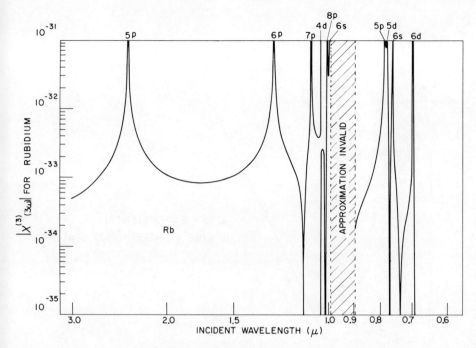

FIG. 3--Nonlinear susceptibility for third
harmonic generation in Rb.

breaking of the phase-matching condition resulting from very small
absorption of the 1.064μ or 3547 Å beams. For example, if the
metal vapor cell is 100 coherence lengths long, then the phase-
matching condition will be broken when about 1% of the Na atoms
are removed from the ground state. Assuming an incident pulse
with an energy density of 3 J/cm², this allows a maximum 1.064μ
absorption of about 10^{-4}. Analysis shows[8] that as a result of
this condition on the breaking of phase-matching, that the square
of the maximum number of Na atoms which may be employed varies inversely as the incident energy density and cell length. At incident energy densities of several joules/cm² and cell lengths of
about a meter, the optimum metal vapor pressure will be in the
range of a few Torr. In Fig. 4 we note that the conversion efficiency is still increasing as the square of the incident power
density, and thus the interruption of phase-matching has not yet
begun.

Considerable effort has gone into determining the best cell
configuration for metal vapor inert gas systems of this type. Two
alternatives are a closed, completely heated sidearm cell, and a
concentric double heat pipe oven.[9,10] In the sidearm cell

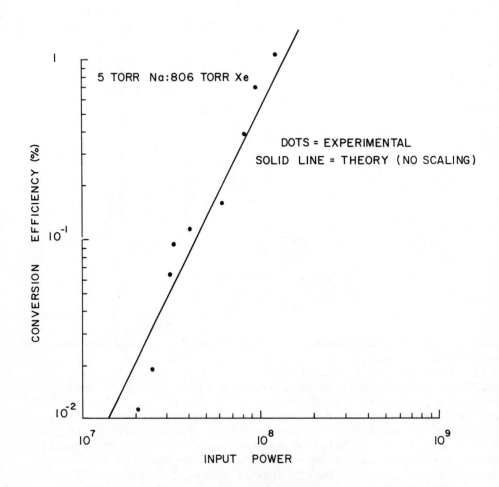

FIG. 4--Conversion efficiency vs. input power for tripling of
     1.064μ radiation. The active metal vapor zone in this
     experiment was 40 cm.

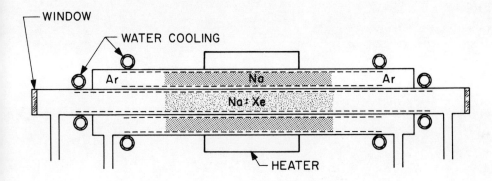

FIG. 5--Schematic of double heat pipe oven.

configuration the metal vapor is in contact with heated sapphire
windows, and its vapor pressure is determined by the coldest point
in the cell. In our first experiments with this type of cell we
have found that it is very difficult to vary the vapor pressure of
the metal vapor sufficiently slowly. The output cell window has
also been scarred by the 1.064μ beam. Figure 5 is a schematic of
the double concentric heat pipe oven which we are presently using.
The outer oven operates in a heat pipe mode and provides a constant
temperature zone for the inner oven. Using this technique we have
now obtained in excess of 75 coherence lengths of phase-matched Na.
The principle problem is Na condensation at the boundary layer between the hot Na-Xe mixture and the cold Xe buffer gas which pro-
tects the input and output windows. Considerable experimental effort will be necessary to eliminate this problem.

Optimization to achieve maximum conversion efficiency involves
first increasing the incident power density to the maximum value
allowed by multiple-photon ionization or avalanche breakdown. The
metal vapor density is then increased to a maximum value determined
by the incident energy density and the cell length. Once in this
regime, the conversion efficiency will only increase linearly with
increasing cell length.[8]

III.   VUV GENERATION

A summary of experiments which have thus far been performed to
extend the frequency tripling technique into the vacuum ultraviolet
is shown in Fig. 6. In early experiments we employed a phase-matched
mixture of Cd and Ar in the ratio 1 part Cd to 25 parts Ar to gener-

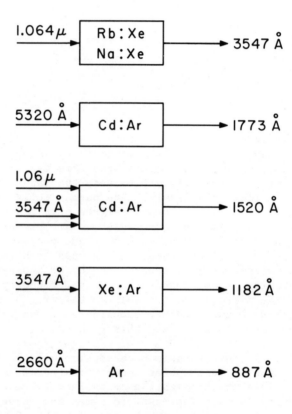

FIG. 6--Summary of experiments for generation of vacuum ultraviolet radiation.

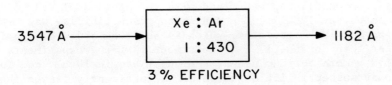

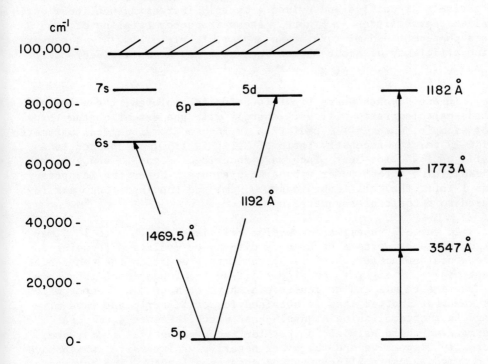

FIG. 7--Generation of 1182 Å radiation in a phase-matched mixture
        of Xe and Ar.

ate 1773 Å.[11] We estimated $\chi^{(3)}$ to be about 2×10^{-34} esu/atom for tripling 5320 Å radiation. This nonlinearity is quite high for this region of the spectrum and, noting the relatively low absorption of 5320 Å by Cd, this system should lead to an efficient and convenient source of 1773 Å radiation. In a second experiment, one photon of 1.06µ radiation was mixed with two photons of 3547 Å to yield an output at 1520 Å. The nonlinear susceptibility was measured to be 2×10^{-33} esu/atom, which is about an order of magnitude larger than for the 1773 Å experiment. Mixing experiments often allow a more optimum use of the upper atomic levels and lead to nonlinear susceptibilities which are significantly larger than those obtainable in direct tripling experiments.

Once the generated wavelength becomes shorter than the longest resonance line of the inert gases, the inert gases themselves become negatively dispersive and evidence sharply increasing nonlinear optical susceptibilities.[7] Figure 7 shows frequency tripling of 3547 Å in a phase-matched mixture of Xe and Ar, to yield 1182 Å. A conversion efficiency of about 3% has been obtained at an incident power density at the focus of about 6.3×10^{12} watts/cm^2.

In experiments where tight focusing is employed, the necessary phase-matching ratio will vary sharply with the extent of the focus. For example, when 3547 Å radiation is focused to a confocal parameter of 0.25 cm, the necessary ratio of Xe:Ar is 1:50, as opposed to a ratio of 1:430 for near plane wave focusing. Figure 8 shows normalized 1182 Å output power versus Ar pressure. Here, the Xe pressure was 1 Torr, the cell length was 0.95 cm, and the input beam was focused to a confocal parameter of 2.1 cm.

In certain cases, a conversion efficiency of 10^{-4} to 10^{-3} may be obtained by using a negatively dispersive media and focusing to a confocal parameter which is approximately equal to the coherence length.[8,12] A tight focus allows off-angle, fundamental frequency k vectors to mix and to generate a third harmonic k vector which is somewhat shorter than is obtained for collinear plane wave mixing, thus accomplishing a quasi-phase match. Figure 9 compares normalized third harmonic output for two hypothetical media, one normally dispersive and the other negatively dispersive, which have the same optical nonlinearity and coherence length. The horizontal axis or cell temperature may instead be interpreted as vapor pressure or atom density of the respective media. Note that the functional form of normalized output power for the negatively dispersive media differs substantially from that of the positive media. For very tight focusing, the output of a normally dispersive media becomes exceedingly small, while that of a negatively dispersive media always exhibits a pronounced peak at the atom density whose coherence length is approximately equal to the confocal parameter of the focus. Curves of this type allow one to distinguish posi-

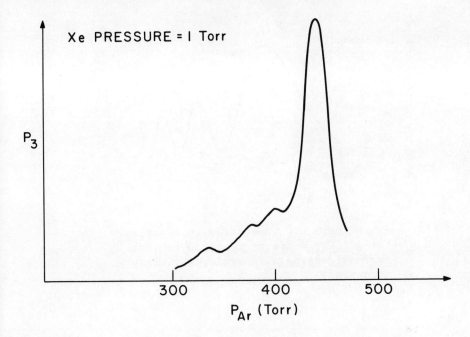

FIG. 8--Phase-matched 1182 Å output vs. argon pressure.

tively dispersive media from negatively dispersive media, and also, by measuring their periodicity, to determine the coherence length of the media. A direct and difficult measurement of refractive index is thus not necessary.

To date, the shortest wavelength generated in these experiments was obtained by tripling 2660 Å radiation to yield 887 Å. The non-linear media was Ar at a pressure of several Torr. Since LiF is no longer transparent at this wavelength, differential pumping was used between the cell exit and the entrance to the McPherson monochromator. The observed efficiency was only 10^{-7}, and the signal was too low to allow a determination of whether Ar is positively or negatively dispersive for this process.

IV.   SOFT X-RAY GENERATION

To extend nonlinear optical techniques into the soft x-ray region of the spectrum it is of interest to consider the use of higher

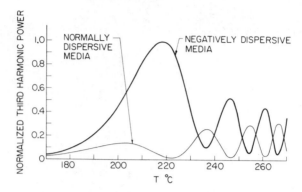

FIG. 9--Third harmonic generation in negatively
        dispersive media as compared to that in
        normally dispersive media, for tight
        focusing.

order nonlinear optical polarizabilities which might allow fifth,
seventh, or higher order harmonic generation, and allow one to
move in bigger steps through the frequency spectrum. A recent
analysis[13] has examined the relative magnitude of high-order
polarizations, subject to the condition that the applied electric field strength not exceed the multi-photon absorption or
ionization limit.

     Analysis shows that if the generated frequency is sufficiently
close to an upper level of the atom that this level determines both
the coherence length and also the multi-photon absorption limit on
the incident applied intensity, then the conversion efficiency is
independent of the order of the nonlinear optical polarizability
employed, and also of the oscillator strengths and positions of
the intermediate levels. If intermediate levels have smaller
oscillator strengths or resonant denominators, the incident applied field is allowed to increase to yield the same conversion
efficiency.

     Results of these calculations are shown in Table I. The first
two examples are concerned with generation of VUV radiation in Xe,
while the latter three are concerned with the generation of soft
x-rays in singly ionized Li. Ionization will be accomplished by

TABLE I--Limiting power density and conversion efficiency for some higher-order nonlinear process.

Process	Specie	$(P/A)_{max.}$ (W/cm^2)	Efficiency (Percent)
3×5320 Å $\rightarrow 1773$ Å	Xe	1.9×10^{12}	.08
5×5320 Å $\rightarrow 1064$ Å	Xe	1.9×10^{12}	.05
5×1182 Å $\rightarrow 236$ Å	Li^+	1.7×10^{15}	.002
7×1182 Å $\rightarrow 169$ Å	Li^+	1.7×10^{15}	.004
15×2660 Å $\rightarrow 177$ Å	Li^+	3.5×10^{15}	4×10^{-7}

the incident laser pulse. At ion densities of approximately 10^{18} ions/cm^3 recombination times are several nanoseconds, and each atom need be ionized only once during the incident pulse. Due to the tight focus, this will require about 10^{-6} of the incident pulse energy.

Note, for instance, that the seventh order process 7×1182 Å $\rightarrow 169$ Å has a somewhat higher predicted conversion efficiency than the fifth order process 5×1182 Å $\rightarrow 236$ Å. The conversion efficiencies in this table are those obtainable in a single coherence length, and might be improved upon by means of phase-matching techniques.

V. HOLOGRAPHY IN THE VACUUM ULTRAVIOLET

One of the interesting applications of short wavelength radiation is holographic microscopy of sub-micron specimens. Radiation generated by a nonlinear optical technique has approximately the same coherence properties as that of the fundamental frequency

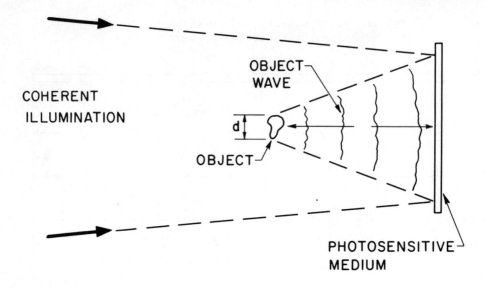

$$\text{FAR FIELD DISTANCE} = d^2/\lambda$$
$$d = 1000\,\text{Å}$$
$$\lambda = 300\,\text{Å}$$
$$\text{FAR FIELD} = 0.3\mu$$

FIG. 10--Schematic of holographic technique for application at
soft x-ray wavelengths.

laser source, and should be immediately useful for recording high-
quality holograms. Our general approach to the microscopy problem
will be to record a far-field or Fraunhofer hologram[14,15] of a
small object onto a grainless photosensitive media, and to then
use an electron microscope and computer techniques to read-out the
hologram and reconstruct the object.[16] Figure 10 shows the general

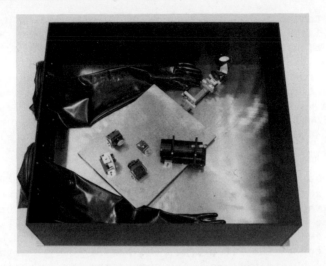

FIG. 11--Photo of holographic apparatus.

$\vdash$ Iμ $\dashv$

1591 Å FRINGES
PRODUCED IN PMMA
BY 1182 Å RADIATION

FIG. 12--Photo of 1591 Å fringes produced with harmonically generated 1182 Å beam.

approach for taking a Fraunhofer hologram. For sub-micron particles the far-field distance may itself be very small, and spacings
on the order of a micron between the object and the photosensitive
surface are sufficient to place the object in the far-field.   In
the far-field the object does not cast a shadow and thus the focused illumination acts both as the object illumination wave and
the reference wave. The input laser beam may be tightly focused,
and an additional reference beam is not required. This technique
also has the advantage of not requiring large f number high precision optics, or Fresnel lenses.

As a first step in the holographic program we have used the
harmonically generated beam of 1182 Å radiation to construct a
holographic grating on a Polymethyl methacrylate (PMMA) substrate.
A photo of the holographic set-up is shown in Fig. 11.   The 1182 Å
radiation is generated in a phase-matched Xe-Ar cell, and enters
the box adjacent to the LiF prism. The box is covered with a plexiglass top and purged with He. The components are conveniently
manipulated using the gloves shown in the photo. Figure 12 shows
a scanning electron microscope read-out of holographically produced
fringes with a spacing of 1591 Å. More recently, fringes with a
spacing of 836 Å have been constructed. Fringe construction of this
type allows the spatial frequency response of different recording
media to be examined and may also have application to sub-micron
fabrication.

ACKNOWLEDGEMENTS

The authors gratefully acknowledge many helpful discussions
with E. A. Stappaerts, and thank B. Yoshizumi for help with the
experimental apparatus.

REFERENCES

1. S. A. Akhmanov, A. I. Kovrigin, A. S. Piskarskas, and R. V.
    Khokhlov, JETP Letters 2, 141 (1965).

2. R. W. Wallace, Opt. Commun. 4, 316 (1971).

3. D. J. Bradley, J. V. Nicholas, and J. R. D. Shaw, Appl. Phys.
    Letters 19, 172 (1971).

4. A. G. Akmanov, S. A. Akhmanov, B. V. Zhdanov, A. I. Kovrigin,
    N. K. Podsotskaya, and R. V. Khokhlov, ZhETF Pis. Red. 10,
    244 (1969).

5. S. E. Harris and R. B. Miles, Appl. Phys. Letters 19, 385 (1971).

6. J. F. Young, G. C. Bjorklund, A. H. Kung, R. B. Miles, and S. E. Harris, Phys. Rev. Letters 27, 1551 (1971).

7. A. H. Kung, J. F. Young, and S. E. Harris, Appl. Phys. Letters 22, 301 (1973).

8. R. B. Miles and S. E. Harris, IEEE J. Quant. Elect. QE-9, 470 (1973).

9. C. R. Vidal and J. Cooper, J. Appl. Phys. 40, 3370 (1969).

10. C. R. Vidal and M. N. Hessel, J. Appl. Phys. 43, 2776 (1972).

11. A. H. Kung, J. F. Young, G. C. Bjorklund, and S. E. Harris, Phys. Rev. Letters 29, 985 (1972).

12. J. F. Ward and G. H. C. New, Phys. Rev. 185, 57 (1969).

13. S. E. Harris, Phys. Rev. Letters 31, 341 (1973).

14. J. B. DeVelis, G. B. Parrent, Jr., and B. J. Thompson, J. Opt. Soc. Amer. 56, 423 (1966).

15. J. W. Goodman, private communication.

16. J. W. Goodman and R. W. Lawrence, Appl. Phys. Letters 11, 77 (1967).

PARAMETRIC OSCILLATORS

Robert L. Byer

Applied Physics Dept., Stanford University

Stanford, California 94305

INTRODUCTION

Parametric oscillators and parametric mixing and sum generation devices offer the widest tuning range of known coherent sources. This paper reviews the present state of parametric devices with a bias toward their use in spectroscopy. In addition to tuning range, bandwidth and bandwidth narrowing methods are considered. The discussion includes the properties of new infrared nonlinear materials that permit parametric tuning over the entire infrared spectrum. Finally, application of parametric oscillators to spectroscopy is illustrated and predictions of future tuning range and output performance is discussed.

LiNbO$_3$ optical parametric oscillators are now in wide use. The properties of parametric oscillators have been reviewed by Harris[1] and recently by Smith[2] and Byer.[3] Parametric oscillator sources cover a spectral range from 0.4 μ to 0.8 μ with ADP[4], 0.6 μ to 3.5 μ with LiNbO$_3$[5], 2 μ and 9.6 μ to 11 μ in CdSe[6] and 1.22 μ to 8.5 μ with proustite.[7] LiNbO$_3$ is the best developed parametric oscillator and is the only oscillator available commercially.

If we include mixing and sum generation in our general definition of parametric sources, then these three frequency processes satisfy the energy and momentum conservation conditions

$$\omega_3 = \omega_2 + \omega_1 \tag{1}$$

and

$$k_3 = k_2 + k_1 \qquad (2)$$

The interaction takes place in the nonlinear material with a gain or conversion efficiency given by

$$G = \Gamma^2 \ell^2 = \left(\frac{2 \, \omega_o^2 \, d^2}{\pi \, n_o^2 \, n_3 \, \epsilon_o \, c^3} \right) P_{30} \, \ell \, k_o (1 - \delta^2)^2 \, \bar{h}(B, \xi) \qquad (3)$$

where ω_o is the degenerate angular frequency, d is the nonlinear coefficient in mks units

$$d \left(\frac{m}{V} \right) = \left(\frac{3 \times 10^4}{4 \, \pi} \right)^{-1} d \, (cm/statvolt)$$

P_{30} is the pump power, ℓ the crystal length,

$$k_o = \frac{2\pi \, n_o}{\lambda_o} \qquad (4)$$

and

$$(1 - \delta^2) = \omega_1 \omega_2 / \omega_o^2 \qquad (5)$$

is the degeneracy factor, and $\bar{h}(B, \xi)$ is the Boyd and Kleinman[9] focusing factor.

For parametric oscillators and mixing experiments, the gain reduces to

$$G(\ell) = \Gamma^2 \ell^2 \, \text{sinc}^2 \left(\frac{\Delta k \ell}{2} \right) \qquad (6)$$

in the low gain limit and to

$$G(\ell) = \frac{1}{4} \exp 2\Gamma\ell \tag{7}$$

in the high gain limit. Here $\text{sinc}^2(\Delta k\ell/2)$ is the phase mismatch factor which leads to an expression for the bandwidth of the interaction. For sum generation the conversion efficiency is the same in the low conversion limit but varies as $\sin^2 \Gamma\ell$ in the high conversion limit.

Equation (3) shows that focusing determines the mixing conversion efficiency and parametric gain through the focusing parameter $\bar{h}(B,\xi)$ which describes coupling between gaussian modes as a function of the double refraction parameter

$$B = \frac{1}{2} \rho (\ell k_o)^{\frac{1}{2}} \tag{8}$$

and the focusing parameter

$$\xi = \ell/b \tag{9}$$

where the confocal parameter

$$b = w_o^2 k_o = \frac{w_o^2 \, 2\pi \, n_o}{\lambda_o} \tag{10}$$

Here w_o is the gaussian beam electric field radius such that the power in the gaussian beam is related to the peak intensity by

$$P = I_o(\pi \, w_o^2/2) \tag{11}$$

The double refraction parameter is a function of the double refraction angle ρ given by

$$\tan \rho = \frac{n_o^2}{2} \left[\frac{1}{n^e(\theta)^2} - \frac{1}{(n_o^o)^2} \right] \sin 2\theta \tag{12}$$

where θ is the propagation direction with respect to the optic axis and we have assumed $n^e < n^o$ or a negative unixial crystal.

In terms of the aperture length

$$\ell_a = \frac{w_o \sqrt{\pi}}{\rho} \tag{13}$$

the double refraction parameter can be written as

$$B = \frac{\sqrt{\pi}}{2} \ell/\ell_a \, \xi^{-\frac{1}{2}} \tag{14}$$

For parametric interactions $\bar{h}(B,\xi)$ can usually be approximated by one of its limiting forms. In the near field approximation with negligible double refraction

$$\bar{h}(B,\xi) \to \xi \qquad \left(\xi < 0.4 \, , \, \xi < \frac{1}{6B^2} \right) \tag{15}$$

For confocal focusing where $\xi = 1$

$$\bar{h}(0,1) \to 1 \tag{16}$$

which corresponds to the near field focusing limit first discussed by Boyd and Ashkin.[10] In fact, the maximum value of $\bar{h}(B,\xi) = \bar{h}_{mm}(0,2.84) = 1.068$ for $\xi = 2.84$ instead of $\xi = 1$. However, practical considerations usually limit focusing such that $\xi < 1$.

When double refraction is important the gain reduction factor is closely approximated by the expression[9]

$$\bar{h}_{mm}(B) \approx \frac{1}{1 + (4B^2/\pi)} \tag{17}$$

which can be rewritten as

$$\bar{h}_{mm}(B) \approx \frac{1}{1 + (\ell/\ell_{eff})} \tag{18}$$

where the effective interaction length is given by

$$\ell_{eff} = \frac{\lambda_o}{2\,n_o\,\rho^2} \tag{19}$$

In the limit of strong double refraction where $\ell/\ell_{eff} > 1$ the focusing parameter becomes

$$\bar{h}_{mm}(B) \to \ell_{eff}/\ell \qquad (B^2/4 > \xi > 2/B^2) \tag{20}$$

so that the conversion efficiency as a function of input power is independent of crystal length. In addition, large double refraction maintains $\bar{h}_{mm}(B)$ constant over a wide range of ξ . Thus to minimize crystal damage problems ξ can be chosen at the least value consistent with maximum $\bar{h}_{mm}(B)$. This value is given by

$$\xi > 2/B^2 \tag{21}$$

which yields a corresponding focal spot size

$$w_o \lesssim \frac{1}{2\sqrt{2}}\,\rho\,\ell \tag{22}$$

and corresponding area

$$\pi\,w_o^2/2 \lesssim \frac{\pi}{16}\,\rho^2\ell^2 \tag{23}$$

Double refraction due to non-collinear phasematching results in a serious reduction of $\Gamma^2 \ell^2$ for power limited pump conditions. For example, for LiNbO$_3$ at room temperature, phasematched at $\theta = 43°$ for a 1.06 μm pump source, $\rho = 0.037$ radians and $B = 4.7\, \ell^{\frac{1}{2}}$. The gain is reduced by $\pi/4B^2 = \ell/\ell_{eff}$ which in this case is 28 times for a 1 cm crystal and 140 times for a 5 cm crystal compared to the 90° phasematched case. For this case ℓ_{eff} is only .36 mm and ξ can vary between $5.5 > \xi > 0.09$ without affecting the gain. This corresponds to a confocal parameter variation between 11 cm and .2 cm. For experimental ease and minimum incident intensity at the LiNbO3 crystal the 11 cm confocal focusing would be used. The pump intensity varies as ℓ^2 in agreement with Eq. (23).

For high power pumping conditions where more than adequate pump power is available, the minimum focal area may be dictated by the crystal damage intensity. Most semiconductor materials show surface damage at near 1J/cm^2. For other nonlinear materials this energy density may approach 4J/cm2.[11] For focusing determined by incident pump energy density with areas larger than that given in Eq. (23), the double refraction gain reduction is correspondingly smaller. In this case, 90° phasematching maintains only the advantages of larger pump acceptance angle and larger effective nonlinear coefficient.

Table I lists selected nonlinear materials and their properties of interest in parametric interactions. Of interest for infrared generation is LiNbO3 proustite, CdSe and the four chalcopyrite crystals AgGaS$_2$, AgGaSe$_2$, ZnGeP$_2$ and CdGeAs$_2$. The characteristics of tunable sources based on these materials are considered in detail by Byer.[3]

EXTENDED INFRARED TUNING

Of particular interest to spectroscopists is the extended tuning range available from parametric devices. In addition, bandwidth and techniques for bandwidth narrowing and ultimately single frequency operation are important.

There has been considerable progress recently in extending the tuning range of parametric oscillators. However, to date the most reliable parametric oscillator uses LiNbO$_3$ at 90° phasematching with temperature tuning. This source covers,with good efficiency and operating stability,the spectral region from 0.6 μ to 3.5 μ with the second harmonic of a Q-switched Nd:YAG laser pump source operating at .532 μ , .561 μ and .659 μ . Figure 1 shows the experimental tuning curve for the .659 μ pump wavelength Also shown are the wavelengths generated by internal SHG and sum

TABLE I

Nonlinear Coefficient, Figure of Merit, and Gain for Nonlinear Crystals

MATERIAL (point group pump wavelength)	$d \times 10^{12}$ (m/V)	n_o n_e	$n_e - n_o$	θ_m	ρ	$d_{eff_{12}} \times 10$	$d^2_{eff}/n^3_o n_3 \times 10^{24}$	$\ell(\rho)_{eff}$ (cm)	$\Gamma^2\ell^2$ 1 Watt	$\Gamma^2\ell^2$ 1 MW/cm²	I_{burn} (MW/cm²)	Transmission Range (μm)
CdGeAs$_2$ ($\bar{4}$2m) $\lambda_p = 5.3$ μm	$d_{36} = 236$[a,b]	3.51 3.59	+.086	II 55° I 35°	.021 .021	193 (d sin θ) 212 (d sin 2θ)	861 1039	.34	3.8×10^{-4} 4.6×10^{-4}	.033 .040	20 - 40	2.4 - 17
ZnGeP$_2$ ($\bar{4}$2m) $\lambda_p = 1.83$ m	$d_{36} = 75$[c]	3.11 3.15	+.038	II 90° I 62°	0.0 0.01	d_{36} 62.2 (d sin 2θ)	187 127	$\ell = 1$ cm .59	7.1×10^{-3} 1.8×10^{-3}	.21 .05	> 4	.7 - 12
AgGaSe$_2$ ($\bar{4}$2m) $\lambda_p = 1.83$ m	$d_{36} = 33$[d]	2.62 2.58	-0.32	I 55° I 90°	.01 0.0	27 (d sin θ) d_{36}	42 63.4	.71 $\ell = 1$ cm	5.52×10^{-4} 1.98×10^{-3}	.02 .07	> 10	.73 - 17
CdSe (6 mm) $\lambda_p = 1.83$ μm	$d_{31} = 19$[e]	2.45 2.47	+.019	90°	0.0	d_{31}	24	$\ell = 2$ cm	1.3×10^{-3}	.09	60	.75 - 25
AgGaS$_2$ ($\bar{4}$2m) $\lambda_p = .946$ μm	$d_{36} = 12$[f]	2.42 2.36	-.054	I 64° I 90°	0.17 0.0	10.8 (d sin θ) d_{36}	9.2 11	.14 $\ell = 1$ cm	2.3×10^{-4} 2.3×10^{-3}	.013 .045	12 - 25	.60 - 13
Ag$_3$AsS$_3$ (3m) $\lambda_p = 1.06$ μm	$d_+ = 11.6$[g]	2.76 2.54	-.223	30°	0.078	d_+	8.2	.007	9.0×10^{-6}	.011	12 - 40	.60 - 13
LiIO$_3$ (6)	$d_{31} = 7.5$[h]	1.85 1.72	-1.35	23°	.071	3.04 (d sin θ)	1.88	.008	5.6×10^{-6}	5.5×10^{-3}	125	.31 - 5.5
LiNbO$_3$ (3m) $\lambda_p = .532$	$d_{31} = 6.25$[i]	2.24 2.16	-.081	90°	0.0	d_{31}	3.88	$\ell = 5$ cm	2.1×10^{-2}	1.28	50 - 140	.35 - 4.5
ADP ($\bar{4}$2m) $\lambda_p = .266$	$d_{36} = .57$[j,k]	1.53 1.48	-.0458	90°	0.0	d_{36}	.100	$\ell = 5$ cm	2.9×10^{-3}	.131	> 1000	.20 - 1.1

References to Table I

a) Byer, R.L., Kildal, H., and Feigelson, R.S. (1971), Appl. Phys. Letts. 19, 237.
b) Boyd, G.D., Beuhler, E., Stortz, F.G., and Wernick, J.H. (1972a), IEEE J. Quant. Electr. QE-8, 419.
c) Boyd, G.D., Beuhler, E., Stortz, F.G., (1971a), Appl. Phys. Letts. 18, 301.
d) Boyd, G.D., Kasper, H., and McFee, J.H. (1971c), IEEE J. Quant. Electr. QE-7, 563.
e) Herbst, R.L. and Byer, R.L. (1971), Appl. Phys. Letts. 19, 527.
f) Boyd, G.D. Kasper, H., and McFee, J.H. (1971c), IEEE J. Quant. Electr. QE-7, 563.
g) Feichtner, J.D., and Roland, G.W. (1972), Appl. Optics. 11, 993.
h) Campillo, A.J., and Tang, C.L. (1971), Appl. Phys. Letts. 19, 36.
i) Byer, R.L. and Harris, S.E. (1968), Phys. Rev. 168, 1064.
j) Bjorkholm, J.E. and Siegman, A.E. (1967), Phys. Rev. 154, 851.
k) Francois, G.E. (1966), Phys. Rev. 143, 597.

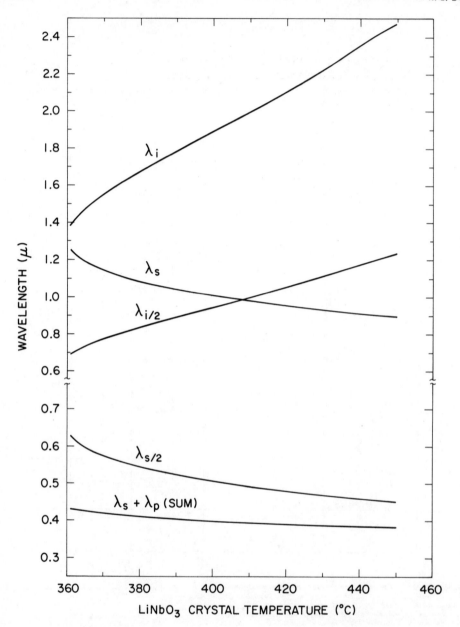

FIG. 1--Experimental tuning curves for a 0.659 μm pumped LiNbO₃ parametric oscillator. Also shown are the up-converted wavelengths obtained with an internal LiIO₃ crystal.

generation using LiIO$_3$. This singly resonant oscillator typically operates at 30% conversion efficiency with a gain bandwidth of 1.5 to 2 cm^{-1} . Peak powers of 300 W at average powers of 30 mW are typical operating parameters. Although historically parametric oscillators have operated with small excess gain, the gain of LiNbO$_3$ parametric oscillators may be so high that they may find application as single pass parametric amplifiers. For example, recently a single pass gain of 60 db was measured experimentally for a .650 μ pumped LiNbO$_3$ crystal 5 cm long. The amplifier was used to amplify a 1 mW 1.15 μ HeNe laser source to 1 kW peak power. The 1 kW output pulse represented 10% of the input pump power. As expected the amplifier showed temporal pulse narrowing of the amplified 1.15 μ HeNe source to one third of the 60 nsec pump pulse length. Simultaneous spatial narrowing was also observed. The potential applications for such a high gain amplifier include amplification of low power narrow bandwidth diode laser sources for saturation spectroscopy, and amplification of infrared sources prior to detection by infrared detectors. Investigations along these lines is continuing.[12]

Although LiNbO$_3$ is transparent to 5 μ parametric oscillation becomes increasingly more difficult beyond 3.5 μ due to decreasing gain. Recently infrared parametric oscillation has been demonstrated in CdSe[6] and proustite.[7,13,14] However, neither of these materials are available with adequate crystal length for convenient oscillator operation. An additional limitation to infrared oscillator operation is the lack of a "fused-silica" type mirror substrate material and high optical quality dielectric coatings. These difficulties have combined such that infrared generation by mixing appears to be the more favorable route to a widely tunable infrared source.

The output power obtained by mixing is

$$\left(\frac{P_1}{P_2}\right) = \frac{\omega_1}{\omega_2} \; \Gamma^2 \ell^2 \; \text{sinc}^2 \left(\frac{\Delta k \ell}{2}\right) \qquad (24)$$

Thus, though visible dye lasers may be convenient tunable sources, they are not the optimum pump source for mixing for two reasons: the reduction in mixing efficiency by the Manley-Rowe factor, and the transparency requirements in both the infrared and visible for the nonlinear material. For infrared generation by mixing, the LiNbO$_3$ parametric oscillator more closely fits the source requirements.

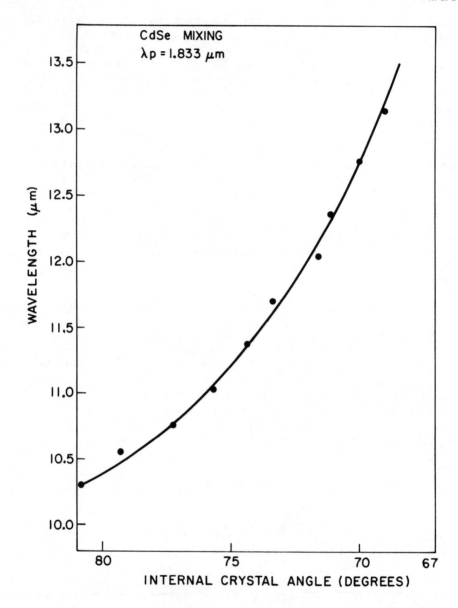

FIG. 2--Measured difference output in CdSe. The tunable pump
source was two LiNbO$_3$ oscillators operating within a
single cavity pumped by 0.659 μm of a doubled Nd:YAG
laser. One oscillator was tuned while the second was
held at 1.833 μm.

We have recently performed two mixing experiments to illustrate
the advantages of widely tunable infrared generation by mixing.
Figure 2 shows the wavelengths obtained by mixing in CdSe. The
tunable pump source was two LiNbO$_3$ oscillators operating within
the same optical cavity pumped by .659 μ . One crystal was tuned
while the other was held fixed at 1.833 μ . The mixer phasematched
between 9.6 μ and 25 μ . However, the detector response cutoff
at 13.5 μ .

CdSe is unique in that it has a wide transparency range,
.75 μ - 25 μ and 70 μ to submillimeter wavelengths, combined
with a large nonlinear coefficient and high crystal quality. The
small birefringence of CdSe allows phasematched far infrared
generation as shown in Fig. 3.

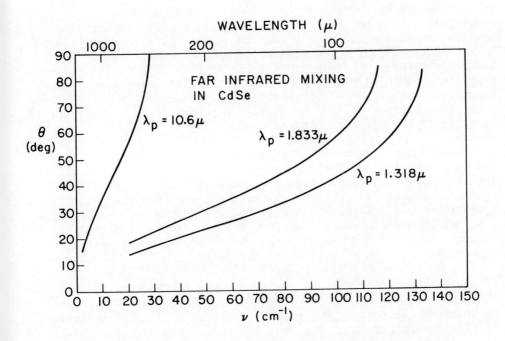

FIG. 3--Calculated far infrared phasematching in CdSe.

However, the birefringence is inadequate for phasematching across
the entire near infrared. For this spectral region crystals belonging to the chalcopyrite group appear particularly useful.

The nonlinear and phasematching properties of the ternary
II - IV - V_2 and I - III - IV_2 chalcopyrite ($\overline{4}2$ m point group)
crystals have been determined. At this time four of the thirty
compounds appear particularly useful for infrared nonlinear optical
applications. These crystals are $AgGaS_2$[15], $AgGaSe_2$[16], $ZnGeP_2$[17]
and $CdGeAs_2$[18]. The properties of these crystals including phase-
matching are reviewed in reference 3. Table I lists their nonlinear
properties and transparency ranges. In general, the chalcopyrites
have a very high nonlinearity.

We have concentrated our efforts on $AgGaSe_2$ and $CdGeAs_2$[19]
due to their extended infrared transparency .7 - 18 μ and
2.3 - 18 μ and unique phasematching properties. In addition,
$AgGaSe_2$ shows useful crystal growth properties and is available
in reasonable quality crystals over 1 cm^3 in volume.[20]

Figure 4 shows the 7 μ to 15 μ tuning range generated by
mixing in $AgGaSe_2$. The output bandwidth is less than 2 cm^{-1}
reflecting the bandwidth of the .659 μ pumped $LiNbO_3$ parametric oscillator source. The unique features of this experiment
are its simplicity and rapid, continuous tuning. The experimental
arrangement consists of collinear geometry with a Q-switch Nd:YAG
laser operating at 1.32 μ internally doubled with $LiIO_3$ to
generate .659 μ. The .659 μ pumps the singly resonant $LiNbO_3$
parametric oscillator. The oscillator output mixes in the $AgGaSe_2$
with the collinear 1.32 μ with phasematching achieved by proper
rotation of the $AgGaSe_2$ crystal.

Continuous tuning was achieved by temperature scanning the
$LiNbO_3$ parametric oscillator at $1^{\circ}C/min$ and synchronously rotating
the $AgGaSe_2$ crystal to phasematch between 7.5 μ and 12 μ
within 10 minutes scan time. The scanning was repeatible with
approximately a 10% peak to peak fluctuation in output pulses and
a signal to noise greater than 1000 in the HgCdTe detector. A
demonstration polystyrene spectrum was scanned to illustrate the
repeatibility of this coherent spectrometer.

The above description of tuning ranges illustrates that it is
now possible with only three nonlinear crystals and a Q-switched
Nd:YAG laser source to generate .6 μ to 3.5 μ by parametric
oscillation in $LiNbO_3$, 3 μ to 18 μ by mixing in $AgGaSe_2$ and
10 μ to 25 μ and 70 μ to 1000 μ by mixing in CdSe. The
next important question in spectroscopic application of these

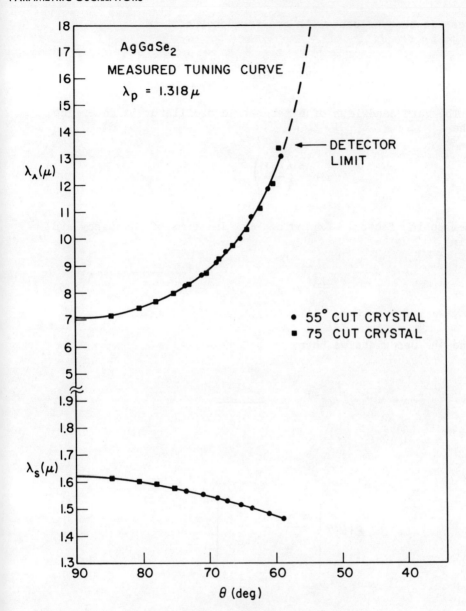

FIG. 4--7 μ - 15 μ output in AgGaSe2 pumped by 1.318 μ mixed
against a .659 μ pumped LiNbO₃ parametric oscillator.

sources is that of bandwidth and frequency control.

BANDWIDTH

The gain bandwidth of a parametric oscillator is determined by the

$$\operatorname{sinc}^2\left(\frac{\Delta k\ell}{2}\right)$$

phasematching factor. Expanding Δk in terms of frequency and letting

$$\frac{\Delta k\ell}{2} = \pi \tag{25}$$

define the bandwidth we have

$$\Delta\nu\,(\mathrm{cm}^{-1}) = \frac{1}{c\,\beta_{12}\,\ell} \tag{26}$$

where

$$\beta_{12} = \left|\frac{\partial k_1}{\partial\omega_1} - \frac{\partial k_2}{\partial\omega_2}\right|$$

$$= \frac{1}{c}\left[n_2 - n_1 + \lambda_1\frac{\partial n_1}{\partial\lambda_1} - \lambda_2\frac{\partial n_2}{\partial\lambda_2}\right]$$

$$\approx \frac{2\Delta n}{c} \tag{27}$$

Thus the gain bandwidth is approximately given by

$$\Delta \nu \, (\text{cm}^{-1}) \sim \frac{1}{2 \Delta n \ell} \qquad (28)$$

where Δn is the crystal birefringence. Figure 5 shows the measured gain bandwidth of a .659 μ pumped LiNbO3 singly resonant oscillator.

Similar to other coherent sources, the parametric oscillator can be frequency narrowed to oscillate in a single axial mode. The frequency narrowing methods include thermally tuned etalon,[21] tilted etalon,[22] Smith and the duel Smith interferometer[23] and birefringent filter.[24] The first three methods have been demonstrated with good results. For example, Wallace[21] reports that a .92 cm^{-1} free spectral range, finesse of 7.8 , thermally tuned etalon effectively controls the oscillator bandwidth to a single axial mode with long term frequency stability of 30 MHz or 0.001 cm^{-1} . The commercial LiNbO3 parametric oscillator is available with this frequency control option and has been used in a series of experiments.[25] The use of a thermally tuned etalon is dictated by the small parametric oscillator cavity spot size which does not permit tilted etalon use.[26] The disadvantage of the thermal etalon is the slow thermal time constant and difficulty in stabilizing the etalon at the frequency of interest.

These thermally tuned etalon difficulties can be overcome by using a beam expanding telescope within the oscillator cavity to permit the use of a tilted etalon. Recently experiments were performed to demonstrate this frequency narrowing method using both refractive and reflective optics to expand the beam from 100 μ to 600 μ .[22] Figure 6 shows the tuning capability achieved with a 2.85 cm^{-1} free spectral range, finesse of 4.0 tilted etalon. The bandwidth was less than the 0.5 cm^{-1} spectrometer resolution. Based on the previous work with thermal etalons, bandwidth narrowing should be less than 0.1 cm^{-1} and approaching single frequency operation. The 2.85 cm^{-1} interval could be rapidly scanned in a convenient and repeatable manner illustrating this advantage of the tilted etalon. The use of a beam expanding telescope is required for the present confocally focused low power LiNbO3 oscillators. Operation of higher power oscillators dictates a larger oscillator cavity mode radius so that the beam expander would not be necessary.

LiNbO$_3$ SRO BANDWIDTH

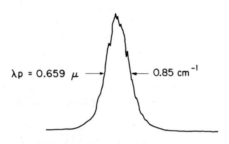

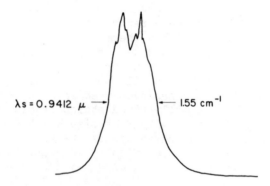

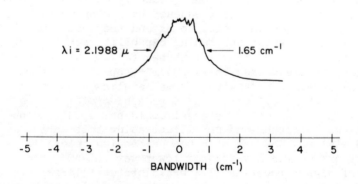

BANDWIDTH (cm^{-1})

FIG. 5--Measured bandwidth for a 0.659 μm pumped LiNbO$_3$ SRO.
The calculated linewidth for the 5 cm crystal is
1.60 cm^{-1} .

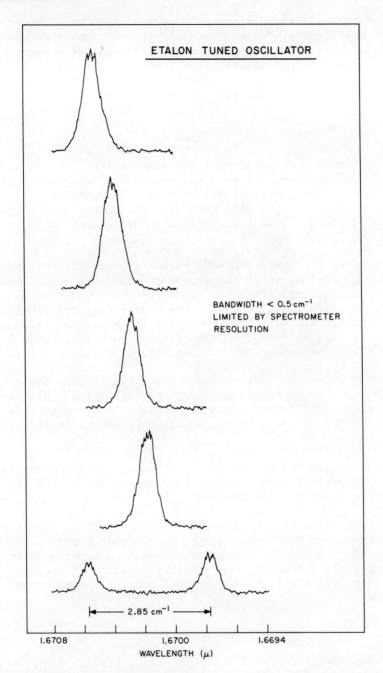

FIG. 6--Tilted etalon tuned .659 μ pumped LiNbO$_3$ parametric
         oscillator. The etalon had a 2.85 cm^{-1} free spectral
         range and a finesse of four.

Although birefringent elements have been proposed for use in parametric oscillators,[24] they have not yet been demonstrated. However experiments are presently in progress and based upon positive results, a combination birefringent filter and tilted etalon may provide the most convenient frequency narrowing method for future parametric oscillator devices.

APPLICATIONS TO SPECTROSCOPY

Spectroscopic applications of tunable coherent sources covers a wide range of chemical, biological and physical systems. Rather than review the range of applications to which parametric oscillators have been applied, I will briefly describe our work which is related to remote air pollution monitoring using tunable infrared sources.[27,28]

Using a .532 μ pumped $LiNbO_3$ parametric oscillator operating in the 2.3 μ region, we have measured pressure broadening and absorption coefficients of the $V'' = 0 \to V' = 2$ CO overtone transition.[29] Figure 7 illustrates the observed absorption spectrum taken by continuously scanning the oscillator wavelength with a resolution of 0.1 cm^{-1} . The detection system used is a dual differential boxcar integrator followed by ratio electronics.

Figure 8 shows the measured absorption coefficient and cross section vs CO pressure for the overtone transition. Similar data was taken for nitrogen pressure boradened CO up to 1 atm pressure. Extending the results to higher pressures is illustrated by Fig. 9.

Using the observed absorption spectra, the pressure broadening in CO is determined to be 0.62 $\times$ 10^{-3} cm^{-1} $Torr^{-1}$ for the $V'' = 0 \to V' = 2$ transition for the P(2) - P(4) rotational levels.

The above discussion illustrates just one application of parametric sources. As the tuning range, output energy and bandwidth improve a wider use of parametric devices can be expected. In particular, applications to long path spectroscopy,[30] chemical transfer rate spectroscopy[25] and optical pumping spectroscopy[31] can be expected.

FUTURE DEVELOPMENTS

If I may speculate for a moment, I would like to describe a potential coherent infrared source that looks particularly promising The system is based on a Nd:YAG oscillator-amplifier pump source operating Q-switched at 1.06 μ . This pump source is capable of

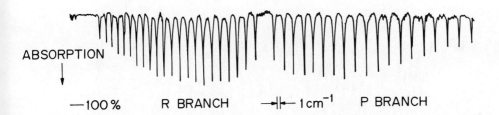

CO OVERTONE SPECTRUM

$V'' = 0 \longrightarrow V' = 2$ 600 Torr

FIG. 7--CO overtone spectrum taken with a LiNbO$_3$ parametric
   oscillator at .1 cm^{-1} resolution at a 1 m absorption
   path length.

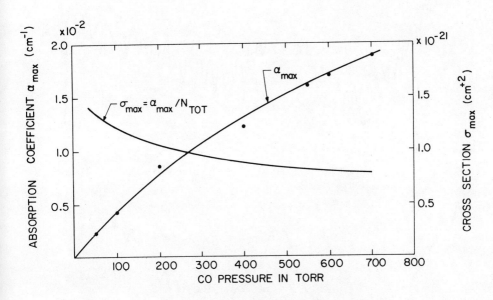

FIG. 8--Absorption coefficient and cross sections at peak of
   P(3) line in V = 0 → V = 2 transition of CO .

CO OVERTONE SPECTRUM
$V'' = 0 \longrightarrow V' = 2$

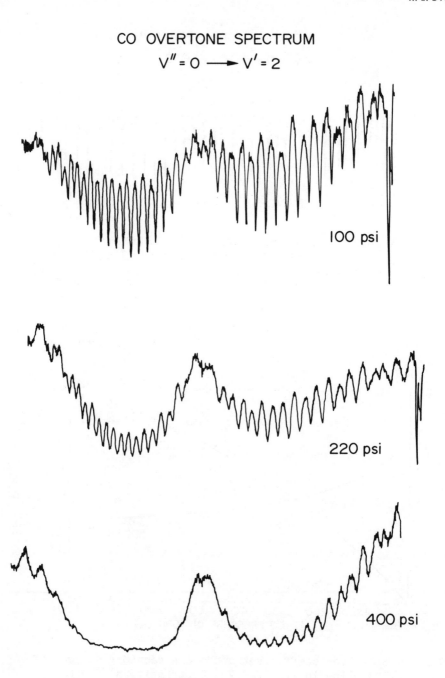

100 psi

220 psi

400 psi

FIG. 9--Pressure broadening of the CO overtone transition.

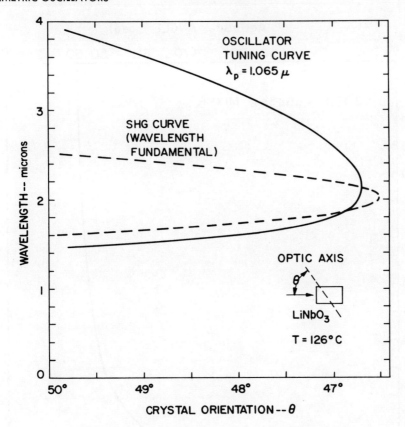

FIG. 10--Angle tuning curve for a   1.06 μ LiNbO₃ parametric
         oscillator.

400 mJ output at up to 40 pps.[32] Figure 10 shows the computed
tuning curve for a LiNbO₃ parametric oscillator directly pumped
by 1.06 μ .[33,34] The output of this oscillator provides an ideal
match to the phasematching properties of AgGaSe₂. Figure 11
illustrates the resulting tuning range achieved by mixing. Of
particular interest from the device standpoint, is rapid tuning by
crystal rotation and the requirement for only one set of mirrors
reflecting between 1.6 μ and 2.1 μ for the LiNbO₃ singly
resonant oscillator. These phasematching curves illustrate the
potential for a high energy continuously tunable 1.6 μ to 18 μ
source based on well developed Nd:YAG laser and LiNbO₃ oscillator technology.

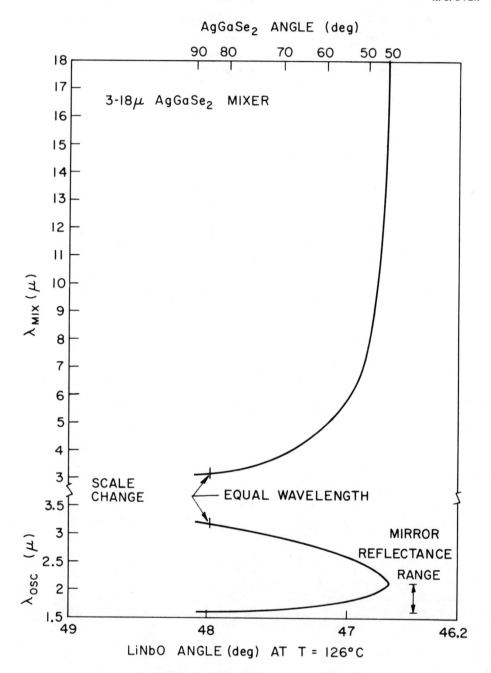

FIG. 11--3 - 18 μ AgGaSe$_2$ mixing source pumped by a LiNbO$_3$ singly resonant parametric oscillator.

In conclusion, parametric oscillator and mixing sources are now finding their way out of the laboratory and into an increasing number of spectroscopic applications. Their convenient operating characteristics, relatively high peak power and adequate bandwidth control coupled with a very wide tuning range assure increased future application as tunable coherent sources.

ACKNOWLEDGEMENT

Dr. R.L. Herbst has carried out a number of the experiments discussed in this paper. I want to acknowledge his contributions especially in the CdSe and $AgGaSe_2$ mixing work.

REFERENCES

1. S.E. Harris, "Tunable Optical Parametric Oscillators", Proc. IEEE, 57, 2096, (1969).

2. R.G. Smith, "Optical Parametric Oscillators", vol. 4 of Advances in Lasers, ed by A.K. Levine and A.J. De Maria, (to be published).

3. R.L. Byer, "Optical Parametric Oscillators", Treatise in Quantum Electronics, eds. H. Rabin and C.L. Tang, (1973), (to be published).

4. J.M. Yarborough and G.A. Massey, "Efficient High Gain Parametric Generation in ADP Continuously Tunable Across the Visible Spectrum", Appl. Phys. Lett. 18, 438, (1971).

5. R.W. Wallace, "Stable, Efficient Optical Parametric Oscillators Pumped with Doubled Nd:YAG", Appl. Phys. Lett. 17, 497, (1970).

6. R.L. Herbst and R.L. Byer, "CdSe Infrared Parametric Oscillator", Appl. Phys. Lett. 21, 189, (1972).

7. D.C. Hanna, B. Luther-Davies, R.C. Smith, "Singly Resonant Proustite Parametric Oscillator Tuned from 1.22 μm to 8.5 μm", Appl. Phys. Lett. vol. 22, 440, (1973).

8. R.W. Wallace and S.E. Harris, "Extending the Tunability Spectrum", Laser Focus, Nov. 1970, p.42; also Chromatix Inc., Mountain View, California.

9.   G.D. Boyd and D.A. Kleinman, "Parametric Interaction of Focused
     Gaussian Light Beams", J. Appl. Phys. $\underline{39}$, 3597, (1968).

10. G.D. Boyd and A. Ashkin, "Theory of Parametric Oscillator
     Threshold with Single-Mode Optical Masers and Observation of
     Amplification in $LiNbO_3$", Phys. Rev. $\underline{146}$, 187, (1966).

11. J.F. Ready, <u>Effects of High Power Laser Radiation</u>, Academic
     Press, New York, 1971.

12. R.L. Byer and R.L. Herbst, "High Gain $LiNbO_3$ Parametric
     Amplifiers", (to be published).

13. E.O. Ammann and J.M. Yarborough, "Optical Parametric
     Oscillation in Proustite", Appl. Phys. Lett. $\underline{17}$,
     233, (1970).

14. D.C. Hanna, B. Luther-Davies, H.N. Rutt and R.C. Smith,
     "Reliable Operation of a Proustite Parametric Oscillator",
     Appl. Phys. Lett. $\underline{20}$, 34, (1972).

15. D.S. Chemla, P.J. Kupecek, D.S. Robertson and R.C. Smith,
     "Silver Thiogallate, a New Material with Potential for
     Infrared Devices", Optic Comm. I, 29, (1971).

16. G.D. Boyd, H.M. Kaspar, J.H. McFee and F.G. Stortz, "Linear
     and Nonlinear Optical Properties of some Ternary Selenides",
     IEEE, J. Quant. Elect. $\underline{QE-8}$, 900, (1972).

17. G.D. Boyd, E. Beuhler and F.G. Stortz, "Linear and Nonlinear
     Optical Properties of $ZnGeP_2$ and CdSe ", Appl. Phys. Lett.
     $\underline{18}$, 301, (1971).

18. R.L. Byer, H. Kildal and R.S. Feigelson, "$CdGeAs_2$ - A New
     Nonlinear Crystal Phasematchable at 10.6 μm", Appl. Phys.
     Lett. $\underline{19}$, 237, (1971).

19. H. Kildal, R.F. Begley, M.M. Choy and R.L. Byer, "Efficient
     Second and Third Harmonic Generation of 10.6 μm in $CdGeAs_2$",
     J. Opt. Soc. of Am. $\underline{62}$, 1398, (1972).

20. H. Kildal, (private communication).

21. R.W. Wallace, IEEE Conf. on Laser Application, Washington, D.C.
     (1971).

22. R.L. Herbst and R.L. Byer, "Line Narrowing of a $LiNbO_3$ Para-
     metric Oscillator Using a Tilted Etalon", (to be published).

23. J. Pinard and J.F. Young, "Interferometric Stabilization of
     an Optical Parametric Oscillator", Optic Comm. $\underline{4}$, 425, (1972).

24. S.E. Harris, (private communication).

25. S.R. Leone and C.B. Moore, "Optical Parametric Oscillator
     Pumped Vibrational Energy Transfer Studies of Hydrogen Chloride",
     J. Opt. Soc. of Am. $\underline{62}$, 1358, (1972).

26. M. Hercher, "Tunable Single Mode Operation of Gas Lasers
     Using Intracavity Etalons", Appl. Optics, $\underline{8}$, 1103, (1969).

27. H. Kildal and R.L. Byer, "Comparison of Laser Methods for
     the Remote Detection of Atmospheric Pollutants", Proc.
     IEEE, $\underline{59}$, 1644, (1971).

28. R.L. Byer and M. Garbuny, "Pollutant Detection by Absorption
     Using Mie Scattering and Topographical Targets as Retro-
     reflectors", Applied Optics, $\underline{12}$, 1496, (1973).

29. M. Garbuny, T. Henningsen and R.L. Byer, "Experimental
     Studies for Remote Gas Detection by Parametric Laser Tuning",
     Spring Meeting Optical Society of America, Denver, Colo.
     March, 1973.

30. V. Vali, R. Goldstein and K. Fox, "Very Long-Path Absorption
     Cell for Molecular Spectroscopy", Appl. Phys. Lett. $\underline{22}$, 391,
     (1973).

31. T.Y. Chang and O.R. Wood, "Optically Pumped N_2O Laser", Appl.
     Phys. Lett. $\underline{22}$, 93, (1973).

32. M. Yarborough, (private communication).

33. E.O. Ammann, J.D. Foster, M.K. Oshman and J.M. Yarborough,
     "Repetively Pumped Parametric Oscillator at 2.13 μm",
     Appl. Phys. Lett. $\underline{15}$, 131, (1969).

34. E.O. Ammann, J.M. Yarborough and J. Falk, "Simultaneous
     Optical Parametric Oscillation and Second Harmonic Generation",
     J. Appl. Phys. $\underline{42}$, 5618, (1971).

A TUNABLE INFRARED COHERENT SOURCE FOR THE 2 TO 25 MICRON REGION

AND BEYOND

James J. Wynne, Peter P. Sorokin* and J. R. Lankard*

IBM Thomas J. Watson Research Center

Yorktown Heights, New York 10598

ABSTRACT

A novel method of generating coherent, widely tunable infrared radiation is described. The method uses a four-wave parametric mixing process in alkali metal vapors. The tuning range presently spans the 2 to 25 micron region. The possibilities of extending the tuning range are discussed.

***

By exploiting a resonantly enhanced four-wave parametric mixing process in alkali metal vapors, a novel source of coherent, tunable infrared radiation has been developed. The technique is characterized by a broad, smooth tuning range, and a spectrally narrow output. In its present embodiment which utilizes optical mixing of two dye lasers simultaneously pumped by a 100 kW nitrogen laser, the output is in the form of 5 nsec long pulses with a repetition rate of up to 100 pps. The infrared output power presently ranges from 100 mW at 2 μ to about .1 mW at 25 μ. The linewidth of the infrared is governed by that of the dye lasers and is approximately .2 cm^{-1}.

Unlike three-wave parametric difference mixing techniques which have been used since 1963 to produce coherent infrared [1], the present method does not require an acentric solid with the inherent problems of opacity and difficulty in phase matching in some regions of the infrared. By utilizing a third order nonlinear

* Partially supported by the Army Research Office, Durham, N.C.

response one is able to employ vapors which are centrosymmetric as
the nonlinear mixing media. Although the nonlinearity is of higher
order and the vapors are far less dense than solids, advantage is
taken of an enormous resonant enhancement to make the process
competitive in efficiency with second order difference mixing in
solids. The ease of preparing indestructible samples, transparent
throughout the infrared and easily phase matchable, encourages us
to think that the problems which have prevented the development of
a practical widely tunable coherent infrared source may now be
overcome.

Figure 1 depicts the experimental arrangement used in our
infrared generation. The output beams from two independently
tunable dye lasers are combined into a colinear beam which is
focused into a heat pipe oven containing alkali metal atoms. Each
dye laser is equipped with a beam expander and diffraction grating
so as to produce spectrally narrow ($\sim$ .1 cm^{-1}) output [2]. The
heat pipe oven [3] is a device which allows one to produce a
homogeneous high temperature region of alkali metal vapor with a
cool zone of buffer gas (such as helium) separating the hot vapor
from the optically transparent windows. Thus there is no problem
with window fogging or corrosion and the vapor region is completely

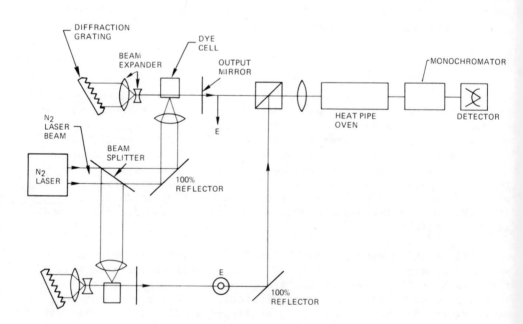

Figure 1.   Experimental block diagram.

accessible to visible light as well as infrared provided that the
windows are transparent in the appropriate regions of the spectrum.

The physical concepts behind the resonant enhancement of the
nonlinearity are very simple. One of the dye lasers is tuned to
emit at a frequency ν_L. Specializing to the case of potassium as
the nonlinear mixing medium, ν_L is tuned to the vicinity of the
4s-5p resonance lines (see Fig. 2). This laser has two important

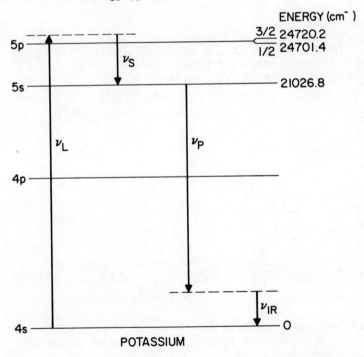

Figure 2.   Energy level diagram for potassium.

functions, to be detailed in what follows. If the power of the laser is sufficiently high and its linewidth sufficiently narrow, it acts as a pump for stimulated electronic Raman scattering (SRS) with coherent Stokes light being emitted at the frequency $\nu_S = \nu_L - \nu_{5s-4s}$. In practice, with powers as low as 100 W and a $\sim .1\ cm^{-1}$ laser linewidth, SRS is observable from potassium at a pressure of 10 torr. With 2 kW laser power, coherent Stokes light is observed over a range of 500 cm^{-1} as ν_L is tuned from the low frequency side of the 5p resonances to the high frequency side. The Stokes intensity shows minima when ν_L is tuned to coincide with the $5p_{3/2}$ and $5p_{1/2}$ lines. These minima correspond to a loss mechanism due to the optical pumping of the $5p_{3/2}$ and $5p_{1/2}$ levels followed by a transition to the 3d levels, with coherent emission as 3.14 and 3.16 μ, respectively. The maximum Stokes intensity is observed when ν_L is tuned to a value $\sim 75\ cm^{-1}$ above the $5p_{3/2}$ resonance.

These characteristics of coherent Stokes generation are typical of all of the alkali metals. A careful study was made of the transmission of laser light through a heat pipe oven containing 5.7 torr of rubidium vapor. The results are shown in Fig. 3 where it is seen that a sizable amount of input light is

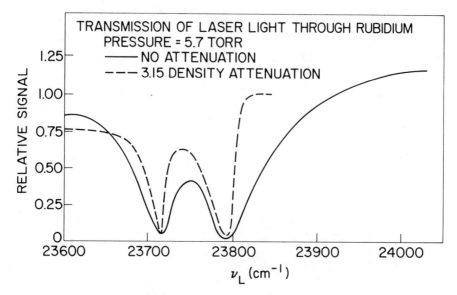

Figure 3.   The solid line is the measured transmission with $\sim$ 5 KW input power. The dotted line is the measured transmission with $\sim$ 3.5 W input power.

lost due to a mechanism that is highly nonlinear, namely SRS. The strongly attenuated laser beam showed linear loss due to the absorption of the 6p resonance lines of rubidium. When the total power of the laser was allowed to enter the cell, the transmission was greatly decreased in the regions where SRS was strong. (In rubidium Stokes generation takes place with the atoms making a transition from 5s to 6s.)

The second function of the beam at ν_L is to mix with ν_S and the beam from the second dye laser with frequency ν_p. The third order nonlinear response of the vapor produces a nonlinear polarization at the frequency

$$\nu_{IR} = \nu_L - \nu_S - \nu_P \tag{1}$$

(see Fig. 2). The response is resonantly enhanced because of the proximity of ν_L to the 5p resonance lines and because $\nu_L-\nu_S$ is exactly resonant with the 4s-5s transition. The dominant term in the third order nonlinear susceptibility is [4]

$$\chi^{(3)} = \frac{iN}{h^3} \frac{<4s|\mu|4p><4p|\mu|5s><5s|\mu|5p><5p|\mu|4s>}{(\nu_{4p-4s}-\nu_P)(\nu_{5p-4s}-\nu_L)\,\Gamma} \tag{2}$$

Here Γ is the linewidth of the laser at frequency ν_L, this linewidth being considerably broader than the natural atomic linewidths. The bracketed expressions are dipole matrix elements. N is the density of atoms. A simple calculation using the known matrix elements shows that for potassium with a density of 10^{17} atoms/cc, with $\nu_{5p-4s}-\nu_L = 50$ cm^{-1}, $\Gamma = .1$ cm^{-1}, and $\nu_{4p-4s}-\nu_P = 5000$ cm^{-1} (i.e. $\nu_P = 18000$ cm^{-1}), $\chi^{(3)} = 6 \times 10^{-10}$ esu which is as large as $\chi^{(3)}$ of almost any solid [5].

The resonantly enhanced nonlinear response fulfills one of the conditions for efficient generation of light through a parametric mixing process. The other important condition which must be satisfied is phase matching. In order for the nonlinear polarization $P^{(3)}(\nu_{IR})$ to radiate efficiently, it must be phase matched with the free wave at ν_{IR}. This condition is expressed by the equation

$$\vec{k}_{IR} = \vec{k}_L - \vec{k}_S - \vec{k}_P \tag{3}$$

The largest volume of interaction is achieved with colinear beams for which Eq. (3) reduces to

$$n_{IR}\nu_{IR} = n_L\nu_L - n_S\nu_S - n_P\nu_P, \tag{4}$$

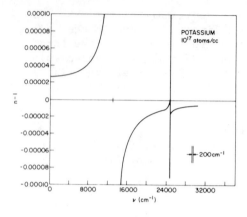

Figure 4
a) Index of refraction of pure potassium as a function of frequency. The curves are drawn for the case of no damping.

b) Expanded view of 200 cm^{-1} region around the 5p resonances.

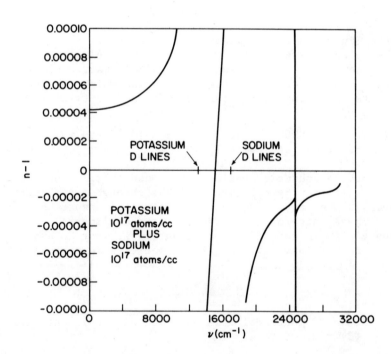

Figure 5. Index of refraction for a mixture of equal parts of sodium and potassium.

where n_i is the index of refraction at frequency ν_i. Subtracting
Eq. (1) from Eq. (3) results in

$$n'_{IR}\nu_{IR} = n'_L\nu_L - n'_S\nu_S - n'_P\nu_P \tag{5}$$

Here $n' = 1-n$ and represents the contribution of the metal vapor to
the index of refraction. In the absence of dispersion, Eq. (5) is
always satisfied. However due to the normal dispersion of the
metal vapor, Eq. (5) is, in general, not satisfied. To see
how to satisfy phase matching it is helpful to refer to Fig. 4,
which plots n' vs ν for potassium. If ν_L is on the high frequency
side of the 5p resonance, n'_L is negative. For a given setting of
ν_L, there is one value of ν_P in the region between the 4p and 5p
resonances where Eq. (5) is satisfied. As ν_P is scanned to larger
values, resulting in longer wavelength infrared generation, ν_L
must be fine tuned to larger values to maintain phase matching.
Eventually one is forced to tune ν_L to such large values that SRS
ceases. Using pure potassium vapor, Sorokin et al. [6]
successfully tuned $\lambda_{IR} = \nu_{IR}^{-1}$ from 2 to 4 μ. In rubidium infrared
was generated from 2.9 to 5.4 μ.

In the pure vapors one is limited by the necessity of phase
matching by fine tuning ν_L. However an obvious alternative is to
alter the linear dispersion characteristics of the metal vapor.
Figure 5 shows what happens when sodium is added to potassium.

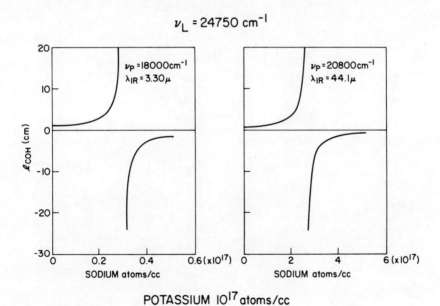

Figure 6. Coherence length for mixtures of potassium and sodium.

The sodium D line resonances alter the linear dispersion
characteristics of the vapor. The dominant effect is to make the
value of n_p' more negative for a given value of ν_p. The index of
refraction changes at the other frequencies of interest but these
changes are relatively small. Thus as one tunes ν_p to higher
values, more sodium is added to keep n_p' sufficiently negative to
maintain phase matching. Figure 6 plots the coherence length,
$\ell_{coh} = \pi/(k_{IR} + k_P + k_S - k_L)$, as a function of sodium atom density
for two different values of ν_p (and λ_{IR}).

In order to establish a homogeneous vapor mixture of sodium
and potassium with independent control of the sodium and potassium
pressures, a concentric heat pipe as described by Vidal and Hessel
[7] was constructed. With the outer pipe charged with pure sodium
and with potassium and sodium in the inner pipe, infrared was
generated out to 25 μ. Table I lists the frequencies and
wavelengths required for infrared generation at various settings
in the infrared. To cover the range from ν_p = 16500 cm^{-1} to
20627 cm^{-1} the dye mixtures given by Stokes et al. [8] were used.
It is seen in Table I that only a short jump is now required to
achieve infrared generation out to 100 μ and beyond.

TABLE I

POTASSIUM

$$\nu_P + \nu_{IR} = \nu_{5s-4s} \sim 21027 \text{ cm}^{-1}$$

ν_P (cm^{-1})	λ_P (Å)	ν_{IR} (cm^{-1})	λ_{IR} (μ)
16500	6061	4527	2.21
17000	5882	4027	2.48
17500	5720	3527	2.84
18000	5556	3027	3.30
18500	5405	2527	3.96
19000	5263	2027	4.93
19500	5128	1527	6.50
20000	5000	1027	9.73
20500	4878	527	18.98
20600	4854	427	23.42
21000	4762	27	373

Equation (2) indicates that the nonlinear polarization is slowly varying with ν_p and therefore with ν_{IR}. The infrared power generation is proportional to the square of the nonlinear polarization. Solutions of Maxwell's equations for the phase matched generation of light via a parametric mixing process, in the limit where a small fraction of the pump light is converted, show that the power generated is proportional to the square of the frequency of the generated light [9]. Thus, in the case of constant input power with phase matched infrared generation, the infrared power should go as the infrared frequency squared. This decreases the efficiency of infrared generation at longer wavelengths. However, by using a stronger pump the infrared power may be enhanced significantly. The nonlinear polarization is proportional to the product of the electric fields at ν_L, ν_S, and ν_p, and thus the generated infrared power will be proportional to the product of the powers at ν_L, ν_S, and ν_p. Therefore the infrared power is expected to be proportional to the cube of the power of the nitrogen laser which pump the dye lasers. With the availability of 10 MW nigrogen lasers [10], one expects 10^6 times more infrared power unless saturation effects limit the conversion. With more pumping power the spectral output of the dye lasers can be narrowed while maintaining sufficient power for efficient infrared generation. In view of the complete infrared transparency of the alkali metal vapors, it is felt that infrared generation from 1 to 500 μ, with sufficient power for spectroscopy, will be attained by this method.

REFERENCES

1. Two recent publications on second order difference mixing are C. D. Decker and F. K. Tittel, Appl. Phys. Letters 22, 411 (1973), and R. L. Aggarwal, B. Lax and G. Favrot, Appl. Phys. Letters 22, 329 (1973).
2. T. W. Hansch, Appl. Optics 11, 895 (1972).
3. C. R. Vidal and J. Cooper, J. Appl. Phys. 40, 3370 (1969).
4. The method used to derive this expression is outlined in N. Bloembergen, Nonlinear Optics, (W. A. Benjamin, New York, 1965), pp. 37-41.
5. J. J. Wynne and N. Bloembergen, Phys. Rev. 178, 1295 (1969).
6. P. P. Sorokin, J. J. Wynne and J. R. Lankard, Appl. Phys. Letters 22, 342 (1973).
7. C. R. Vidal and M. M. Hessel, J. Appl. Phys. 43, 2776 (1972).
8. E. D. Stokes, F. B. Dunning, R. F. Stebbings, G. K. Walters and R. D. Rundel, Optics Commun. 5, 267 (1972).
9. See, for example, Ref. 3, p. 82.
10. B. Godard, 1973 IEEE/OSA Conference on Laser Engineering and Applications, Paper 4.5.

OSCILLATION ON THE ULTRAVIOLET BOUND-FREE CONTINUA OF DIATOMIC MOLECULAR XENON AND MOLECULAR KRYPTON[*]

C. K. Rhodes and P. W. Hoff

University of California, Lawrence Livermore Laboratory

Livermore, California 94550

I. INTRODUCTION

The initial proposal for the use of molecular bound-free transitions in laser systems was advanced by Houtermans [1] in 1960 in relation to the continua of H_2 and Hg_2. In spite of this early suggestion, only relatively recently has genuine stimulated emission been observed on transitions of this type on the ultraviolet molecular continua of xenon[2] at ~ 1722 Å and krypton[3] at ~ 1457 Å. We note that the rare gas molecular continua represent a subset of the much larger class of bound-free systems. Some additional members of this more extensive group are the examples[4-7] Zn_2, Cd_2, Hg_2, HeH, NeH, and LiXe. Among the simplest systems representing the general properties of molecular bound-free transitions, and for which there exists a considerable literature is He_2. A partial energy level diagram representative of this class of molecular systems is illustrated in Fig. (1). The essential property is that some of the upper states have a potential minimum at an internuclear separation R_o for which the ground state curve is strongly repulsive. Provided that a kinetic mechanism exists for population of the upper bound states, the repulsive character of the ground level enhances the tendency for the production of large population inversions in the Franck-Condon region around R_o.

---

[*] This work was performed under the auspices of the United States Atomic Energy Commission.

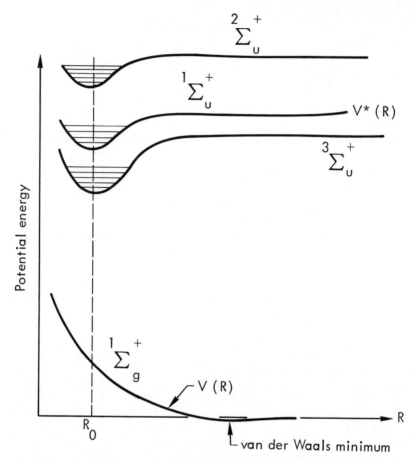

Figure 1

Plot of typical potential energy for homonuclear
rare gas molecules versus internuclear separation
(R) illustrating the ground $^1\Sigma_g^+$ state the first
excited $^3\Sigma_u^+$ and $^1\Sigma_u^+$ states, and the $^2\Sigma_u^+$ ionic
core state.

Although the rare gases are normally considered to comprise
one of the simplest classes of materials, a complicated chain
comprised of several processes contributes to the kinetic scheme
which leads to neutral molecular formation. We illustrate briefly
with the example of xenon, some of the processes that are operative
when high pressure rare gases are excited with relativistic
electron beams. A more complete analysis is contained elsewhere.[9]

They are the following:

$$e^- + Xe \rightarrow e^- + Xe^* \qquad (1)$$

$$e^- + Xe \rightarrow e^- + e^- + Xe^+ \qquad (2)$$

$$Xe^* + Xe + Xe \rightarrow Xe_2^* + Xe \qquad (3)$$

$$Xe^+ + Xe + Xe \rightarrow Xe_2^+ + Xe \qquad (4)$$

$$Xe_2^+ + e^- \rightarrow Xe^* + Xe \qquad (5)$$

$$Xe_2^* \rightarrow \gamma + Xe + Xe \qquad (6)$$

$$Xe^{**} + Xe \rightarrow Xe^* + Xe \qquad (7)$$

$$Xe_2^*(v) + Xe \rightarrow Xe_2^*(v') + Xe \qquad (8)$$

$$Xe^{**} + e^- \rightarrow Xe^* + e^- \qquad (9)$$

$$\gamma + Xe_2^* \rightarrow Xe_2^+ + e^- \qquad (10)$$

$$Xe_2^* + Xe_2^* \rightarrow Xe_2^+ + e^- + Xe + Xe \qquad (11)$$

$$Xe^* + e^- \rightarrow Xe^+ + e^- + e^- \qquad (12)$$

At around the 1 MeV electron energies of interest, the primary electrons lose energy mainly through excitation and ionization of the xenon medium.[10] In this connection, it is important to note that the combination of reactions (3), (4), and (5) channels both excited and ionized atoms into the excited dimer state.[11] The rates for these processes are taken as 2.5×10^{-32} cm^6/sec, $(3.57 \pm 0.17) \times 10^{-31}$ cm^6/sec[13], and $(1.4 \pm 0.1) \times 10^{-6}$ cm^3/sec[14], respectively. Process (11) is an important loss mechanism of molecular dimers whose influence is felt most severely at high dimer densities. Clearly this will establish a relationship between radiative efficiency and the level of excitation.

II. MOLECULAR XENON SYSTEM

The essential observations unequivocally demonstrating the presence of a coherent stimulated process are the measurement of spectral line narrowing (a necessary but insufficient condition), the observation of spatial coherence and the resulting spatially directed beam, the existence of a sharp oscillator threshold, a time dependence of the directed output radiation radically different from the spontaneous emission observed below threshold, and a dip in the spontaneous emission viewed perpendicular to the laser axis. The experimental apparatus giving rise to laser

emission at 1722 ± 1 Å in relativistic electron beam excited
material over a range of xenon gas pressures near 200 psia is
illustrated in Fig. 2. A high pressure gas cell with internal
mirrors and mode-limiting optical apertures was attached to a
Febetron 705 relativistic electron beam generator which emits a
100 nsec pulse in a 10 kA beam with a nominal energy of 1.5 MeV
over a 2 cm diameter aperture. The electron beam current density
penetrating the diaphragm and incident on the gas was ∿300 A/cm²
as measured by a calibrated Faraday cup. The optical cavity consisted of two 1-meter radius of curvature mirrors separated by
5 cm. The mirrors were fabricated from highly polished MgF_2 substrates which were Al coated and MgF_2 overcoated. The reflectance
and transmittance of the mirrors at 1700 Å were measured to be 80%
and 8% respectively. The 2 mm diameter intracavity apertures
limited the field of view of the diagnostic systems essentially to
be volume contained in the cylinder defined by the apertures. The
spectral composition of the radiation along the optical axis was
detected by an 0.75 meter Seya-Namoika spectrograph. The temporal
characteristics were observed using a photodiode with a 2 ns risetime. Spatial propagation studies were performed by replacing the
photodiode with SWR film mounted on a variable length line-of-
sight pipe. The spontaneous radiation emitted perpendicular to
the cavity axis was monitored using an 0.82 meter Czerny-Turner
spectrometer with a typical bandpass of 5 Å about a center wavelength of 1715 Å

Figure 3 illustrates the typical spectral output of the laser
for conditions above and below threshold along with a calibration
spectrum. The spectral linewidth (measured photographically)
narrowed from ∿160 Å (as observed without mirrors) to a half-width
of ∿15 Å with the optical cavity. The absorptive feature at
∿1707 Å, appearing only in the stimulated spectrum, is very probably due to the MgI resonance transition arising from damage of
the MgF_2 mirror overcoat during the laser pulse. In addition to
the spectral narrowing, marked changes in the time dependence of
the directed output radiation were observed for conditions above
and below the oscillation threshold with a sharp temporal pulse of
∿10 ns being characteristic of conditions above threshold. In
addition, the divergence of the resulting beam was determined to
be 5 mrad in agreement with the cavity parameters. The pressure
threshold for oscillation was ∿125 psia.

III. MOLECULAR KRYPTON SYSTEM

The molecular continuum of krypton originating from the bound-
free transition of the excited krypton dimer is centered near
1500 Å. As an initial approximation, the molecular structure and
kinetic properties of the krypton system should strongly resemble

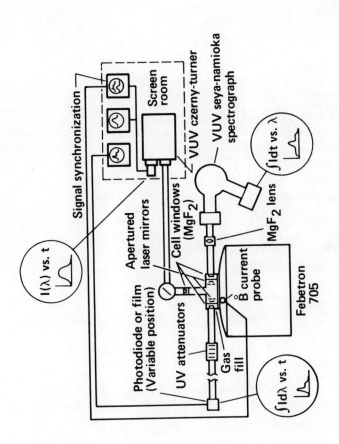

Figure 2

Schematic of the experimental apparatus illustrating the three diagnostic systems:
Seya-Namoika spectrograph for time integrated spectral determinations, a photodiode
to record the directed emission and a Czerny-Turner spectrometer equipped with a
photomultiplier to view the spontaneous emission side light.

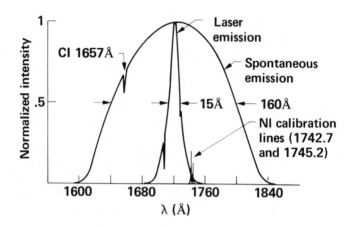

Figure 3

Time integrated spectral output of the xenon oscillator (linear
vertical scale) from spectrograph for (i) conditions below
threshold and (ii) conditions above threshold showing NI calibration lines.   Two absorption lines appear at 1707 Å and 1726 Å.
Above threshold the emission narrows to ∿15 Å at ∿ 1720 Å.

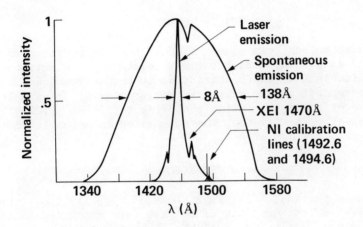

Figure 4

Time integrated spectral output of the high pressure krypton
oscillator (linear vertical scale) showing NI calibration lines
for (i) conditions below threshold and (ii) conditions above
threshold. Above threshold the transition narrows to ∿8 Å at
1457 ± 1 Å. Minor absorption bands tentatively attributed to
the $X^1\Sigma^+ - a'^1\Sigma_u - (\Delta v = 0,1)$ transition of N_2 present in the
line of sight pipe are observed in addition to the xenon reso-
nance absorption at 1470 Å. See M. Ogawa and Y. Tanaka, J. Chem.
Phys. 32, 754 (196?) for N_2 absorption data.

the xenon case. This similarity is reflected experimentally in
the observation[3] of stimulated emission in high pressure krypton
at 1457 ± 1 Å with a linewidth of 8 Å as illustrated in Fig. (4).
The experimental arrangement was identical to that used in the
xenon studies described above. The threshold pressure was ~ 250
psia above which the emission occurred in an approximately 8 ns
pulse. Due to a small but unavoidable impurity of xenon in the
krypton sample, the xenon 1470 Å resonance line is a prominent
absorptive feature in the spectrum.

IV. SUMMARY

Oscillation has been observed in high pressure xenon and
krypton gas excited by a high current relativistic electron beam.
The radiation in both cases attributed to bound-free transitions
of the corresponding homonuclear molecular dimers, Kr_2^* and Xe_2^*.
In the case of xenon, the stimulated line occurs at ~ 1722 with a
15 Å width while for krypton, the stimulated emission is centered
at 1457 Å with a width of 8 Å.

REFERENCES

1. F. G. Houtermans, Helv, Phys. Acta $\underline{33}$, 933 (1960). Other
    analyses of the continua of H_2 have been discussed in the
    following: For the H_2 $a^3\Sigma_g^+ \rightarrow b^3\Sigma_u^+$ transition see A. J. Palmer
    J. Appl. Phys. $\underline{41}$, 438 (1970); C. V. Heer, J. Appl. Phys. $\underline{41}$,
    1875 (1970). For the molecular continuum of Hg_2 see
    D. A. Leonard, J. C. Keck, and M. M. Litvak, Proc. IEEE $\underline{51}$,
    1785 (1963); R. J. Carbone and M. M. Litvak, J. Appl. Phys.
    $\underline{39}$, 2413 (1968); D. C. Lorents, R. M. Hill, and
    D. J. Eckstrom, Molecular Metal Laser, Stanford Research
    Institute Report, November 27, 1972; R. M. Hill,
    D. J. Eckstrom, D. C. Lorents, and H. H. Nakano, Measure-
    ments of Negative Gain for Hg_2 Continuum Radiation, to be
    published; D. J. Eckstrom, R. M. Hill, D. C. Lorents, and
    H. H. Nakano, Collisional Quenching and Radiative Decay of
    the Mercury Excimer, to be published.

2. N. G. Basov, V. A. Danilychev, and Yu. M. Popov, Sov. J.
    Quant. Elec. $\underline{1}$, 18 (1971); H. A. Koehler, L. J. Ferderber,
    D. L. Redhead, and P. J. Ebert, Appl. Phys. Letters $\underline{21}$, 198
    (1972); P. W. Hoff, J. C. Swingle, and C. K. Rhodes,
    Demonstration of Temporal Coherence, Spatial Coherence, and
    Threshold Effects in the Molecular Xenon Laser, UCRL-74665,
    Opt. Commun., to be published; J. B. Gerardo and
    A. Wayne Johnson, High Pressure Xenon Laser at 1730 Å, to be
    published, Preliminary calorimetric studies were reported in
    A. C. Kolb, N. Rostoker, R. White, K. Boyer, R. Jensen,

P. Robinson, and A. Sullivan, Bull. Am. Phys. Soc. <u>17</u>, 1031 (1972). Line narrowing with unexplained structure has also observed, R. Jensen, private communication.

3. P. W. Hoff, J. C. Swingle, and C. K. Rhodes, <u>Observation of Stimulated Emission from High Pressure Krypton and Argon/Xenon Mixtures</u>, UCRL-74715, Appl. Phys. Letters, to be published.

4. For data on the continua of Zn_2, Cd_2, and Hg_2, as well as a general discussion of the early work on continuous spectra, see Wolfgang Finkelnburg, <u>Kontinuierliche Spektren</u> (Springer-Verlag, Berlin, 1938).

5. HeH is discussed in C. A. Slocomb, W. H. Miller and H. F. Schaefer III, J. Chem. Phys. <u>55</u>, 926 (1971).

6. The potential curves of HeH are considered in V. Bondybey, P. K. Pearson, and H. F. Schaefer III, J. Chem. Phys. <u>57</u>, 1123 (1972).

7. The LiXe molecule is discussed in W. F. Baylis, J. Chem. Phys. <u>51</u>, 2665 (1969).

8. The molecular properties of the He_2 system are discussed in Marshall L. Ginter and Rubin Battino, J. Chem. Phys. <u>52</u>, 4469 (1970) and further references cited therein. Recent experimental data on He_2 are described by W. A. Fitzsimmons in <u>Atomic Physics 3</u>, edited by S. J. Smith and G. K. Walters (Plenum Press, New York, 1973) p. 477.

9. E. V. George and C. K. Rhodes, UCRL-74516, <u>Kinetic Model of Ultraviolet Inversions in High Pressure Rare Gas Plasmas</u>, Appl. Phys. Letters, to be published. Another analysis has been given by D. C. Lorents and R. E. Olson, <u>Excimer Formation and Decay Processes in Rare Gases</u>, Stanford Research Institute Report, December 1972. Also see Charles K. Rhodes, <u>Review of Ultraviolet Lasers</u>, UCRL-74816, to be published.

10. Martin J. Berger and Stephen M. Seltzer, <u>Tables of Energy Losses and Ranges of Electrons and Positrons</u>, N65-12506 (NASA, Washington, D.C., 1964).

11. Of course, there are many possible dimer states. For a discussion of the structure of Xe_2 see R. S. Mulliken, J. Chem. Phys. <u>52</u>, 5170 (1970).

12. R. Boucique and P. Mortier, J. Phys. D3, 1905 (1970).

13.   D. Smith, A. G. Dean, and I. C. Plumb, J. Phys. <u>B5</u>, 2134
      (1972).

14.   H. J. Oskam and V. R. Mittelstadt, Phys. Rev. <u>132</u>, 1445
      (1963).

SPECTROSCOPY II

ULTRAHIGH RESOLUTION SATURATED ABSORPTION SPECTROSCOPY[*]

Ch. Bordé[†] and J. L. Hall[‡]

Joint Institute for Laboratory Astrophysics

National Bureau of Standards and University of Colorado

Boulder, Colorado 80302

In this paper we report our progress in the search for super narrow lines, our present state of the art in the line shrinking business and what looks like the way towards still higher resolution based on our current knowledge of the broadening mechanisms.

It is conventional wisdom that the natural linewidth is the only fundamental limit to ultrahigh resolution in saturated absorption spectroscopy. Still it appeared that an actual experimental attempt to achieve such ultrahigh resolution might teach us about some other realistic limitations.

The test molecule we chose was, of course, methane. Its narrow natural linewidth (<100 Hz)[1] and easy wavelength coincidence[2] with the He-Ne laser at 3.39 μm have been noted before. The high resolution potential has motivated development of several appropriate laser devices for such a study. Actually methane may not be the most favorable case for obtaining very narrow lines because it is a light molecule, but it has remarkable features:

    (1) It shows a well resolved magnetic hyperfine structure,[3] providing an unambiguous natural frequency scale.
    (2) Precisely because it is a light molecule, it helps the study and the understanding of all the broadening factors

[*]Contribution of National Bureau of Standards, not subject to copyright.

[†]Supported by CNRS; also NATO Fellow, 1973.

[‡]Staff member, National Bureau of Standards, Boulder, Colorado.

that decrease with increasing mass--these are the transit-
time broadening, the second-order Doppler shift and
broadening, and mainly the recoil structure.
(3) A careful study of the CH_4 lines at 3.39 μm is interesting
in connection with their possible applications as a wavelength or frequency standard. [4]

To discuss the linewidth problem in a detailed way, a good
starting point may be the two equations that the molecules must
satisfy simultaneously to be able to interact with two waves having
the same frequency ω_L but opposite wave vectors $\pm k_L$:

$$
\begin{cases}
\omega_L = \omega_0 + \underline{k}_L \cdot \underline{v} \\
\\
\omega_L = \omega_0 - \underline{k}_L \cdot \underline{v} \ .
\end{cases}
\tag{1}
$$

For resonance, the laser frequency ω_L should be equal to the resonant frequency of the molecule ω_0 shifted by the first-order
Doppler effect. The well-known solution to this problem is
$\omega_L = \omega_0$ and $\underline{k}_L \cdot \underline{v} = 0$. In fact each one of these quantities contains
some uncertainty which ultimately appears as a broadening mechanism
for the saturation resonance.

For ω_L this is the laser spectral width. The basic requirement for high resolution spectroscopy is to have a source whose
frequency stability is appreciably better than the desired resolution.

The molecular absorption frequency ω_0 is affected by the
finite natural lifetime. This is the only fundamental limit in
saturated absorption spectroscopy, for the methane lines we estimate it to be less than 100 Hz. ω_0 is also affected by the collisions and by the Zeeman effect due to the earth's magnetic field.
These two broadening causes can be reduced to a negligible value.
Internal interactions may also split ω_0 into several resonant frequencies which, if unresolved, will give an extra inhomogeneous
character to the line. For example, this was the case before we
resolved the methane hyperfine structure.

The last term describes the first-order Doppler broadening.
To obtain sub-Doppler resolution in conventional spectroscopy, the
experimenter may constrain the function $\underline{k} \cdot \underline{v}$ using molecular beam
techniques. In the present saturated absorption experiments we
rely on the light both to select the $\underline{v}$ and to probe the selected
molecules.

If there is a distribution in the spatial frequencies, that is, if the waves are misaligned or are not perfectly plane, there is going to be a residual first-order Doppler broadening in saturated absorption spectroscopy.

Thus the success of laser saturated spectroscopy is not only due to the laser's very narrow spectral distribution of ω_L but also to its ability to produce a wave with a very narrow distribution of k_x k_y k_z. How well one can do in saturated absorption spectroscopy will clearly depend on how well one can produce a monochromatic wave and on how well one can produce a plane wave.

To achieve the necessary frequency stability it is absolutely essential to electronically control the laser frequency. The most powerful concept known to us is called frequency offset locking[5],[6] whereby one acquires good frequency stability with one laser device and transfers the stability to the laser used to explore the molecular resonances. Our present apparatus is shown in Fig. 1. Its reference laser has a 5-power internal telescope which reduces the reference methane saturation peak width to 50 kHz HWHM.

FREQUENCY OFFSET-LOCKED LASER SPECTROMETER

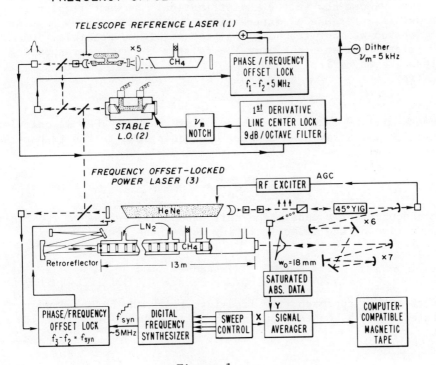

Figure 1

A "solid ingot" local oscillator laser, phase locked 5 MHz red from the methane line, provides the (offset) output frequency of the reference system. Finally, the power laser is phase locked relative to the stable local oscillator using a tunable frequency offset to scan the studied line. To achieve this controlled tuning, the beat frequency between the last two lasers can be varied by digital request to a frequency synthesizer which is in this second heterodyne offset loop. Thus one has both frequency stepping capability and very high frequency stability, along with a natural and stable correspondence being made between offset frequency and channel number of the signal averager.

The power laser's frequency stability is approximately $3 \times 10^{-13} \tau^{-1/2}$ for an averaging time τ in the range $10^{-2} < \tau < 10^2$ sec. At shorter times the noise is characterized by a linewidth which we estimate to be about 500 Hz. At longer times the frequency stability fails to improve beyond 3×10^{-14} (3Hz rms) for $10^2 < \tau < 10^4$ sec, due to changes in the systematic errors which limit the present reproducibility to about $\pm$ 300 Hz. We think those offsets are partly due to the intensity-dependent shifts to be discussed later.

Let us come now to the $\bar{k}$ distribution. We want it as narrow as possible. Because of diffraction, to obtain a narrow angular distribution we want the beam diameter as big as possible. An appropriate Fourier transformation relates the two corresponding widths. We may write

$$\Delta k_\perp \; \Delta x \simeq 1 \quad ,$$

where Δx will be taken as the beam waist radius. Thus we calculate, using Eq. (1), a residual Doppler broadening due to the transverse velocity

$$\Delta \omega = \underline{\Delta k}_\perp \cdot \underline{v} \simeq \frac{1}{w_0} v \quad \text{diffraction} .$$

We can equivalently understand this broadening as due to the finite transit time of the molecule flying across the beam:

$$\Delta \omega \simeq \frac{1}{\Delta t} \simeq \frac{v}{w_0} \quad \text{transit time} .$$

So with either approach, the linewidth times the beam radius is expected to be a constant having the dimensions of a velocity: we write

$$\Delta \nu_{HWHM} \times w_0 = k_v .$$

This relation is illustrated on Fig. 2. For methane we find

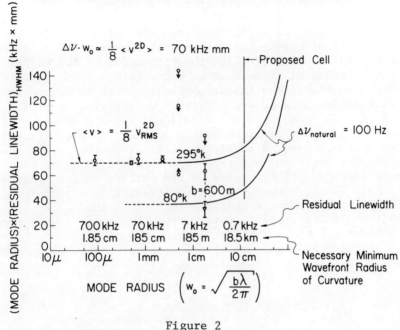

Figure 2

$k_\nu \simeq 1/8(2kT/M)^{1/2}$ which is 70 kHz mm at room temperature. Liquid nitrogen narrows the lines by about a factor of two. In our last experiment two successive reflective telescopes were used to expand the beam by a factor of 42. However, due to imperfect cell straightness, the effective beam had an 18 mm mode radius, which gives us about 4 kHz contribution to the linewidth at room temperature.

In the pursuit of narrow saturation resonances we must also beware of wave front deviations from the ideal beam considered above: Of course there are many ways that laser beams can have more angular content than is imposed by their fundamental diffraction limit. One can easily observe such effects as back surface reflections in mirrors, imaging aberrations, beam deformation by the air path, and non-ideal output mode structure of the laser. Similarly the retroreflected probe must be parallel, colinear and have matched (planar) wavefronts. To achieve this alignment condition, we use a "cat's eye" retroreflector as illustrated in Fig. 1. The 100 μW return beam is steered to the detector by polarization optics: a Faraday rotator (45° YIG) and a Rochon prism comprise this optical directional coupler. The cell has a 5 cm diameter, is 13 m long and filled with CH_4 at pressures around 100 μTorr or less.

The part of the CH_4 spectrum that one can cover with the He-Ne laser at 3.39 μm is represented in Fig. 3. This is a conventional spectrum of the Coriolis fine structure P(7) of ν_3.[7]

The ν_3 vibration has the carbon atom vibrating relative to the tetrahedral frame of hydrogen. Tetrahedral symmetry is sufficiently high that the three perpendicular vibration modes of the carbon against the tetrahedral frame must be degenerate. Thus, in the excited vibrational state a linear combination can be formed in which a circulation is evident. There is a resulting angular momentum $\underline{L}$ that is coupled to the rotational angular momentum of the frame $\underline{R}$ to give the total angular momentum $\underline{J}$. Some of the degeneracy within a given J manifold is lifted in the ground state by centrifugal distortion, but this term is small compared with the Coriolis perturbation that appears in the excited state. These effects result in six components for the transition of interest, P(7). The laser emission is centered in close coincidence to the F component at 2947.912 cm^{-1}. This is the line that we have first studied without any special trick.[6] With permanent magnets Zeeman shifting the laser emission, we have also studied the E component.[8]

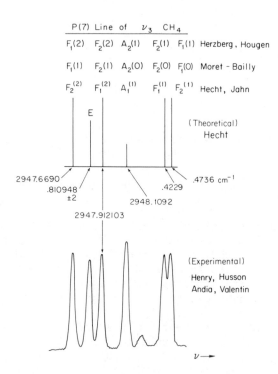

Figure 3

With an electromagnet we observed the A component, and with superconducting solenoids one probably could cover all six Coriolis components. In Fig. 3 we also show the most popular notations for these Coriolis components.[9],[10] The notation adopted here is that of Herzberg[11] and Hougen.[12] The important thing is that the F components correspond to levels having a total nuclear spin $I = 1$ for the four hydrogens, whereas $I = 0$ for the E line.

Figure 4 shows a typical derivative spectrum for the $F_2^{(2)}$ component. Whereas the spectrum for the E line is a single line, this spectrum exhibits three strong lines. On the low frequency side there is a weaker and broader line, and on the 15 times magnified picture two additional very weak lines are visible. The intensities are in the approximate ratios $1:(1/20):(1/20)^2$.

As noted above the two levels of the observed $F_2^{(2)}$ transition have a total nuclear spin $I = 1$. Thus, there are three combinations to be formed coupling $\underline{I}$ and $\underline{J}$ to form $\underline{F} = \underline{I} + \underline{J}$. Corresponding to $J = 7$ in the ground state, the F levels will be 6, 7, and 8 and corresponding to $J = 6$ in the excited state the F levels will be 5, 6, and 7.

The degeneracy of these levels is removed by magnetic interaction terms in the Hamiltonian. For the ground state these are: a scalar and a tensor spin-rotation interaction and a spin-spin interaction between the protons; for the excited state one expects also a spin-vibration interaction with both scalar and tensor parts.[13]

From the theory of the relative intensities in a multiplet we expect three main diagonal lines for which $\Delta F = \Delta J = -1$, with intensities which increase with F. So we identify the + 9.7 kHz line as the $8 \rightarrow 7$ transition, the central line at −1.7 kHz as $7 \rightarrow 6$ and the low −12.8 kHz one as $6 \rightarrow 5$. Besides these lines one expects two much weaker satellite lines with $\Delta F = 0$ that we identify with the two weak low frequency lines.

In addition to these usual components, saturated absorption spectroscopy provides a new resonance each time two transitions share a common level. This is a distinctive feature of saturated absorption spectroscopy. These three level resonances are called "Doppler-generated level crossings." They occur half way between the two usual transitions and have an intensity proportional to the geometrical mean of the two parent lines. We have four of these represented in Fig. 4. To reproduce the spectrum near −38 kHz with these crossings we have to assume that the −72.0 kHz peak is the $6 \rightarrow 6$ and that the −91.2 kHz peak is the $7 \rightarrow 7$. The transitions indicated in the energy level diagram are arranged in the same order as the spectral lines, with higher absolute frequency to the right.

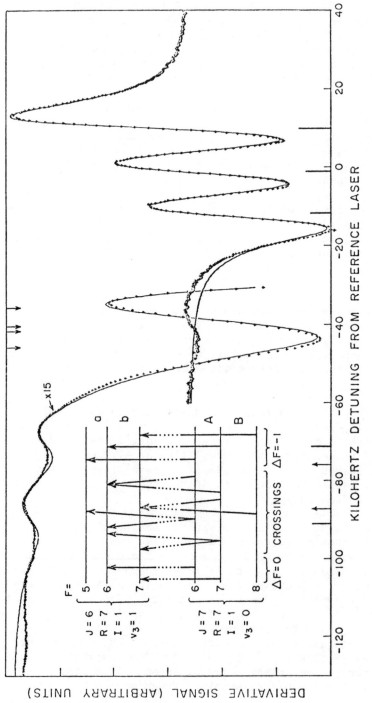

Figure 4.   Composite saturated absorption hyperfine spectrum: experimental data ×, least-squares fit ——.   T = 77°K.   Main triplet: methane pressure 70 μTorr, modulation 2 kHz peak-to-peak at 600 Hz.   Fitted Lorentz width = 5.8 kHz HWHM.   Low frequency spectrum taken at lower resolution, ~9.8 kHz HWHM.   The unobserved ΔF = +1 line and its crossings are not illustrated on the energy level diagram.

With this interpretation we find the values of the energy intervals given in Table I. The experimental values have an uncertainty of $\pm$ 1.5 kHz arising primarily from measurement of the weak lines. These experimental values are to be compared with those recently calculated by Jon Hougen[13],[14] for the ground state, using the constants derived from the magnetic resonance experiments by Yi, Ozier and Ramsey.[15] The agreement is just within the uncertainties of both experiments. As pointed out by Dr. Hougen,[14] it is not obvious that the agreement should be any better: the hyperfine coupling constant may be J dependent because of the centrifugal distortion of the molecule. This experiment also demonstrates that one can have access to knowledge of the hyperfine splittings of a vibrationally excited state.

The least-squares fit of the main components in the spectrum of Fig. 4 gave us a linewidth HWHM equal to 5.8 kHz. Recently, this was a typical result for a room temperature spectrum. To account for the linewidth, we must consider effects omitted in Eqs. (1) which are no longer negligible. These are the second-order Doppler shift and the recoil effect. Both come out of a proper relativistic energy and momentum balance. In the laboratory frame we may write

$$\hbar\omega = \sqrt{P_b^2 c^2 + (Mc^2 + E_b)^2} - \sqrt{P_a^2 c^2 + (Mc^2 + E_a)^2}$$

(2)

$$\hbar\underline{k} = \underline{P}_b - \underline{P}_a \quad .$$

Here let the "b" subscript refer to the upper level. An expansion to the order $1/c^2$ yields shifted absorption and emission resonance frequencies

$$\omega_{absorption} = \omega_o + \underline{k} \cdot \underline{v}_a - \omega_o \frac{v^2}{2c^2} + \frac{\hbar\omega^2}{2Mc^2}$$

(3)

$$\omega_{emission} = \omega_o + \underline{k} \cdot \underline{v}_b - \omega_o \frac{v^2}{2c^2} - \frac{\hbar\omega^2}{2Mc^2} \quad .$$

Table I. Energy level differences illustrated in Fig. 4.

Level	Interval	Experiment[*]	Calculation[†]	pure Ca $\underline{I} \cdot \underline{J}$[‡]
$v_3 = 1$	a	59.2 kHz		62.4 kHz
	b	89.5		72.8
$v_3 = 0$	A	70.3	68.4 kHz	72.8
	B	100.9	99.7	83.2

[*]Ref. 3   [†]Ref. 14   [‡]Calculated using Ca = 10.4 kHz (Ref. 15)

The v^2 term leads to a second-order Doppler shift of ≈ 0.5 Hz/°C for methane.[16] Since we deal with a thermal distribution of velocities, this shift will also be accompanied by a broadening effect. The last term is the recoil frequency shift, which leads to an absorption/emission frequency difference $\hbar\omega^2/Mc^2 = 2.16$ kHz[17] for the CH_4 line of interest. We can rewrite Eqs. (1) expressing the saturated absorption resonance condition,

$$
\begin{matrix}
\text{lower} \\ \text{state} \\ \text{resonance}
\end{matrix}
\left\{
\begin{aligned}
\omega_L &= \omega_o + \underline{k}\cdot\underline{v}_a - \omega_o \frac{v^2}{2c^2} + \frac{\hbar\omega^2}{2Mc^2} \\[2mm]
\omega_L &= \omega_o - \underline{k}\cdot\underline{v}_a - \omega_o \frac{v^2}{2c^2} + \frac{\hbar\omega^2}{2Mc^2}
\end{aligned}
\right.
$$

$$
\begin{matrix}
\text{upper} \\ \text{state} \\ \text{resonance}
\end{matrix}
\left\{
\begin{aligned}
\omega_L' &= \omega_o + \underline{k}\cdot\underline{v}_b - \omega_o \frac{v^2}{2c^2} - \frac{\hbar\omega^2}{2Mc^2} \\[2mm]
\omega_L' &= \omega_o - \underline{k}\cdot\underline{v}_b - \omega_o \frac{v^2}{2c^2} - \frac{\hbar\omega^2}{2Mc^2}
\end{aligned}
\right. \quad .
$$

$$(4)$$

There are two saturation resonance peaks now, for $\underline{k}\cdot\underline{v}_a = 0$ and $\underline{k}\cdot\underline{v}_b = 0$ with the corresponding resonance frequencies

$$
\omega_L = \omega_o - \omega_o \frac{v^2}{2c^2} + \frac{\hbar\omega^2}{2Mc^2} \quad .
$$

and

$$(5)$$

$$
\omega_L' = \omega_o - \omega_o \frac{v^2}{2c^2} - \frac{\hbar\omega^2}{2Mc^2} \quad .
$$

This double peak structure has been previously predicted by Kol'chenko, Rautian, and Sokolovskii.[18]

To the two sets of Eqs. (4), there are two corresponding sets of lines on the ω vs. v plot of Fig. 5. The upper pair of lines corresponds to the ground state. From the figure one can see that the population peaks created in the excited state cross each other for a laser frequency lower than that for which the holes cross each other. Note that for a three-level resonance only one level is involved--that is the common level. So there we expect only a red or a blue component depending on whether we have a common upper or lower level.

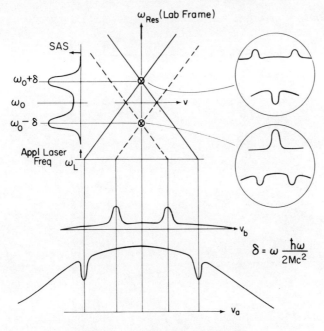

RECOIL – INDUCED DOUBLET

Figure 5

The recoil structure has not yet been observed in saturated
absorption spectroscopy, but our highest resolution line shapes are
consistent with the existence of the doublet.   Figure 6 illustrates
the best resolution we have obtained by cooling the cell to liquid
nitrogen temperature.   In view of the time, expense and labor that
will precede experimental data of adequate resolution to display
the recoil doublet, perhaps we may be forgiven a brief attempt to
"hyper-analyze" the existing data.   The least-squares fit in Fig.
6 gave a 4.14 kHz linewidth but the fit is not very good--the experimental data are straighter at line center than either a Lorentz or
a Gaussian.   The line wings are Lorentzian, but decrease faster
than the Lorentzian which fits the peak-to-peak separations.
These differences are more apparent if one imposes a fit at the
derivative peaks rather than allowing the unweighted least-squares
fit.   Modulation broadening[19] can give a somewhat similar effect,
but accounts for only 1/15 of the present effect.   The line shape
in the free-flight regime has been shown to be accurately
Lorentzian.[20]

Thus it was interesting to try to fit the spectrum of Fig. 6
with each component being a (recoil) doublet.   A common splitting
and height ratio were adopted as additional fitting parameters in
the least-squares fitting program.   Miraculously the program

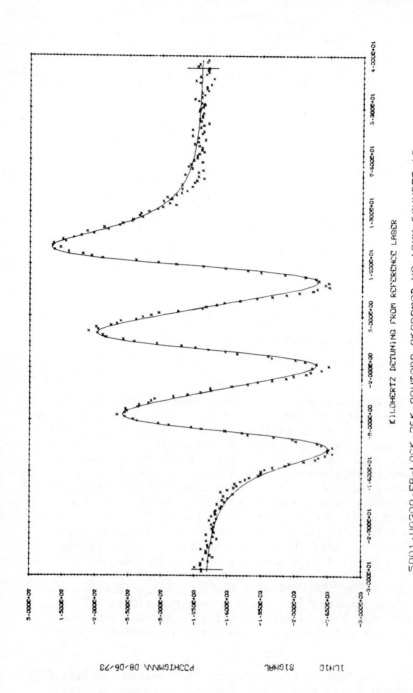

Figure 6. Lorentz least-squares fit to highest resolution data.

converged, giving a 55/45 height ratio and a splitting of (2.22±.04) kHz! Starting the initial height ratio at 10:1 or the splitting at 1/2 or 2 times the preceding value gave convergence to the same final splitting. The doublet fit reduced the variance of the residuals from 2.9 times the expected white noise variance to only 1.26 times this value. Figure 7 shows the doublet fit to the data of Fig. 6.

Two other high resolution data sets were available. Fits to both were also significantly improved, and converged to doublet splittings of (2.03 ± .05)kHz and (2.20 ± .02)kHz with height ratios of 60/40 and 51/49. The low frequency peak intensity is favored above 50/50 by 3, 5 and 1 standard deviations in the three cases. All these quoted errors are on a "one standard deviation" basis and contain no allowance for systematic effects. We conclude that our best data like a doublet representation but that an increment in resolution will be necessary to remove all doubts.

To make a comparison with conventional spectroscopy, the present linewidth is 3 kHz, which corresponds to 10^{-7} wave numbers. We have thus demonstrated at 3000 wave numbers a resolving power of 2.4×10^{10}, where we understand zero second derivative as the resolution condition.

Since we have achieved the diameter-scaling linewidth contribution (1.9 kHz) at liquid nitrogen temperature, it is interesting to estimate the other linewidth contributions:

1)	natural linewidth	<100 Hz
2)	second-order Doppler broadening	≃100 Hz
3)	pressure broadening at 30 μTorr	≃300 Hz
4)	residual earth's field Zeeman broadening	≃100 Hz
5)	saturation broadening	≤200 Hz
6)	modulation broadening	≃300 Hz
7)	laser spectral width	≃500 Hz

Most of these broadening contributions are resistant to further important reductions, but it still appears that another major improvement can be obtained by increasing the cell aperture. A cell of 38 cm aperture, 14 m length using internal reflective optics is being prepared. The potential increment in resolution should make possible a careful study of the line shapes of the two recoil peaks. It is interesting that if the two-state lifetimes are not equal, the peak heights of the doublet will be different. A dependence of the relative strengths of the two lines on experimental param-

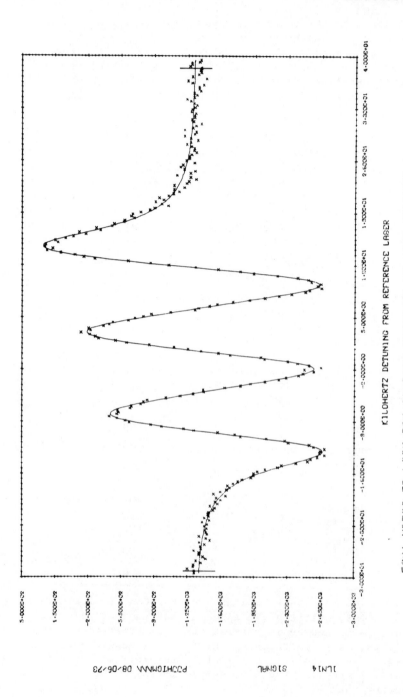

Figure 7.   Lorentz doublets fit to data of Fig. 6.

eters will certainly affect the frequency reproducibility achievable with saturated absorption frequency standards. At low resolution such differential shifts may be avoided by choice of a regime in which both levels are dominated by the same relaxation mechanisms.

Another important example of the effect of unresolved structure on the frequency reproducibility of a stabilized laser is of course also provided by the presence of hyperfine structure in methane. This effect is illustrated in Fig. 8 where we have a synthetic derivative of the peak versus frequency for different saturation parameters. A linewidth of 50 kc was assumed corresponding to our reference telescope laser. Since the intensity of the different components is a non-linear function of the optical field, the apparent zero crossing of the derivative of the unresolved line is shifted towards low frequencies as the saturation increases.[21] This means an intensity-dependent red shift of a laser stabilized at the apparent center of this line. The consequence is that long term stability using the F component will require careful stabilization of the intensity, or sufficient resolution to separate the structure. Alternatively we can use the E Coriolis component which is free of this type of hyperfine structure.

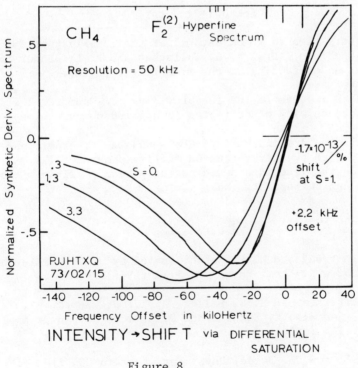

Figure 8

Since the reference laser at present does not have such an intensity stabilizer, we were not able to make a reliable measurement of the second-order Doppler shift upon cooling the long cell from room temperature to 77°K. In the absence of a proper theory which relates the signal contribution to the molecular transverse velocity, it is not obvious what mean velocity to use in the second-order Doppler term of Eq. (5). In particular, in the transit-limited domain at low pressures, we may expect the effective transverse velocity to increase with the laser power.

In this paper we have discussed the several factors which limit our present spectral resolution to 2.4×10^{10}. Some effects such as the methane hyperfine structure have been elucidated; the hyperfine energy splittings of an excited state have been obtained. With the higher resolution expected in the near future, it should be interesting to further investigate the recoil structure as well as the Doppler-generated level-crossing resonances. Ramsey double excitation methods may prove an attractive alternative to simple diameter magnification for exceptionally long-lived molecules. Perhaps storage techniques for ions--eventually neutrals--can be used to obtain dramatically lengthened interaction times.

Acknowledgments

The authors wish to thank several of their colleagues for valuable discussions. Drs. A. C. Gallagher and P. L. Bender have sharpened our understanding of the free-flight regime.

We thank Dr. Judah Levine for providing some numerical filters and for assistance with other topics in applied computerology.

Especially we are indebted to Chela Kunasz for expert and cheerful help with the least-squares fitting program. The glimpse inside the linewidth would not have been possible without her skillful and enthusiastic help.

References

1.   R. B. Armstrong and H. L. Welsh, quoted by J. T. Yardley and C. B. Moore, J. Chem. Phys. 49, 1111 (1968).

2.   H. J. Gerritsen and M. E. Heller, Appl. Opt. Suppl. #2, 73 (1965). See also K. Sakurai and K. Shimoda, Japan J. Appl. Phys. 5, 744 (1966).

3.   J. L. Hall and Ch. Bordé, Phys. Rev. Letters 30, 1101 (1973).

4. In their June, 1973, Paris meeting, the Comité Consultatif pour la Définition du Mètre recommended a wavelength value for the methane P(7) F component, $\lambda = 3392231.4 \times 10^{-12}$ m. They also recommended a value for a certain iodine transition at .633 µm and for the speed of light, c = 299,792,458 m/sec. The assigned fractional uncertainties of these quantities is $\pm 4 \times 10^{-9}$.

5. J. L. Hall, IEEE J. Quant. Electron. QE4, 638 (1968).

6. R. L. Barger and J. L. Hall, Phys. Rev. Letters 22, 4 (1969).

7. Spectrum from L. Henry, N. Husson, R. Andia, and A. Valentin, J. Mol. Spectry. 36, 511 (1970).

8. J. A. Magyar and J. L. Hall, Bull. Am. Phys. Soc. 17, 67 (1972); and J. A. Magyar, to be published.

9. J. Moret-Bailly, J. Mol. Spectry 15, 344 (1965).

10. K. T. Hecht, J. Mol. Spectry. 5, 355 and 390 (1960).

11. G. Herzberg, Electronic Spectra of Polyatomic Molecules (Van Nostrand, Princeton, N.J., 1967), p. 102.

12. J. T. Hougen, J. Chem. Phys. 39, 358 (1963).

13. J. T. Hougen, private communication.

14. J. T. Hougen, 28th Symposium on Molecular Structure and Spectroscopy, The Ohio State University, Columbus, Ohio, June, 1973; and to be published.

15. P. N. Yi, I. Ozier, and N. F. Ramsey, J. Chem. Phys. 55, 5215 (1971).

16. S. N. Bagaev and V. P. Chebotayev, ZhETF Pis. Red. 16, 614 (1972) [JETP Lett. 16, 433 (1972)].

17. The recoil splitting quoted in Ref. 3, 2.4 kHz, is unfortunately in error. The proper splitting is 2.44×10^{-11} fractional shift, which is 2.16 kHz.

18. A. P. Kol'chenko, S. G. Rautian, and R. I. Sokolovskii, ZhETF 55, 1864 (1968) [Sov. Phys. - JETP 28, 986 (1969)].

19. R. L. Smith, J. Opt. Soc. Am. 61, 1015 (1971); also H. Wahlquist, J. Chem. Phys. 35, 1708 (1961).

20. J. L. Hall, in <u>Atomic Physics 3</u> edited by S. J. Smith and
     G. K. Walters (Plenum, New York, 1973), pp. 615–646.

21. J. L. Hall and Ch. Bordé, Proceedings of the 27th Annual
     Symposium on Frequency Control, Cherry Hill, N. J. (June 1973).

SPEED OF LIGHT FROM DIRECT LASER FREQUENCY AND WAVELENGTH MEASUREMENTS: EMERGENCE OF A LASER STANDARD OF LENGTH

K. M. Evenson, F. R. Petersen,
and J. S. Wells
Institute for Basic Standards
National Bureau of Standards
Quantum Electronics Division
Boulder, Colorado 80302

Abstract

Recent frequency and wavelength measurements of a methane stabilized laser yield a value of the speed of light 100 times less uncertain than the previously accepted value. Various possibilities using lasers as radiation sources for a new length standard are discussed. One possibility is to fix the value of the speed of light in the redefinition of the meter.

143

The speed of light is one of the most interesting and important of the fundamental (dimensioned) constants of nature.[1] It enters naturally into ranging experiments, such as geophysical distance measurements which use modulated electromagnetic radiation, and astronomical measurements such as microwave planetary radar and laser lunar ranging. Basically, very accurately measured delay times for electromagnetic waves are dimensionally converted to distance by use of the light propagation speed. Recent experiments have set very restrictive limits on any possible speed dependence on direction[2] or frequency.[3] Another interesting class of applications involves the speed of propagating waves in a less obvious manner. For example, the conversion between electrostatic and electromagnetic units involves the constant, c, as does the relativistic relationship between the atomic mass scale and particle energies.

With the perfection of highly reproducible and stable lasers, their wavelength-frequency duality becomes of wider interest. We begin to think of lasers as frequency references for certain kinds of problems such as optical heterodyne spectroscopy.[4] At the same time, we use the wavelength aspect of the radiation, for example, in precision long-path interferometry.[5] Also since the fractional uncertainty in the frequency of these lasers is approximately one part in 10^{11}, the duality implies that the radiation must have a similar wavelength characteristic, i. e., $\Delta\lambda/\lambda \approx 10^{-11}$ or about 100 times better than the current length standard. Hence, the stabilized laser must be considered to have tremendous potential in wavelength as well as frequency standards applications and perhaps in both.[6]

It has been clear since the early days of lasers that this wavelength-frequency duality could form the basis of a powerful method to measure the speed of light. However, the laser's optical frequency was much too high for conventional frequency measurement methods. This fact led to the invention of a variety of modulation or differential schemes, basically conceived to preserve the small interferometric errors associated with the short optical wavelength, while utilizing microwave frequencies which were still readily manipulated and measured. These microwave frequencies were to be modulated onto the laser output or realized as a difference frequency[7] between two separate laser transitions. Indeed, a proposed major long-path interferometric experiment[8] based on the latter idea has been made obsolete by the recent high-precision direct frequency measurement.[9] An ingenious modulation scheme, generally applicable to any laser transition, has recently successfully produced an improved value for the speed of light.[10] While this method can undoubtedly be perfected further, its differential nature leads to limitations which are not operative in direct frequency measurements.

The product of the frequency and wavelength of an electromagnetic wave is the speed of propagation of that wave. For an accurate determination of both of these quantities, the source should be stable and monochromatic and should be at as short a wavelength as possible. At shorter optical wavelengths the accuracy of the wavelength measurement increases. A suitable source of such radiation is the methane-stabilized He-Ne laser[11] at 3.39 μm (88THz). Direct frequency measurements were recently extended to this frequency[12] and subsequently refined[9] to the present accuracy of 6 parts in 10^{10}. The wavelength of this stabilized laser has been compared[13-16] with the krypton-86 length standard to the limit of the usefulness of the length standard (approximately 4 parts in 10^9). The product of the measured frequency and the wavelength yields a new, definitive value for the speed of light, c.

The previously accepted value[17] of c was similarly determined by measuring the frequency and wavelength of a stable electromagnetic oscillator; however, it oscillated at 72 GHz (more than 1000 times lower in frequency than in the case of the present measurements). The 100-fold improvement in the presently reported measurement comes from the increased accuracy possible in the measurement[13] of the shorter wavelength.

A block diagram in figure 1 illustrates the entire cesium to methane frequency chain. The three saturated-absorption-stabilized lasers are shown in the upper right-hand section, the transfer chain oscillators are in the center column, and the cesium frequency standard is in the lower right-hand corner. The He-Ne and CO_2 lasers in the transfer chain were offset locked,[11] that is, they were locked at a frequency a few megahertz different from the stabilized lasers. This offset-locking procedure produced He-Ne and CO_2 transfer oscillators without the frequency modulation used in the molecular-stabilized lasers. The measurements of the frequencies in the entire chain were

made in three steps shown on the right-hand side, by using
standard heterodyne techniques previously described.[12, 18-20]

Conventional silicon point-contact harmonic generator-
mixers were used up to the frequency of the HCN laser. Above
this frequency, tungsten-on-nickel diodes were used as harmonic
generator-mixers. These metal-metal diodes required 50 or
more mW of power from the lasers to obtain optimum signals.
The 2-mm-long, 25 μm-diam tungsten antenna, with a sharpened
tip which lightly contacted the nickel surface, seemed to couple
to the radiation in two separate manners. At 0.89 and 10.7 THz
it acted like a long wire antenna,[21, 22] while at 29-88 THz its
conical tip behaved like one-half of a biconical antenna.[22] Conventional detectors were used in the offset-locking steps.

The methane-stabilized He-Ne laser used in these experiments is quite similar in size and construction to the device
described by Hall.[23] The gain tube was dc excited,
and slightly higher reflectivity mirrors were employed. The
latter resulted in a higher energy density inside the resonator
and consequently a somewhat broader saturated absorption.
Pressure in the internal methane absorption cell was about 0.01
Torr (1 Torr = 133.3N/m^2).

The two 1.2-m-long CO_2 lasers used in the experiments
contained internal absorption cells and dc-excited sealed gain
tubes. A grating was used on one end for line selection, and
frequency modulation was achieved by dithering the 4-m-radius-
of-curvature mirror on the opposite end. CO_2 pressure in the
internal absorption cell was 0.020 Torr. The laser frequency
was locked to the zero-slope point on the dip in the 4.3 μm fluorescent radiation.[24] The 0.89-, 10.7-, and 88-THz transfer
lasers were 8-m-long linearly polarized cw oscillators with
single-mode output power greater than 50 mW. The Michelson
HCN laser has been described.[25] The H_2O laser used a double-
silicon-disk partially transmitting end mirror, and a 0.5-mil
polyethylene internal Brewster-angle membrane polarized the
laser beam. The 8-m He-Ne laser oscillated in a single mode
without any mode selectors because of a 4-Torr pressure with a
7:1 ratio of helium to neon. This resulted in a pressure width
approximately equal to the Doppler width, and the high degree of
saturation allowed only one mode to oscillate.

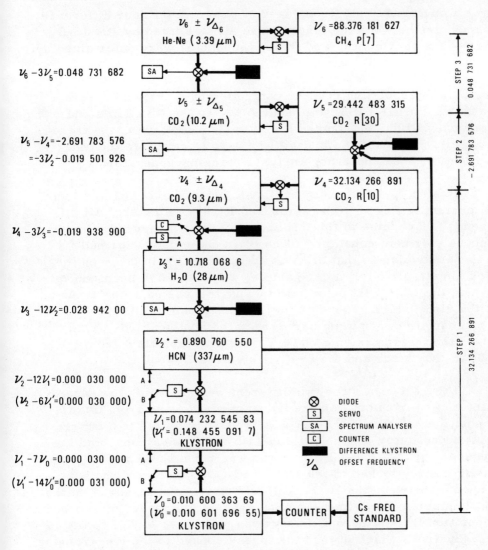

Figure 1. Stabilized Laser Frequency Synthesis Chain. All frequencies are given in THz; those marked with an asterisk were measured with a transfer laser oscillator tuned to approximate line center.

Conventional klystrons used to generate the four difference frequencies between the lasers were all stabilized by standard phase-lock-techniques, and their frequencies were determined by cycle counting at X-band.

An interpolating counter controlled by a cesium frequency standard, the AT (NBS) time scale[26, 27] in the NBS Time and Frequency Division, counted the 10.6-GHz klystron in the transfer chain. This same standard was used to calibrate the other counters and the spectrum-analyzer tracking-generator.

In step 1, a frequency synthesis chain was completed from the cesium standard to the stabilized R(10) CO_2 laser. All difference frequencies in this chain were either measured simultaneously or held constant. Each main chain oscillator had its radiation divided so that all beat notes in the chain could be measured simultaneously. For example, a silicon-disk beam splitter divided the 10.7-THz beam into two parts: one part was focused on the diode which generated the 12th harmonic of the HCN laser frequency, the remaining part irradiated another diode which mixed the third harmonic of 10.7-THz with the output from the 9.3 μm CO_2 laser and the 20-GHz klystron.

Figure 1 shows the two different ways in which the experiment was carried out. In the first scheme (output from mixers in position A), the HCN laser was frequency locked to a quartz crystal oscillator via the 148- and 10.6-GHz klystrons, and the frequency of the 10.6-GHz klystron was counted. The H_2O laser was frequency locked to the stabilized CO_2 laser, and the beat frequency between the H_2O and HCN lasers was measured on the spectrum analyzer. In the second scheme (output from mixers in position B), the 10.6-GHz klystron was phase locked to the 74-GHz klystron, which in turn was phase locked to the free-running HCN laser. The 10.6-GHz klystron frequency was again counted. The free-running H_2O laser frequency was monitored relative to the stabilized CO_2 laser frequency. The beat frequency between the H_2O and HCN lasers was measured as before on the spectrum analyzer.

In step 2, the difference between the two CO_2 lines was measured. The HCN laser remained focused on the diode used in step 1, which now also had two CO_2 laser beams focused on it.

The sum of the third harmonic of the HCN frequency, plus a micro-wave frequency, plus the measured rf beat signal is the difference frequency between these two CO_2 lines. The two molecular-ab-sorption-stabilized CO_2 lasers were used directly, and the rela-tive phase and amplitudes of the modulating voltages were adjusted to minimize the width of the beat note. The beat note was again measured on a combination spectrum analyzer and tracking-gen-erator-counter. The roles of the CO_2 lasers were interchanged to detect possible systematic differences in the two laser-stabili-zation systems.

In step 3, the frequency of the P(7) line in methane was measured relative to the 10. 18 µm R(30) line of CO_2. Both the 8-m 3. 39 µm laser and the CO_2 laser were offset locked from saturated-absorption-stabilized lasers and thereby not modulated. The 10- to 100-MHz beat note was again measured either on a spectrum analyzer and tracking generator, or in the final meas-urement when the S/N ratio of the beat note was large enough (about 100), directly on a counter.

The measurements were chronologically divided into four runs, and values for each of the steps and for ν_4, ν_5, and ν_6 were obtained by weighting the results of all runs inversely pro-portional to the square of the standard deviations. The largest uncertainty came from step 1; a recent measurement by NPL[28] gave a value of the R(12) line which was only 2×10^{-10} (7kHz) different from the number obtained by adding the R(12) - R(10) difference[29] to the present R(10) value. Thus, the first step of the experiment has been verified.

The final result is:

Molecule	Line	λ	Frequency
$^{12}C^{16}O_2$	R(30)	10. 18 µm	29. 442 483 315 (25) THz
$^{12}C^{16}O_2$	R(10)	9. 33	32. 134 266 891 (24)
$^{12}C H_4$	P(7)	3. 39	88. 376 181 627 (50)

The numbers in parentheses are 1-standard-deviation-type
errors indicating uncertainties in the last two digits.

In a coordinated effort, the wavelength of the 3.39 μm
line of methane was measured with respect to the [86]Kr 6057 Å
primary standard of length by R. L. Barger and J. L. Hall.[13]
Using a frequency-controlled Fabry-Perot interferometer with a
pointing precision of about 2×10^{-5} orders, a detailed search for
systematic offsets inherent in the experiment, including effects
due to the asymmetry of the Kr standard line, was made. Offsets
due to various experimental effects (such as beam misalignments,
mirror curvatures and phase shifts, phase shift over the exit
aperature, diffraction, etc.) were carefully measured and then removed from the data with an uncertainty of about 2 parts in 10^9.
This reproducibility for a single wavelength measurement illustrates the high precision which is available using the frequency-
controlled interferometer.

At the 5th session of the consultative committee on the
definition of the meter (CCDM)[30], results of Barger and Hall as
well as measurements made at the International Bureau of
Weights and Measures[15], the National Research Council[16], and
the National Bureau of Standards at Gaithersburg[14] were all
combined to give "recommended" values for the wavelengths of
this transition of methane and the transition of iodine used to
stabilize the visible He-Ne laser. The recommended values are:

$$\lambda CH_4 \left(P(7), \text{ band } \nu_3 \right) \qquad = 3\ 392\ 231.\ 40 \times 10^{-12} \text{ m,}$$

$$\lambda I_2\ 127 \left(\begin{array}{l} R(127), \text{ band } 11\text{-}5, \\ \text{i component} \end{array} \right) \quad = 632\ 991.\ 399 \times 10^{-12} \text{ m.}$$

These "recommended" values are in agreement with wavelength
measurements to the limits possible with the krypton length
standard (that is, about $\pm 4 \times 10^{-9}$). It is "recommended" that
either of these values be used to make length measurements using
these stabilized lasers in the interim before the meter is
redefined.

Multiplying this recommended wavelength of methane by the measured frequency yields the value for the speed of light:

$$c = 299\ 792\ 458\ \text{m/s}$$

which was also recommended by the CCDM. This value of c is one which should be used in distance measurements where time of flight is converted to length and for converting frequency to wavelength and vice versa.

The fractional uncertainty in this value for the speed of light, $\pm 4 \times 10^{-9}$, arises from the interferometric measurements with the incoherent krypton radiation which operationally defines the international meter. This limitation is indicative of the remarkable growth in optical physics in recent years; the present krypton-based length definition was adopted only in 1960!

This result is in agreement with the previously accepted value of c = 299 792 500 (100) m/s and is about 100 times more accurate. As mentioned above, a recent differential measurement of the speed of light has been made by Bay, Luther, and White;[10] their value is 299 792 462 (18) m/s, which is also in agreement with the presently determined value.

As a result of the recommendations made by CCDM, two different definitions of a new length standard must be considered. First, we can continue as before with separate standards for the second and meter but with the meter defined as the length equal to $1/\lambda$ wavelengths in vacuum of the radiation from a stabilized laser instead of from a ^{86}Kr lamp. Either the methane-stabilized[11, 23] He-Ne laser at 3. 39 µm (88 THz) or the I_2-stabilized[31] He-Ne laser at 0. 633 µm (474 THz) appear to be suitable candidates. The 3. 39 µm laser is already a secondary frequency standard in the infrared, and hopefully direct measurements of the frequency of the 0. 633 µm radiation will give the latter laser the same status in the visible. The 3. 39 µm laser frequency is presently known to within 6 parts in 10^{10}, and the reproducibility and long term stability have been demonstrated to be better by more than two orders of magnitude. Hence, frequency measurements with improved apparatus in the next year or two are expected to reduce this uncertainty to a few parts in 10^{11}. A new value of the speed of light with this accuracy would thus be achievable if the standard of length were redefined to be this laser.

Alternately, one can consider defining the meter as a specified fraction of the distance light travels in one second in vacuum (that is, one can fix the value of the speed of light). The meter would thus be defined in terms of the second and, hence, a single unified standard would be used for frequency, time, and length. What at first sounds like a rather radical and new approach to defining the meter is actually nearly one hundred years old. It was first proposed by Lord Kelvin in 1879.[32] With this definition, the wavelength of all stabilized lasers would be known to the same accuracy with which their frequencies can be measured. Stabilized lasers would thus provide secondary standards of both frequency and length for laboratory measurements with the accuracy being limited only by the frequency reproducibility and measurability and the long term frequency stability. It should be noted that an adopted nominal value for the speed of light is already in use for high-accuracy astronomical measurements; thus, now there are two different standards of length in existence: one for terrestrial measurement and one for astronomical measurements. A definition which fixes c and unites these two values of c would certainly be desirable from a philosophical point of view.

Independent of which type of definition is chosen we believe that research on simplified frequency synthesis chains bridging the microwave-optical gap will be of great interest, as will refined experiments directed toward an understanding of the factors that limit laser optical frequency reproducibility. No matter how such research may turn out, it is clear that ultraprecise physical measurements made in the interim can be preserved through wavelength or frequency comparison with a suitably stabilized laser such as the 3.39 μm methane device.

Frequencies are currently measurable to parts in 10^{13}, and hence the over-all error of about six parts in 10^{10} for the frequency measurement represents a result which can be improved upon. The experiment was done fairly quickly to obtain frequency measurements of better accuracy than the wavelength measurements; this was easily done. It should be possible to obtain considerably more accuracy by using tighter locks on the laser. For example, the 8-m HCN laser has recently been phase locked[33] to a multiplied microwave reference which currently determines the HCN laser line widths. An improved microwave reference could be a superconducting cavity stabilized oscillator[34] for best stability in short term (narrowest line width) coupled with a primary cesium beam standard for good long term stability.

The relative ease with which these laser harmonic signals were obtained in the second round of frequency measurements indicates that the measurement of the frequencies of visible radiation now appears very near at hand. Such measurements should greatly facilitate one's ability to accurately utilize the visible and infrared portion of the electromagnetic spectrum.

The authors wish to acknowledge B. L. Danielson and G. W. Day who participated in the measurement of the frequency of the methane stabilized He-Ne laser and to R. L. Barger and J. L. Hall who first stabilized the laser and measured its wavelength. We also express our gratitude to D. G. McDonald and J. D. Cupp whose work with the Josephson junction was a parallel effort to ours in trying to achieve near-infrared frequency measurements. We note here that experiments using the Josephson junction to measure laser frequencies are continuing, and may well lead to better methods for near-infrared frequency synthesis in the future.

References

1. The interested reader will find a useful, critical discussion of the speed of light in K. D. Froome and L. Essen, The Velocity of Light and Radio Waves (Academic Press, New York, 1969).

2. T. S. Jaseja, A. Javan, J. Murray, and C. H. Townes, Phys. Rev. 133, A 1221 (1964), using infrared masers; D. C. Champeney, G. R. Isaak, and A. M. Khan, Phys. lett. 7, 241 (1963), using Mössbauer effect.

3. B. Warner and R. E. Nather, Nature (London) 222, 157 (1969), from dispersion in the light flash from pulsar NP 0532, obtain $\Delta c/c \leq 5 \times 10^{-18}$ over the range $\lambda = 0.25$ to 0.55 μm.

4. E. E. Uzgiris, J. L. Hall, and R. L. Barger, Phys. Rev. Lett. 26, 289 (1971).

5. J. Levine and J. L. Hall, J. Geophys. Res. 77, 2595 (1972).

6. Donald Halford, H. Hellwig, and J. S. Wells, Proc. IEEE 60, 623 (1972).

7. J. L. Hall and W. W. Morey, Appl. Phys. Lett. 10, 152 (1967).

8. J. Hall, R. L. Barger, P. L. Bender, H. S. Boyne, J. E. Faller, and J. Ward, Electron Technol. 2, 53 (1969).

9. K. M. Evenson, J. S. Wells, F. R. Petersen, B. L. Danielson, and G. W. Day, Appl. Phys. Lett. 22, 192 (1973).

10. Z. Bay, G. G. Luther, and J. A. White, Phys. Rev. Lett. 29, 189 (1972).

11. R. L. Barger and J. L. Hall, Phys. Rev. Lett. 22, 4 (1969).

12. K. M. Evenson, G. W. Day, J. S. Wells, and L. O. Mullen, Appl. Phys. Lett. 20, 133 (1972).

13. R. L. Barger and J. L. Hall, Appl. Phys. Lett. 22, 196 (1973).

14. R. D. DeSlattes, H. P. Layer, and W. G. Schweitzer (paper in preparation).

15. P. Giacomo, results presented at the 5th session of the Comite Consultatif pour la Definition du Metre, BIPM, Sevres, France, 1973.

16. K. M. Baird, D. S. Smith, and W. E. Berger, Opt. Comm., 7, 107 (1973).

17. K. D. Froome, Proc. Roy. Soc., Ser. A247, 109 (1958).

18. L. O. Hocker, A. Javan, D. Ramachandra Rao, L. Frenkel, and T. Sullivan, Appl. Phys. Lett. 10, 5 (1967).

19. K. M. Evenson, J. S. Wells, L. M. Matarrese, and L. B.
     Elwell, Appl. Phys. Lett. 16, 159 (1970).

20. K. M. Evenson, J. S. Wells, and L. M. Matarrese, Appl.
     Phys. Lett. 16, 251 (1970).

21. L. M. Matarresse and K. M. Evenson, Appl. Phys. Lett.
     17, 8 (1970).

22. Antenna Engineering Handbook, edited by Henry Jasik
     (McGraw-Hill, New York, 1961), Chap. 4 and 10.

23. J. L. Hall in Esfahan Symposium on Fundamental and
     Applied Laser Physics, edited by M. Feld and A. Javan
     (Wiley, New York, to be published).

24. Charles Freed and Ali Javan, Appl. Phys. Lett. 17, 53
     (1970).

25. K. M. Evenson, J. S. Wells, L. M. Matarrese, and D. A.
     Jennings, J. Appl. Phys. 42, 1233 (1971).

26. D. W. Allan, J. E. Gray, and H. E. Machlan, IEEE Trans.
     Instrum. Meas. IM-21, 388 (1972).

27. H. Hellwig, R. F. C. Vessot, M. W. Levine, P. W.
     Zitzewitz, D. W. Allan, and D. Glaze, IEEE Trans.
     Instrum. Meas. IM-19, 200 (1970).

28. T. G. Blaney, C. C. Bradley, G. J. Edwards, D. J. E.
     Knight, P. T. Woods, and B. W. Jolliffe, Nature, (to
     be published, 1973).

29. F. R. Petersen, D. G. McDonald, J. D. Cupp, and B. L.
     Danielson, "Accurate Rotational Constants, Frequencies,
     and Wavelengths from $^{12}C^{16}O_2$ Lasers Stabilized by
     Saturated Absorption," Proc. of the Laser Spectroscopy
     Conference, Vail, Colorado (June 1973).

30. Comite Consultatif pour la Definition du Metre, 5th session,
     Rapport, (Bureau International des Poids et Mesures,
     Sevres, France, 1973).

31. G. R. Haines and C. E. Dahlstrom, Appl. Phys. Lett. 14, 362 (1969); G. R. Haines and K. M. Baird, Metrologia 5, 32 (1969).

32. W. F. Snyder, IEEE Trans. on Instr. and Meas. IM-22, 99 (1973).

33. J. S. Wells and Donald Halford, NBS Tech. Note #620 May (1973).

34. S. R. Stein and J. P. Turneaure, Electron Lett., 8, 431 (1972).

HIGH RESOLUTION SPECTROSCOPY WITH TUNABLE DYE LASERS

Herbert Walther

I. Physikalisches Institut der Universität zu
Köln, Köln
Federal Republic of Germany

If tunable dye lasers are used for the spectroscopic investigation of electronically excited atoms
and molecules two groups of methods are available which
yield a resolution comparable to the natural width of
the investigated transitions. To the first group belong
the well known methods of radio frequency spectroscopy
as for example the optical double resonance method and
furthermore the purely optical techniques such as the
level crossing method and the observation of quantum
beats. For this group of methods a narrow spectral distribution of the exciting light is not needed, in fact
a broad banded excitation is even more favorable. The
advantage of the use of tunable dye lasers for these experiments, compared to an excitation with classical light
sources, is that a higher population of the excited states
may be achieved. This results in a better signal to
noise ratio and in a more accurate measurement. Furthermore, levels can now be populated which could not be
reached by the excitation with discharge lamps since the
corresponding transition probabilities are too small.
In addition a step-wise excitation is now feasible.

The second group of high resolution experiments requires a bandwidth of the spectral distribution of the
laser which should be smaller than the natural linewidth.
Experiments of this type are those performed by saturation spectroscopy or by means of well collimated atomic
beams. In the latter case the atomic velocity with respect to the direction of the exciting laser light is
reduced, so that the absorption width of the beam is

comparable to the natural width of the excited transition. This last group of experiments cannot be performed using excitation by classical discharge lamps.
The advent of the narrow banded tunable dye lasers was
essential for the application of these methods.

     In this report experiments of both types recently
performed in our laboratory in Köln will be described.
The first experiment has been performed together with
Hartmut Figger and Shü-chuan Cha. The purpose of this
experiment was the investigation of the hyperfine-split-
ting of the $3^2P_{3/2}$ level of NaI by the use of the level
crossing method.

     For the excitation of the atoms a nitrogen laser
pumped dye laser was used. Since the fluorescent light
was observed a certain time interval after the excitation by the laser pulse the signal observed was determined only by those atoms which remained in the excited
state for a longer time than the average. Therefore
the linewidth observed was several times smaller than
the natural width. Experiments using this principle
were performed several years ago in connection with
Mössbauer effect studies e.g. [1,2] as well as in atomic
spectroscopy [3,4]. One of these experiments which was
performed by Copley, Kibble and Series [4] was also a
level crossing experiment at the $3^2P_{3/2}$ level of NaI.
In their set-up the light of a sodium discharge lamp
pulsed by a Kerr cell shutter was used to excite the
atoms. Since in this way only a small fraction of atoms
were excited the maximum delay time possible for the observation of the signal was about 2.5 lifetimes. Using
a pulsed dye laser, however, the excited state can be
saturated so that after eight or ten lifetimes the
10^{-4} - 10^{-3} part of the atoms is still in the excited
state; this is about the same amount as can be excited
if classical discharge lamps are used.

     The experimental set-up for our measurements is
shown in Fig. 1. The dye laser is similar to that described by Hänsch [5]. The grating was an echelle grating with 79 grooves per mm and a blaze angle of 63.4
degrees. The 39th order of the grating was used. The
laser beam was expanded by the beam expander so that a
surface of the grating of about 5 cm diameter was illuminated. The nitrogen laser, an Avco Model C5000, was
normally run at a repetition rate of about 300 Hz. Rhodamine 6G in a 10^{-3} molar solution in ethanol was the dye.
The pulse duration and the peak power of the dye laser
pulse were about 5 nsec and 3 kW respectively. The
length of the dye laser cavity was about 40 cm.

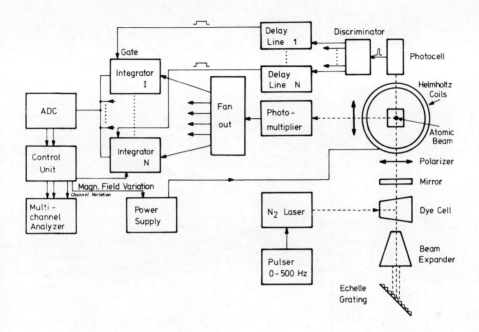

Fig. 1

Free atoms of an atomic beam were investigated.
The direction of polarization of the laser light was
perpendicular to the external magnetic field, which was
produced by a pair of Helmholtz coils. The fluorescent
light of the Na-atoms was observed through a linear polarizer the polarization direction of which was also
perpendicular to the magnetic field.

The laser light was monitored by a photocell whose
signal triggered a discriminator. The output of the
discriminator was split into four different delay lines
producing the gate signals which open the signal inte-
grators so that the signal is integrated for four different delay times with an integration time of 100 nsec
in each case. The integrator signal is fed to a voltage
to frequency converter and stored in a multichannel analyzer operating in the multiscaling mode. The channel
variation was synchronized to the magnetic field variation. If the signal is measured with a delay of eight
lifetimes it is necessary to average the signal for a-
bout one hour in order to obtain an appropriate signal-
to-noise ratio.

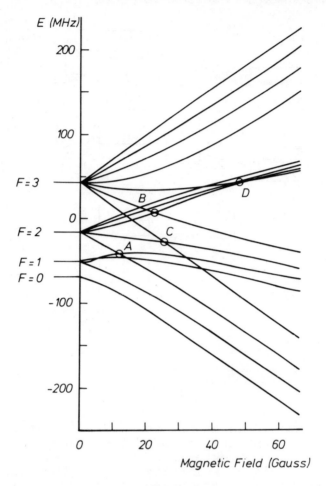

Fig. 2

Fig. 2 shows the level scheme of the hyperfine-
structure of the $3^2P_{3/2}$ level of Na^{23}. The four hyperfine levels are split into 16 Zeeman sublevels. Up to
60 Gauss four level crossings marked by A,B,C,D with
$\Delta m = 2$ may be observed. The crossing D belongs to two
levels which intersect at a very small angle, which implies that this crossing is very broad and therefore
difficult to observe. The crossings A,B,C are hidden
by the zero field crossing signal (Hanle signal) in a
usual level crossing experiment. Therefore a better
resolution is desirable. In our experiment it was even
possible to isolate crossings B. Although B and C were
separated from each other by about 2 Gauss.

In a delayed level crossing experiment the lineshape of the crossing signal is no longer a Lorentzian.
A bell shaped minimum is observed at the crossing point
and weak intensity oscillation at the wings of the signal. The distance between the side-minima decreases
and their relative amplitude increases if a larger delay time is used. These line shapes are discussed in
detail in Ref. [4] and especially in Ref. [6]. At certain delay times the level crossing signals at higher
fields may coincide with the side-minima of the zero
field crossing. By the appropriate choice of the parameters, however, the higher crossing can be isolated.
This is shown in Fig. 3. In this figure the position
of crossing A is marked by a dotted line. In order to

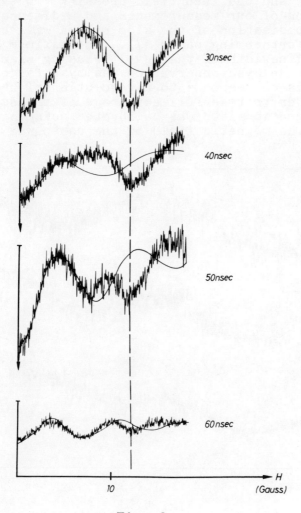

Fig. 3

show the influence of the side-minimum on the zero field
crossing the calculated line shape of the zero field
crossing signal is shown as a solid line. It can be
seen that with a delay time of somewhat more than 50 nsec
the 12 Gauss crossing lies between two side-minima of the
zero field crossing. The signal measured for a delay of
40 nsec shows that the undulation of the zero crossing
and crossing A may even cancel.

Some measurements obtained for crossing B are shown
on Fig. 4. The position of the crossing B at about 22
Gauss can clearly be located. The second minimum at the
right side changes with the delay time demonstrating
that there is still an influence from the undulation
of crossing A and the zero field crossing. Details of
the evaluation of our measurements and a discussion of
a possible apodisation of the side-minima will be presented in a forthcoming paper [7]. The maximum reduction of the linewidth of the level crossing signal obtained by the delayed observation was by a factor of
nearly 6 (This corresponds to a resolution of about 3
MHz). In order to take full advantage of the small linewidth the magnetic field was calibrated using a Rb mag-
netometer. The magnetic field of the earth was care-

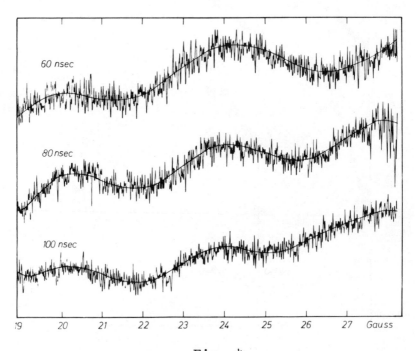

Fig. 4

fully compensated.

The results obtained for the position of the cross-ings are compiled in Table 1. Table 2 shows the hyper-fine constants evaluated from the position of the level crossing signals compared to other measurements. There is a good agreement between the different results. The measurements of Ackermann [8] have been performed using the double resonance method in a strong external mag-netic field.

Summarizing we can say that the technique applied here is useful for obtaining higher resolution in level crossing or in double resonance experiments. More accu-rate results than in an usual experiment can be expected especially in those cases where we have an overlap of the signals within the natural line width.

Table 1: Observed level crossing signals

Designation of the crossings (F, m ; F', m')	Position (Gauss)
A: (2,-2; 1,0) B: (3,-2; 2,0)	12.42 (5) 22.10 (10)

Table 2: Results for the hyperfine constants

$(3^2P_{3/2}$ level of $Na^{23})$

Authors	A (MHz)	B (MHz)
Ackermann [8]	18.7 (4)	3.4 (4)
Copley et al. [4]	18.5 (4)	3.0 (6)
Figger et al. [7] (this paper)	18.62 (8)	3.14 (19)

In the second part of this paper some of the high resolution work we performed using cw dye lasers will be reviewed. This work was performed together with Wolfgang Hartig, Rudolf Schieder and Hermann Hartwig.

With a cw dye laser, which was similar to that described by Hercher and Pike [9], a rather good long term stability on the order of 25 MHz/min was obtained. This enabled us to lock the output frequency of the dye laser to a specific hyperfine transition of the Na D lines. To reduce the Doppler width a well collimated atomic beam was used. The collimation ratio was 1:550, corresponding to a Doppler width of 2.5 MHz when the atoms are viewed at right angles to the beam. Since the natural width of the Na D lines is 10 MHz the absorption width of the beam was essentially determined by the natural linewidth. Details of the set-up and the high resolution experiments, which have been performed on the Na D lines are described in Ref. [10].

The set-up used for the frequency stabilization is shown in Fig. 5. It is the more or less standard set-up employed for the frequency control of gas lasers.

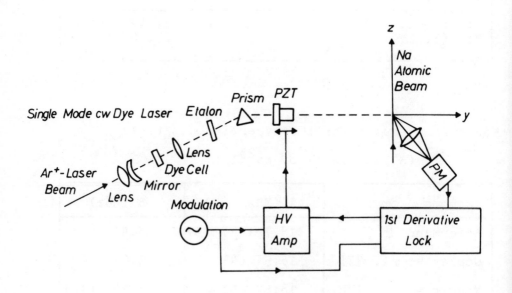

Fig. 5

The voltage applied to the piezoelectric drive of the flat mirror is modulated by 1.5 kHz. The first derivative of the fluorescence signal is used as the error signal to modify the average cavity length.

The intensity of the laser was also stabilized using an electro optical filter. Details on this control circuit are given in [10].

In order to measure the frequency stability of the laser the dc part of the fluorescent light was recorded as a function of time. From the intensity variation, the frequency deviation $\Delta \nu$ of the laser averaged for a certain time τ has been derived. The results obtained are $\Delta \nu$ = 2 MHz for τ = 10^{-1} sec and $\Delta \nu$ = 200 kHz for τ = 4 sec. The unity gain bandwidth for the control circuit was about 2 kHz. Since the frequency noise of the free running laser is as much as 50 kHz, a further improvement of the short term stability of the dye laser can be expected if a higher bandwidth for the control circuit is used.

For high precision spectroscopic investigations using tunable dye lasers the wavelength variation of the laser must be measured as the laser is tuned over the structure of the lines under investigation. This may be done using a highly stable optical cavity. In our experiments another approach was used. The principle of the method is shown in Fig. 6. Two dye lasers are pumped by the same Ar^+-laser, both of which are stabilized to different transitions whose frequency difference is to be measured. This can then be accurately determined by measuring the beat note between the two monomode lasers.

We first used this set-up in an exploratory experiment in order to measure the splitting between two hyperfine components of the sodium D_1 line. For this purpose the two lasers were stabilized using different regions of the same atomic beam [11]. In our opinion the most accurate measurement of the splitting between the two transitions can be achieved with this method since both lasers are stabilized and the absolute determination of the beat frequency can be performed very accurately.

The technique described here can, of course, not be used if the two lines are not well resolved. For such a case the set-up shown in Fig. 7 was considered. The first dye laser is stabilized to a particular transition and the second is locked to the first one via the beat

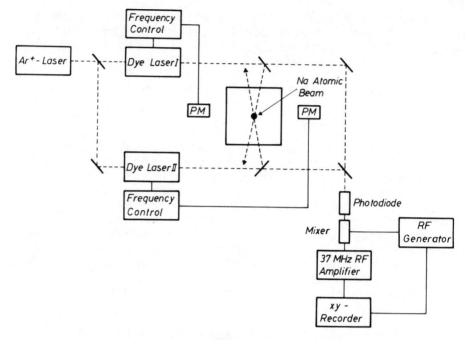

Fig. 6

signal. In this way a certain frequency difference between both lasers is maintained. By changing the output frequency of the rf generator, the frequency of laser
II is changed by the same amount as the rf generator.
The frequency variation is therefore known very precisely. This is in principle the technique used by Barger
and Hall in connection with the methane stabilized He-Ne
laser [12].

    In the last part of this paper some of the measure-
ments performed with a new dye laser set-up built in our
laboratory will be reported. The laser configuration is
shown in Fig. 8, it is in principle similar to the one
first used by Dienes, Ippen and Shank [13]. In order
to provide high thermal and mechanical stability the
prism, intracavity mirror (R= 100 mm), dye cell, and
end mirror (R= 10 mm) are mounted in appropriate holes
of a solid block of invar. This block as well as the
plane end mirror and the intracavity etalon are fixed
on a solid table which is also made of invar. The total
cavity length of the laser is 62 cm corresponding to a
longitudinal mode distance of 245 MHz. Single mode operation of the laser may be obtained using a solid silica
intracavity etalon with a coating of 30% reflectivity.

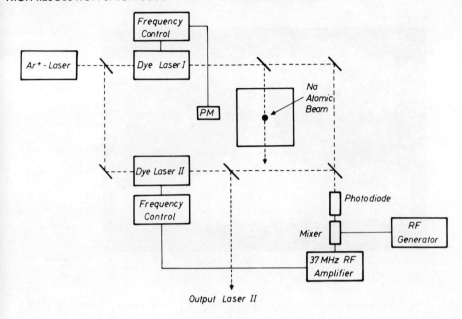

Fig. 7

In order to get information on the coherence length of the laser light, the beat signal between two different longitudinal modes was investigated, which gives a rough

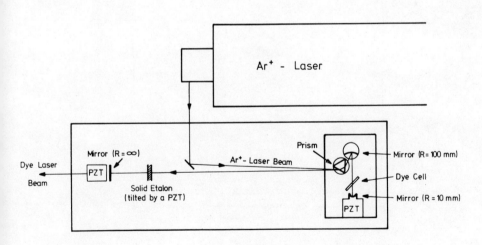

Fig. 8

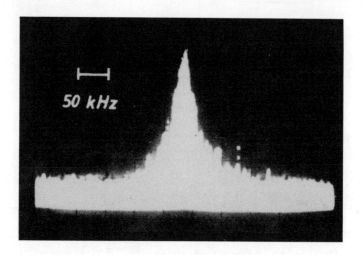

Fig. 9

value for the natural linewidth of the laser. This beat
signal is shown in Fig. 9. For this measurement about
three signal transients of a frequency analyser are stored
on a storage oszilloscope. The linewidth observed is
about 50 kHz.

If the linewidth is measured by means of a Fabry
Perot the influence of microphonics, optical path variations in the dye solutions and other environmental disturbances become evident. Fig. 10 shows such an inter-

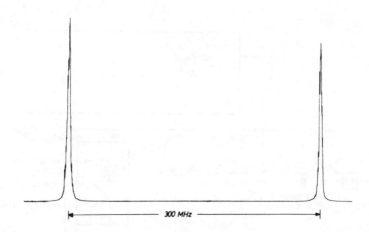

Fig. 10

ferometric linewidth measurement. The Fabry Perot had
a free spectral range of 300 MHz. The linewidth resulting from this recording is 2.8 MHz. This corresponds to
a linewidth of the dye laser of roughly 0.5 MHz, since
the resolution limit of the Fabry Perot was 2.4 MHz.
This is the smallest linewidth which has been observed
so far with a dye laser.

To record the structure of one of the signals shown
in Fig. 10 a time of about 10 sec is needed. The linewidth observed therefore represents the linewidth of the
laser including the frequency jitter averaged over a time
of 10 sec. Since these frequency variations can be reduced by a fast frequency control, a further reduction
of the linewidth to the range of 50 kHz may be feasible.

The support of the Deutsche Forschungsgemeinschaft
is gratefully acknowledged.

REFERENCES

1. R.E. Holland, F.J. Lynch, G.J. Perlow, S.S. Hanna
    Phys. Rev. Lett. 4, 181 (1960).
2. W. Neuwirth, Z. Physik 197, 473 (1966).
3. I.J. Ma, G. zu Putlitz, G. Schütte, Z. Physik 208,
    276 (1968).
4. G. Copley, B.P. Kibble and G.W. Series, J. Phys. B
    1, 724 (1968).
5. T.W. Hänsch, Appl. Opt. 11, 895 (1972).
6. R.C. Hilborn, R.L. deZafra, JOSA 62, 1492 (1972).
7. U. Figger, H. Walther, Z. Physik to be published
    (1973).
8. H. Ackermann, Z. Physik 194, 251 (1966).
9. M. Hercher, H.A. Pike, Opt. Comm. 3, 65 (1971).
10. W. Hartig, H. Walther, Appl. Phys. 1, 171 (1973).
11. W. Hartig, H. Walther, unpublished material
12. R.L. Barger and J.L. Hall, Appl. Phys. Lett. 22,
    196 (1973).
13. A. Dienes, E.P. Ippen and Ch. V. Shank, IEEE Journ.
    of Quant. Electr. QE8, 388 (1972).

PRECISION HETERODYNE CALIBRATION

C. Freed, D. L. Spears and R. G. O'Donnell

MIT Lincoln Laboratory,[*] Lexington, Massachusetts

A. H. M. Ross, MIT Research Laboratory of Electronics[**]

Cambridge, Massachusetts

ABSTRACT

The frequencies of $^{12}C^{18}O_2$, $^{13}C^{16}O_2$ and $^{13}C^{18}O_2$ isotope lasers have been measured to better than 3 MHz (0.0001 cm^{-1}) by comparison with a $^{12}C^{16}O_2$ reference laser. Heterodyne techniques were used to generate 139 difference frequencies in a liquid-nitrogen cooled HgCdTe photodiode. Microwave frequency counter measurements of the difference frequencies were then used to calculate new values for the band centers and rotational constants of the rare isotopes. It is shown that an additional thousandfold improvement in accuracy and a greatly extended spectral range can be readily achieved.

INTRODUCTION

This paper will concentrate on new results relating to CO_2 isotope lasers[1] and high frequency HgCdTe varactor photodiodes.[2] We have obtained and calibrated hundreds of CO_2 isotope laser transitions, most of which have never been observed before.

Isotope laser action was first reported by Wieder and McCurdy[3] in $^{12}C^{18}O_2$, by Jacobs and Bowers[4] in $^{13}C^{16}O_2$, and by Siddoway[5] in $^{14}C^{16}O_2$. The lasing transitions reported in these papers were measured with conventional grating spectrometers, with accuries no better than 1500 MHz (0.05 cm^{-1}). To our knowledge, $^{13}C^{18}O_2$ lasers

*This work was sponsored by the Department of the Air Force and the Advanced Research Projects Agency of the Department of Defense.
**This work was sponsored in part by NSF Grant GK-37979X.

have never been previously reported. By contrast, the absolute
frequencies and vibrational-rotational constant given in this preliminary paper are accurate within about 3 MHz (0.0001 cm^{-1}); this
accuracy can be further improved by at least another factor of 1000
by repeating these experiments with the lasers stabilized to the
center frequency of the lasing transitions. In previously published
results[6,7] it was shown that CO_2 lasers can be frequency stabilized
in any lasing transition by using the standing-wave saturation resonances in a low-pressure, room-temperature, pure CO_2 absorber via
the intensity changes observed in the entire collisionally-coupled
spontaneous emission band at 4.3 μm between the upper and the ground
state. It was also demonstrated that the frequency shift due to
changes in pressure is very small in CO_2, typically much less than
100 Hz/mTorr. By utilizing the above frequency stabilization technique, Petersen et al have recently determined[8] the rotational
constants[9] and frequencies of most of the common $^{12}C^{16}O_2$ laser transitions[9] to within a few kHz.[10] Thus it is clear that the frequency stability and resettability together with the availability
of hundreds of lasing transitions uniquely endow the CO_2 system for
either direct use in high resolution spectroscopy or as precision
calibration bench marks in heterodyne spectroscopy with tunable
lasers.

EXPERIMENT

Grating controlled lasers with a 150-cm semiconfocal cavity
configuration and an approximately 120-cm long, 1.2-cm diameter,
active discharge were used. The design, stability, and TEM_{00q} (7)
mode characteristics of these lasers were previously described.[7]
The estimated round-trip loss due to coupling through the dielectri-
cally-coated output mirrors, 0-order grating loss, diffraction and
scatter was about 20-25%. This loss limited the observation of
lasing transitions to those which have small signal single pass
gains greater than about 10% per meter under sealed-off cw operating
conditions. Some enhancement of gain was produced, however, by
cooling the discharge tubes with refrigerated alcohol near -60°
Centigrade instead of water at room temperature. Under these conditions about 80 lasing transitions were observed for each of the
CO_2 isotopes we have measured. Figure 1 graphically illustrates
these lasing transitions and the spectral region they occupy.

In this paper the components of a Fermi multiplet are labelled
with Roman subscripts in order of decreasing energy. It has been
determined[11,12,13] recently that the [10^00, 02^00] level is to be
identified with the unperturbed 02^00 level in $^{12}C^{16}O_2$, $^{12}C^{18}O_2$, and
$^{13}C^{18}O_2$, and with the unperturbed 10^00 level in $^{13}C^{16}O_2$. Note that
this identification is the reverse of the traditional in $^{12}C^{16}O_2$.
Figure 1 illustrates this notation and the 330 lasing transitions
obtained with the lasers used in these experiments; however, a very
large number of additional laser lines were obtained since then with
lasers of more recent design.

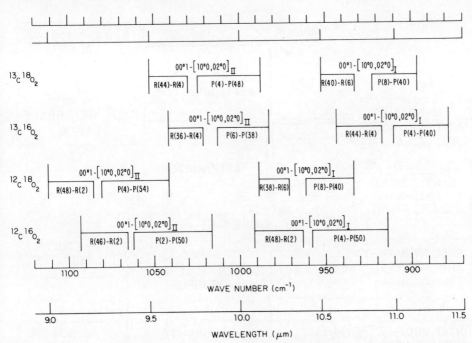

Fig. 1.   The 330 Lasing Transitions Obtained with the Lasers Used
in These Experiments.

A mixture of CO_2, N_2, Xe, He and H_2 with partial pressures
of 2.3, 3.0, 1.1, 8.3 and 0.15 Torr was used in the lasers and we
did not try to optimize either the laser fill or the output coupling
for any given lasing band. Under these conditions the power outputs
varied from about 30% in the weak $^{13}C^{16}O_2$ $00^01 - [10^00, 02^00]_{II}$
R-branch to about 90% in the strong $00^01 - [10^00, 02^00]_{II}$ P-branch
of $^{12}C^{18}O_2$, when compared to the corresponding branches of the
$^{12}C^{16}O_2$ reference laser. The relative strengths of each lasing
branch can be also ascertained from Figure 1, by counting and comparing the number of lasing transitions observed in each branch of
the CO_2 isotopes.

Optical heterodyne techniques were used to generate beat frequencies between the rare and the abundant isotope laser lines.
Figure 2 shows a block diagram of the experimental apparatus. All
the radio frequency and microwave devices were standard laboratory
instruments.

The high speed, high quantum efficiency HgCdTe photodiodes we
used were previously described by Spears, Harman, and Melngailis.[14]
The external quantum efficiencies for these detectors at 77°K were
typically 40% over the 9-11 μm region with zero bias low frequency
detectivities $D* \simeq 1.1 \times 10^{10}$ cm $Hz^{1/2}$ W^{-1}. The diodes used in

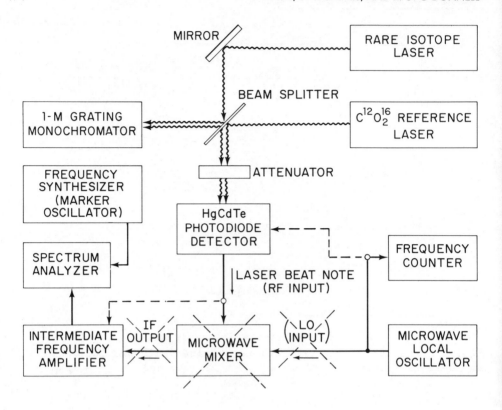

Fig. 2.   Block Diagram of the Experimental Apparatus.

these experiments were in the form of circular mesas 200 μm in
diameter and displayed a 3 dB roll-off in the vicinity of 1000 MHz
with reversed biased operation into 50 Ω coaxial components.

In the initial phase of our measurements, separate microwave
mixers were employed as indicated by the solid lines in Figure 2.
In this mode of operation we obtained about 20 dB signal-to-noise
ratio at 20 GHz with a 10 KHz noise bandwidth.

We have also successfully achieved mixing of the microwave
local oscillator with the CO_2 laser beats directly in the HgCdTe
photodiodes, as indicated by the dashed lines in Figure 2.   The
mixing of the microwave signals closely corresponded to varactor
diode behavior.[15]   In this varactor – photodiode mode of operation,
details of which have been described elsewhere,[2] the detector output at the intermediate frequency was linearly proportional to both
laser and to microwave local oscillator power.   Figure 3 shows a
typical spectrum analyzer display of a laser beat signal at 21.3 GHz,

which resulted from varactor-photodiode mixing. The approximately 45 dB signal-to-noise ratio in Figure 3 was obtained with 0.5 mW of $^{12}C^{18}O_2$ laser power of the (attenuated) $00^01 - [10^00, 02^00]_I$ P(24) transition, 0.6 mW of $^{12}C^{16}O_2$ laser power of the (attenuated) $00^01 - [10^00, 02^00]_I$ P(16) transition, and was somewhat limited by the ∿1 mW power available from the 20.9 GHz microwave local oscillator into the varactor-photodiode. For all beat frequencies we have measured, the signal-to-noise ratio was more than adequate with a 10 kHz spectrum analyzer noise bandwidth.

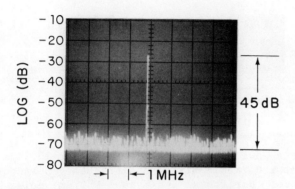

Fig. 3. 21,296 MHz Beat Frequency of the $^{12}C^{18}O_2$ Laser $00^01 - [10^00, 02^00]_I$ Band P(24) and the $^{12}C^{16}O_2$ Laser $00^01 - [10^00, 02^00]_I$ Band P(16) Lines. Power Levels into Photodiode: $^{12}C^{18}O_2$ Laser, 0.5 mW; $^{12}C^{16}O_2$ Laser, 0.6 mW. The Horizontal Scale is 1 MHz/cm; the Noise Bandwidth is 10 kHz.

Microwave pump harmonic mixing was used to detect millimeter-wave laser beats.[2] The four different laser beats in the 40 to 60 GHz range shown in Fig. 4 were detected by mixing the 3rd or 4th harmonics of the microwave pump with the laser beat. For example, to detect the 43,429 GHz laser beat a microwave signal at 14.343 GHz was fed into the diode to obtain the 0.400 GHz intermediate frequency. The signal-to-noise ratios in the four cases shown ranged from 12 to 29 dB. A fall-off in quantum efficiency of the photodiode at 11.1 μm was responsible for the smaller signals observed in the upper half of Figure 4.

The highly efficient, low noise varactor mixing in HgCdTe photodiodes enabled us to easily detect 60 GHz heterodyne beats of CO_2 lasers with less than a milliwatt of laser and only a few milliwatts of microwave power. With the use of waveguides and/or narrow band synchronous detection schemes, much higher optical heterodyne

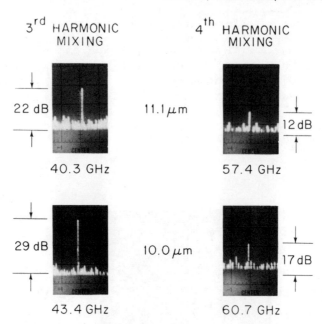

Fig. 4. Millimeter-wave CO_2 laser beat signals at 10.0 μm and 11.1 μm Detected by Varactor Harmonic Mixing and Frequency Downconversion in a HgCdTe photodiode. Horizontal Scale is 1 MHz/cm; noise bandwidth is 10 kHz.

frequencies and very much larger signal-to-noise ratios should be obtainable in a straightforward manner. Thus, varactor-photodiodes provide a very simple and effective means of precisely measuring and monitoring infrared laser line separations, since the 200 μm diameter active area is easily brought into alignment and does not require focusing of the laser beams.

Although we could observe and measure laser beats well beyond the 40–60 GHz separating adjacent CO_2 laser transitions, within the scope of this work we have only utilized 139 difference frequencies below 21 GHz out of the several hundred lasing transitions we observed in the three rare isotope lasers.

In the experiments, the two lasers were tuned to line center by observing the detuning at which oscillation ceased, and setting the tuning midway between these values. This procedure and the stability of the lasers was generally sufficient to measure the frequency differences with an accuracy of a few MHz, as shown by the result in Tables I through V. The difference frequencies we have measured are listed in Tables I, II, III and were used to obtain the vibrational-rotational constants listed in Table IV as further discussed in the next section of this paper.

Table I.

Difference Frequencies Measured Between $^{12}C^{18}O_2$ and $^{12}C^{16}O_2$ Transitions

ISOTOPE LINE			REFERENCE LINE		MEASURED F(I)-F(R)	MEAS.-CALC.
I	P(24)	I	P(16)		21296.0	-2.3
I	P(22)	I	P(14)		19810.2	-1.9
I	P(20)	I	P(12)		18259.4	-0.5
I	P(18)	I	P(10)		16642.1	-0.0
I	P(16)	I	P(8)		14960.0	0.9
I	P(14)	I	P(6)		13213.0	1.9
I	P(12)	I	P(4)		11401.0	2.2
I	R(6)	I	R(14)		-8904.6	-5.9
I	R(8)	I	R(16)		-11336.5	5.4
I	R(10)	I	R(18)		-13840.9	-1.7
I	R(12)	I	R(20)		-16389.6	-0.0
I	R(14)	I	R(20)		19490.4	4.1
I	R(16)	I	R(22)		16080.7	-0.5
I	R(18)	I	R(24)		12616.8	-7.3
I	R(20)	I	R(26)		9120.0	3.5
I	R(22)	I	R(28)		5561.0	1.2
I	R(24)	I	R(30)		1950.7	-4.9
I	R(26)	I	R(32)		-1694.6	0.0
I	R(28)	I	R(34)		-5385.5	3.5
I	R(30)	I	R(36)		-9123.0	2.9
I	R(32)	I	R(38)		-12900.0	3.2
I	R(34)	I	R(38)		14845.0	1.1
I	R(34)	I	R(40)		-16722.3	-3.2
I	R(36)	I	R(40)		10197.7	-3.0
I	R(38)	I	R(42)		5519.0	1.2
II	P(54)	II	P(28)		4010.0	0.1
II	P(52)	II	P(26)		2981.6	-7.5
II	P(50)	II	P(24)		2212.6	-1.3
II	P(48)	II	P(22)		1691.8	7.2
II	P(46)	II	P(20)		1409.2	8.2
II	P(44)	II	P(18)		1369.3	6.0
II	P(42)	II	P(16)		1563.0	-8.0
II	P(40)	II	P(14)		2018.5	-5.5
II	P(38)	II	P(12)		2716.0	-5.6
II	P(36)	II	P(10)		3669.0	5.7
II	P(34)	II	P(8)		4852.0	3.7
II	P(32)	II	P(6)		6276.7	0.9
II	P(30)	II	P(4)		7945.0	0.4
II	P(24)	II	R(2)		-8424.2	0.7
II	P(22)	II	R(4)		-5404.0	-1.2
II	P(20)	II	R(6)		-2147.5	-0.9
II	P(18)	II	R(8)		1343.8	2.2
II	P(16)	II	R(10)		5053.0	-6.7
II	P(14)	II	R(12)		9003.0	-2.3
II	P(12)	II	R(14)		13171.5	-4.4
II	P(10)	II	R(16)		17563.5	-5.4
II	P(8)	II	R(20)		-16436.3	-0.9
II	P(6)	II	R(22)		-10801.4	9.8
II	P(4)	II	R(24)		-4970.0	4.9
II	R(2)	II	R(32)		222.4	1.6
II	R(4)	II	R(34)		7364.0	10.6
II	R(6)	II	R(36)		14681.1	5.5
II	R(6)	II	R(38)		-16888.4	-1.4
II	R(8)	II	R(40)		-8620.8	-7.8
II	R(10)	II	R(42)		-161.2	1.1
II	R(12)	II	R(44)		8453.5	-7.1
II	R(12)	II	R(46)		-20066.0	-2.8

STANDARD DEVIATION= 4.8 MHz

Table II.

Difference Frequencies Measured Between $^{13}C^{16}O_2$ and $^{12}C^{16}O_2$ Transitions

ISOTOPE LINE		REFERENCE LINE		MEASURED F(I)-F(R)	MEAS.-CALC.
I	R(4)	I	P(48)	19995.0	0.9
I	R(6)	I	P(46)	765.0	-1.9
I	R(8)	I	P(44)	-18279.0	-1.7
I	R(12)	I	P(42)	5909.7	1.9
I	R(14)	I	P(40)	-13377.5	4.5
I	R(18)	I	P(38)	8755.8	-0.8
I	R(20)	I	P(36)	-10813.2	-2.6
I	R(24)	I	P(34)	9230.3	-0.1
I	R(26)	I	P(32)	-10650.0	-1.9
I	R(30)	I	P(30)	7237.3	-0.8
I	R(32)	I	P(28)	-12986.0	1.7
I	R(36)	I	P(26)	2681.3	1.1
I	R(38)	I	P(24)	-17931.0	0.2
I	R(40)	I	P(24)	16586.0	1.2
I	R(42)	I	P(22)	-4553.0	-1.7
II	P(38)	I	R(32)	-10176.8	0.2
II	P(36)	I	R(34)	18289.3	2.6
II	P(36)	I	R(36)	-14842.0	1.0
II	P(34)	I	R(38)	14366.0	-0.5
II	P(34)	I	R(40)	-17202.0	-5.5
II	P(32)	I	R(42)	12772.0	-0.5
II	P(32)	I	R(44)	-17203.0	0.2
II	P(30)	I	R(46)	13544.0	2.4
II	R(4)	II	P(46)	12060.0	0.9
II	R(6)	II	P(44)	-7834.0	0.4
II	R(10)	II	P(42)	14401.0	-1.9
II	R(12)	II	P(40)	-6669.0	1.0
II	R(16)	II	P(38)	11781.0	-2.4
II	R(18)	II	P(36)	-10433.0	3.6
II	R(22)	II	P(34)	4279.3	-1.2
II	R(24)	II	P(32)	-19052.0	-4.0
II	R(26)	II	P(32)	16289.3	0.0
II	R(28)	II	P(30)	-8002.0	4.0
II	R(32)	II	P(28)	374.0	0.9
II	R(36)	II	P(26)	6122.1	-1.3

STANDARD DEVIATION=   2.4 MHz

Table III.

Difference Frequencies Measured Between $^{13}C^{18}O_2$ and $^{12}C^{16}O_2$ Transitions

ISOTOPE LINE		REFERENCE LINE		MEASURED F(I)-F(R)	MEAS.-CALC.
I	P(16)	I	P(48)	6980.0	0.3
I	P(14)	I	P(46)	-10571.0	-1.7
I	P(10)	I	P(44)	17132.0	-0.6
I	P(8)	I	P(42)	-900.0	-0.0
I	R(6)	I	P(32)	10019.0	0.4
I	R(8)	I	P(30)	-9224.5	0.2
I	R(12)	I	P(28)	8976.0	0.0
I	R(14)	I	P(26)	-10823.1	0.3
I	R(18)	I	P(24)	4761.0	1.0
I	R(20)	I	P(22)	-15617.6	2.2
I	R(22)	I	P(22)	18328.6	-0.7
I	R(24)	I	P(20)	-2699.1	-1.7
I	R(28)	I	P(18)	8179.0	0.2
I	R(30)	I	P(16)	-13475.2	-0.9
I	R(32)	I	P(16)	16985.8	1.1
I	R(34)	I	P(14)	-5347.4	0.1
I	R(38)	I	P(12)	668.0	-0.2
II	P(48)	I	R(40)	3393.0	1.5
II	P(46)	I	R(44)	-1502.5	-9.9
II	P(44)	I	R(48)	-3732.5	9.6
II	P(14)	II	P(50)	-298.2	-3.1
II	P(12)	II	P(48)	-20075.5	-0.0
II	P(8)	II	P(46)	4521.0	1.1
II	P(6)	II	P(44)	-15741.0	1.6
II	R(6)	II	P(36)	4077.5	1.1
II	R(8)	II	P(34)	-17320.0	2.1
II	R(12)	II	P(32)	-1016.4	-1.2
II	R(16)	II	P(30)	13488.4	0.2
II	R(18)	II	P(28)	-8848.2	-2.4
II	R(22)	II	P(26)	3358.0	1.7
II	R(24)	II	P(24)	-19355.0	-1.5
II	R(26)	II	P(24)	13786.5	-3.8
II	R(28)	II	P(22)	-9408.7	2.0
II	R(32)	II	P(20)	-1206.0	0.1
II	R(36)	II	P(18)	5284.8	2.5
II	R(38)	II	P(16)	-18702.7	-2.7
II	R(40)	II	P(16)	10080.6	1.4
II	R(42)	II	P(14)	-14345.4	0.0
II	R(44)	II	P(14)	13212.0	-0.2

STANDARD DEVIATION= 3.0 MHz

CALCULATION OF THE VIBRATIONAL-ROTATIONAL
CONSTANTS AND LASING TRANSITION FREQUENCIES

The measured difference frequencies were fitted[1] in a least-squares sense to the expansion of the line frequencies

$$f = f_o + B_u [J'(J'+1) - J(J+1)] - (B_\ell - B_u) J(J+1)$$

$$- D_u[J'^2(J'+1)^2 - J^2(J+1)^2] + (D_\ell - D_u) J^2(J+1)^2$$

$$+ H_u[J'^3(J'+1)^3 - J^3(J+1)^3] - (H_\ell - H_u) J^3(J+1)^3$$

$$- L_u [\ldots \tag{1}$$

where $J' = J - 1$ for a P(J) line and $J' = J + 1$ for a R(J) line. The $^{12}C^{16}O_2$ reference frequencies were computed from the known constants. Because the laser bands of each isotope share the upper level, all measurements of each isotope were used simultaneously to determine the constants in Eq. (1). In analyzing our data we used the constants recently determined by Petersen, et al[8] for the $^{12}C^{16}O_2$ reference lines. Inclusion of the terms $H_u J^3(J+1)^3$ in the reference frequencies was essential in reducing the computed band center errors in Table IV. With the Bridges and Chang constants,[9] which were originally used to analyze our data, several band centers had errors up to 30 MHz, due principally to the large J-values of many of the reference lines in Tables I, II and III.

Errors in the derived constants were calculated from the estimated standard deviation of the present observations and from the reported errors in the $^{12}C^{16}O_2$ reference constants;[5, 6] for simplicity of calculation, all observations and reference constant errors were assumed to be normally distributed and statistically independent.

The standard deviation of the measurements for each isotope was calculated according to

$$\sigma^2 = \sum_{i=1}^{N} \left[f_{measured}(i) - f_{calc.}(i) \right]^2 \Big/ (N-m) \tag{2}$$

Where N is the number of line pairs measured and m is the number of coefficients to be fitted. Within the accuracy of our measurements, the H_v were found to be indeterminate; the calculated errors in every case were of the same order or larger than the constants themselves. We therefore fitted only the eight constants listed in Table IV for each isotope (giving m = 8 in Eq. (2)).

Table IV

Vibrational-Rotational Constants of CO_2 Isotope Laser Transitions

Isotope\\Constants in MHz	$^{12}C^{16}O_2$ [5, 6]	$^{12}C^{18}O_2$	$^{13}C^{16}O_2$	$^{13}C^{18}O_2$
$V^{(0)}_{001 - I}$	28 808 813.8 $\pm$ 0.025	28 988 595.0 $\pm$ 2.5	27 383 788.8 $\pm$ 1.5	27 838 550.1 $\pm$ 1.6
$V^{(0)}_{001 - II}$	31 889 960.2 $\pm$ 0.024	32 489 194.0 $\pm$ 1.4	30 508 658.1 $\pm$ 1.8	30 785 883.1 $\pm$ 1.3
B_{001}	11 606.2072 $\pm$ 0.0003	10 315.6761 $\pm$ 0.0523	11 610.2920 $\pm$ 0.0496	10 319.0659 $\pm$ 0.0293
$B_I - B_{001}$	91.362504 $\pm$ 0.000027	99.33095 $\pm$ 0.00910	73.26310 $\pm$ 0.00440	84.37119 $\pm$ 0.00673
$B_{II} - B_{001}$	100.157620 $\pm$ 0.000025	72.95927 $\pm$ 0.00434	109.20516 $\pm$ 0.00481	79.87460 $\pm$ 0.00465
D_{001}	39.8825 ($\pm$ 0.0019) X 10^{-4}	32.714 ($\pm$ 0.607) X 10^{-4}	40.366 ($\pm$ 0.022) X 10^{-4}	31.272 ($\pm$ 0.078) X 10^{-4}
$D_I - D_{001}$	-5.42091 ($\pm$ 0.00033) X 10^{-4}	-3.8018 ($\pm$ 0.0876) X 10^{-4}	-3.9448 ($\pm$ 0.0305) X 10^{-4}	-4.3814 ($\pm$ 0.0457) X 10^{-4}
$D_{II} - D_{001}$	7.23304 ($\pm$ 0.00036) X 10^{-4}	3.4527 ($\pm$ 0.0355) X 10^{-4}	7.5763 ($\pm$ 0.0331) X 10^{-4}	4.9325 ($\pm$ 0.0214) X 10^{-4}
H_{001}	5.07 ($\pm$ 0.54) X 10^{-10}			
$H_I - H_{001}$	5.212 ($\pm$ 0.012) X 10^{-9}			
$H_{II} - H_{001}$	6.469 ($\pm$ 0.015) X 10^{-9}			

It is the nature of a least-squares polynomial fit that spectral regions from which there is data are accurately represented at the expense of other regions. In particular, we have no direct measurements of the $00^01 - [10^00, 02^00]_I$ band P branch of $^{13}C^{16}O_2$, which lies beyond the P(50) line of the long wavelength end of the $^{12}C^{16}O_2$ laser spectrum. The calculated frequencies in this branch, therefore, might exhibit errors substantially larger than those within the $^{12}C^{16}O_2$ bands. To investigate the overall accuracy of our fitted constants, we have measured difference frequencies between the $^{13}C^{16}O_2$ and $^{13}C^{18}O_2$ long wavelength band P branches. The measured and calculated (from the constants of Table IV) frequencies are compared in Table V.

There is clearly a systematic error of about 2 MHz, which is not excessive compared to the other uncertainties. In Table IV we have therefore adopted the constants calculated using only the $^{13}C^{16}O_2 - ^{12}C^{16}O_2$ differences. By utilizing the constants given in Table IV, we have finally calculated[1] the transitions listed in Tables VI through XI for the $00^01 - [10^00, 02^00]_I$ and $00^01 - [10^00, 02^00]_{II}$ bands of $^{12}C^{18}O_2$, $^{13}C^{16}O_2$ and $^{13}C^{18}O_2$.

Table V

Difference Frequencies Measured Between $^{13}C^{16}O_2$ and $^{13}C^{18}O_2$ Transitions

$^{13}C^{16}O_2$ LINE	$^{13}C^{18}O_2$ LINE	MEASURED f (13-16) - f (13-18)	MEAS. - CALC.
I P(20)	I P(40)	+14388	0.0
I P(18)	I P(38)	+11824	+0.8
I P(16)	I P(36)	+9370	+2.4
I P(14)	I P(34)	+7021	+1.9
I P(12)	I P(32)	+4779	+3.2
I P(10)	I P(30)	+2639	+3.3
I P(8)	I P(28)	+598	+1.3
I P(6)	I P(26)	−1340	+3.0

Mean Meas. - Calc. = +2.0 MHz

Standard Deviation = 1.1 MHz

Table VI

Calculated Frequencies of the $^{12}C^{18}O_2$ Laser $00^01 - [10^00, 02^00]_I$ Band

LINE	FREQ. (MHZ.)	WV.N. (INV.CM.)	WV.L.(NM.)
P (60)	27384896.2	913.461808	10947.36519
P (58)	27450167.2	915.639012	10921.33457
P (56)	27514599.2	917.788234	10895.75964
P (54)	27578196.0	919.909592	10870.63347
P (52)	27640960.8	922.003201	10845.94932
P (50)	27702897.0	924.069171	10821.70071
P (48)	27764007.7	926.107604	10797.88132
P (46)	27824295.9	928.118602	10774.48505
P (44)	27883764.4	930.102258	10751.50599
P (42)	27942415.9	932.058663	10728.93842
P (40)	28000253.1	933.987901	10706.77681
P (38)	28057278.1	935.890053	10685.01579
P (36)	28113493.5	937.765194	10663.65020
P (34)	28168901.1	939.613395	10642.67502
P (32)	28223503.0	941.434719	10622.08541
P (30)	28277301.1	943.229230	10601.87671
P (28)	28330297.0	944.996981	10582.04439
P (26)	28382492.1	946.738024	10562.58410
P (24)	28433888.0	948.452405	10543.49163
P (22)	28484485.8	950.140166	10524.76294
P (20)	28534286.5	951.801341	10506.39411
P (18)	28583291.3	953.435964	10488.38137
P (16)	28631500.8	955.044059	10470.72112
P (14)	28678915.7	956.625650	10453.40986
P (12)	28725536.5	958.180752	10436.44425
P (10)	28771363.5	959.709378	10419.82107
P (8)	28816397.1	961.211535	10403.53724
P (6)	28860637.1	962.687224	10387.58981
P (4)	28904083.6	964.136444	10371.97594
P (2)	28946736.4	965.559186	10356.69293
V (0)	28988595.0	966.955439	10341.73820
R (0)	29009226.3	967.643626	10334.38316
R (2)	29049892.7	969.000110	10319.91627
R (4)	29089763.3	970.330052	10305.77171
R (6)	29128837.4	971.633422	10291.94733
R (8)	29167113.8	972.910186	10278.44106
R (10)	29204591.4	974.160305	10265.25095
R (12)	29241268.9	975.383734	10252.37519
R (14)	29277144.8	976.580426	10239.81204
R (16)	29312217.5	977.750325	10227.55988
R (18)	29346485.3	978.893374	10215.61722
R (20)	29379946.1	980.009508	10203.98263
R (22)	29412598.1	981.098659	10192.65484
R (24)	29444438.9	982.160755	10181.63264
R (26)	29475466.2	983.195716	10170.91494
R (28)	29505677.6	984.203460	10160.50076
R (30)	29535070.5	985.183899	10150.38920
R (32)	29563641.9	986.136939	10140.57947
R (34)	29591389.0	987.062485	10131.07088
R (36)	29618308.8	987.960432	10121.86286
R (38)	29644398.0	988.830673	10112.95490
R (40)	29669653.2	989.673097	10104.34660
R (42)	29694071.0	990.487586	10096.03769
R (44)	29717647.6	991.274018	10088.02795
R (46)	29740379.4	992.032267	10080.31728
R (48)	29762262.2	992.762199	10072.90568
R (50)	29783292.1	993.463680	10065.79324
R (52)	29803464.7	994.136567	10058.98015
R (54)	29822775.8	994.780715	10052.46669
R (56)	29841220.7	995.395971	10046.25324
R (58)	29858794.8	995.982180	10040.34028

Table VII

Calculated Frequencies of the $^{12}C^{18}O_2$ Laser $00^01 - [10^00, 02^00]_{II}$ Band

LINE	FREQ. (MHZ.)	WV.N. (INV.CM.)	WV.L. (NM.)
P (60)	30991733.6	1033.772949	9673.30399
P (58)	31049505.3	1035.700007	9655.30553
P (56)	31106768.3	1037.610094	9637.53153
P (54)	31163518.1	1039.503064	9619.98126
P (52)	31219750.4	1041.378773	9602.65396
P (50)	31275461.1	1043.237082	9585.54884
P (48)	31330646.1	1045.077855	9568.66511
P (46)	31385301.5	1046.900962	9552.00192
P (44)	31439423.5	1048.706277	9535.55845
P (42)	31493008.4	1050.493679	9519.33381
P (40)	31546052.9	1052.263051	9503.32713
P (38)	31598553.4	1054.014280	9487.53750
P (36)	31650506.7	1055.747257	9471.96399
P (34)	31701909.8	1057.461878	9456.60568
P (32)	31752759.6	1059.158045	9441.46159
P (30)	31803053.3	1060.835662	9426.53076
P (28)	31852788.2	1062.494639	9411.81219
P (26)	31901961.6	1064.134889	9397.30489
P (24)	31950571.2	1065.756331	9383.00783
P (22)	31998614.6	1067.358887	9368.91998
P (20)	32046089.7	1068.942484	9355.04028
P (18)	32092994.3	1070.507054	9341.36768
P (16)	32139326.6	1072.052533	9327.90109
P (14)	32185084.8	1073.578862	9314.63943
P (12)	32230267.2	1075.085984	9301.58160
P (10)	32274872.3	1076.573849	9288.72646
P (8)	32318898.6	1078.042411	9276.07290
P (6)	32362345.1	1079.491628	9263.61978
P (4)	32405210.4	1080.921462	9251.36594
P (2)	32447493.7	1082.331880	9239.31022
V (0)	32489194.0	1083.722854	9227.45143
R (0)	32509825.4	1084.411041	9221.59552
R (2)	32550650.0	1085.772804	9210.02991
R (4)	32590890.1	1087.115070	9198.65824
R (6)	32630545.3	1088.437826	9187.47931
R (8)	32669615.4	1089.741062	9176.49187
R (10)	32708100.2	1091.024776	9165.69469
R (12)	32745999.6	1092.288966	9155.08653
R (14)	32783313.9	1093.533638	9144.66611
R (16)	32820043.4	1094.758799	9134.43217
R (18)	32856188.3	1095.964464	9124.38343
R (20)	32891749.3	1097.150651	9114.51859
R (22)	32926726.9	1098.317381	9104.83634
R (24)	32961122.1	1099.464680	9095.33538
R (26)	32994935.7	1100.592581	9086.01437
R (28)	33028168.8	1101.701118	9076.87197
R (30)	33060822.6	1102.790331	9067.90685
R (32)	33092898.4	1103.860265	9059.11764
R (34)	33124347.7	1104.910968	9050.50297
R (36)	33155322.1	1105.942494	9042.06146
R (38)	33185673.2	1106.954900	9033.79171
R (40)	33215453.1	1107.948247	9025.69233
R (42)	33244663.5	1108.922603	9017.76190
R (44)	33273306.7	1109.878038	9009.99899
R (46)	33301385.0	1110.814627	9002.40216
R (48)	33328900.6	1111.732450	8994.96997
R (50)	33355856.2	1112.631591	8987.70094
R (52)	33382254.3	1113.512139	8980.59361
R (54)	33408097.9	1114.374187	8973.64648
R (56)	33433389.7	1115.217831	8966.85806
R (58)	33458132.9	1116.043174	8960.22684

Table VIII

Calculated Frequencies of the $^{13}C^{16}O_2$ Laser 00^01 - $[10^00, 02^00]_I$ Band

LINE	FREQ. (MHZ.)	WV. N. (INV.CM.)	WV. L. (NM.)
P (60)	25720614.1	857.947330	11655.72716
P (58)	25784819.5	860.088993	11626.70385
P (56)	25848396.5	862.209693	11598.10668
P (54)	25911348.6	864.309550	11569.92886
P (52)	25973679.3	866.388678	11542.16376
P (50)	26035391.8	868.447186	11514.80500
P (48)	26096489.3	870.485180	11487.84635
P (46)	26156974.8	872.502757	11461.28183
P (44)	26216851.0	874.500013	11435.10560
P (42)	26276120.7	876.477036	11409.31204
P (40)	26334786.3	878.433910	11383.89568
P (38)	26392850.2	880.370715	11358.85125
P (36)	26450314.7	882.287524	11334.17365
P (34)	26507181.8	884.184406	11309.85791
P (32)	26563453.4	886.061425	11285.89928
P (30)	26619131.3	887.918640	11262.29313
P (28)	26674217.1	889.756104	11239.03501
P (26)	26728712.3	891.573867	11216.12059
P (24)	26782618.1	893.371971	11193.54572
P (22)	26835935.7	895.150456	11171.30639
P (20)	26888666.2	896.909354	11149.39871
P (18)	26940810.3	898.648695	11127.81897
P (16)	26992368.8	900.368502	11106.56357
P (14)	27043342.2	902.068793	11085.62903
P (12)	27093731.0	903.749581	11065.01204
P (10)	27143535.4	905.410785	11044.70940
P (8)	27192755.4	907.052678	11024.71801
P (6)	27241391.0	908.674988	11005.03495
P (4)	27289442.0	910.277798	10985.65737
P (2)	27336908.1	911.861097	10966.58256
V (0)	27383788.8	913.424866	10947.80794
R (0)	27407009.3	914.199421	10938.53242
R (2)	27453010.5	915.733854	10920.20346
R (4)	27498424.2	917.248694	10902.16870
R (6)	27543249.6	918.743906	10884.42594
R (8)	27587485.3	920.219450	10866.97309
R (10)	27631130.0	921.675281	10849.80818
R (12)	27674182.2	923.111350	10832.92931
R (14)	27716640.4	924.527601	10816.33473
R (16)	27758502.6	925.923975	10800.02276
R (18)	27799767.0	927.300406	10783.99183
R (20)	27840431.4	928.656825	10768.24046
R (22)	27880493.6	929.993156	10752.76730
R (24)	27919951.2	931.309320	10737.57106
R (26)	27958801.6	932.605230	10722.65056
R (28)	27997042.2	933.880798	10708.00473
R (30)	28034670.0	935.135926	10693.63257
R (32)	28071682.0	936.370516	10679.53319
R (34)	28108075.2	937.584461	10665.70577
R (36)	28143846.2	938.777652	10652.14962
R (38)	28178991.4	939.949971	10638.86409
R (40)	28213507.4	941.101300	10625.84867
R (42)	28247390.3	942.231512	10613.10290
R (44)	28280636.2	943.340477	10600.62644
R (46)	28313241.1	944.428059	10588.41900
R (48)	28345200.7	945.494116	10576.48041
R (50)	28376510.7	946.538504	10564.81058
R (52)	28407166.5	947.561071	10553.40949
R (54)	28437163.4	948.561662	10542.27722
R (56)	28466496.7	949.540114	10531.41394
R (58)	28495161.3	950.496263	10520.81990

Table IX

Calculated Frequencies of the $^{13}C^{16}O_2$ Laser $00^01 - [10^00, 02^00]_{II}$Band

LINE	FREQ. (MHZ.)	WV.N. (INV.CM.)	WV.L. (NM.)
P (60)	28729368.7	958.308582	10435.05212
P (58)	28800186.5	960.670807	10409.39303
P (56)	28870277.5	963.008793	10384.12118
P (54)	28939632.7	965.322233	10359.23514
P (52)	29008243.2	967.610834	10334.73339
P (50)	29076100.5	969.874310	10310.61437
P (48)	29143196.3	972.112385	10286.87645
P (46)	29209522.7	974.324793	10263.51795
P (44)	29275071.8	976.511278	10240.53713
P (42)	29339836.4	978.671592	10217.93222
P (40)	29403809.4	980.805500	10195.70139
P (38)	29466983.8	982.912774	10173.84275
P (36)	29529353.3	984.993196	10152.35439
P (34)	29590911.6	987.046559	10131.23435
P (32)	29651652.7	989.072664	10110.48062
P (30)	29711571.0	991.071323	10090.09116
P (28)	29770661.1	993.042358	10070.06390
P (26)	29828918.0	994.985599	10050.39672
P (24)	29886336.9	996.900887	10031.08747
P (22)	29942913.3	998.788073	10012.13398
P (20)	29998643.1	1000.647016	9993.53402
P (18)	30053522.2	1002.477587	9975.28536
P (16)	30107547.1	1004.279665	9957.38572
P (14)	30160714.5	1006.053139	9939.83281
P (12)	30213021.4	1007.797908	9922.62428
P (10)	30264465.0	1009.513882	9905.75779
P (8)	30315042.8	1011.200977	9889.23095
P (6)	30364752.8	1012.859122	9873.04135
P (4)	30413593.0	1014.488256	9857.18656
P (2)	30461561.1	1016.088325	9841.66411
V (0)	30508658.1	1017.659286	9826.47153
R (0)	30531878.7	1018.433841	9818.99815
R (2)	30577664.2	1019.961083	9804.29565
R (4)	30622575.2	1021.459152	9789.91669
R (6)	30666611.4	1022.928040	9775.85872
R (8)	30709772.7	1024.367749	9762.11913
R (10)	30752059.6	1025.778288	9748.69533
R (12)	30793472.6	1027.159677	9735.58466
R (14)	30834012.7	1028.511947	9722.78448
R (16)	30873680.9	1029.835138	9710.29210
R (18)	30912478.9	1031.129298	9698.10480
R (20)	30950408.3	1032.394487	9686.21987
R (22)	30987471.2	1033.630774	9674.63455
R (24)	31023670.1	1034.838236	9663.34607
R (26)	31059007.4	1036.016963	9652.35161
R (28)	31093486.2	1037.167052	9641.64836
R (30)	31127109.7	1038.288610	9631.23346
R (32)	31159881.3	1039.381755	9621.10404
R (34)	31191805.0	1040.446614	9611.25719
R (36)	31222884.8	1041.483324	9601.68999
R (38)	31253125.0	1042.492030	9592.39947
R (40)	31282530.5	1043.472890	9583.38266
R (42)	31311106.4	1044.426069	9574.63654
R (44)	31338857.0	1045.351742	9566.15807
R (46)	31365789.0	1046.250094	9557.94418
R (48)	31391907.7	1047.121322	9549.99177
R (50)	31417219.4	1047.965628	9542.29770
R (52)	31441730.4	1048.783228	9534.85881
R (54)	31465447.5	1049.574346	9527.67190
R (56)	31488377.7	1050.339214	9520.73374
R (58)	31510528.3	1051.078078	9514.04107

Table X

Calculated Frequencies of the $^{13}C^{18}O_2$ Laser $00^01 - [10^00, 02^00]_I$ Band

LINE	FREQ. (MHZ.)	WV.N. (INV.CM.)	WV.L. (NM.)
P(60)	26288296.3	876.883173	11404.02771
P(58)	26350130.2	878.945727	11377.26676
P(56)	26411234.4	880.983946	11350.94464
P(54)	26471613.4	882.997972	11325.05432
P(52)	26531271.2	884.987940	11299.58901
P(50)	26590211.5	886.953980	11274.54212
P(48)	26648438.3	888.896216	11249.90726
P(46)	26705955.1	890.814769	11225.67828
P(44)	26762765.2	892.709750	11201.84920
P(42)	26818871.9	894.581267	11178.41426
P(40)	26874278.2	896.429424	11155.36788
P(38)	26928987.1	898.254315	11132.70467
P(36)	26983001.2	900.056031	11110.41941
P(34)	27036323.1	901.834659	11088.50708
P(32)	27088955.2	903.590277	11066.96282
P(30)	27140899.7	905.322960	11045.78194
P(28)	27192158.7	907.032776	11024.95992
P(26)	27242734.0	908.719787	11004.49241
P(24)	27292627.4	910.384051	10984.37521
P(22)	27341840.4	912.025619	10964.60427
P(20)	27390374.3	913.644538	10945.17571
P(18)	27438230.5	915.240846	10926.08578
P(16)	27485409.8	916.814581	10907.33089
P(14)	27531913.3	918.365769	10888.90760
P(12)	27577741.6	919.894435	10870.81258
P(10)	27622895.2	921.400597	10853.04268
P(8)	27667374.5	922.884266	10835.59484
P(6)	27711179.6	924.345450	10818.46619
P(4)	27754310.8	925.784150	10801.65393
P(2)	27796767.7	927.200360	10785.15543
V(0)	27838550.1	928.594071	10768.96817
R(0)	27859188.2	929.282485	10760.99051
R(2)	27899957.9	930.642416	10745.26567
R(4)	27940051.6	931.979797	10729.84633
R(6)	27979468.4	933.294599	10714.73039
R(8)	28018207.2	934.586786	10699.91589
R(10)	28056266.8	935.856318	10685.40097
R(12)	28093645.8	937.103147	10671.18389
R(14)	28130342.6	938.327221	10657.26303
R(16)	28166355.5	939.528483	10643.63687
R(18)	28201682.6	940.706868	10630.30402
R(20)	28236321.8	941.862308	10617.26318
R(22)	28270270.9	942.994728	10604.51316
R(24)	28303527.5	944.104048	10592.05288
R(26)	28336089.0	945.190181	10579.88138
R(28)	28367952.5	946.253036	10567.99780
R(30)	28399115.4	947.292516	10556.40135
R(32)	28429574.4	948.308518	10545.09140
R(34)	28459326.2	949.300934	10534.06738
R(36)	28488367.6	950.269649	10523.32884
R(38)	28516694.8	951.214544	10512.87542
R(40)	28544304.2	952.135494	10502.70688
R(42)	28571191.8	953.032368	10492.82306
R(44)	28597353.5	953.905029	10483.22391
R(46)	28622785.1	954.753335	10473.90948
R(48)	28647482.1	955.577138	10464.87991
R(50)	28671440.0	956.376286	10456.13546
R(52)	28694653.9	957.150619	10447.67646
R(54)	28717118.9	957.899973	10439.50337
R(56)	28738830.0	958.624178	10431.61672
R(58)	28759781.9	959.323057	10424.01715

C. FREED, D. L. SPEARS, AND R. G. O'DONNELL

Table XI

Calculated Frequencies of the $^{13}C^{18}O_2$ Laser $00^01 - [10^00, 02^00]_{II}$ Band

LINE	FREQ. (MHZ.)	WV.N. (INV.CM.)	WV.L. (NM.)
P (60)	29264563.4	976.160754	10244.21435
P (58)	29323757.1	978.135245	10223.53509
P (56)	29382410.3	980.091706	10203.12685
P (54)	29440517.0	982.029934	10182.98899
P (52)	29498071.1	983.949734	10163.12079
P (50)	29555067.2	985.850918	10143.52152
P (48)	29611499.7	987.733302	10124.19038
P (46)	29667363.2	989.596711	10105.12655
P (44)	29722652.8	991.440974	10086.32916
P (42)	29777363.6	993.265927	10067.79729
P (40)	29831490.7	995.071412	10049.52999
P (38)	29885029.6	996.857280	10031.52628
P (36)	29937976.1	998.623384	10013.78513
P (34)	29990325.9	1000.369587	9996.30549
P (32)	30042075.2	1002.095757	9979.08626
P (30)	30093220.1	1003.801766	9962.12632
P (28)	30143757.0	1005.487497	9945.42451
P (26)	30193682.6	1007.152835	9928.97965
P (24)	30242993.6	1008.797674	9912.79050
P (22)	30291687.1	1010.421913	9896.85583
P (20)	30339760.2	1012.025459	9881.17435
P (18)	30387210.2	1013.608222	9865.74476
P (16)	30434034.8	1015.170121	9850.56572
P (14)	30480231.6	1016.711082	9835.63588
P (12)	30525798.7	1018.231034	9820.95385
P (10)	30570734.0	1019.729916	9806.51821
P (8)	30615036.0	1021.207672	9792.32753
P (6)	30658703.1	1022.664250	9778.38035
P (4)	30701734.0	1024.099607	9764.67517
P (2)	30744127.7	1025.513706	9751.21048
V(0)	30785883.1	1026.906516	9737.98476
R (0)	30806521.2	1027.594930	9731.46102
R (2)	30847317.9	1028.955761	9718.59080
R (4)	30887474.9	1030.295254	9705.95561
R (6)	30926991.8	1031.613398	9693.55383
R (8)	30965868.7	1032.910191	9681.38381
R (10)	31004105.6	1034.185637	9669.44390
R (12)	31041702.9	1035.439746	9657.73242
R (14)	31078660.9	1036.672534	9646.24766
R (16)	31114980.5	1037.884023	9634.98789
R (18)	31150662.4	1039.074243	9623.95138
R (20)	31185707.7	1040.243229	9613.13635
R (22)	31220117.6	1041.391022	9602.54101
R (24)	31253893.7	1042.517671	9592.16355
R (26)	31287037.5	1043.623229	9582.00213
R (28)	31319550.9	1044.707757	9572.05489
R (30)	31351435.7	1045.771322	9562.31997
R (32)	31382694.3	1046.813998	9552.79546
R (34)	31413329.1	1047.835862	9543.47943
R (36)	31443342.5	1048.837002	9534.36995
R (38)	31472737.4	1049.817510	9525.46505
R (40)	31501516.6	1050.777483	9516.76274
R (42)	31529683.5	1051.717027	9508.26101
R (44)	31557241.2	1052.636252	9499.95782
R (46)	31584193.2	1053.535277	9491.85112
R (48)	31610543.4	1054.414224	9483.93882
R (50)	31636495.6	1055.273225	9476.21883
R (52)	31661453.9	1056.112414	9468.68900
R (54)	31686022.5	1056.931935	9461.34719
R (56)	31710005.9	1057.731937	9454.19123
R (58)	31733408.9	1058.512575	9447.21889

CONCLUSIONS

It was shown that rare CO_2 isotopes can provide a many-fold expansion of the already highly useful spectral range of CO_2 lasers. In terms of number of lasing transitions, power output, gain, stability, and sealed-off cw operation characteristics, these three rare isotope lasers are generally similar to the commonly used $^{12}C^{16}O_2$ lasers. Since sealed-off CO_2 laser operating life times of over 10,000 hours have been reported by a number of laboratories, the additional cost of a few Torr-liters of isotope required for a properly designed laser is not significant.

The vibrational-rotational constants and the lasing transition frequencies of the $00^01 - [10^00, 02^00]_I$ and $00^01 - [10^00, 02^00]_{II}$ bands of $^{12}C^{18}O_2$, $^{13}C^{16}O_2$, and of the previously unreported $^{13}C^{18}O_2$, have also been obtained. The calculated frequencies are accurate to about $\pm$ 3 MHz [0.0001 cm^{-1}], in the spectral regions measured.

It should be emphasized that substantially improved data can be obtained in a straightforward way, but with increased complexity of experimental apparatus. Previous work has shown[6,7,8] that CO_2 lasers can be stabilized to within a few kHz of the natural molecular frequencies by the saturation resonance technique using pure, low pressure CO_2 as a reference.

We would like to point out that, instead of using the $^{12}C^{16}O_2$ lines as a secondary reference, the rotational constants of each isotope could be obtained directly by measuring rotational line separations and determine the band centers by two comparisons against the $^{12}C^{16}O_2$ lines measured to high accuracy by Evenson, et al.[10] We are also certain that by using stabilized lasers sufficiently accurate measurements can be obtained to allow determination of additional higher order coefficients of the polynomial expansion (1), with a resultant improvement of all predicted line frequencies.

Finally, similar experiments and calculations can be carried out for other CO_2 isotopes as well. Such work is currently underway and the results will be submitted for publication in the near future. Also, with the sole exception of the saturation resonance stabilization technique using the 4.3 μm spontaneous emission, all other aspects of this work can be easily extended to the molecular CO laser system. Stable, sealed-off CO laser operation has been previously demonstrated.[16] The Lamb-dip in CO lasers[17] could be used to set the laser transitions within 1 MHz of line center. The HgCdTe photodiodes we used are just as useful at CO wavelengths. Indeed, we did obtain microwave frequency beats between CO laser transitions which are equivalent to the ones shown in Figures 3 and 4. By comparison with selected, doubled CO_2 transitions[18] the entire CO laser spectrum may be also utilized as a secondary frequency standard. Thus CO_2 and CO lasers can readily provide more than a thousand precise calibration frequency or wavelength benchmarks in the 5 to 12 μm portion of the infrared spectrum.

ACKNOWLEDGEMENT

We are especially appreciative to E. J. Carell for preparing the design details of the CO_2 lasers.

REFERENCES

1. C. Freed, A. H. M. Ross and R. G. O'Donnell, "Determination of Laser Line Frequencies and Vibrational-Rotational.Constants of the $^{12}C^{18}O_2$, $^{13}C^{16}O_2$, and $^{13}C^{18}O_2$ Isotopes from Measurements of CW Beat Frequencies with Fast HgCdTe Photodiodes and Microwave Frequency Counters," to be published in the Feb. 1974 issue of J. Mol. Spectry.

2. D. L. Spears and C. Freed, "HgCdTe Varactor/Photodiode Detection of CW CO_2 Laser Beats Beyond 60 GHz," Appl. Phys. Lett., Vol. 23, pp. 445-447, 15 Oct. 1973.

3. I. Wider and G. B. McCurdy, "Isotope Shifts and the Role of Fermi Resonance in the CO_2 Infrared Laser," Phys. Rev. Lett., Vol. 16, pp. 565-567, 28 March 1966.

4. G. B. Jacobs and H. C. Bowers,"Extension of CO_2 Laser Wavelength Range with Isotopes," J. Appl. Phys., Vol. 38, pp. 2629-2693, May 1967.

5. J. C. Siddoway, "Calculated and Observed Laser Transitions using $^{14}C^{16}O_2$", J. Appl. Phys., Vol. 39, pp. 4854-4855, Sept. 1968.

6. C. Freed and A. Javan, "Standing Wave Saturation Resonances in the 10.6μ Transitions Observed in a Low-Pressure Room-Temperature Absorber Gas," Appl. Phys. Lett., Vol. 17 pp. 53-56,15 July 1970.

7. C. Freed, "Designs and Experiments Relating to Stable Lasers," Proceedings of the Frequency Standards and Metrology Seminar, University Laval, Quebec, Canada, pp. 226-261, September 1, 1971.

8. F. R. Petersen, D. G. McDonald, F. D. Cupp, and B. L. Danielson, "Rotational Constants for $^{12}C^{16}O_2$ from Beats Between Lamp-Dip-Stabilized Laser Lines," Phys. Rev. Lett., Vol. 31, No. 9, pp. 573-576, 27 August 1973. Also discussed in a preceding paper.

9. T. J. Bridges and T. Y. Chang, "Accurate Rotational Constants of $^{12}C^{16}O_2$ from Measurements of CW Beats in Bulk GaAs between CO_2 Vibrational-Rotational Laser Lines," Phys. Rev. Lett., Vol. 22, pp. 811-814, 21 April 1969.

10. K. M. Evenson, J. S. Wells, F. R. Petersen, B. L. Danielson and G. W. Day, "Accurate Frequencies of Molecular Transitions used In Laser Stabilization: the 3.39-μm transition in CH_4 and the 9.33- and 10.18-μm transitions in CO_2," Appl. Phys. Lett., Vol. 22, pp. 192-195, 15 Feb. 1973.

11. G. Amat and M. Pimbert, "On Fermi Resonance in Carbon Dioxide," J. Mol. Spectry., Vol. 19, pp. 278-290, 1965.

12. H. R. Gordon and T. K. McCubbin, Jr., "The 2.8 Micron Bands of CO_2," J. Mol. Spectry., Vol. 19, pp. 137-154, 1966.

13. H. E. Howard-Lock and B. P. Stoicheff, "Raman Intensity Measurements of the Fermi Diad ν_1, $2\nu_2$ in $^{12}CO_2$ and $^{13}CO_2$," J. Mol. Spectry., Vol. 37, pp. 321-326, 1971.

14. D. L. Spears, T. C. Harman and I. Melngailis, "High Quantum Efficiency HgCdTe Photodidodes at 10.6 μm," to be published in the Proceedings of the IRIS Detector Specialty Group Meeting, held in Washington, D. C., 13-15 March 1973.

15. P. Penfield and R. P. Rafuse, "Varactor Applications," the M.I.T. Press, 1962.

16. C. Freed, "Sealed-off Operation of Stable CO Lasers," Appl. Phys. Lett., Vol. 18, pp. 458-461, 15 May, 1971.

17. C. Freed and H. A. Haus, "Lamb Dip in CO Lasers," IEEE J. Quantum Electron., Vol. QE-9, pp. 219-226, Feb. 1973.

18. R. S. Eng, H. Kildal, F. C. Mikkelsen and D. L. Spears, "Determination of Absolute Frequencies of $^{12}C^{16}O$ and $^{13}C^{16}O$ Laser Lines." Submitted for publication to Appl. Phys. Lett.

ABSORPTION SPECTROSCOPY FROM SELECTIVELY EXCITED ATOMIC SINGLET LEVELS

D.J. Bradley, P. Ewart, J.V. Nicholas and
J.R.D. Shaw

Department of Pure and Applied Physics
The Queen's University, Belfast BT7 1NN, Northern Ireland

ABSTRACT

Employing high power, narrow band, frequency tunable dye lasers, the $^1P_1^0$ levels of Ba I and Mg I have been selectively excited. The excited state absorption series $^1P_1^0 - ^1D_2$ have been measured and extended. Series perturbations by $^1D_2$ levels of Ba I have shown lines of the 6s 6p $^1P_1^0$ - 6s $(n + 1)$s $^1S_0$ series and terms of the auto-ionization series 6s 6p $^1P_1^0$ - 5d ns $^1D_2$ have also been observed and assigned. A strong auto-ionization resonance in Mg I at $\lambda 300.9$ nm has been recorded, with a measured photo-ionization cross-section of $\sim 5 \times 10^{16}$ cm^{-2}, in good agreement with calculations.

INTRODUCTION

The new possibilities for selective atomic and molecular spectroscopy, opened up by the development of organic dye lasers as high power, frequency tunable, nanosecond and picosecond light sources[1,2] permitted, for example, the detection of resonant optical-frequency Stark effects[3] and the investigation of self-induced transparency and linear dispersion delays in potassium vapour[4]. The high energy, high power pulses obtainable from dye lasers have been employed for the selective excitation of triplet calcium[5], doublet potassium[6] and singlet magnesium[7] atomic levels. Previous observations of absorptions from excited levels of atomic metal vapours have been from low lying levels, thermally populated at normal oven temperatures, or from lower metastable levels excited in a shock tube[8]. It is now possible with dye lasers to substantially populate selected higher levels, and

193

stimulated electronic Raman scattering[9] and photo-ionization[7,10] from the selectively excited levels of rubidium, potassium and magnesium have already been investigated experimentally.

Absorption spectra are normally employed for the detection of higher series members and auto-ionization levels, and the capability of obtaining in absorption transitions to levels not reached from the ground state, greatly extends the study of series perturbations and auto-ionization levels. There is the added practical advantage that levels above the normal ionization limit can be investigated without working in the experimentally difficult vacuum ultra-violet spectral region, since absorption takes place from levels well above the ground state.

EXPERIMENTAL

General Considerations

To saturate the $6s^2 \; ^1S_0 - 6s \; 6p \; ^1P_1^o$ transition of Ba I requires a laser operating at $\lambda 553.5$ nm. The saturating laser power density, P_S, for a homogeneously broadened line must satisfy the condition

$$P_S \;\; >> \;\; \frac{h\nu}{\sigma_1 T_1} \quad\quad\quad\quad\quad\quad (1)$$

where σ_1 is the cross-section for the $6s^2 \; ^1S_0 - 6s \; 6p \; ^1P_1^o$ transition, and T_1 is the lifetime of the $6s \; 6p \; ^1P_1^o$ level. Taking the values $T_1 = 8$ nsec and $\sigma_1 \simeq 2 \times 10^{-11}$ cm^2[11] gives $P_S >> 2.5$ watts cm^{-2}. Thus saturation is easily achieved with a pulsed dye laser. If the absorption from the selectively excited level to higher atomic levels is to be recorded photographically then a minimum number density, N, of excited atoms per unit area will be required in the vapourizing furnace. The value of N will depend upon the cross-section of the excited state absorption transition and on the signal-to-noise capability of the photographic plate. An absorption of 5% for a spectral line with a peak cross-section of 10^{-17} cm^2, corresponding to a strong auto-ionization transition, requires $N = 5 \times 10^{15}$ cm^{-2}. If the oven is long enough to absorb all of the selectively exciting laser beam, the laser power density required to produce an excitation density N per cm^2 is

$$P_N \;\simeq\; \frac{N \; h\nu}{T_1} \quad\quad\quad\quad\quad\quad (2)$$

We have neglected resonant imprisonment of radiation which would tend to reduce the laser power density requirement. For the above Ba I transition, equation (2) gives the value $P_N \simeq 0.25$ Mwatt cm^{-2} for $N = 5 \times 10^{15}$ cm^{-2}.

Barium I

The general experimental arrangement for Barium is shown in
Figure 1. A single transverse mode, Nd:glass oscillator, actively
Q-switched with a Pockels cell was used with amplifiers to produce
a 10 joule, 30 nsec pulse with a uniform beam cross-section.
Frequency doubling in a phase-matched ADP crystal produced a 1 joule,
22 nsec pulse at λ530 nm. Part of the second harmonic beam
(filtered from the fundamental frequency) was used to transversely
pump[12] the dye laser cell containing a 2.5×10^{-3} M solution of
2' 7' dichlorofluorescein in ethanol, with 1% of ammonia added.
The 4 cm long, 1 cm^2 dye cell had 10° wedged windows to avoid sub-
cavities forming in the laser resonator, which was terminated by a
single quartz parallel plate reflector and a 100% reflectivity
concave mirror of 1 m radius of curvature. Frequency narrowing and
tuning was achieved by the insertion of two, optically contacted,
Fabry-Perot etalons with gaps of 5 μm and 50 μm, respectively. A
beam expanding telescope in the laser cavity prevented damage to
the dielectric coatings of the tuning etalons. The dye laser output spectral bandwidth was <0.1 nm for a 20 nsec, 60 mj pulse,
corresponding to a pump power density of 10 MW cm^{-2} in the 0.3 cm^2
cross-sectional area beam with a divergence of 3 milliradians,
half-angle. Rotation of the two etalons permitted coarse and fine
tuning, respectively, of the dye laser wavelength[13].

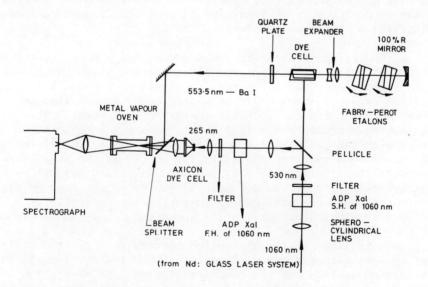

Figure 1   Experimental arrangement for absorption spectroscopy
from selectively excited barium vapour.

A pellicle beam-splitter reflected ~20% of the neodymium second harmonic beam into another frequency doubling ADP crystal to generate the fourth harmonic frequency at λ265 nm. The ultra-violet light was filtered from the remaining λ530 nm light and focussed into a cell, containing a fluorescent dye to produce a background continuum covering the appropriate spectral region. Because of the short excited state lifetime of Ba I and the 20 nsec duration of the pumping dye laser pulse it is necessary to employ dyes with sufficiently short fluorescent lifetimes. We have used several scintillator dyes to cover the spectral range of interest. Coincidence, to within 1 nsec, of the dye laser pumping pulse and the fluorescent background continua is achieved by adjusting the optical path lengths. The scintillator dye was contained in a highly polished aluminium axicon reflector which efficiently collects fluorescence from the line source due to saturation pumping by the λ265 nm beam. A lens focuses the background continuum inside the metal vapour oven. A beam splitter, transmitting the fluorescent light and reflecting the dye laser beam, ensures good spatial overlap of the background light with the region of excited atoms in the oven.

The metal vapour oven was a simple stainless steel tube of 2 cm internal diameter and a metal vapour path length of 35 cm. The oven was heated with a resistance coil wound around the outside of the tube. After initial baking and evacuation, argon gas was admitted to a pressure of ~100 torr, to act as a dephasing buffer gas and to prevent the quartz end windows becoming covered with barium deposits. The oven was operated at a temperature of 850°C giving a barium vapour density of 10^{16} atoms per cm^3.

The absorption spectra were recorded with a 1 m Czerny-Turner spectrograph, with a 1200 lines/mm plane grating blazed at λ750 nm used in second order where the reciprocal dispersion was 0.38 nm/mm. A spectral range of 20 nm could be photographed at one time and this was also the typical useful bandwidth of the scintillator dyes. A single exposure was sufficient to obtain an absorption spectrum with the 3000 ASA Polaroid used for testing and lining up purposes. With a spectrograph slit width of 50 μm, up to 8 laser pulses were required to obtain an exposure on Ilford HP3 or FP4 plates suitable for microdensitometry. Tuning of the pumping dye laser to resonance was obtained by comparing the laser output spectrum with a resonance absorption spectrum of the barium vapour.

About 150 lines of the absorption spectra from excited states of barium have been detected and measured over the wavelength range λ320 nm to λ475 nm. To remove confusion with absorption lines from the ground state and impurities, spectra were recorded with and without the pumping laser beam. Besides absorptions from the excited singlet level 6s 6p $^1P_1^o$, many transitions of similar strengths were seen from the metastable levels, 6s 5d $^1D_2$,

6s 5d $^3D_{3,2,1}$ and 6s 6p $^3P^o_{2,1,0}$, which lie below the selectively excited level. The relative strengths of the metastable states absorption lines decreased with decreasing argon gas pressures and this was used to distinguish them from the 6s 6p $^1P_1^o$ absorptions. Over 60 of the observed lines could not be ascribed to known transitions[14,15]. Many of these lines have been assigned to absorptions from the 6s 6p $^1P_1^o$ level and the series 6s 6p $^1P_1^o$ - 6s nd $^1D_2$ has been extended from n = 9 up to n = 41. Figure 2 shows microdensitometer traces of these absorption spectra with the section n = 23 to n = 41 shown expanded. (See Table 1)

Only two members of the series 6s 6p $^1P_1^o$ - 6s (n + 1)s ^1S_o were observed because the higher members are very nearly degenerate with the 6s nd $^1D_2$ series. Perturbations in the quantum defect of the 6s nd $^1D_2$ series occurred for n = 26 and n = 27 due to a transition at λ420.39 nm which has been assigned to the 5d 7d $^1D_2$ level. This perturbation shifts the two lines of the 6p $^1P_1^o$ - nd $^1D_2$ series sufficiently to reveal the n = 26 and n = 27 absorption lines of the 6s 6p $^1P_1^o$ - 6s (n + 1)s ^1S_o series as shown in Figure 2. As expected, the lines of the latter series are not perturbed by the 5d 7d $^1D_2$ level. Finally we have recorded the series 6s 6p $^1P_1^o$ - 5d ns $^1D_2$ including the auto-ionization lines corresponding to n = 9, 10 and 11. (See Table 2)

Magnesium I

While Bacher showed in 1933[16] that the interaction of $3p^2$ $^1D_2$ with the 3s nd $^1D_2$ series of Mg I could account for the location of the 3s 3d ^1D term below the corresponding triplet level 3s 3d ^3D, the $^1D_2$ and ^1S_o displaced terms of the $3p^2$ configuration have not been previously observed. The presence of $3p^2$ $^1D_2$, ^1S_o in the continuum will affect the dielectronic recombination rate and the ultra-violet radiation balance in, for example, astrophysical sources. Calculations of the photo-ionization cross-section from the lowest excited singlet level 3s 3p $^1P_1^o$ have been carried out[17], using the approach of Henry and Lipsky[18]. The photo-ionization cross-section for the $3p^2$ ^1S_o resonance was found to have maximum values, at λ297.7 nm, of 12.9 x 10^{-16} cm^2 in the dipole-velocity approximation.

Transitions from the Mg I $3s^2$ ^1S_o ground state to ^1S_o, $^1D_2$ states are forbidden and we have employed selective excitation and absorption to the experimental detection and measurement of the calculated 3s 3p $^1P_1^o$ - $3p^2$ ^1S_o resonance (Figure 3).

The lifetime of the 3s 3p $^1P_1^o$ level is 2 nsec[11]. To record photographically an excited state absorption of the predicted cross-section requires a laser power density of ~1 MW cm^{-2} at λ285.2 nm. The experimental arrangement was similar to that employed for barium.

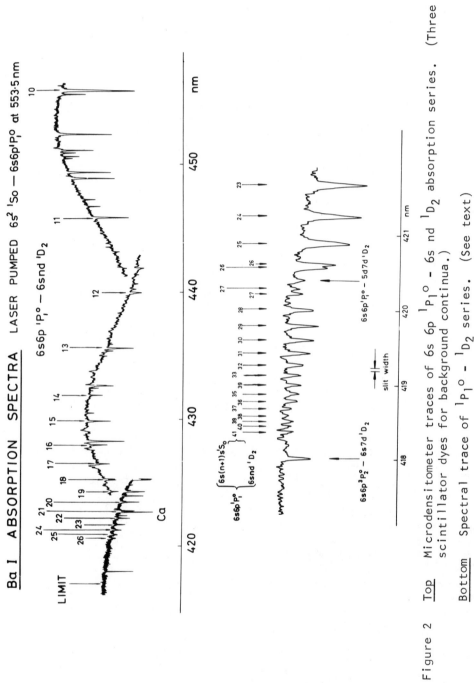

Figure 2   Top   Microdensitometer traces of 6s 6p $^1P_1^o$ – 6s nd $^1D_2$ absorption series. (Three scintillator dyes for background continua.)

Bottom   Spectral trace of $^1P_1^o$ – $^1D_2$ series. (See text)

TABLE   1

n	λ (air) nm	ν (vacuum) cm^{-1}	n*
10	455.709	21937.7	7.339
11	445.842	22423.2	8.410
12	439.997	22721.0	9.355
13	435.661	22947.2	10.333
14	432.301	23125.5	11.366
15	429.900	23254.7	12.344
16	428.020	23356.8	13.324
17	426.517	23439.1	14.311
18	425.294	23506.5	15.306
19	424.299	23561.7	16.297
20	423.467	23607.9	17.292
21	422.772	23646.8	18.288
22	422.192	23679.2	19.264
23	421.688	23707.5	20.258
24	421.258	23731.7	21.242
25	420.880	23753.1	22.243
26	420.61	23768.4	23.05
27	420.21	23791.2	24.44
28	419.981	23803.9	25.33
29	419.763	23816.3	26.30
30	419.567	23827.4	27.28
31	419.394	23827.2	28.23
32	419.23	23846.4	29.22
33	419.09	23854.6	30.20
34	418.94	23862.9	31.30
35	418.83	23869.3	32.24
36	418.73	23875.1	33.16
37	418.63	23880.6	34.11
38	418.53	23886.5	35.23
39	418.45	23890.9	36.14
40	418.39	23894.6	36.97
41	418.31	23899.0	38.02

TABLE   2

n	λ (air) nm	ν (vacuum) cm^{-1}	n*
8	451.067	22163.4	3.83
9	402.81	24818.6	4.77
10	380.63	26264.6	4.69
11	367.17	27227.5	6.73

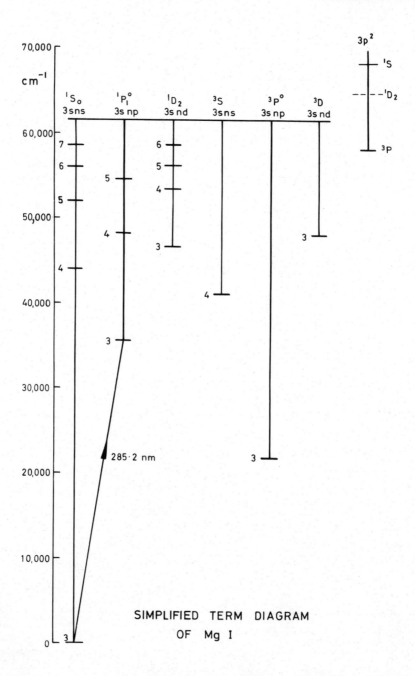

SIMPLIFIED TERM DIAGRAM
OF Mg I

Figure 3

The transversely pumped rhodamine 6G laser, had an output spectral
bandwidth of 0.08 nm in a 22 nsec, 120 mj pulse, corresponding to
a power density of 18 MW cm^2 in the 0.3 cm^2 cross-sectional area
beam. Beam divergence was 3 milliradians, half-angle. The dye
laser operating frequency, was doubled in a second phase-matched
ADP crystal to give a 15 mJ, 15 nsec pulse (3.3 MW cm^{-2}), of
spectral width 0.04 nm at the $3s^2$ $^1S_0$ - $3s$ $3p$ $^1P_1^o$ transition
wavelength. Magnesium was kept molten in a reservoir in the
stainless steel oven and the 10 cm metal vapour path length was
heated to 650°C. Argon at a pressure of a few hundred torr was
used as a buffer gas. To permit accurate comparison of the
recorded background continua with and without selective excitation
of the magnesium vapour, the intensities of the background pulses
were monitored with a co-axial photodiode and recorded on a fast
oscilloscope.

Absorption spectra were recorded from λ400 nm to λ290 nm. The
series $3s$ $3p$ $^1P_1^o$ - $3s$ nd $^1D_2$ was obtained (Figure 4) in absorption
for the first time, up to n = 24 (ionization limit at λ375.58 nm),
using the spectrograph in second order with a slit width of 50 μm.
This extends the series from n = 13, previously seen in emission

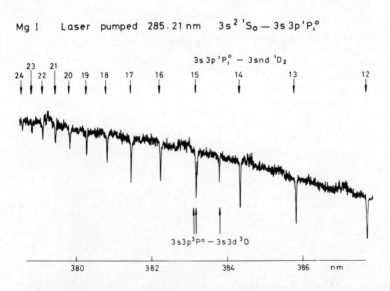

Figure 4   Microdensitometer trace showing higher terms of the
$3s$ $3p$ $^1P_1^o$ - $3s$ nd $^1D_2$ absorption series obtained from
selectively excited Mg I. Triplet absorption lines
arising from $^1P_1^o$ - $^3P^o$ intersystem crossing are also
present.

AUTOIONIZATION IN Mg I   Laser pumped $3s^2{}^1S_0 - 3s\,3p\,{}^1P_1^o$

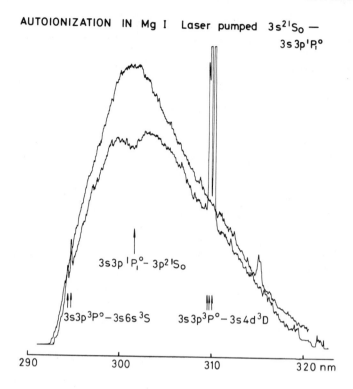

Mg I $3s\,3p\,{}^1P_1^o - 3p^2\,{}^1S_0$ ABSORPTION

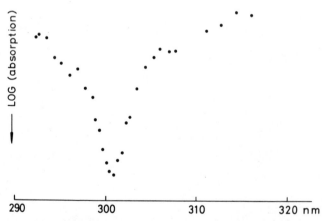

Figure 5 Top Microdensitometer traces of the indole continuum
              with and without selective excitation of the
              magnesium vapour.

         Bottom Absorption profile of $3s\,3p\ {}^1P_1^o - 3p^2\ {}^1D_2$
                 autoionization line.

spectra[19]. Figure 5 shows the indole fluorescence continuum, with and without excited magnesium in the oven, recorded in first order with a 200 μm slit width. Strong absorption at λ300.9 nm from the auto-ionization transition 3s 3p $^1P_1^0$ - $3p^2$ $^1S_0$ can be clearly seen near the peak of the continuum. There is good agreement with the predicted resonance at λ297.7 nm[17]. By quantitative microdensitometry a linewidth of 3.5 nm (FWHM) was measured from the absorption line profile of figure 4, obtained by subtracting the normalized background intensity profile. A peak cross-section of 5 x 10^{-16} cm^2 was estimated from the spectral densities and the excited state population determined by the laser flux. Both of these experimental values are in excellent agreement with those calculated. Figure 4 also shows two sets of triplet emission multiplets. The selectively exciting pumping beam at λ285.2 nm will also cause ionization from the 3s 3p $^1P_1^0$ state and recombination can leave the atom in one of the excited triplet states. We have recorded emission lines terminating in the 3s 3p $^3P^0$ level from 3s ns ^3S, 3s nd ^3D and $3p^2$ ^3P levels. When the argon buffer gas pressure was increased to 600 Torr from the 300 Torr pressure normally used these emission lines generally disappeared and often appeared in absorption. This indicates that inter-system crossing is then occurring during the lifetime of the background pulse. Experiments are in progress to measure the 3s 3p $^1P_1^0$ - $3p^2$ $^1S_0$ photo-ionization cross-section photoelectrically with a second tunable dye laser probe.

ACKNOWLEDGEMENTS

The authors are glad to acknowledge useful discussions with Professor W.R.S. Garton who also kindly made available the use of a plate measuring machine. We wish to thank the Science Research Council for the provision of equipment and technical support.

REFERENCES

1. D.J. Bradley, A.J.F. Durrant, G.M. Gale, M. Moore and P.D. Smith, J. Quantum Electron. QE-4, 707 (1968).

2. D.J. Bradley, M.H.R. Hutchinson, H. Koetser, T. Morrow, G.H.C. New and M.S. Petty, Proc. Roy. Soc. Lond. A 328, 97 (1972).

3. D.J. Bradley, Appl. Optics 8, 1957 (1969).

4. D.J. Bradley, G.M. Gale and P.D. Smith, Nature 225, 719 (1970).

5. T.J. McIlrath, Appl. Phys. Lett. 15, 41 (1969).

6. D.J. Bradley, G.M. Gale and P.D. Smith, J. Phys. B: Atom. Molec. Phys. 3, L11 (1970).

7. D.J. Bradley, P. Ewart, J.V. Nicholas and J.R.D. Shaw, International Quantum Electronics Conference, Digest of Technical Papers, 58-59.

8. W.R.S. Garton, W.H. Parkinson and E.M. Reeves, Proc. Phys. Soc. 80, 860 (1962).

9. D.J. Bradley, G.M. Gale and P.D. Smith, J. Phys. B: Atom. Molec. Phys. 4, 1349 (1971).

10. R.V. Ambartzumian and V.S. Letokhov, Appl. Optics 11, 354 (1972).

11. A. Lurio, Phys. Rev. 136, A376 (1964).

12. D.J. Bradley, J.V. Nicholas and J.R.D. Shaw, Appl. Phys. Lett. 19, 480 (1971).

13. D.J. Bradley, W.G.I. Caughey and J.I. Vukusic, Opt. Commun. 4, 150 (1971).

14. H.N. Russell and C.E. Moore, J. Research N.B.S. 55, 299 (1955).

15. W.R.S. Garton and K. Codling, Proc. Phys. Soc. 75, 87 (1960).

16. R.F. Bacher, Phys. Rev. 43, 264 (1933).

17. D.G. Thompson, unpublished.

18. R.J.W. Henry and L. Lipsky, Phys. Rev. 153, A51 (1967).

19. A.R. Striganov and N.S. Sventitskii, Tables of Spectral Lines of Neutral and Ionized Atoms (Plenum, New York, 1968), p.245.

SPECTROSCOPY OF HIGHLY EXCITED LEVELS IN ALKALI ATOMS USING A

CW TUNABLE DYE LASER*

S. Svanberg†

Columbia Radiation Laboratory, Department of Physics

Columbia University, New York, New York 10027

The alkali atoms have a particularly simple electronic structure, and a very detailed theoretical understanding of their properties could perhaps be expected. Although a single valence electron is responsible for the gross features in the energy level
diagram, excitations of core electrons can have a significant
influence on such atomic parameters as fine and hyperfine coupling
constants. With modern many-body computational techniques, such
effects can be studied, and the alkali metals form natural test
cases for the new theoretical methods.

Until recently, accurate experimental hyperfine structure
studies of excited states in alkali atoms have been confined to
the sequences of ^2P levels, which are readily available for observation through direct excitation with resonance radiation.
Detailed studies of P states have revealed several effects of core
electron excitations. These include magnetic core polarization,
electric quadrupole shielding, and possible deviations from
expected Landé g_J factors.(1) Clearly, a corresponding knowledge
of the properties of non-P states is of great interest for the
understanding of the various atomic interactions.

It is well known that the D and F state fine structure inter-

*Work supported in part by the Joint Services Electronics Program
(U. S. Army, U. S. Navy, and U. S. Air Force) under Contract
DAAB07-69-C-0383, in part by the Air Force Office of Scientific
Research under Contract AFOSR-72-2180, and in part by the Swedish
Natural Research Council.

†Permanent address: Department of Physics, Chalmers University of
Technology, Fack, S-40220 Göteborg, Sweden.

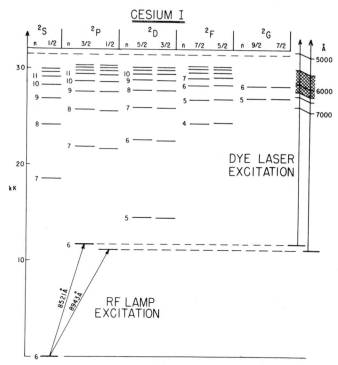

Fig. 1.   Energy level diagram for Cs, with wavelengths relevant
for two-step excitation indicated.

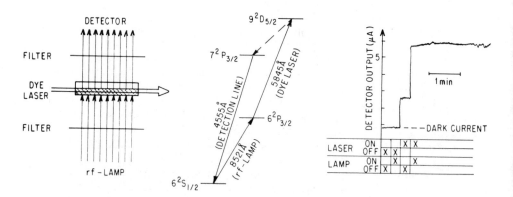

Fig. 2.   Scattering cell arrangement and observation of light
released after the two-step excitation process, exemplified by
the case of the 9 $^2D_{5/2}$ state of Cs.

vals are often anomalously small and even inverted, and recent
experimental work has shown that similar large anomalies occur in
the D-state hyperfine structure.(2,3) Studies of non-P states
have been obstructed by the difficulty in producing such states.
Archambault et al. (4) have populated D states by electron bom-
bardment and could make rough estimates of the hyperfine structure
of the 5 $^2D_{5/2}$ state in Na^{23} and the 9 $^2D_{5/2}$ state in Cs^{133}. With
the recently introduced cascade decoupling (5) and cascade rf
spectroscopy methods,(2) the Columbia group has investigated the
hyperfine structure of several S and D levels in the alkali
atoms.(6) In these experiments the states are produced by cascading from a P level excited by a conventional rf lamp. However,
for reasons of intensity, this method is limited to comparatively
low-lying S and D levels, as the absorption oscillator strengths
in the S-P transition sequence decrease very rapidly with
increasing P-state energy, and conventional lamps are not very
efficient sources for the high resonance lines.

Recently we have reported a two-step excitation method which
allows us to get access to more highly excited non-P states.(7)
In the first step the strong D_1 and D_2 lines from an rf lamp are
used to transfer atoms from the ground S state to the first
excited P states. In the second step, the intense tunable radiation from a cw dye laser, operating with rhodamine 6G, is used to
excite atoms from one of the P levels to a highly excited S or D
level. From such levels, P and F levels are also populated in the
cascade decay. Absorption of light by atoms in the 5 $^2P_{3/2}$ level
of Rb has earlier been demonstrated by Bradley et al. (8), utilizing high power pulsed dye lasers. Further, a two-step excitation of the 3 ^2D term in lithium has been used in a fine-structure
investigation by Smith and Eck.(9) In this case the light for the
excitation was provided in a discharge in the studied alkali vapor.
The excitation method we have utilized is particularly useful,
since it can be used in conjunction with conventional optical
double resonance (ODR) and level crossing (LC) methods to investigate the hyperfine and fine structures of a large number of excited
states in the alkali atoms.

The principles of the two-step excitation method are shown in
Figs. 1 and 2, where the explicit case of cesium is chosen. Figure
1 displays the cesium level diagram with the D_1 and D_2 lines indicated. The cw operation region for the tunable laser, using rhodamine 6G as a dye, is indicated on the wavelength scale, starting
from the first excited ^2P levels. In Fig. 2 the scattering cell
arrangement and the observation of fluorescent light released after
a two-step excitation of the 9 $^2D_{5/2}$ Cs level is illustrated. Here
the laser was tuned to the 6 $^2P_{3/2}$ - 9 $^2D_{5/2}$ transition. From the
recording it is seen that with the filters we used there is no
direct leakage of lamp light, whereas a substantial laser light
leakage into the detection channel is evident. Our experimental
setup is based on the apparatus described in Refs. 2 and 6.

A powerful rf lamp was constructed for producing the light for the
first excitation step. The beam of D_1 and D_2 light enters the
resonance cell in the direction of the external magnetic field,
which is also the detection direction. The laser beam, which has
a diameter of about 3 mm, passes through the cell at right angles
to the field direction. We use a Spectra Physics model 370
tunable dye laser, which is pumped by the 5145-Å line, or the all
line output of a Spectra Physics model 165 argon ion laser. In
multimode operation the dye laser gives up to 300 mW output power
for an input power of 2 W (5145 Å). A bandwidth of about 0.3 Å
is obtained. With an intracavity etalon assembly consisting of
uncoated quartz discs, operation in a few or even a single mode
could be achieved with a substantial loss of output power. Some
of our measurements were made using such a narrow-line operation.
Fluorescent light released in the decay of the highly excited
state is detected by an RCA C31000F photomultiplier. The rf field
required in the ODR measurements is produced in a cavity, tuned to
about 147 MHz.   Lock-in detection is used in the experiments,
and the signals are stored in a PDP-8/S computer, which also
controls a linear magnetic field sweep.
     In the two-step excitation process it is possible to take
advantage of radiation trapping of the D_1 and D_2 resonance lines.
In the multiple scattering process an incoming photon can excite
several ground-state atoms into a ^2P state so that more P state
atoms will be available for laser excitation. This is very
valuable, as the lifetime of the P states is only about 30 nanoseconds. In this way we have achieved adequate populations for
ODR and LC spectroscopy of the 9, 10, 11 $^2S_{1/2}$ and the 8, 9, 10
^2D levels in Cs and the 8, 9 ^2S and 6, 7 ^2D levels in Rb. The
interaction region between the atoms and the two light beams is
about 0.5 cm^3 and contains at the usual operating temperature
($100^{\circ}C$ for Cs, $120^{\circ}C$ for Rb) about 10^{13} atoms. The number of
detected photons per second in a solid angle of about 0.15 sr is
$10^6 - 10^8$, depending on which particular state is studied and
what detection line is used.
     We will now briefly describe some ODR and LC measurements
which we have performed utilizing the two-step excitation
method.(7,10) The LC technique was used to measure the hyperfine
structure of the 6 and 7 $^2D_{3/2}$ states in Rb^{87}. Because of the
multiple scattering of the D lines the different sublevels of the
5 $^2P_{3/2}$ and the 5 $^2P_{1/2}$ states, respectively, have approximately
equal populations. With the tunable laser, σ transitions to the
$^2D_{3/2}$ state are induced from one of these P states, and the decay
down to the other one is observed. To avoid direct leakage of the
detection line from the rf lamp, a Schott RG 10 colored glass
filter is used in the exciting beam to block all lines except the
infrared ones. An interference filter with a half-width of 50 Å
is used to isolate the detection line from the stray light from
the strong laser beam. This cannot be done completely, as the

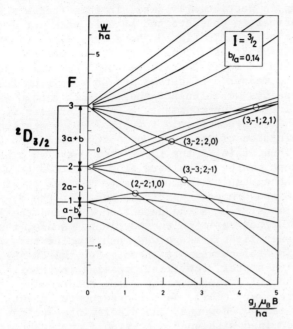

Fig. 3. Energy level diagram for the hyperfine structure of a $^2D_{3/2}$ level in Rb^{87} (nuclear spin 3/2). Dimensionless units are used for energy and magnetic field. The $\Delta m \doteq 2$ level crossings are indicated.

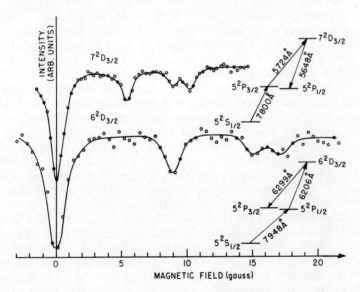

Fig. 4. Level-crossing signals for the 6 and 7 $^2D_{3/2}$ states in Rb^{87}. The sampling time for each of the curves is about 1.5 h.

wavelength separation is particularly small in this case. A
rotating linear polarizer in the detected beam is used for lock-in
detection of the LC signals. In Fig. 3 the energy level diagram
for the $^2D_{3/2}$ states of Rb^{87} is shown. Out of the four $\Delta m = 2$
crossings only the first three are suitable for detection. In
Fig. 4 examples of the measured LC curves are shown. From the
positions of the crossings we calculate the magnetic dipole and
electric quadrupole interaction constants a and b. So far we have
performed a precision measurement only for the 7 $^2D_{3/2}$ state.
This allows a determination of the quadrupole interaction constant
to a 15% accuracy. Thus, this is the first case where an alkali
metal quadrupole moment can be calculated from D state data and
can be compared with the results from extensive P-state studies.(1)

Corresponding LC measurements were also made for the 8, 9,
and 10 $^2D_{3/2}$ states in Cs^{133}. As the detection line and the laser
excitation line are further apart on the wavelength scale, the
suppression of the laser light leakage was much simpler than for
Rb. In Fig. 5 experimental curves are shown. For the Cs $^2D_{3/2}$
states as for the $^2P_{3/2}$ states, the first crossing is resolved,
whereas the second and third ones overlap. The quadrupole
coupling constants are very small, as the quadrupole moment of
Cs^{133} is only -3 millibarn. However, the dipole interaction
constants can be determined with good precision from the crossing
field values.

The $^2D_{3/2}$ states can also be studied with ODR in the Paschen-
Back region for the hyperfine structure. As an example, the
signal structure obtained for the 9 $^2D_{3/2}$ state in Cs^{133} is shown
in Fig. 6. The same geometry was used, and the rf field was
100 per cent square wave modulated for lock-in detection. Eight
signals of approximately equal intensity are expected around the
center of gravity (C.G.), corresponding to $g_J = 0.8$. The structure
obtained is unresolved, with a total width closely agreeing with
the one inferred from the more accurate LC results.

As the transition between a $^2D_{5/2}$ state and a $^2P_{1/2}$ state is
forbidden and it is not possible to use the laser transition for
the detection due to the inevitable stray light, a different
technique was used for the measurement of the highly excited
$^2D_{5/2}$ states in Cs and Rb.   These states decay with a branching
ratio of about 15% into the second excited $^2P_{3/2}$ level, and the
wavelength for the transition from this state to the ground state
(4555 Å for Cs, 4202 Å for Rb) can be well isolated from both
exciting lines. The diagram in the center of Fig. 2 is an example
of this two-step fluorescence scheme. Schott BG 3 and BG 18
colored glass filters were useful in further suppressing the background light. The LC method is not very applicable in this case,
as the coherence at the non-zero-field crossing points is strongly
degraded in the passage through the second state.(11) Thus, we
used the ODR technique. Unpolarized light was used in the first
excitation step, whereas the laser light polarization was adjusted

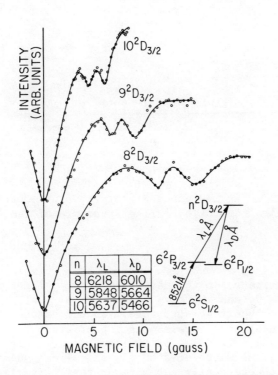

Fig. 5. Level-crossing signals for the 8, 9, and 10 $^2D_{3/2}$ states in Cs^{133}. Total sampling time for each curve is about 1 h.

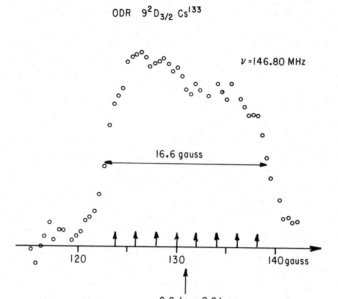

Fig. 6. Optical double resonance signal structure obtained in the
Paschen-Back region for the hyperfine structure of the 9 $^2D_{3/2}$
state in Cs^{133}.

to give π excitations. In the detection step, σ light was
observed. The 8, 9, and 10 $^2D_{5/2}$ levels in Cs^{133} all turn out to
have a very small hyperfine structure. In the Paschen-Back region
an unresolved signal structure consisting of eight closely spaced
signals is obtained. Examples of the experimental curves are
shown in Fig. 7. The width of a single signal component is
calculated from theoretically estimated natural lifetimes and is
displayed for each state in the figure. From the half-widths of
the signal structures obtained by extrapolation to zero rf power,
only rough values for the dipole coupling constants can be
obtained.

In Fig. 8 corresponding ODR curves are shown for the 6 and 7
$^2D_{5/2}$ states in Rb^{87}. As can be seen, a better resolution is
obtained for the four Paschen-Back region signals, and the dipole
coupling constants can be extracted with reasonable accuracy.

In Cs the ^2F levels from 4 ^2F to 7 ^2F can be populated by
cascading from the ^2D levels excited in these experiments. For
example, the 8 $^2D_{5/2}$ and 8 $^2D_{3/2}$ levels have a 2% branching ratio
into the 5 $^2F_{7/2}$ and the 5 $^2F_{5/2}$ levels, respectively. The decay
of the F states to the 5 $^2D_{5/2}$ and 5 $^2D_{3/2}$ levels can be detected

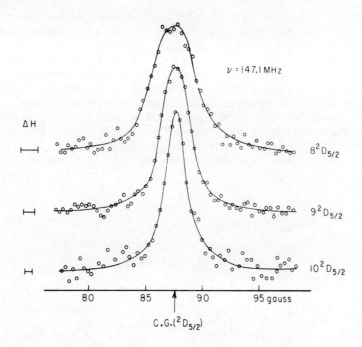

Fig. 7. Optical double resonance signals for the 8, 9, and 10
$^2D_{5/2}$ states in Cs^{133}.

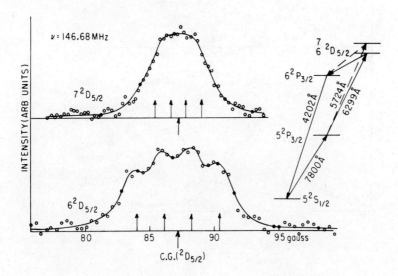

Fig. 8. Optical double resonance signals for the 6 and 7 $^2D_{5/2}$
states in Rb^{87}.

at 8079 and 8016 Å. We have observed unresolved ODR signals in
these F states, as shown in Fig. 9. In these experiments resonances in the feeding D states are also obtained. The 5 ^{2}F
doublet is inverted, with a fine structure splitting of only about
4410 MHz. Thus, at the fields used, the beginning decoupling of
the fine structure will broaden the signal in addition to the
broadening due to the hyperfine structure. Only rough upper
limits for the hyperfine structure constants can be estimated.
We have performed similar experiments on the 6 ^{2}F doublet in Cs^{133}.

        In all the ODR and LC measurements discussed here, only the
absolute values of the coupling constants, but not the signs,

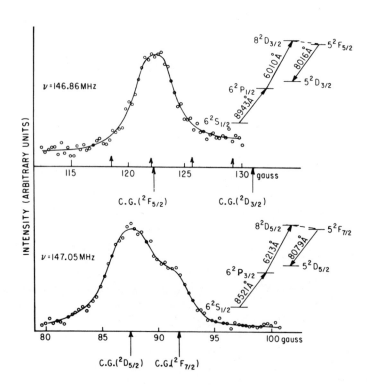

Fig. 9. Optical double resonance curves with signals from the
5 ^{2}F$_{7/2}$ and 5 ^{2}F$_{5/2}$ levels in Cs^{133}. Sampling time for each of
the curves is about 1 h.

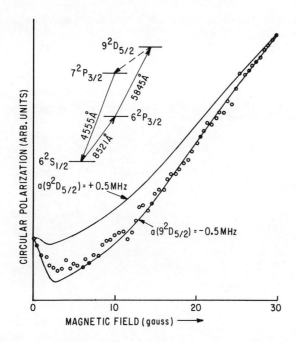

Fig. 10. Experimental and theoretical decoupling curves for the 9 $^2D_{5/2}$ states of Cs[133]. A negative a-factor is suggested.

could be determined. The positions of resonances and level crossings are the same for normal or inverted hyperfine structures. The signs of the coupling constants can be determined by decoupling measurements.[2] Such measurements are now being performed for the states studied in this work. We use circularly polarized laser light, irradiated onto the atoms in the same direction as the first step excitation light. In addition to determining the sign, an improvement in the absolute values is anticipated for the $^2D_{5/2}$ and ^2F states. As an example, an experimental decoupling curve for the 9 $^2D_{5/2}$ state in Cs[133] is shown in Fig. 10. The lines are theoretical curves calculated for the two a-values compatible with the ODR results. Clearly, the better fit is obtained for the negative a-factor. A negative a-value for this state cannot be explained by a one-electron picture of the alkali atom. However, similar effects, presumably due to core polarization, have been observed for lower $^2D_{5/2}$ states in alkali atoms.[2,3]

We summarize our present results for the hyperfine structure
of highly excited alkali states in Table I.

Natural lifetimes of excited states can be obtained from the
half-widths of level crossing and optical double resonance
signals. Much information about the radiative properties of the
highly excited states can be obtained in the experiments described
here. Particularly strong signals yielding lifetime values are

TABLE I. Results for hyperfine coupling constants a and b,
obtained in two-step excitation experiments.

Element	State	Abs. value of a (MHz)	Abs. value of b (MHz)
Rb^{87}	$6\ ^2D_{3/2}$	7.72(20)	0.6(4) b/a>0
	$6\ ^2D_{5/2}$	3.6(7)	
	$7\ ^2D_{3/2}$	4.53(3)	0.26(4) b/a>0
	$7\ ^2D_{5/2}$	2.2(5)	
Cs^{133}	$8\ ^2D_{3/2}$	3.98(12)	
	$8\ ^2D_{5/2}$	0.9(4)	
	$9\ ^2D_{3/2}$	2.37(3)	
	$9\ ^2D_{5/2}$	0.5(2)	
	$10\ ^2D_{3/2}$	1.52(3)	
	$10\ ^2D_{5/2}$	0.4(2)	
	$5\ ^2F_{5/2}$	< 0.7	
	$5\ ^2F_{7/2}$	< 1.0	
	$6\ ^2F_{5/2}$	< 1.0	
	$6\ ^2F_{7/2}$	< 1.0	

obtained in Hanle-effect experiments. Such experiments can also be performed when two-step fluorescence is observed, whereas the non-zero-field crossings have a very low detection intensity, as discussed above. Examples of Hanle curves, observed in two-step fluorescence, are shown in Fig. 11.

In order to obtain accurate lifetime values, great care must be taken to obtain results free from systematic errors. The highly excited levels have comparatively long lifetimes and are quite sensitive to collision broadening. Thus, extrapolations of measured half-widths to low atomic densities have to be performed. Further, the influence of the laser light intensity should be studied.

In this paper some examples of possible spectroscopic measurements of highly excited Rb and Cs states using a two-step excitation

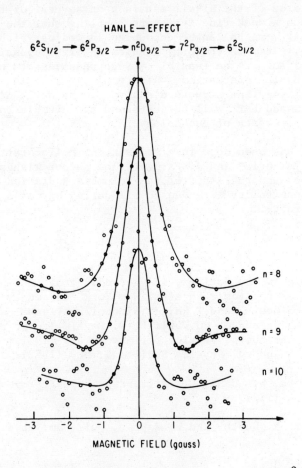

HANLE—EFFECT

$6^2S_{1/2} \rightarrow 6^2P_{3/2} \rightarrow n^2D_{5/2} \rightarrow 7^2P_{3/2} \rightarrow 6^2S_{1/2}$

n = 8

n = 9

n = 10

MAGNETIC FIELD (gauss)

Fig. 11. Hanle-effect signals for the 8, 9, and 10 $^2D_{5/2}$ states in Cs[133], observed in two-step fluorescence following two-step excitation.

method in conjunction with a tunable dye laser have been given. Obviously, the method can be extended to many other states in the alkalies and also to other elements. For the alkali atoms, S states, directly populated, and highly excited P states, populated in the cascade decay from higher lying S and D states, can be studied with ODR. With the new dyes now available for cw operation, the usable wavelength region has been extended to both longer and shorter wavelengths.(12) Using shorter wavelengths one quickly comes into a region of very closely spaced levels, which converge toward the ionization limit. (See Fig. 1.) Very systematic studies along the different sequences of levels seem possible, and even photoionization experiments can be performed. Apart from hyperfine structure, narrow fine structures can also be studied by ODR, LC, and anti-level-crossing techniques. Among the D and F states of the alkali metals even several of the lower states have such a weak fine structure coupling that the orbital and spin angular momenta can be decoupled in ordinary laboratory electromagnets. In this context, studies of off-diagonal hyperfine structures seem to be feasible. Such measurements would yield very valuable information, which would greatly increase the reliability of hyperfine structure analyses. Measurements along the lines discussed here will be performed at Columbia University and Chalmers University of Technology.

This work has been done in a stimulating collaboration with W. Happer, P. Tsekeris, and J. Farley. The author wishes to thank Professor Happer and the staff of the Columbia Radiation Laboratory for the warm hospitality extended to him during a year's stay at Columbia University.

References

1. See e.g. G. zu Putlitz, in Proceedings of the International Conference on Atomic Physics, New York, edited by B. Benderson, B. W. Cohen, and F. M. J. Pichanick (Plenum, New York, 1969); D. Feiertag and G. zu Putlitz, Z. Physik (to be published); G. Belin and S. Svanberg, Physica Scripta 4, 269 (1971); S. Rydberg and S. Svanberg, Physica Scripta 5, 209 (1972).

2. R. Gupta, S. Chang, C. Tai, and W. Happer, Phys. Rev. Letters 29, 695 (1972).

3. C. Tai, R. Gupta, and W. Happer, Bull. Am. Phys. Soc. 18, 121 (1973).

4. Y. Archambault, J. P. Descourbes, M. Priou, A. Omont, and J. C. Pébay-Peyroula, J. Phys. Radium 21, 677 (1960).

5. S. Chang, R. Gupta, and W. Happer, Phys. Rev. Letters 27, 1036 (1971).

6. R. Gupta, S. Chang, and W. Happer, Phys. Rev. A 6, 529 (1972); L. K. Lam, R. Gupta, and W. Happer, Bull. Am. Phys. Soc. 18, 121 (1972), and works to be published (Phys. Rev. A).

7. S. Svanberg, P. Tsekeris, and W. Happer, Bull. Am. Phys. Soc. 18, 611 (1973); S. Svanberg, P. Tsekeris, and W. Happer, Phys. Rev. Letters 30, 817 (1973).

8. D. J. Bradley, G. M. Gale, and R. D. Smith, J. Phys. B: Proc. Phys. Soc., London 3, L11 (1970).

9. R. L. Smith and T. G. Eck, Phys. Rev. A 2, 2179 (1970).

10. S. Svanberg, P. Tsekeris, and W. Happer, Phys. Rev. A (to be published).

11. C. Tai, R. Gupta, and W. Happer (to be published).

12. See e.g. S. A. Tuccio, K. H. Drexhage, and G. A. Reynolds, Opt. Commun. 7, 248 (1973).

TUNABLE LASERS II

HIGH RESOLUTION TUNABLE INFRARED LASERS*

A. Mooradian

Lincoln Laboratory, Massachusetts Institute of Technology

Lexington, Massachusetts 02173

INTRODUCTION

This paper describes some recent developments in the technology of tunable sources which operate in the near to middle infrared regions. Included among these are external cavity controlled and hydrostatic pressure tuned semiconductor lasers, spin-flip Raman lasers, and nonlinear mixing using tunable and quasitunable laser sources. The spectral linewidth and tuning range of these devices will be discussed with emphasis on their application for high resolution spectroscopy.

SEMICONDUCTOR LASERS

Cavity Controlled Semiconductor Lasers

Semiconductor diode lasers have been used for high resolution spectroscopic studies of gases in the middle and near infrared more extensively than any other presently available tunable infrared laser. The output of these diodes when operated well above threshold usually occurs in many different modes. One of the modes which lies close to the absorption line of interest is selected by a spectrometer and then fine tuned by changing the current through the diode. The change of current through the diode alters the I^2R heating and hence the refractive index in the mode volume. The frequency tuning for each mode is typically about one cm^{-1} with unfilled frequency gaps of the same order. Another diode laser with a different cavity length at the same temperature would have shifted cavity mode frequencies and provide a partial overlap.

---

*This work was sponsored by the Department of the Air Force, the Atomic Energy Commission, and the National Science Foundation.

Initial studies of the mode control properties of a semiconductor laser have been made with the use of a grating controlled external cavity
GaAs laser[1,2,3] in order to improve the usefulness of the spectral output
of these devices. The spectral properties of both a 77 K continuously operating[2] and a room temperature pulsed[3] external cavity grating controlled GaAs diode laser have been studied. With only a grating as one of
the cavity elements, output powers of 17 milliwatts CW in a single axial
mode have been extracted[2] from a diode at 77 K operating just above
threshold. A single mode output power of 60 milliwatts CW (double ended)
has been reported[4] from a GaAs diode laser operating just above threshold near liquid helium temperature. More recently,[3] a grating controlled
room temperature GaAs diode laser has operated with a spectral width of
less than 0.2 Å and a peak output power of nearly three watts. It is interesting to note that the same diode operating with a 100% reflector in place
of the grating produced 3.5 watts in a bandwidth of over 40 Å. The output
intensity at the uncoated face of the diode in these experiments was close
to the value where catastrophic degradation of GaAs diode lasers usually
occurs. It was deduced from these experiments that the GaAs spontaneous linewidth is almost entirely homogeneously broadened. Figure 1

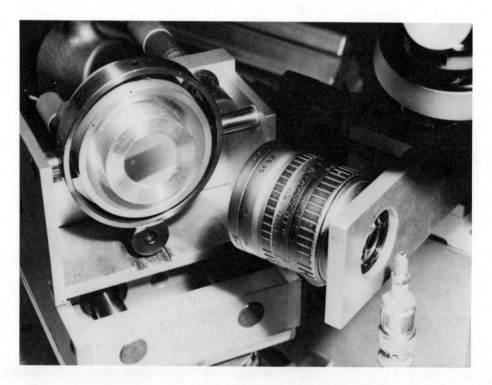

Fig. 1.   Photograph of a room temperature external cavity gallium
           arsenide diode laser.

shows a photograph of this device. Only one end of the diode is antireflection coated. Figure 2 shows the tuning characteristics of this device. By
varying both the grating position and the temperature of the diode, it would
be possible to get a tunable source from 8500 to 9000 Å using only GaAs
diode lasers. For temperatures below 77 K, output powers of up to one
watt CW in a single frequency should be possible. A $Pb_xSn_{1-x}Te$ diode
laser has been made to lase[5] with an output near 10 μm using an external
cavity consisting of a single concave mirror. This device operated near
liquid helium temperature with only one end of the diode antireflection
coated.

The spectral output of these grating controlled semiconductor diode
lasers occurs in a number of axial modes for pumping levels much above
threshold. Additional intercavity mode control elements would be necessary to limit the output to a single mode. One of the advantages of an external cavity controlled semiconductor laser is the reduced frequency
chirp for a given mode during pulsed operation. The frequency chirp over
a "raw" diode is reduced by the ratio of the optical length of the diode to
the overall length of the cavity. This reduction can be as much as a factor
of one hundred for typical devices, however, the smaller mode spacing in

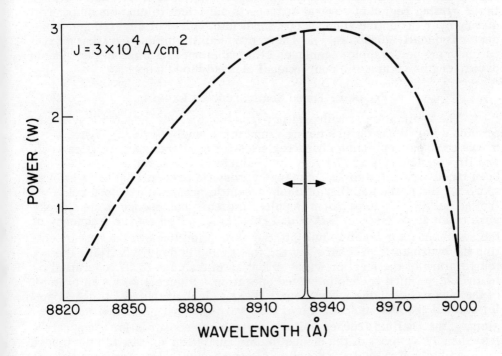

Fig. 2.   Output of the external cavity, grating controlled pulsed
gallium arsenide diode laser shown in Fig. 1.

long external cavities makes it more difficult to achieve single frequency
operation. For a number of spectroscopic applications, single frequency
operation may not be necessary or desired. An appropriate external cavity device operated pulsed could have its modes chirped enough to fill in
the spectral envelope defined by the external cavity grating dispersion and
the mode width defined by the junction. Such an envelope width could be
significantly less than 0.1 cm^{-1} if the diode were not driven very far
above threshold. Tuning would effectively be continuous for spectroscopic
lines that were broader than the envelope width.

   The external cavity controlled diode laser is a hybrid device. In
the future a distributed feedback grating fabricated along the active gain
region of the diode will provide oscillation at a single wavelength given by

$$\lambda = \frac{2n(T)d}{M} \tag{1}$$

where d is the spacing of the grating, n(T) is the temperature dependent
refractive index, and M is an odd integer. Optically pumped distributed
feedback GaAs laser action has been demonstrated in GaAs at 77 K.[6] An
important feature of Eq. (1) is that the temperature dependent refractive
index $n(T) \sim n_o + (dn/dT)\Delta T$ provides a mechanism for continuously tuning the resonant condition without any mode jumps. A temperature change
in the grating region of several degrees is sufficient to tune the output frequency over the entire spontaneous bandwidth typical of infrared semiconductor (typically 20-50 cm^{-1}). The reproducibility of such grating fabrication using holographic techniques eliminates uncertainties in the random
cavity lengths obtained in conventional cleaved diode lasers.

Pressure Tuned Semiconductor Lasers

   The application of hydrostatic pressure to semiconductor lasers can
provide a very broad gross tuning range for a single device. Figure 3
shows the band gap tuning characteristics for a number of binary lead salt
and III-V compounds at 77 K. A lead selenide diode laser at 77 K has
been tuned[7] from 7.5 to 22 μm using hydrostatic pressure up to 14 kilobars.
More recently, the stability of the hydrostatic pressure has been controlled sufficiently to perform Doppler limited spectroscopy using a pressure tuned diode laser of GaAs[7] and PbS_xSe_{1-x}. The course frequency of
the semiconductor laser is tuned to the absorption lines of interest by setting the hydrostatic pressure and the fine tuning is done by varying the
temperature of the junction with current as in the case of CW operation or
by the chirp which occurs in pulsed operation. Figure 4 shows an example[8]
of the latter case using a GaAs diode laser to look at the real time, Doppler limited absorption spectrum of the $6s^2S_{1/2} \rightarrow 6p^2P_{3/2}$ transition in cesium. Since helium freezes solid at these high pressures for temperatures
less than 77 K, most of the diodes to date have been operated in the pulsed
mode with chirp rates low enough to sweep through a Doppler linewidth in
a time longer than the response time of the detector. Development of infrared diodes which operate continuously at 77 K or higher will greatly

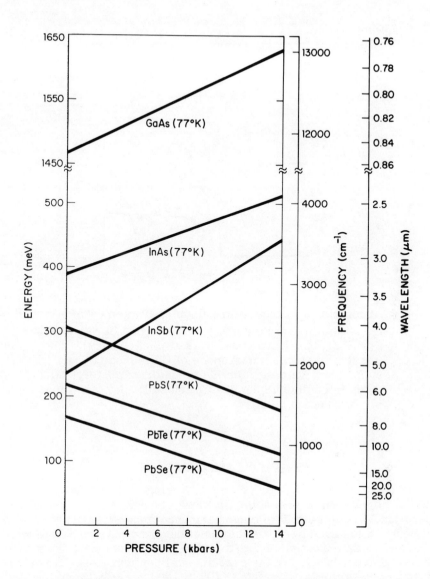

Fig. 3.   Energy gap variation as a function of hydrostatic pressure
           at 77 K for some binary semiconductor compounds.

facilitate the usefulness of hydrostatic pressure tuning.   Because of the
broad band wavelength coverage using hydrostatic pressure tuning, only
the binary semiconductor compounds and one or two alloy semiconductors
would be necessary to cover the wavelength range from 2 to 35 μm.

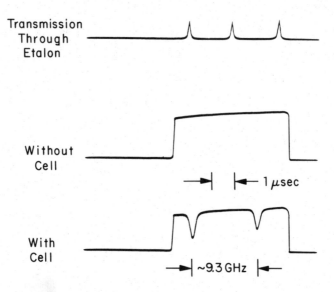

Fig. 4.   Absorption spectrum of the Doppler limited $6s^2S_{1/2} \rightarrow 6p^2P_{3/2}$
           transition in cesium using a 77 K chirped GaAs diode laser
           ($\lambda = 8521.1$ Å, hydrostatic pressure = 160 bars). Top: trans-
           mission of chirped axial mode of GaAs diode through the free
           spectral range of a Fabry-Perot interferometer. Middle:
           mode intensity without absorption cell. Bottom: mode inten-
           sity with cesium cell.

SPIN-FLIP RAMAN LASERS

There have been a number of recent developments in the understanding and operation of InSb spin-flip Raman lasers. One of these is
the operation of a gradient field tuned permanent magnet system[9] in
which tuning is achieved by a precision motion of the magnetic field gradient normal to the direction of the laser beam in the crystal. Figure 5
shows a schematic drawing of the experimental configuration. A permanent magnet whose field varies from 1100 to 1800 gauss between the ta-
pered pole pieces is moved vertically by a precision electro-mechanical
drive. This mechanical tuning of the spin-flip laser output is both stable
and reproducible enough to record Doppler limited spectra. Figure 6
shows the tuning of five Stokes and two anti-Stokes components as a function of magnet position. The mechanical tuning range is sufficient to completely cover the spacing of the rotational lines of the CO laser in the
range between 5.1 to 6.5 μm while two small modulation coils on the magnet provide a fine tuning capability. For the longer wavelengths, continuous operation becomes more difficult as the threshold increases and the
available CO laser power decreases. The amount of field variation over

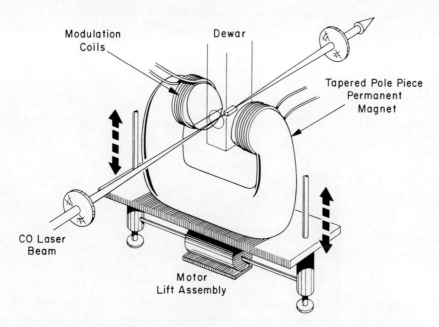

Fig. 5.   Diagram of a field gradient permanent magnet spin-flip
           Raman laser system.

the mode diameter does not produce any serious broadening of the spontaneous linewidth. This permanent magnet system provides a simpler,
less costly, and potentially more stable spin-flip laser.

The spontaneous spin-flip Raman linewidth is an important parameter in helping to determine not only the laser threshold but the mode tuning characteristics. Recently, [10] linewidths as small as 200 MHz have
been observed using a small signal gain technique. A more detailed experimental and theoretical analysis of this lineshape as a function of magnetic field[11] and temperature[12] has been reported using this technique. A
sample of InSb was pumped with a CO laser at frequency ω_1 below threshold but with enough gain over the sample length to produce about a 10%
amplification at frequency $\omega_2 = \omega_1 - g\beta H$. The intensity of a weak colinear
probe beam at ω_2 was measured as the magnetic field was swept. A typical small signal gain spectrum at ω_2 is shown in Fig. 7 while Fig. 8
shows the variation of measured spontaneous linewidth with magnetic
field. The most striking feature of the latter figure is that in low concentration samples, the linewidth which is very close to Lorentzian actually
increases with magnetic field at least up to 10 kG. This in part contributes to the low threshold observed for these samples at low magnetic
fields. A threshold power of less than 5 milliwatts has been reported[10]
in diameter of about 100-200 μm using a 5.3 μm pump line from a CO laser. This threshold power increases for a fixed pump photon energy as

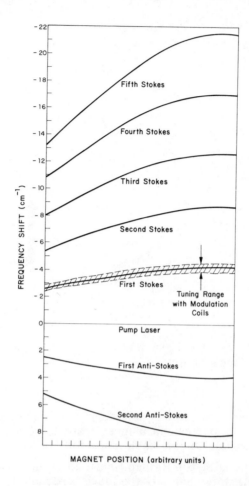

Fig. 6.   Tuning characteristics of the field-gradient permanent-
magnet InSb spin-flip Raman laser as a function of
magnet position.

the magnetic field increases because the empty upper spin state moves away from the pump laser with which it is strongly resonant as well as the linewidth increasing. Threshold powers of a watt or more can occur for pump wavelengths longer than 6-6.5 μm. Continuous operation using a CO_2 laser should be possible in a properly designed InSb spin-flip laser crystal with a pump power of only a few watts.

The five Stokes components and the two anti-Stokes components of Fig. 6 are not all due to a stimulated Raman process.[11] A strongly res-onant nonlinear four-wave parametric mixing process accounts for most

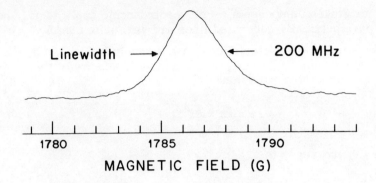

Fig. 7.   Small signal gain of an InSb spin-flip Raman amplifier. Here,
n = 8 x 10^{14} cm^{-3}, $\omega_1 - \omega_2 = 4.17$ cm^{-1}, and T = 2 K.

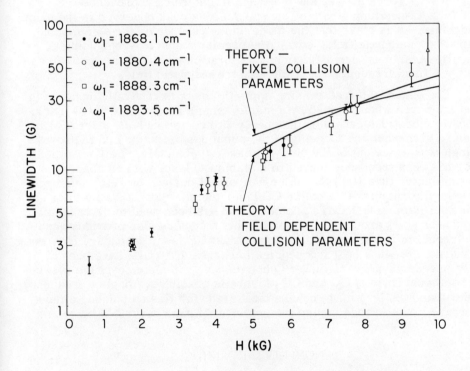

Fig. 8.   Linewidth of spontaneous spin-flip Raman scattering as a
function of magnetic field for an InSb sample with
n = 8 x 10^{14} cm^{-3} at T = 2 K. Theoretical fits are described
in detail in reference 10.

of higher order Stokes components. For spectroscopic applications, however, these parametrically generated lines are still quite useful.

The mode tuning characteristics[13] are strongly influenced by the linewidth of the spontaneous process. The output frequency of the spin-flip laser is given by the expression

$$\nu = \frac{\nu_c \Gamma_s + \nu_s \Gamma_c}{\Gamma_s + \Gamma_c} \quad , \tag{2}$$

where the cavity frequency is

$$\nu_c = \frac{cM}{2n(\nu, H)\ell} \quad , \tag{3}$$

ν_s is the peak of the spontaneous frequency, Γ_c the cavity width, Γ_s the spontaneous width, M is an integer, n is the frequency and field dependent index, and ℓ is the cavity length. The tuning rate of the mode frequency with magnetic field has been measured to vary from 16 MHz/Gauss[12] for a sample with $n = 1 \times 10^{16}$ cm^3 to greater than 60 MHz/Gauss[14] for samples with $n \gtrsim 10^{15}$ cm^{-3} at low magnetic fields which approaches the 67.5 MHz/Gauss tuning rate of the spontaneous peak observed in the same field range. It is clear that the mode tuning depends on the sample concentration, magnetic field, etc. Any precision spectroscopy using a spin-flip laser must use a frequency calibration technique rather than assuming a fixed frequency variation with the magnetic field.

The output of a free running spin-flip laser is generally unstable in frequency even when pumped by a stable, single frequency CW CO laser. Acoustic vibrations can cause frequency jitters as much as several MHz or more even when the focusing of the pump has been adjusted to provide a single frequency out of the spin-flip laser. Recently,[14] a CW spin-flip laser has been frequency locked to a stable CO laser with an absolute stability of better than 100 kHz. An error signal derived by heterodyning the spin-flip output with a stable CO laser together with a stable microwave oscillator in a HgCdTe photodetector has been used to drive small modulation coils around the sample. This technique also provides absolute frequency tuning and calibration against CO laser secondary frequency standards. Because beat frequencies in excess of 60 GHz have been directly measured using HgCdTe[15] photodiodes, complete coverage between the rotational lines of CO and CO_2 lasers is possible. Absolute frequency rather than wavelength can now be measured with this technique, almost anywhere in the infrared using the nonlinear mixing described below.

NON-LINEAR MIXING

Perhaps one of the more potentially useful ways of producing tunable coherent IR radiation over a broad band is sum and difference frequency generation in a non-linear material using a fixed and a tunable frequency source. There have been a number of successful experiments in which a dye laser plus either a ruby or a doubled Nd:YAG laser has been used to generate tunable radiation in the UV-visible-near IR using $LiNbO_3$ as the non-linear crystal. More recently,[16] a stabilized CW dye laser and a CW single frequency argon-ion laser have been mixed in $LiNbO_3$ to generate a CW tunable output from 2-4.5 μm. The frequency stability was about 20 MHz which is adequate for Doppler limited spectroscopy in this region. Output power exceeded one μwatt for 10 m watts and 100 m watts of input dye and argon laser power, respectively. Two of the difficulties associated with using visible lasers to generate IR is the fact that non-linear materials such as $LiNbO_3$ do not stand up well under focused visible-UV radiation and most non-linear materials which are highly transparent in the visible do not transmit much beyond 5 μm.

An alternate approach to generating stable frequency tunable radiation in the infrared is the use of infrared nonlinear materials such as the chalcopyrites.[17,18,19] Some of the advantages of these materials include good IR transmission properties and high nonlinear figure-of-merits. Recently,[19] a CW CO and CO_2 laser have been mixed in $CdGeAs_2$ to generate quasi-tunable infrared radiation in the range from 2.5-17 μm. Figure 9 shows the coverage obtained using just the common isotopes of carbon and oxygen. The output power at the sum or difference frequency is given in the plane wave approximation by

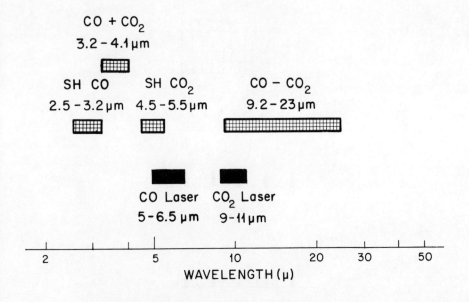

Fig. 9.   Wavelength coverage obtainable from the sum and difference frequencies of a CO and a CO_2 laser. Such a range has been obtained using $CdGeAs_2$ up to the transmission limit of 17 μm. Far infrared frequencies are also possible beyond 40 μm.

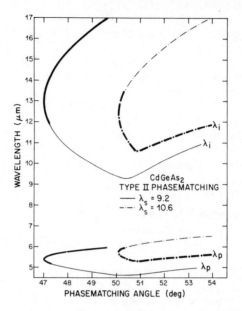

Fig. 10. Phase-matching angle vs wavelength for $CdGeAs_2$ for difference frequency generation. Heavier curves are experimentally obtained data.

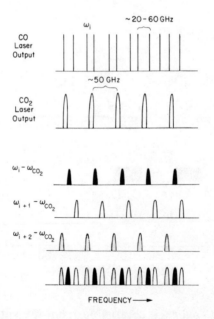

Fig. 11. Frequency coverage obtained by permuting the rotational lines of a CO and a CO_2 laser with a pressure broadened bandwidth.

$$P_{\pm} = \frac{2d^2 \ell^2}{n_{\pm} n_2 n_1 \epsilon_o c^3} (\omega_1 \pm \omega_2)^2 \frac{P_1 P_2}{A_1 + A_2} \tag{4}$$

where P_1 and P_2 are the power densities inside the crystal for the two input waves, A_1 and A_2 are the areas of the two gaussian beams, d is the
non-linear coefficient, ℓ is the length of the crystal, n is the refractive
index at the designated frequency, and ω_1 and ω_2 are frequencies of the
two input waves. Using $CdGeAs_2$, internal conversion efficiencies of 17%
have been measured for second harmonic generation using a pulsed 10 μm
CO_2 laser. For difference frequency generation, an output power at
12.87 μm, for example, of 4 μwatts has been measured from $CdGeAs_2$
using 97 and 1250 m watts CW of CO and CO_2 laser radiation, respectively. Figure 10 shows the phasematching angle for various wavelengths
which have been obtained in $CdGeAs_2$. These nonlinear materials should
be capable of generating several watts average power and hundreds of mil-
lijoules in less than a hundred nanoseconds of tunable or quasi-tunable IR
radiation.

Quite often it is not necessary to have broad-band frequency coverage for a tunable laser, especially when one is only probing the line profile of a pressure or Doppler broadened line. Figure 11 shows how a
large wavelength range can be covered by permuting the lines of a CO laser with the pressure broadened lines of a CO_2 laser. As the technology
of tunable high pressure gas lasers[20] is developed, especially the capillary[21] variety, operation at pressures high enough for complete overlap
of the rotational lines will become routine and continuous tuning will become possible. In addition, the use of CO and CO_2 lasers in a device of
this sort will automatically provide an absolute frequency calibration.
Such a room-temperature device could eliminate the need of a spin-flip
laser for broad band IR spectroscopic applications in the near future.

REFERENCES

1.   H. D. Edmunds and A. W. Smith, IEEE J. Quantum Electron. 6,
      356 (1970).

2.   R. Ludeke and E. P. Harris, Appl. Phys. Lett. 20, 499 (1972).

3.   J. A. Rossi, S. R. Chinn, and H. Heckscher, Appl. Phys. Lett.
      23, 25 (1973).

4.   A. Mooradian, in Laser Handbook, edited by F. T. Arecchi and
      E. O. Schulz-DuBois, (North-Holland Publishing Company, 1972),
      p. 1409.

5.   E. J. Johnson, private communication.

6.   M. Nakamura, A. Yariv, H. W. Yen, S. Somekh, and H. L. Gar-
      vin, Appl. Phys. Lett. 22, 515 (1973); M. Nakamura, H. W. Yen,
      A. Yariv, E. Garmire, S. Somekh, and H. L. Garvin, (to be pub-
      lished in Appl. Phys. Lett.); H. W. Yen, M. Nakamura, E. Gar-
      mire, S. Somekh, and A. Yariv, Optics Commun., 9, 35 (1973).

7.   J. M. Besson, J. F. Butler, A. R. Calawa, W. Paul, and R. H. Rediker, Appl. Phys. Lett. 7, 206 (1965).

8.   A. S. Pine, C. J. Glassbrenner, and J. A. Kafalas, IEEE J. Quantum Electron. 9, 800 (1973).

9.   M. A. Guerra, S. R. J. Brueck, and A. Mooradian, IEEE J. Quantum Electron. (December, 1973).

10. A. Mooradian, in Fundamental and Applied Laser Physics, Proceedings of the Esfahan Symposium, 1971, edited by M. S. Feld, A. Javan, and N. Kurnit, (John Wiley & Sons, 1972), p. 613, addendum.

11. S. R. J. Brueck and A. Mooradian, Optics Commun. 8, 363 (1973).

12. C. S. DeSilets and C. K. N. Patel, Appl. Phys. Lett. 23, 543 (1973).

13. S. R. J. Brueck and A. Mooradian, Appl. Phys. Lett. 18, 229 (1971).

14. S. R. J. Brueck and A. Mooradian (to be published).

15. D. L. Spears and C. Freed, Appl. Phys. Lett. 23, 445 (1973).

16. A. S. Pine (to be published).

17. G. D. Boyd, T. J. Bridges, C. K. N. Patel, and E. Buehler, Appl. Phys. Lett. 21, 553 (1972).

18. R. L. Byer, M. M. Choy, R. L. Herbst, D. S. Chemla, and R. S. Feigelson, (to be published).

19. H. Kildal and J. C. Mikkelsen, (to be published).

20. V. N. Bagratashvili, I. N. Knyazev, Yu. A. Kudryavtsev, and V. S. Letokhov, JETP Letters, (to be published); A. J. Alcock, K. Leopold, and M. C. Richardson, Appl. Phys. Lett. 23, 562 (1973).

21. P. W. Smith, ibid., and R. Abrams, ibid.

ADVANCES IN TUNABLE LEAD-SALT SEMICONDUCTOR LASERS*

I. Melngailis

Lincoln Laboratory, Massachusetts Institute of Technology

Lexington, Massachusetts 02173

ABSTRACT

As a result of recent progress in the materials and device technology of tunable lead-salt diode lasers, including the use of stripe-
geometry and heterostructures, CW output powers as high as 1/3 mW
have been obtained and CW operation has been observed at temperatures
considerably higher than previously. In addition to their already demonstrated capabilities in ultra-high resolution molecular absorption
spectroscopy, the high-power capability should extend the use of these
lasers to additional applications including saturation spectroscopy,
heterodyne radiometry and high sensitivity absorption measurements
with the spectrophone technique.

INTRODUCTION

Tunable diode lasers of the various lead-salt compounds have
already proven to be extremely valuable tools in molecular spectroscopy,
where they have been used in high-resolution absorption measurements
at wavelengths between 3.5 and 11 μm.[1] In these measurements the
diode lasers have proven to be very effective and convenient to use;
effective because of their very narrow linewidth and stability (linewidths
as narrow as 54 kHz with a frequency stability at least an order of
magnitude better than this have been measured in 10 μm PbSnTe diodes[2])
and convenient because they can be very simply and reproducibly tuned,
for example, by changing the diode current, without the necessity of
complex external optics. Wider application of these lasers in spectroscopy has in the past been limited by the relatively small CW output
powers (typically less than 1 mW) and for some of the proposed field

*This work was sponsored by the Department of the Air Force.

applications, such as air pollutant measurements, the need for operation at temperatures below 20°K has posed economic problems. As a result of recent advances in the technology of these lasers very significant progress has been made both in increasing the power level[3, 4] and achieving CW operation at higher temperatures.[5] Present results in terms of CW output power and external quantum efficiency are summarized in Table I for 4.3 μm PbS lasers and PbSnTe in the 10 - 12 μm wavelength range, both operated near liquid helium temperature. The total power is the sum of the emission from both ends of the laser while the single mode power has been measured from one end. By use of the heterostructure CW operation has been observed in PbSnTe diodes in the 10 - 12 μm range at temperatures as high as 65°K and from the results obtained we anticipate that CW operation will be possible at temperatures above 77°K for all lead-salt lasers in the 3 - 12 μm wavelength in the near future.

This paper includes a review of the tuning characteristics, a summary of the technological advances that have led to improved mode quality and increased output power, and a description of preliminary results obtained with heterostructure lasers. A number of new applications made possible by the improved laser performance are also outlined.

TUNABILITY

One of the important features of the lead salts is the possibility to produce lasers at any desired wavelength in a broad range by choosing the appropriate composition of one of the alloys. Figure 1 shows the wavelength ranges covered by the three lead-salt compounds. PbCdS covers the wavelength range between 2.5 and 4 μm, PbSSe between 4 and 8.5 μm, and PbSnTe from 6.5 to 32 μm (beyond the scale limit of Fig. 1). Indicated in the figure are also the wavelengths of absorption lines of various gases, which have so far been investigated in high-resolution absorption measurements.

The wavelength of an individual laser can be tuned either by a magnetic field, hydrostatic pressure or temperature, or by simply changing the diode current which produces a small temperature change.

The current tuning of a PbSnTe laser operating near 10 μm is shown in Fig. 2. The maximum continuous tuning range of a mode which occurs as a result of an index of refraction change with temperature is about 50 GHz ($\sim$ 2 cm^{-1}). The gross tuning which involves jumps from one cavity mode to the next results from a shift of the gain spectrum as the energy gap of the material changes. Even though this type of discontinuous tuning occurs in all diodes, recent results show that continuous coverage (avoiding gaps between adjacent modes) can be achieved.[6] Further flexibility in tuning may be achieved with the use of an external cavity.[7] Suppression of multi-mode operation and extension of the single-mode tuning range might also be achieved in the future by incorporating a distributed feedback structure[8] into the diode laser.

TABLE I

RECENT LASER PERFORMANCE

	Wavelength (μm)	Total CW Power (mW)	Single-Mode Power (mW)	External Q. E. (%)
PbS	4.3	350	52	27
$Pb_{1-x}Sn_xTe$	10-12	28	6	10

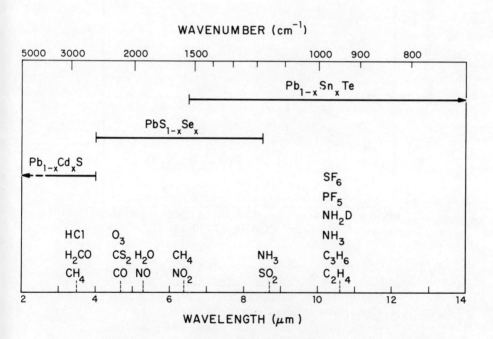

Fig. 1.   Wavelength ranges covered by three lead-salt ternary compounds. Wavelengths of principal absorption lines of various gases are also indicated.

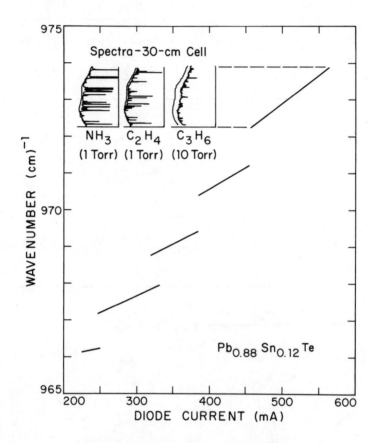

Fig. 2.   Current dependence of mode frequency for a
CW $Pb_{0.88}Sn_{0.12}Te$ diode laser operating near
liquid helium temperature. The inset shows
absorption spectra for three gases. [From
E. D. Hinkley and H. A. Pike, to be published.]

As indicated in the figure by the absorption spectra of three gases, the continuous tuning range is quite adequate for molecular spectroscopy because the Doppler broadened lines are only on the order of 100 MHz wide.

IMPROVED STRIPE-GEOMETRY LASERS

A number of important factors have contributed to the increased external efficiency and CW output power capability of the lead-salt lasers. In the materials area improvements in the closed-tube vapor growth technique[9] have recently made possible the growth of crystals with facets larger than 1 cm^2 with dislocation counts as low as 10^3 cm^2.

In the device area the DC current carrying capability has been greatly increased, lowering the contact resistance by improved contacting techniques and by improved heatsinking. The external efficiency has been increased by improving the optical quality of the end-mirrors,[3] by use of the stripe-geometry[10] to suppress parasitic modes and by using optical coatings on the end mirrors.

The structure of a stripe-geometry laser is shown in Fig. 3. Here an n-region is diffused into a p-type substrate through a stripe-opening in a SiO_2 diffusion mask. In earlier lasers where the junction extended across the entire device the devices frequently operated in parasitic bounce-modes involving all four sides of the laser, resulting in poor mode quality and low external efficiency. In the stripe-geometry structure where bounce-modes are suppressed by the lossy bulk regions along the sides of the active region, operation in a fundamental spatial mode has been observed.

The near-field (mirror illumination) patterns of a liquid helium cooled PbSnTe stripe-laser which emits near 11 μm are shown in Fig. 4a. In the direction parallel to the junction the width of the pattern corresponds approximately to the width of the stripe, while in the direction normal to the junction the spatial extent of the mode appears to be limited by the diffusion length of the injected carriers. Figure 4b shows the far-field patterns for this laser, which are in good quantitative agreement with the near-field patterns. It is interesting to note that the beam is nearly circular, in contrast to GaAs lasers where the beam is generally rectangular because of the much narrower active region in GaAs. The lasers generally operate in 1 to 4 longitudinal modes, which in most cases have been found to have a TE polarization.

In addition to the stripe geometry, a second important innovation has been the use of polishing rather than cleaving to form the reflecting end faces of the laser cavity.[3] It has been determined with the aid of a surface profile monitor that cleaved surfaces in the lead salts contain cleavage steps which are sufficiently deep (a fraction of the wavelength) and sufficiently closely spaced to very seriously degrade the optical quality of the mirrors at the laser wavelength. A technique involving sequential mechanical and etch-polishing was found to yield surfaces of

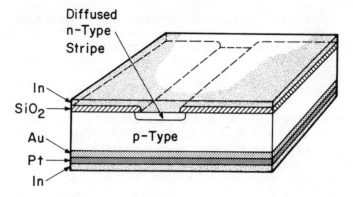

Fig. 3.   Stripe geometry lead-salt laser structure. The stripe
            width is typically 50 μm and the length is about 500 μm.
            [From R. W. Ralston et al., to be published.]

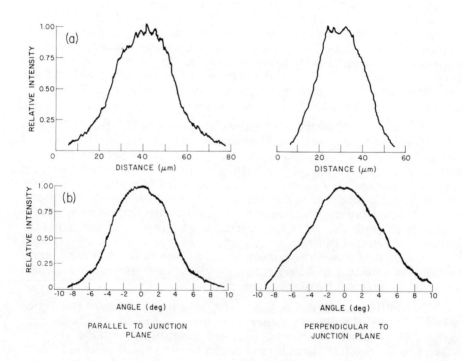

Fig. 4.   (a) Near field and (b) far field patterns of a $Pb_{0.88}Sn_{0.12}Te$
            stripe geometry laser with a 50 μm wide stripe. [From
            R. W. Ralston et al., IEEE J. Quant. Electr. QE-9, 350 (1973).]

adequate flatness.   For some PbSnTe crystals an order of magnitude higher laser output power has been obtained with polished rather than cleaved end faces.

Figure 5 shows the output power vs. current in a PbS stripe laser with polished end faces.   The upper curve is the sum of the powers emitted from the two ends of the laser, the lower curve is the single ended power, and the dotted curve corresponds to emission from one end after a reflective coating was applied to the opposite end face.   From the reflective-coating experiment an internal quantum efficiency of 44% and a loss coefficient of 15 cm^{-1} has been deduced for this laser.   The structure in the curves of Fig. 5 is probably due to changes in the mode structure.   It should be noted that a DC current of 5 A (15,000 A/cm^2) has been applied to this diode without causing noticeable degradation in output power due to heating.   This has been possible because of the low contact resistance (10 mΩ) that has been achieved by a technique involving evaporation of gold on the p-side and evaporation of indium on the n-side, and because of improved heatsinking by mounting the stripe contact directly to a heatsink.

HETEROSTRUCTURE LASERS

Lead-salt heterostructure lasers have been fabricated[5] in an effort to reduce the high-temperature threshold and to ultimately achieve CW operation above 77°K.   As previously demonstrated in GaAs-GaAlAs heterostructure lasers, the confinement of both the injected carriers and of the light wave to an active region which is bounded by epitaxially-grown regions of wider energy gap results in a substantial decrease in laser threshold.   As a first step in developing such structures in the lead salts, a single heterojunction was formed by evaporating a layer of n-type PbTe on a p-type PbSnTe substrate.   A stripe-geometry was formed by masking the surface with SiO_2 and opening stripes in the SiO_2 prior to evaporation.   The evaporation technique is similar to the one previously used in the growth of PbTe films on BaF_2 substrates.

Figure 6 shows the temperature variation of threshold for two homojunction lasers and a single-heterostructure laser.   While the low temperature threshold of the heterostructure laser is comparable to the lowest observed in homojunction devices at 77°K, the threshold is considerably lower than previously observed in homojunctions.   Even though the preliminary devices were not constructed with adequate heat-sinking (the stripe was not in contact with the heatsink) CW operation was observed up to 65°K.   Thus, CW operation at 77°K appears quite feasible with properly mounted diodes.   Also, further threshold reduction can be anticipated in double-heterostructures, which will ultimately be fabricated by epitaxial techniques.

It is also interesting to note that the single heterostructures show an abrupt change of slope in their current-voltage characteristics at threshold.[5]   This phenomenon, which is associated with a decrease in

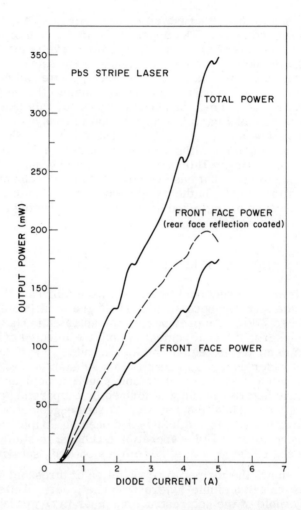

Fig. 5.   CW output power from PbS stripe geometry
             laser with polished end mirrors operating
             at 4.2°K. [From R. W. Ralston et al., to be
             published.]

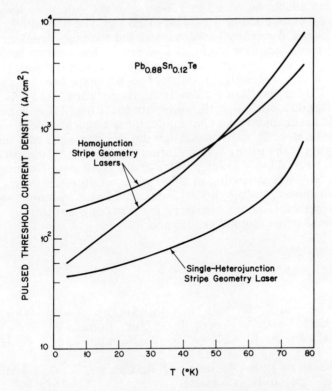

Fig. 6.   Temperature dependence of pulsed threshold current
density for $Pb_{0.88}Sn_{0.12}Te$ diode lasers: a) a single
heterostructure stripe geometry laser, b) a homojunction
stripe geometry laser with unusually low $4.2^{\circ}K$ threshold,
and c) a typical homojunction stripe geometry laser.
[From J. N. Walpole et al., Appl. Phys. Letters,
Dec. 1, 1973.]

minority carrier lifetime, has been predicted but has not been clearly
observed before in diode lasers. The ability to observe this effect is
indicative of highly efficient internal radiative recombination.

APPLICATIONS

In addition to their use in absorption spectroscopy and in a number
of air pollution measurement schemes which have been already demonstrated with the earlier low-power diode lasers,[11] the higher powers
recently obtained should now extend the use of the lasers to new areas
of application. With 10 - 100 mW powers, saturation spectroscopy of gas

vibrational levels should be possible. The use of the lasers as tunable local oscillators in conjunction with wide-band HgCdTe photodiode detectors should enable heterodyne detection and the measurement of spectra of emission from astronomical sources or from warm pollutants such as smokestack effluents or aircraft exhausts. Lasers with single-mode CW powers of a few milliwatts should be adequate for this application, since ideal local-oscillator noise limited operation has been observed with HgCdTe diodes at 10.6 μm with local oscillator powers of 1 - 2 mW.[12] In laser radar and communications a tunable local oscillator could be used to compensate for the Doppler shift of a moving target or a moving transmitter. In pollution monitoring by means of the spectrophone[13] 10 - 100 mW laser powers could yield sensitivities better than 10 ppb, corresponding to absorption coefficients as low as 10^{-6} m^{-1}. High-power tunable diode lasers could also become useful in certain applications in photochemistry for initiating or altering chemical reactions or in isotope separation.

REFERENCES

1. F. A. Blum and K. W. Nill, this volume.
2. E. D. Hinkley and C. Freed, Phys. Rev. Letters 23, 277 (1969).
3. J. N. Walpole, A. R. Calawa, R. W. Ralston, and T. C. Harman, J. Appl. Phys. 44, 2905 (1973).
4. R. W. Ralston, J. N. Walpole, A. R. Calawa, T. C. Harman and J. P. McVittie, to be published.
5. J. N. Walpole, A. R. Calawa, R. W. Ralston, T. C. Harman and J. P. McVittie, Appl. Phys. Letters, Dec. 1, 1973.
6. A. R. Calawa, private communication.
7. J. A. Rossi, S. R. Chinn, and H. Heckscher, Appl. Phys. Letters 23, 25 (1973).
8. M. Nakamura, H. W. Yen, A. Yariv, E. Garmire, S. Somekh and H. L. Garvin, Appl. Phys. Letters 23, 224 (1973).
9. T. C. Harman and J. P. McVittie, Journal of Electronic Materials, to be published.
10. R. W. Ralston, I. Melngailis, W. T. Lindley and A. R. Calawa, IEEE J. Quantum Electronics QE-9, 350 (1973).
11. E. D. Hinkley, Opto-Electronics 4, 69 (1972).
12. B. J. Peyton, A. J. DiNardo, G. M. Kanischak, F. R. Arams, R. A. Lange and E. W. Sard, IEEE J. Quantum Electronics QE-8, 252 (1972).
13. L. B. Kreutzer and C. K. N. Patel, Science 173, 45 (1971).

HIGH PRESSURE WAVEGUIDE GAS LASERS

P. W. Smith

Bell Telephone Laboratories

Holmdel, New Jersey   07733

Waveguide gas lasers are lasers within which the light is guided by a hollow dielectric waveguide which also serves to confine the discharge. Because these lasers operate at high gas pressures and thus have relatively broad oscillation bandwidths, they are of interest as potential sources for laser spectroscopy.

Let us begin by describing the concept of a hollow dielectric waveguide. Figure 1 compares the hollow dielectric waveguide with a conventional dielectric fiber waveguide. A conventional waveguide has the refractive index of the core higher than that of the cladding. Light guidance in the core takes place by total internal

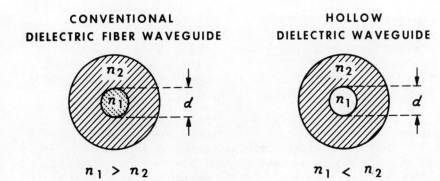

Fig. 1 Comparison of hollow dielectric waveguides and conventional
dielectric waveguides.

reflection, and the diameter of the core, d, can be of the order
of the wavelength of the guided radiation for low-loss guiding.
In a hollow dielectric waveguide, the refractive index of the core
is lower than that of the cladding, and guidance takes place by
grazing incidence reflections. In order to obtain low-loss guiding,
it is usually necessary for d, the diameter of the core, to be of
the order of 100 wavelengths or more. In theory, for sufficiently
straight and smooth hollow dielectric waveguides, however, low-
loss modes can propagate. Marcatili and Schmeltzer[1] showed that,
in theory, for a 2mm-bore glass waveguide, losses at λ = 1μm will
be less than 2dB/km.

The interest for gas lasers stems from the fact that for most
gas lasers, the electron temperature in the positive column of the
gas discharge is determined by the product of the gas pressure and
the discharge tube diameter (Pd). At constant current density,[2]
discharge conditions are reproduced by maintaining Pd constant.
As (for a Doppler-broadened line) the laser gain for constant
excitation is proportional to the number of excited atoms, we have

$$P \propto 1/d \tag{1}$$

and

$$\text{gain} \propto 1/d . \tag{2}$$

Thus small-diameter laser discharge tubes should allow high-pressure
and high-gain laser operation. The limit for conventional lasers
comes from the requirement that the fundamental resonator mode be
able to pass through the laser tube without suffering appreciable
diffraction loss. By using a hollow dielectric waveguide both to
confine the discharge and to guide the laser light, however, we
are no longer limited by free-space propagation diffraction losses,
and much smaller discharge tube diameters become possible.

We have performed experiments to determine the waveguide
losses achievable with selected specially-straightened lengths of
precision-bore pyrex capillary tubes. Using 30-cm lengths of
tubing we measured waveguide losses of less than 10% per meter,
both for 430μm-bore tubes at λ = 0.63μm and for 1000μm-bore tubes
at λ = 10.6μm.

Figure 2 shows the set-up used to make these measurements.
The fundamental-transverse-mode beam from a conventional laser is
matched into the hollow dielectric waveguide with a matching lens.
Experimentally, it is found that the optimum waveguide transmission
occurs when the Gaussian beam waist is at the entrance face of the
waveguide, and the Gaussian beam radius, w_0, is related to the
radius of the waveguide, a, by the relationship

$$w_0 \approx 0.6a . \tag{3}$$

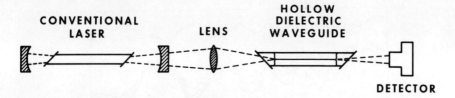

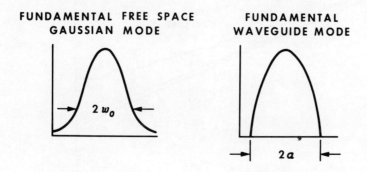

Fig. 2   Matching a free-space Gaussian beam into a hollow dielec-
         tric waveguide.

Abrams[3] has computed the theoretical coupling efficiency for
this situation and has shown that for a matching relationship
somewhat different from Eq. 3, 98% of the incident light can be
coupled into the fundamental waveguide mode.

In order to make a laser, it is necessary to form a resonator. This can be most conveniently done using conventional
laser mirrors - either flat mirrors directly on the ends of the
hollow dielectric waveguide, or curved mirrors spaced from the
ends of the waveguide so that the curvature of the mirror matches
the phase-front curvature of the beam leaving the waveguide. Calculations of mode losses for this type of resonator have been made
by several authors.[3,4,5] Figure 3 shows the round-trip loss as a
function of the Fresnel number for a waveguide laser with external
curved mirrors (see Ref. 6). Note the large loss discrimination
between the lowest-order (EH_{1m}) and higher-order resonator modes.

In 1968, Steffen and Kneubühl[7] introduced hollow dielectric
waveguide modes to explain the transverse mode frequencies they
observed with a submillimeter-wave laser. The first experiments
to clearly demonstrate waveguide gas laser action were made with
a 6328Å He-Ne laser system in early 1971.[8] Since that time waveguide gas laser operation has been observed with a He-Ne system

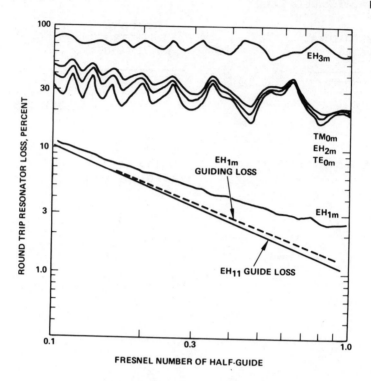

Fig. 3 Computed round-trip resonator loss as a function of resonator Fresnel number for a waveguide laser with external
curved mirrors[6]. The lowest two curves show the loss
within the guide for the lowest-order guide mode (EH_{11})
and the lowest order resonator mode (EH_{1m}).

at $3.39\mu m$[9], the CW CO_2 system at $10.6\mu m$[10-14], and the CO system at
$5\mu m$[15]. Recently, experiments have been reported with a He-Xe
system at $3.5\mu m$[16], and an optically-pumped far-infrared laser in
the range of $70-200\mu m$[17].

The first waveguide gas laser[8] was made with a specially-
selected 20-cm length of $430\mu m$-bore pyrex capillary tubing. The
resonator consisted of two 30-cm mirrors spaced 29.5 cm from the
ends of the waveguide. The free-space resonator (two 30-cm
mirrors spaced by 79 cm) is unstable and has no low-loss modes.
Using a mixture of He and Ne in the discharge tube, laser action
was obtained at $6328\mathring{A}$ and it was determined[8] that the total round-
trip loss in the resonator was less than 1%. This low-loss
operation was obtained because of the waveguiding action of the
capillary tube used to confine the discharge. Figure 4 illustrates

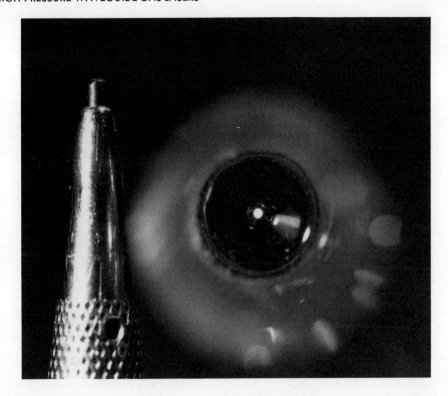

Fig. 4   A cross-sectional view of the waveguide laser tube used
         in Ref. 8. The bright spot in the center is the bore of
         the tube, and the tip of a mechanical pencil is shown for
         size comparison.

the small size of this laser tube by comparing the bore with the
lead in a mechanical pencil. The intensity distribution of output beam of this waveguide gas laser was similar to that of a
conventional fundamental transverse mode laser as shown in Fig. 5.

     The optimum gas pressure for the 6328Å He-Ne waveguide laser
described here was about 7 Torr. At this pressure the pressure-
broadened homogeneous linewidth is of the order of the Doppler
broadening. The gain curve is thus partially homogeneously
broadened. Figure 6 shows the output frequency spectra observed
under these conditions. The frequency spectrum was bistable and
alternated between the two spectra shown. At higher gas filling
pressures, only the single-frequency spectrum was observed. When
the length of the laser resonator was changed, the single-frequency
output would only shift by one half of the longitudinal mode
spacing from line center before the adjacent mode closer to line

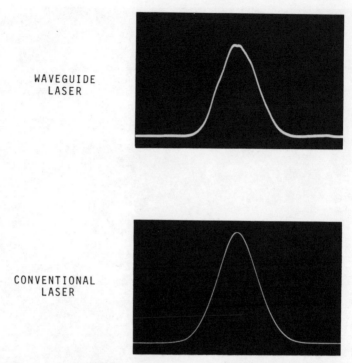

Fig. 5   Measurements of the intensity distribution in the far
         field output beam of a waveguide laser, and a conventional
         fundamental-transverse-mode laser.

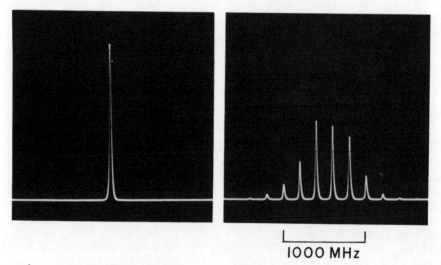

Fig. 6   Output frequency spectra observed with a high-pressure
         He-Ne waveguide laser.[8]

center would begin to oscillate, and the original mode would be suppressed. This strong mode competition, for which oscillation only takes place on the resonator mode closest to line center and all other modes are suppressed, is due to the homogeneous nature of the gain curve broadening.

This pressure broadening is disadvantageous for the He-Ne laser, however, for as the gas pressure is increased above ~10 Torr, the width of the gain profile begins to increase linearly with pressure. Under these conditions the peak gain will no longer increase linearly with pressure, as predicted by Eqs. 1 and 2, but will remain constant. As the peak gain obtained at this pressure with the 6328Å He-Ne system is not very large (2.7dB/m) this limits the interest in a 6328Å He-Ne waveguide laser.

We will now turn our attention to some other waveguide laser systems. Figure 7 illustrates the effect of the dispersion associated with a laser transition on the frequencies of the laser modes. For most lasers the deviation of the laser frequency

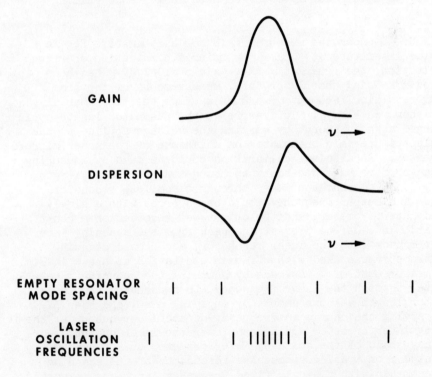

Fig. 7   The dispersion associated with the laser gain will cause the laser oscillation frequencies to be different from the frequencies of the modes of the passive resonator.

spacings from the empty resonator mode spacings is small. Near
line center, the laser frequency spacing, $\Delta\nu$ is given by

$$\Delta\nu = (c/2L)(1+\beta)^{-1} \qquad (4)$$

where

$$\beta = K(\ell/L)(cg/\Delta\nu_G) \qquad (5)$$

Here $c/2L$ is the empty resonator mode spacing; K is a numerical
factor that depends on the shape of the gain curve and has a
value of ~ 0.2; ℓ is the length of the laser medium and L the
length of the resonator; g is the saturated gain, and $\Delta\nu_G$ is the
width of the gain curve. From Eqs. 4 and 5 we find that for most
gas lasers $\beta < 0.01$. For the He-Xe laser transition, however,
the combination of high gain and narrow linewidth gives $\beta \sim 3$ for
conventional He-Xe lasers.[18] By using a high-gain waveguide He-
Xe laser, we have recently observed much larger mode pulling
effects.[16]

Figure 8 shows the measured gain for a 250μm-bore He^3-Xe^{136}
waveguide laser tube at 3.5μm. By using a combination of rf and
dc excitation, the remarkably high gain of 1000 dB/m has been
achieved.[16] Using this laser tube, measurements of the mode
spacing reduction factor, β, were made using the experimental
set-up shown in Fig. 9. To measure β, the resonator length could
be changed a known amount by varying the voltage applied to the
piezoelectric ceramic translator on which one of the laser mirrors
was mounted. An absolute calibration could be made by measuring
the change in voltage necessary to change the resonator length
by $\lambda/2$, i.e., to tune the laser through a complete resonator
period. The output frequency could be observed with a slowly
scanned interferometer, and the output was recorded on a chart
recorder. In order to prevent feedback from the scanning interferometer back into the laser resonator, a confocal scanning
interferometer was used with off-axis excitation as shown in Fig.
9. A measurement of β was made by scanning the scanning interferometer through the laser frequency, then changing the laser
resonator length a known amount by applying a voltage step ΔV,
and observing the change in frequency of the laser output as the
scanning interferometer again scans over the laser output. The
frequency scale of the scanning interferometer output can be
calibrated from tracing successive resonances of the scanning
interferometer with a single fixed frequency input and measuring
the interferometer length with a ruler.

The lower part of Fig. 9 shows a typical result. The output

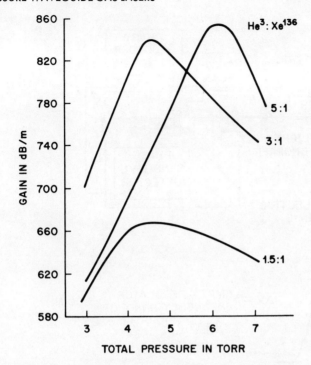

Fig. 8   Gain of a 250μm-bore dc-excited He^3-Xe^{136} laser as a
function of gas pressure and mixture.[16]

frequency was observed to shift only 23 MHz when the resonator
frequency was changed by 800 MHz, which indicates a β of 34. The
maximum β measured at low power levels was β = 50. This agrees
very well with the theoretical value of β = 48 computed from Eq. 5
using parameters appropriate for this experiment.[16] It was also
observed that under these conditions only the single laser mode
closest to line center would oscillate, and mode switching would
occur as the laser was tuned away from line center. Thus we have
achieved a single-frequency laser oscillator which, without any
external electronics, would always oscillate within ±(1/100)(c/2L)
of the center of the net gain curve.

A number of groups are currently studying CW CO_2 waveguide
lasers[10-14] operating at 10.6μm. Figure 10 shows the construction
of a typical flowing-gas He-N_2-CO_2 waveguide laser. The tube uses
a conventional Kovar cathode and platinum tab anode, and has
provisions for circulating a coolant around the waveguide discharge section of the tube. Figures 11 and 12 show the results
obtained with this tube.[14] The gas pressure plotted is the average

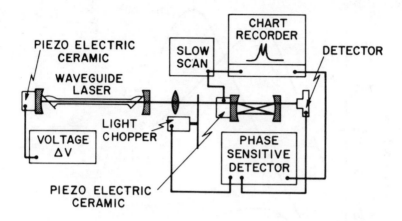

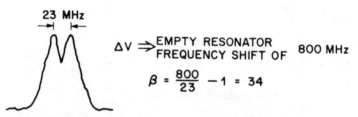

Fig. 9 Schematic diagram of the apparatus used to measure dis-
        persive mode pulling in a waveguide He-Xe laser.[16] The
        lower figure shows how the mode spacing reduction factor,
        β, is computed from the experimental data.

of the inlet and outlet pressures. This has been shown to be a
good approximation to the actual pressure in the waveguide region.[21]
Note the improvement in gain and output power that results from
flowing the gas through the laser tube. Using a cooled 5-cm
pyrex bore discharge tube, gains as high as 37 dB/m have been
measured.[10] Laser tubes of this type have been operated at
pressures as high as 1/2 atmosphere.[10] Abrams and Bridges have
reported on the operation of sealed-off waveguide CO_2 lasers.[12,19]

As a demonstration of the enhanced bandwidth capabilities of
these high-pressure waveguide CO_2 lasers, mode-locking experiments
were performed using an internal, Brewster's angle, germanium
acoustic light modulator. Figure 13 shows that with the highest
gas pressure (and thus largest gain-linewidth) operation, pulses
as narrow as 3 ns could be obtained.[20] Using a shorter resonator,
Abrams has achieved mode-locked pulsewidths of 2 ns.[21] The
narrowest pulses previously achieved with a conventional mode-
locked CW CO_2 laser have been approximately 20 ns in duration.

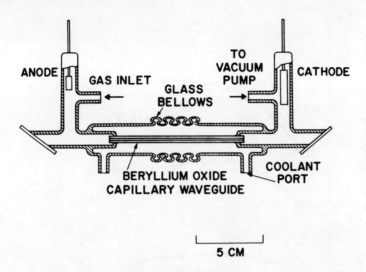

Fig. 10 A beryllium oxide waveguide CO_2 laser tube.[14]   The waveguide bore is 1 mm.

The solid line in Fig. 13 is a theoretical curve computed
using the theory of Kuizenga and Siegman for the mode-locked
pulsewidth of a homogeneously-broadened laser with internal modulation.[22]   They find that the pulsewidth, $\Delta\tau$, is related to the
gain linewidth, $\Delta\nu_G$, by the equation,

$$\Delta\tau = k/\Delta\nu_G^{1/2} \tag{6}$$

where k is group of parameters which depend on the experimental
set-up. For these experiments we compute $k = 9.8 \times 10^5$ sec^{-1}
in good agreement with the value $k = 8.1 \times 10^5$ sec^{-1} which gives
the best fit with the data in Fig. 13.

Recently, Provorov and Chebotaev[23] have described experiments
with a 2.5-mm bore CO_2 waveguide laser. They reported CW operation at pressures as high as 1 atmosphere. By using a mixture
of $C^{12}O_2^{16}$, $C^{13}O_2^{16}$ and $C^{14}O_2^{16}$ isotopes, they compute that gain
bandwidths of ~15 GHz can be achieved at many points in the
spectral region between 9 and 11μm.

The last type of waveguide laser that we will discuss is a

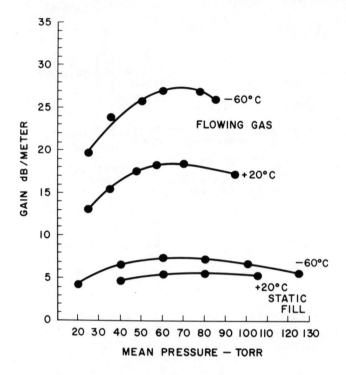

Fig. 11 Measured gain for the waveguide CO_2 laser tube of Fig. 10.

high-pressure transverse-discharge (TEA) waveguide laser. Figure
14 illustrates the construction of this type of waveguide laser.

     A hollow waveguide of 1-mm square cross-section and 6-cm
length is made from slabs of fused quartz and metal in such a way
that multiple transverse discharges from a large number of elec-
trodes can be formed inside the waveguide. The cathode, consisting
of a solid block of polished copper, comprises one wall of the
waveguide, and the anodes, thin (2 μm) copper strips coated on a
fused quartz substrate, make up the opposite waveguide wall. The
remaining waveguide walls are formed by fused quartz slabs whose
edges have been polished. The ends of the waveguide can be
closed by NaCl Brewster's angle windows positioned so as to provide minimum loss for linearly-polarized modes with the electric
field oriented parallel to the two metal walls of the waveguide.
We have recently performed experiments on this type of waveguide
TEA laser using two 8-cm radius curvature mirrors spaced 8 cm
from the ends of the waveguide to complete the laser resonator.
Pulsed transverse excitation was provided by simultaneously dis-
charging individual capacitors attached to each anode strip. A
mixture of CO_2, N_2, and He entered the laser through an inlet

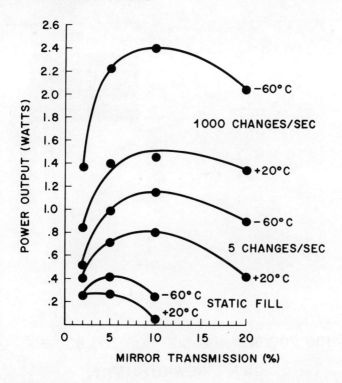

Fig. 12   Measured power output using the waveguide CO_2 laser tube
          of Fig. 10.

hole in the copper cathode, flowed down the waveguide, and exited
through a second hole in the cathode as shown in Fig. 14.

This laser has been operated with filling pressures as high
as one atmosphere.[24]   The peak output power was 100 W with a
pulsewidth of 200 ns.   The laser could be operated at repetition
rates as high as 4 kHz (this limit being set by the external
electronics) with no discernible decrease in gain or output power.
In fact, as the repetition rate was raised above ∼ 400 Hz the
discharge appeared more diffuse, and time and amplitude jitter in
the output were greatly reduced.   This marked change in discharge
character at higher repetition rates is attributed to "preionization" of the laser medium by the previous discharge pulse.   The
ionization remaining in the discharge medium from preceding pulses
conditions the medium so that subsequent electrical pulses have a
higher probability of initiating a diffuse glow discharge.

A major problem associated with TEA lasers is that of ensur-
ing that their discharges be uniform (no temperature or index
gradients) and stable (minimum tendency to degenerate into a

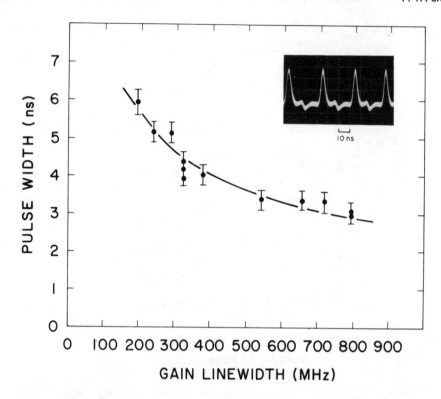

Fig. 13   Mode-locked pulsewidth as a function of laser gain
          bandwidth for an internally mode-locked waveguide CO_2
          laser.[20]   The insert shows a typical pulse train.   The
          small blips between the main pulses are due to cable
          reflections.

narrow constricted arc).   In this and other respects the waveguide
TEA laser has advantages over conventional TEA lasers.   Firstly,
due to the waveguiding action of the walls, the device is insensitive to bright arcs which tend to deflect the laser beam in conventional TEA laser resonators - thereby causing high losses.
Secondly, due to preionization from previous discharge pulses a
more diffuse and homogeneous discharge is insured.   Finally, due
to its compactness, excitation voltage requirements are reduced
by 1 to 2 orders of magnitude from conventional TEA lasers.
Because of the unique properties of this device it should be
possible to operate at extremely high repetition rates and at
pressures much higher than one atmosphere.   Work is proceeding
in these two directions.

     Operation of CO_2 lasers at high pressures is of particular
interest for spectroscopy applications because at a pressure of

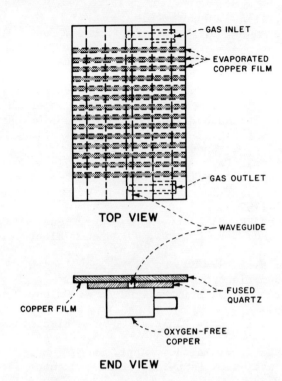

Fig. 14   Schematic diagram of a waveguide TEA laser.

~7 atmospheres, the pressure-broadened gain curves corresponding
to the various rotational-vibrational transitions (P_{20}, P_{22}, etc.)
overlap to give almost continuous gain from 9 - 11μm.

At present, longitudinal discharge CW waveguide CO_2 lasers
have been operated at pressures as high as 1 atmosphere - giving
a tuning range of ~4GHz. In the future, using a mixture of isotopes, tuning ranges of ~15GHz should be achieved.[23] With pulsed
TEA waveguide CO_2 lasers, 1 atmosphere, 4GHz operation has been
achieved,[24] and we are presently attempting to extend operation to
the 10 atmosphere range where continuous tuning from 9 - 11μm
should be possible.

REFERENCES

1.   E. A. J. Marcatili and R. A. Schmeltzer, Bell System Tech. J.
     43, 1783 (1964).
2.   See, for example, C. S. Willett, "Handbook of Lasers", CRC
     Press, Cleveland, 1971, p. 183.

3. R. L. Abrams, IEEE J. Quantum Electron. QE-8, 838 (1972).
4. A. N. Chester and R. L. Abrams, Appl. Phys. Lett. 21, 576
    (1972).
5. J. J. Degnan and D. R. Hall, IEEE J. Quantum Electron. QE-9,
    to be published.
6. R. L. Abrams and A. N. Chester, Presented at the 1973 Spring
    Meeting of the Optical Society of America, Paper Th D16: to
    be published. The author would like to thank Dr. Abrams for
    permission to use Fig. 3 prior to its publication.
7. H. Steffen and F. K. Kneubühl, Phys. Lett. 27A, 612 (1968).
8. P. W. Smith, Appl. Phys. Lett. 19, 132 (1971).
9. P. W. Smith and P. J. Maloney, unpublished results.
10. T. J. Bridges, E. G. Burkhardt, and P. W. Smith, Appl. Phys.
    Lett. 20, 403 (1972).
11. R. E. Jensen and M. S. Tobin, Appl. Phys. Lett. 20, 509 (1972).
12. R. L. Abrams and W. B. Bridges, J. Quantum Electron., to be
    published.
13. J. J. Degnan, H. E. Walker, J. H. McElroy, and N. McAvoy,
    IEEE J. Quantum Electron. QE-9, 489 (1973).
14. E. G. Burkhardt, T. J. Bridges, and P. W. Smith, Opt. Comm.
    6, 193 (1972).
15. R. Yusek and G. Lockhart, IEEE J. Quantum Electron. QE-9, 694
    (1973).
16. P. W. Smith and P. J. Maloney, Appl. Phys. Lett. 22, 667 (1973).
17. D. T. Hodges and T. S. Hartwick, IEEE J. Quantum Electron.
    QE-9, 695 (1973).
18. Lee Casperson and Amnon Yariv, Appl. Phys. Lett. 17, 259 (1970).
19. R. L. Abrams, the following paper at this conference.
20. P. W. Smith, T. J. Bridges, E. G. Burkhardt and O. R. Wood,
    Appl. Phys. Lett. 21, 470 (1973).
21. R. L. Abrams, private conversation.
22. D. J. Kuizenga and A. E. Siegman, IEEE J. Quantum Electron.
    QE-6, 694 (1970).
23. A. S. Provorov and V. P. Chebotaev, Doklady Akademii Nauk
    USSR 208, 318 (1973).
24. P. W. Smith, P. J. Maloney, and O. R. Wood II, Appl. Phys.
    Lett., to be published.

WIDEBAND WAVEGUIDE CO_2 LASERS*

Richard L. Abrams

Hughes Research Laboratories

Malibu, California 90265

INTRODUCTION

The development[1] of the He:Ne waveguide gas laser and its extension[2] to small, high pressure, sealed-off waveguide CO_2 lasers allows us to take advantage of the CO_2 pressure broadened gain bandwidth for mode locking and frequency tuning experiments. As the operating pressure is increased above 10 Torr, the linewidth increases from the 50 MHz Doppler broadening at a rate of ~5.3 MHz/Torr. At pressures above several hundred Torr, it is clear that tuning ranges in the GHz region are possible and nanosecond mode-locked pulses may be produced. CO_2 lasers with these characteristics have been shown to be useful in laser spectroscopy and are essential for CO_2 laser communications.

In this paper we review briefly what may be expected from sealed-off waveguide CO_2 lasers as the discharge diameter is decreased (with a corresponding increase in operating pressure). Gain, output power, and efficiency data are presented for several waveguide CO_2 laser devices. We then present recent results on mode locking and frequency tuning experiments designed to take advantage of the increased gain bandwidth available from these high pressure lasers.

---

*The research reported in this paper was sponsored in part by the Air Force Cambridge Research Laboratories and the Advanced Research Projects Agency, monitored by the Office of Naval Research.

WAVEGUIDE CO_2 LASER CHARACTERISTICS

Waveguiding allows us to extend measurements of laser medium characteristics to smaller diameter discharges without being limited by the high optical losses that would attend simple free-space diffraction-limited wave propagation in such tubes. Konyukhov[3] has given the most logical development of the similarity laws applicable to CO_2 lasers. His results were derived for a Doppler broadened molecular laser, but have been extended by Abrams and Bridges[2] for the pressure broadened case of interest here. Briefly, the waveguide CO_2 laser will have the following characteristics as the discharge tube diameter D is decreased:

- The optimum gas pressure increases
- Tube current decreases as D
- The tube voltage increases as 1/D
- The gain remains constant
- The power per unit length remains constant
- The efficiency remains constant
- The saturation flux increases as $1/D^2$.

Of course, the above assumes that the optical losses are independent of discharge tube diameter. This assumption is less true for smaller diameters in a waveguide laser than for a conventional laser, but eventually breaks down for a waveguide laser also. For a discussion of waveguide losses for various materials, see Ref. 2.

It is clear from the above discussion that power per unit volume is not useful as a scaling parameter in the same sense that it is used in TEA lasers, for example. High values of P/V obtained from small diameter lasers cannot be scaled to large volumes. Power per unit length is the proper quantity for scaling waveguide CO_2 lasers (or, for that matter, all wall-cooling-dominated molecular lasers).

Measurements of gain, output power, and efficiency on sealed-off capillary bore CO_2 lasers were reported in Ref. 2, and will be reviewed briefly here. Fig. 1 shows the measured gain versus total pressure for an 18 cm x 1.5 mm ID BeO laser tube. The tube had an all metal-ceramic construction with CdTe Brewster windows epoxied on. For mixtures rich in CO_2, the gain peaks at lower pressures. As the He concentration is raised, the gain peaks at higher pressures, but has a lower value. Mixtures which are very lean in CO_2 show a gain almost independent of pressure, implying that the inversion ratio is also independent of pressure for these mixtures. Extraction of laser power depends also on the saturation intensity which has been shown to increase as the square of the pressure.[2] The maximum available power for the tube described here occurs at a total gas pressure of about 150 Torr and significant output should be attainable at a pressure of 300 Torr.

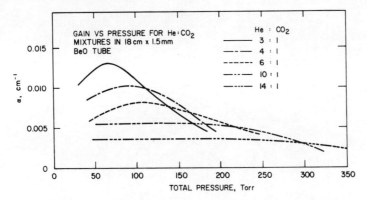

Fig. 1. Gain versus total pressure for various sealed-off He:CO$_2$
         mixtures in an 18 cm x 1.5 mm BeO tube.

We have made output power measurements with a variety of gas
mixtures and pressures for the 18 cm x 1.5 mm ID tube discussed
above as well as for a 7.5 cm x 1.5 mm square tube (which we call
a channel laser). These results are discussed below.

CIRCULAR WAVEGUIDE LASER

The 18 cm x 1.5 mm ID BeO tube was inserted into a resonator*
formed by a 23 cm radius of curvature total reflector spaced 22 cm
from one end of the laser and a 14 cm radius of curvature, 95%
reflecting mirror positioned 10 cm from the opposite end. Laser
output power versus total pressure for the various He:CO$_2$ mixtures
is shown in Fig. 2. The peak laser output clearly occurs at pressures well above the optimum pressure for peak gain, demonstrating
the effect of increasing saturation parameter with pressure.

We have seen[2] that high pressure operation can be realized
with helium-rich mixtures and that discharge voltage can be reduced
by adding Xe. In Fig. 3 the laser output power and discharge
voltage versus total pressure are shown for a He:CO$_2$:Xe mixture of
14:1:0.25. The maximum power is reduced because the gain is reduced
to the point where losses are important, but laser action is demonstrated out to 320 Torr. For these latter measurements, the 95%
mirror was replaced with a 98% reflecting mirror with the same radius
of curvature.

----

*The design of optimum resonators for waveguide lasers is discussed
 in the papers by Abrams and Chester.[4]

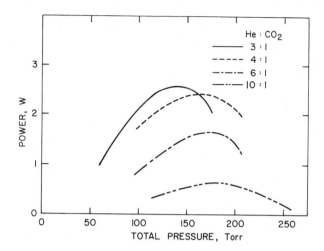

Fig. 2.   Output power versus total pressure for He:CO_2 mixtures in
          18 cm x 1.5 mm BeO tube.

Recent measurements with the same laser using higher quality
CdTe windows gave a power output of 4.0 Watts at 150 Torr with a
gas mixture of He:CO_2:N_2:Xe = 4:1:0.5:0.25. Under these conditions
the laser efficiency was 7.1% comparing favorably with larger
diameter tubes of the same gain as we expect from the scaling laws.
This corresponds to a power per unit length of 0.22 W/cm of length.

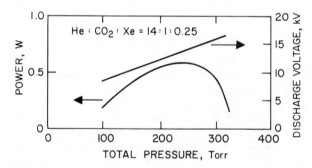

Fig. 3.   Laser output power and discharge voltage versus total
          power for mixture of He:CO_2:Xe = 14:1:0.25. For these
          measurements, laser output mirror reflectivity is 98%.

CHANNEL LASER

Hollow waveguide lasers may also be operated in non-circular
guides. In particular, we have experimented with square waveguides
machined in Al-300 alumina. The advantages of this structure include
ease of manufacture and a convenient geometry for internal modulator
applications. An exploded view of such a 7.5 cm long channel laser
is shown in Fig. 4. One mirror is totally reflecting and the other
has 3% transmission. The mirrors are both flat. The center electrode
is hollow and acts both as cathode and gas filling port. The pins
at either end are the anodes.

The output power versus total pressure for several $He:CO_2$ gas
mixtures in the three-inch channel laser is shown in Fig. 5. With
a mixture of $He:CO_2:N_2:Xe$ = 4:1:0.5:0.25, an output power of 1.4 W
at 6% efficiency was obtained at 150 Torr and 0.3 W was observed at
350 Torr. Thus a power per unit length of 0.19 W/cm was achieved.
This compares with 0.22 W/cm obtained from the longer 18 cm x 1.5 mm
ID laser.

It is clear from the above results that reasonably large amounts
of cw power at good efficiency are available from rather small sealed-
off waveguide CO_2 lasers. Operating pressures up to 320 Torr have
been demonstrated where the pressure broadened linewidth is greater
than 1.8 GHz. In the next sections we discuss experiments designed
to exploit this large bandwidth.

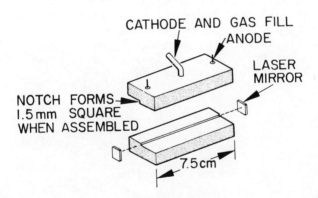

Fig. 4. Exploded view of three-inch 1.4 Watt channel laser.

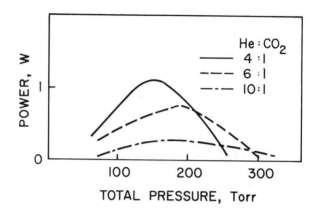

Fig. 5.   Laser output power versus total pressure for He:CO_2
mixtures in 7.5 cm channel laser.

FREQUENCY TUNING EXPERIMENTS

Frequency tuning of the waveguide CO_2 laser is performed by
varying the length of the laser resonator which tunes the resonator
mode over one free spectral range. The tuning range is thus limited
to c/2L in addition to limitations set by the laser bandwidth and the
tendency of the CO_2 laser to change transitions as the length is
tuned. A schematic drawing of our laser is shown in Fig. 6. The
150 ℓ/mm diffraction grating (97% efficiency) is mounted on a bender
bimorph and moves linearly when a voltage is applied to the bimorph.
An applied voltage of 22 V is sufficient to translate the bimorph
5.3 µm or one free spectral range. The gas discharge region is contained in a 9.5 cm x 1.5 mm ID BeO tube which is conductively cooled
to a heat sink. It is terminated at the opposite end by a 97%
reflecting output mirror. The tube is statically filled with 150 Torr
of a He:CO_2:N_2:Xe = 4:1:0.5:0.25 mixture.

Fig. 6.   Schematic drawing of tunable laser.

Figure 7 shows the output of the tunable laser on P(20) of the 10.6 μm transition as a function of cavity length (controlled by the bimorph voltage). The peak output is 0.6 W. The free spectral range is 1500 MHz and we see that the laser tunes over 620 MHz which is many times the Doppler width of 50 MHz, but a little less than the 800 MHz pressure broadened linewidth. Fixed losses including waveguide loss, mirror output coupling, and grating loss are presently limiting the output tuning range. In the future we expect to increase the tuning range above 1.5 GHz.

We have performed a heterodyne experiment where the laser was mixed with a conventional CO_2 laser stabilized to line center on P(20) (of the 10.6 μm transition). The IF beam was observed on a spectrum analyzer. The beat frequency was tunable over ±310 MHz and appeared as a narrow spike (<100 kHz) on the spectrum analyzer, typical of conventional CO_2 laser beats. The frequency stability of the lasers was limited by power supply ripple (∿1%) and the poor acoustic environment of the laboratory. There is every indication that waveguide CO_2 lasers will exhibit the same frequency stability that has been demonstrated for conventional CO_2 lasers.[5]

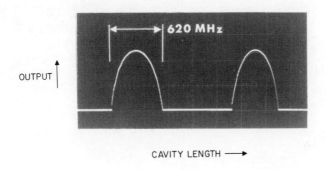

Fig. 7. Laser output versus cavity length.

MODE-LOCKING EXPERIMENT

Mode locking of a waveguide CO_2 laser was first demonstrated by Smith et al[6] in a flowing gas system. They produced a cw train of mode-locked pulses with a 3 nsec pulse width. We report here cw mode locking of a sealed-off waveguide CO_2 laser using a germanium acoustooptic loss modulator similar to that used for mode locking a TEA CO_2 laser.[7]

The experimental arrangement is shown in Fig. 8. The optical
length of the resonator is 83 cm resulting in a mode-locking frequency of 180 MHz. The 18 cm x 1.5 mm ID BeO tube is the same laser
used for gain and power measurements earlier in this paper. It is
filled with 150 Torr of the optimized He, CO_2, N_2, Xe mixture. With
the germanium acoustooptic loss modulator in the cavity, the laser
output power drops to 0.8 W because of the high insertion loss of
the modulator (6%). When 0.3 W of 90 MHz RF is applied to the modulator, the cw laser power drops to 0.5 W and a train of mode-locked
pulses appears at the output.

The pulses are detected with a S.A.T. HgCdTe photodiode with a
frequency response greater than 500 MHz and displayed on a fast
oscilloscope. The observed pulses are shown in Fig. 9. The pulse
repetition frequency is 180 MHz and the observed pulsewidth is
2.1 nsec. Only three or four modes are being locked due to the high
insertion loss of the modulator. With a lower loss modulator, the
round trip resonator losses would be reduced allowing more modes to
be above threshold and resulting in a corresponding reduction in
pulse width.

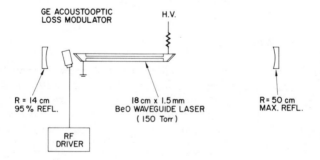

Fig. 8. Experimental arrangement for mode-locking of the waveguide
        CO_2 laser.

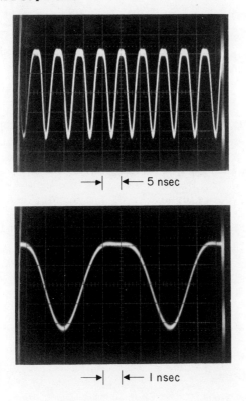

Fig. 9. Observed Mode-locked laser pulses.

SUMMARY

We have shown that efficiency and reasonably high power operation are possible in remarkably small, high pressure, sealed-off
capillary bore CO_2 lasers. The increased bandwidth available in
these lasers, made available by pressure broadening of the laser
transition, allows improved performance in mode locking and frequency
tuning experiments. We have demonstrated frequency tuning tunability
of 620 MHz and have formed mode-locked pulses with a 2.1 nsec pulse
width. In the future we expect to frequency tune a waveguide CO_2
laser over 1.5 GHz and to demonstrate a continuous train of nanosecond mode-locked pulses.

REFERENCES

1.   P. W. Smith, Appl. Phys. Lett. 19, 132 (Sept. 1971).
2.   R. L. Abrams and W. B. Bridges, IEEE J. of Quantum Electronics
     8 (Sept. 1973), (to be published).
3.   V. K. Konyukhov, Soviet Physics-Technical Physics 15, 1283
     (1971).
4.   R. L. Abrams, IEEE J. of Quantum Electronics 8, 838 (1972);
     A. N. Chester and R. L. Abrams, Appl. Phys. Lett. 21, 576
     (1972); R. L. Abrams and A. N. Chester, 1973 Spring Meeting of
     the Optical Society, paper ThD16 (March 1973).
5.   C. Freed, IEEE J. of Quantum Electronics 4, 404 (1968).
6.   P. W. Smith, T. J. Bridges, E. G. Burkhardt, and O. R. Wood,
     Appl. Phys. Lett. 21, 470 (1972).
7.   O. R. Wood, R. L. Abrams, and T. J. Bridges, Appl. Phys. Lett.
     17, 376 (1970).

FREQUENCY STABILIZATION OF A CW DYE LASER

R. L. Barger

Quantum Electronics Division
National Bureau of Standards, Boulder, CO   80302

A large improvement in the frequency stability of a single-mode cw dye laser has been obtained[1] by using a wide-band (0-100 kHz) servo system to lock the dye laser frequency to a transmission fringe of a high-finesse optical cavity. By applying corrections to the optical length of the dye laser cavity, the free-running laser stability of ±50 MHz (for one second averaging time) was reduced to drifts of about 1 - 2 MHz/minute (corresponding to controllable drifts of optical cavity temperature and possibly of optical coupling parameters). The frequency has been made to track these drifts in a single cavity with errors less than 50 kHz rms for short times (20 μsec) and less than 100 Hz for long times (10 sec).

The dye laser whose frequency stabilization is reported here is similar to the three-mirror astigmatically compensated dye laser of Dienes, Ippen and Shank,[2] and uses a quartz-window dye cell. Single-mode operation was obtained by using two intracavity Brewster-angle prisms, two uncoated solid etalons with optical thicknesses of 3 and 19 mm, and a mode-limiting aperture. It was also found that using a very short focal length (R = 1 cm) high-reflectivity mirror next to the Brewster-angle dye cell greatly improved the stability of the single-mode operation, perhaps owing to spatial saturation effects in the dye cell (1 mm thick). By simultaneously tilting both etalons and displacing the flat output mirror, continuous tuning of the single mode could be accomplished over intervals of 1.5 GHz. Typical single-mode output power was 5 mW (with 1 W argon pump power) with approximately 10% amplitude noise (0 to ∿ 100kHz). Continuous free-running operation in a given longitudinal (c/2L   120 MHz) mode was limited to about 1 min.

A much improved dye laser with an Invar cavity has been constructed using a free jet dye stream, rather than a window dye
cell, and a crystal quartz birefringent wavelength tuning element
similar to the laser design reported by Jarrett.[3] Although the
frequency of this laser has not yet been stabilized, the free
running stability is greatly improved (±2 MHz for averaging times
of 1 sec.). It can be tuned single-mode over approximately 10 GHz
using piezoelectrically tuned etalons, with the mode stable for
periods of hours. Single-mode output power of over 150 mW has been
obtained. Amplitude noise is also reduced by more than an order of
magnitude, the remaining noise corresponding almost entirely to the
±1% (at 180 Hz) and ±0.5% (at 100 kHz) amplitude fluctuations of
the argon pump laser.

Due to the relatively poor free-running stability, an optical
frequency discriminator with a high signal-to-noise ratio over very
short times, such as a high-finesse optical cavity, is needed for
servo controlling the frequency. In addition to a high short-term
signal-to-noise ratio, an optical cavity has the added advantage
of being tunable, thus providing the possibility of repeatable con-
trolled-frequency scanning. However, owing to thermal drifts and
acoustic disturbances of an optical cavity, the long-term frequency
stability of such a frequency discriminator is limited. Thus a
system is now being assembled to stabilize this reference cavity
itself by using frequency offset locking[4] from a saturated-absorption-stabilized HeNe/CH_4 laser at 3.39 μm.[5] The frequency discriminator used was a 17.8-cm Invar-spaced cavity enclosed in a sealed
tank and optically isolated from the laser with a λ/4 plate and
linear polarizer. The cavity has a finesse of 400 (2.1-MHz FWHM
fringes) and one piezoelectrically controlled mirror. The servo
lock is made by using the side of a transmission fringe with the
zero for the error signal located approximately halfway up the
fringe. In order to avoid mapping intensity fluctuations into the
frequency domain, a fast differencing technique using a separate
reference channel is employed. An alternative approach of servo
controlling the laser intensity has also been demonstrated, using
a linear polarizer and an electro-optic crystal (KDP).

The servo control system is illustrated in Fig. 1. A portion
of the dye laser output is mode matched into the Invar frequency
discriminator, and an error signal is formed by taking the difference
between the reference and transmission fringe channels. This error
is then passed through a variable-gain-integrator and a 3-dB/octave
active filter into a high voltage amplifier. The resulting correction signal is then divided into low- and high-frequency components
by a crossover network. The low-frequency corrections (< 1 kHz) are
applied to a long-excursion (10 μm) piezoelectric driver attached to
the output coupling mirror (resonant frequency of ∿ 18 kHz), while
the high-frequency ones are applied at the opposite end of the dye
laser cavity to a very fast piezoelectrically driven mirror

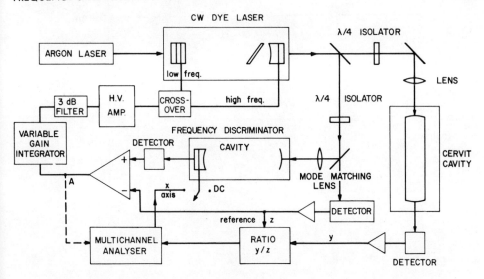

Figure 1. Block Diagram of Servo System.

(resonant frequency $\sim$ 200 kHz). Useful servo gain was evident
beyond 50 kHz. It should be possible to obtain a servo bandwidth
of several MHz by using a fast servo element, such as a resonance
damped dueterated KDP crystal, inside the cavity for phase control.

The frequency stability of the servo controlled laser is best
represented by an Allan variance plot (a plot of σ, the normalized
averaged first differences $\Delta\nu/\nu$ of the laser's absolute frequency,
versus the averaging time τ). Figure 2 shows the results of two
separate tests of the frequency stability for the window dye cell
laser. The crosses are experimental points obtained by measuring
the error signal from the Invar frequency discriminator cavity.
The voltage-to-frequency conversion is done by knowing the fringe
width (2.1 MHz FWHM) and fringe height ($\sim$ 10 V). These points
represent the ability of the servo system to make the laser frequency track the Invar cavity system (including its small instability and drift).

The servo gain increases at 9 dB/octave for frequencies below
the unity gain frequency of approximately 50 kHz, which for a white-
noise spectrum would result in a σ-τ curve falling at 3 dB/octave
($\tau^{-1/2}$).   However, the observed frequency variance increases to a
soft maximum near 0.1 msec, before breaking into a regime near
τ^{-1}. This effect is consistent with the observed weak spectral
noise peak at $\sim$ 2 kHz, which is thought to be due to dye flow
anomalies.

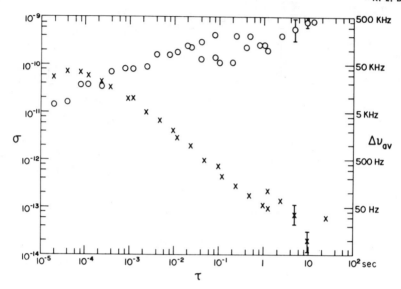

Figure 2. Allan variances, $\sigma[(\Delta\nu)_{av}/\nu]$, of laser frequency for window dye cell laser; τ is the integration time. ×, variance for laser frequency locked to Invar cavity transmission fringe; O, variance for light transmitted by auxiliary cavity with laser frequency locked to Invar cavity. Divergence of the two cases demonstrates relative instabilities of the two discriminator cavities.

The circles in Fig. 2 are the σ-τ points obtained by measuring the frequency deviation of the locked laser relative to a transmission fringe of an auxiliary optically isolated 30-cm Cervit-spaced test cavity. The rapid divergence of the two σ-τ plots for averaging times greater than 1 msec arises partly from relative thermal instability of the two optical cavities, but was mainly due to the fact that one of the cavities was inadvertantly not closed air-tight. The correlation of the two curves for short averaging times shows that the relative stability of the two cavities for short times is comparable to or better than the frequency stability of the dye laser.

A qualitative measure of the absolute stability of the dye laser plus Invar frequency discriminator is illustrated in Fig. 3. Here the saturated absorption signal (first derivative, laser amplitude stabilized) from an iodine hyperfine triplet is displayed as a function of voltage applied to the discriminator cavity's piezoelectric mirror driver. These peaks are typical of

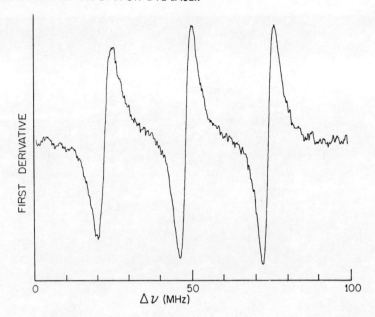

Figure 3.   Saturated absorption signal (first derivative) for an
            iodine triplet near 5975 Å. Dither amplitude, 5 MHz.
            Intensity stabilizer suppressed dither's AM noise by
            66 dB.

the many iodine multiplets we have observed throughout the rhodamine 6G region. With an ∿ 5-MHz dither, the narrowest peak is
approximately 7 MHz FWHM. The scan rate was approximately
1 MHz/sec (100-sec sweep). The spectrum was found to be degraded
because of drifts in the discriminator cavity after averaging for
approximately 3 or 4 min, which thus indicates an absolute frequency drift of about 1.5 MHz/min. This drift could be greatly
reduced by thermal controlling the discriminator cavity and could
be essentially eliminated by locking the cavity length to a
saturated absorption resonance through variable frequency offset
locking techniques.[4] The effect of amplitude stabilization of
the dye laser is illustrated in Fig. 4. Two multi-channel analyzer
traces are shown of an iodine quartet, with the amplitude servo
first open and then closed during the trace. Averaging time per
channel was 40 msec.

    In light of the conference call for "speculation regarding
future work," I will briefly discuss some of the further developments and applications which are underway in our laboratory.
Changes being made in the design of the laser (more stable cavity)

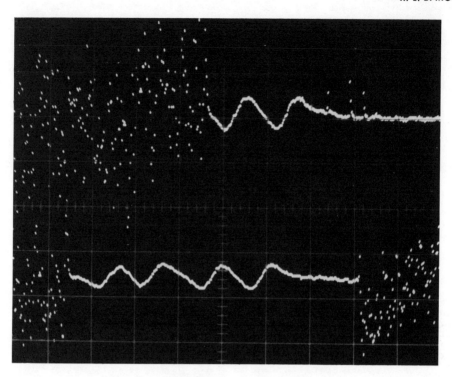

Figure 4. Saturated absorption signal (first derivative) for an
iodine quartet, showing effect of amplitude servo.

and its servo system (gain to higher frequencies using an electro-
optic crystal) should give higher frequency stability by one to
two orders of magnitude. The stability, power, and tuneability
already achieved brings about many new possibilities for accurate
laser spectroscopy measurements throughout most of the visible
spectrum. For instance, one of our goals is to measure a series
of medium accuracy (about 1 in 10^9) easily obtainable molecular
saturated absorption lines to serve as a mesh of secondary wavelength standards in the visible.

Another application making use of the stability and output
power of this laser is the development of a new high signal/noise
ratio technique for detecting saturated absorption involving
long-lived states, for instance the calcium $^1S-^3P$ (lifetime 1/3
msec) and oxygen $^1D-^1S$ (lifetime about 1 sec) transitions, where
the transition probability is too low for satisfactory detection
by measurement of the optical absorption or fluorescence intensity.
In this technique, the laser beam is absorbed in an atomic or
molecular beam, and the saturation is detected by observing the

changes produced in the atomic beam intensity for one of the energy
levels involved. The levels can be separately detected by using
level-discriminating hot wire or Auger effect detectors, or by
using magnetic or electric field deflection with subsequent non-
level-discriminating detection. Very high signal/noise ratio
should be obtained since essentially only the atoms which absorb
light are detected (with detector efficiencies which can approach
unity) and high atomic beam intensities can be used.

Another application underway is an accurate measurement of
the Rydberg constant using the hydrogen Balmer α line. Here, the
cw dye laser will be used to produce saturated absorption in a
beam of metastable hydrogen atoms. With the laser frequency
locked to the saturation peak, the laser wavelength will be
measured using frequency-offset-controlled interferometry techniques.[4] Since the hydrogen atoms will be in a perturbation-
free environment (except for the laser radiation field), and since
the frequency-stabilized cw dye laser will be used together with
well-established accurate interferometry techniques, the accuracy
of the experiment should be limited only by properties of the
Balmer α line. These conditions should result in lower systematic errors[6] than those encountered in the recent excellent work
of Hansch and it is hoped that an accuracy for the Rydberg of
better than a part in 10^9 will be achieved.

The author is pleased to acknowledge the collaboration of
J. L. Hall and M. S. Sorem in the dye laser stabilization work,
of D. A. Jennings in development of our jet stream dye laser,
and of F. Russell Petersen and T. C. English in the atomic beam
saturation investigations.

REFERENCES

1. R. L. Barger, M. S. Sorem and J. L. Hall, Appl. Phys, Lett 22,
    573 (1973).

2. A. Dienes, E. P. Ippen, and C. V. Shank, IEEE J. Quantum
    Electron. QE-8, 388 (1972).

3. S. Jarrett, paper in these proceedings.

4. R. L. Barger and J. L. Hall, Appl. Phys. Lett. 22, (1973).

5. R. L. Barger and J. L. Hall, Phys. Rev. Lett. 22, 4 (1969).

6. T. Hansch, paper in these proceedings.

COHERENT TRANSIENT EFFECTS

COHERENT TRANSIENT STUDY OF VELOCITY-CHANGING COLLISIONS[§]

J. Schmidt,[*][†] Paul R. Berman,[‡] and Richard G. Brewer[*]

*IBM Research Laboratory, San Jose, California 95193

†Physics Department, New York University, New York 10003

ABSTRACT

The effect of velocity-changing collisions on the optical phase memory of coherently prepared molecular gas samples is examined by the method of Stark-pulse switching. Experiments are interpreted through the solution of a transport equation, which extends the earlier Fokker-Planck description known both in NMR and in Dicke line narrowing. The magnitude of a characteristic velocity jump for binary molecular collisions and its cross section are thereby obtained.

It is well known that collisions between molecules can influence their optical lineshape through changes in molecular velocity. Theoretical discussions[1-5] of this problem usually invoke a Brownian motion diffusion model in velocity space that is based on a solution of the Fokker-Planck equation.[6,7] A Doppler or Gaussian lineshape is predicted for low pressure and a Lorentzian profile for high pressure, the width becoming narrower with increasing presure as first recognized by Dicke.[1] To be valid, these treatments imply characteristically small Doppler phase changes $kt\Delta u_{rms} \ll 1$ over the period of observation t where $\vec{k}$ is the propagation vector of light and Δu_{rms} is a characteristic velocity jump, essentially the root mean square change in velocity per collision.

*Work sponsored in part by the U.S. Office of Naval Research under Contract No. N00014-72-C-0153.
†Present address: Kammerlingh Onnes Laboratorium, Leiden, the Netherlands.
‡Research supported in part by a National Science Foundation Institutional Grant to New York University and the U.S. Army Research Office, Durham.
§To appear in Physical Review Letters.

We present here new optical coherent transient effects which reveal the persistence of velocity in dilute molecular gases where low angle elastic scattering is an important dephasing mechanism. The measurements are performed by the recently introduced technique of *Stark-pulse switching*.[8-9] For short observation times, the change in the Doppler phase factor is small, but for longer times it becomes progressively larger so that ultimately $kt\Delta u_{rms} > 1$ and the Fokker-Planck solution then fails. To obtain a proper description for all times, we find it necessary to begin with the more general transport equation using a collision kernel of the form given by Keilson and Storer.[10] Our observations show new features of coherent transient effects not anticipated previously and provide the first measurement of Δu_{rms} as well as the cross section for velocity-changing collisions. By comparison, similar coherent transients in the NMR of condensed media (spin echoes)[11-12] are restricted to the Fokker-Planck diffusion behavior because of the small phase excursions encountered and never enter the long-time regime monitored here.

The great advantage of coherent transient methods is that individual dephasing processes can be isolated and studied with high sensitivity. This is illustrated in Fig. 1 for the ν_3 band line (J,K = 4,3 $\rightarrow$ 5,3) of $C^{13}H_3F$ at 1035.474 cm^{-1}, which shows the decay behavior for a series of two-pulse photon echoes as well as results for two new methods. The first of these is *two-pulse optical nutation* and the second is the *optical analog of the Carr-Purcell*[13] *multiple-pulse spin echoes* (See Fig. 2); both of these, in contrast to the two pulse echo decay, can be made insensitive to collisions which change velocity. Thus, a comparison of these three experiments permits a quantitative examination of momentum-changing collisions.

The two-pulse nutation method gives a measurement of the phenomenological and state independent decay time T_1. A $\pi/2$ pulse initially places the absorbing molecules in a superposition state while the second pulse, which may be a step-function, monitors the recovery of population as a function of delay time. Electronic pulses are derived from a two-pulse generator that periodically Stark switches molecules into resonance with a cw CO_2 laser beam. Molecules that remain in the superposition state at the time of the second pulse will have experienced a rapid inhomogeneous dephasing. Thus, they will not absorb radiation and will go undetected while those molecules that relaxed into the absorbing level are monitored.

The multiple-pulse echo technique, explained below, can also reduce the dephasing caused by velocity-changing collisions to a negligible value and provides an additional measure of T_1. The Stark pulse sequence is shown in Fig. 2. The first two pulses

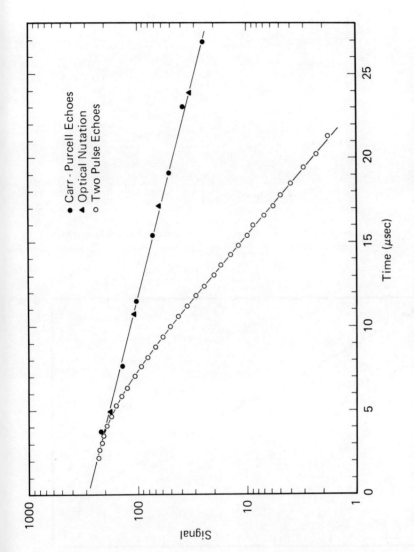

Figure 1. Decay curves of optical coherent transients in $C^{13}H_3F$. Here and in Figs. 2 and 3 the pressure is 0.32 mTorr, the laser beam diameter equals a 1.3 cm Stark spacing, the Stark pulses are 40 volts/cm, and a Stark bias field of 80 volts/cm is added to remove any complication of level degeneracy. The expanded laser beam's power density is about 350 mW/cm². The time axis is the total elapsed time t for each experiment, so that $t = 2\tau$ in a two-pulse echo or $t = 2n\tau$ in an n-pulse echo train.

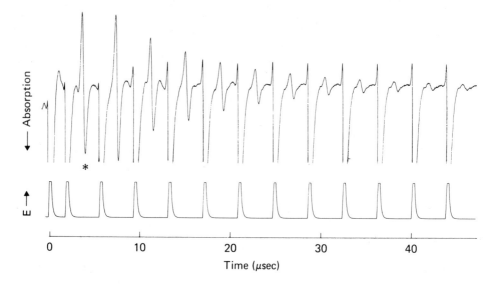

Figure 2. Optical Carr–Purcell echoes in $C^{13}H_3F$. The first echo of the sequence is marked by an asterisk.

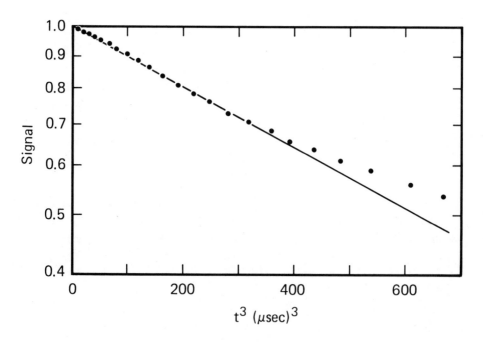

Figure 3. Decay curve of two–pulse photon echoes in the region of short times where the t^3 behavior is evident out to $400(\mu sec)^3$ (replotted from Fig. 1 with the T_1 decay subtracted).

produce an echo which subsequently dephases but is refocussed by the next pulse, the cycle repeating sequentially thereafter. The experiment is closely related to spectral diffusion in NMR, either in solids by fluctuating magnetic fields of neighboring spins[12] or in a liquid by spatial diffusion of spins through an external magnetic field gradient.[11]

Spectral diffusion caused by momentum-changing collisions will not affect T_1 in these experiments provided the velocity jump Δu_{rms} is much less than the velocity width ($\sim 1/k \cdot$ pulse width) sampled, a condition realized in this work. Consequently, T_1 is determined by lifetime limiting processes contained in the $C^{13}H_3F$ pressure dependence

$$\frac{1}{T_1} = [0.115 + 0.076 \ P(mTorr)] \mu sec^{-1} \ . \tag{1}$$

Here, the intercept is due to the transverse molecular transit time across the laser beam and the pressure dependent part yields the relaxation rate to neighboring *rotational states*, the corresponding cross-section being 440Å^2.

In the theoretical description of the photon echo problem, we treat the center of mass motion classically using the assumption that the collisional interaction is state independent for the vibration-rotation levels involved. For binary collisions, the echo amplitude may be derived from the transport equation[2]

$$\frac{\partial \rho_{mn}(\vec{R}, \vec{v}, t)}{\partial t} = -\vec{v} \cdot \vec{\nabla} \rho_{mn}(\vec{R}, \vec{v}, t) - \rho_{mn}(\vec{R}, \vec{v}, t)/T_1$$

$$-\Gamma \rho_{mn}(\vec{R}, \vec{v}, t) + \int d^3 v' W(\vec{v}' \rightarrow \vec{v}) \rho_{mn}(\vec{R}, \vec{v}', t)$$

$$+[H(\vec{r}, \vec{R}, t), \rho(\vec{R}, \vec{v}, t)]_{mn}/(i\hbar) \tag{2}$$

where $\rho_{mn}(\vec{R}, \vec{v}, t)$ is the phase space distribution function for the mn density matrix element. The off-diagonal density matrix elements, which depend on molecular position $\vec{R}$ and velocity $\vec{v}$, give rise to a sample polarization and to an echo radiation pattern derivable from Maxwell's equations. The first term on the rhs of (2) denotes the spatial behavior or convective flow, the second term explicitly expresses the population decay through the constant T_1, the third and fourth terms result from velocity-changing collisions that respectively decrease or increase $\rho_{mn}(\vec{R}, \vec{v}, t)$, and the last term is the quantum mechanical commutator appearing in Schrödinger's equation. Thus, in the absence of the last term, (2) reduces to the linearized Boltzmann

equation. Note that $H(\vec{r},\vec{R},t)$ is the Hamiltonian for the system
excluding collisions, $W(\vec{v}' \to \vec{v})$ is the probability density per
unit time for a collision that takes a molecule from velocity $\vec{v}'$
to $\vec{v}$, and Γ is the rate constant for velocity-changing collisions
given by $\Gamma = \int d^3 v W(\vec{v}' \to \vec{v})$. Finally, there is no term in Eq.
(2) representing "phase-interrupting collisions." Its absence
is consistent with our assumption of a state-independent
collisional interaction.

To solve (2), we assume the Keilson and Storer[10] "weak"
collision kernel

$$W(\vec{v}' \to \vec{v}) = \Gamma[\pi(\Delta u)^2]^{3/2} \exp[-(\vec{v} - \alpha\vec{v}')^2/(\Delta u)^2] . \quad (3)$$

The velocity jump Δu (the rms subscript is dropped hereafter) is
related to the most probable speed u of the thermal equilibrium
distribution by

$$(\Delta u)^2 = (1 - \alpha^2)u^2 \approx 2(1 - \alpha)u^2 \quad (4)$$

and α is a constant very close to but less than unity.

Inserting (3) into (2) leads to an analytic solution for the
echo amplitude under the following assumptions:

(a) The molecular sample is subject to resonant infrared
pulses, a $\pi/2$ pulse at $t = 0$ and a π pulse at $t = \tau$, which are
sufficiently brief that collisions and spatial terms in (2) can
be neglected leaving only the Schrodinger equation. Following
a pulse, the off-resonant condition prevails and all terms in
(2) are needed to describe relaxation and echo formation.

(b) The average velocity of the sample decays slowly, i.e.,
$\Gamma(1-\alpha)\tau \ll 1$ where $\Gamma(1-\alpha)$ is the effective decay rate of the
average velocity. The collision rate Γ is taken to be much less
than the Doppler width, $\Gamma/ku \ll 1$.

(c) The initial velocity distribution at $t = 0$ is the thermal
equilibrium value $\sqrt{\dfrac{1}{\pi u^2}}\, e^{-v^2/u^2}$

We thus obtain the normalized echo amplitude E, evaluated at
$t = 2\tau$,

$$E(t = 2\tau) = \exp[-(t/T_1) - \Gamma t + (4\Gamma/k\Delta u) \int_0^{kt\Delta u/4} e^{-x^2} dx] . \quad (5)$$

This has the limiting form for short times, $kt\Delta u < 1$,

$$E(t = 2\tau) = \exp[-(t/T_1) - \Gamma k^2 (\Delta u)^2 t^3/48] \qquad (6)$$

and for long times, $kt\Delta u > 1$,

$$E(t = 2\tau) = \exp[-(t/T_1) - \Gamma t + (2\Gamma/k\Delta u)\pi^{1/2}] . \qquad (7)$$

Extension of (5) to the Carr-Purcell[13] (n pulse) echo sequence is obtained by simply taking $E(t = 2\tau)$ to the nth power giving

$$E_n(t = 2n\tau) = \exp[-(t/T_1) - \Gamma t + (4n\Gamma/k\Delta u)\int_0^{k\tau\Delta u/2} e^{-x^2} dx] . \qquad (8)$$

For short pulse separations $k\tau\Delta u \ll 1$,

$$E_n(t = 2n\tau) = \exp[-(t/T_1) - \frac{1}{12} \Gamma(k\tau\Delta u)^2 t] . \qquad (9)$$

where the second term on the rhs can be made arbitrarily small by reducing the interval τ so that the T_1 decay dominates. The equivalent of (6) and (9) are well-known in NMR,[11-13] but (5) or (7) have not been proposed previously.

Note that our general expression (5) reduces to the Fokker-Planck solution[12] (6) in the short time limit $kt\Delta u < 1$. In this regime, those molecules that collided suffer small phase changes and contribute to echo formation with the t^3 behavior displayed in (6). For long times, $kt\Delta u > 1$, small excursions in velocity eventually produce large phase changes as time elapses and the echo results largely from those molecules that didn't collide at all, their survival probability being $e^{-\Gamma t}$ as shown in (7). The form of this time behavior is independent of the specific collision kernel used.

Many of the features predicted by Eqs. (5)-(9) are fully verified in our echo measurements. The observed time dependence, shown for example in Fig. 1, is in striking agreement with the linear behavior of (7) for long times and the t^3 behavior of (6) for short times (replotted in Fig. 3). The linear pressure behavior indicated in (5)-(9), through the quantity Γ, is also obeyed experimentally and stands in contrast to the Fokker-Planck solution[12] which in the long time region has a decay parameter with an inverse pressure dependence (Dicke line narrowing effect). Furthermore, the Carr-Purcell measurement, illustrated in Fig. 1, reduces to a T_1 decay for sufficiently short pulse intervals as suggested in (9).

The CH_3F-CH_3F collision parameters

$$\Delta u = 200 \text{ cm/sec}$$

$$\sigma = 580\text{Å}^2 \qquad\qquad (10)$$

were obtained by extrapolating to the low intensity limit where
radiative effects can be neglected.[14] Here, the cross-section
is defined by $\sigma = \Gamma/(Nu\sqrt{2})$. We see that the velocity jump Δu is
about 0.5% of thermal velocity and from (4), $(1-\alpha) = 1.5 \times 10^{-5}$.
It follows that our initial assumptions of $\Gamma(1-\alpha)\tau \ll 1$ and Γ/ku
$\ll 1$ are well justified.

While the Carr-Purcell method may be insensitive to
velocity-changing collisions, it will still respond to other
dephasing processes such as "phase interrupting collisions" that
exhibit a pure exponential decay. However, the fact that the
Carr-Purcell and the two-pulse nutation give the same decay
constant T_1 shows that phase interrupting collisions are
unimportant. This is a reasonable result since the vibrational
transition levels are not expected to shift significantly relative
to each other during a collision and supports our assumption of
a state independent collisional interaction.

In addition, the method of *coherent Raman beats*,[9] which also
is independent of velocity-changing collisions, exhibits a T_1
decay in agreement with the two-pulse nutation measurement and
further confirms that the unusual and prominent decay behavior
observed in a two-pulse echo is due to velocity-changing
collisions.

REFERENCES

1. R. H. Dicke, Phys. Rev. 89, 472 (1953); J. P. Wittke and R.
    H. Dicke, ibid 103, 620 (1956).

2. P. R. Berman and W. E. Lamb, Jr., Phys. Rev. A2, 2435 (1970);
    P. R. Berman, ibid A6, 2157 (1972) and references therein.

3. M. Borenstein and W. E. Lamb, Jr., Phys. Rev. A5, 1311 (1972).

4. S. G. Rautian and I. I. Sobelman, Soviet Physics Uspekhi 9,
    701 (1967); V. A. Alekseev, T. L. Andreeva, and I. I.
    Sobelman, Soviet Physics JETP 35, 325 (1972).

5. L. Galatry, Phys. Rev. 122, 1218 (1961).

6. S. Chandrasekhar, Reviews of Modern Physics 15, 1 (1943).

7. G. E. Uhlenbeck and L. S. Ornstein, Phys. Rev. 36, 823 (1930).

8. R. G. Brewer and R. L. Shoemaker, Phys. Rev. Letters $\underline{27}$, 631
    (1971); ibid Phys. Rev. $\underline{A6}$, 2001 (1972).

9. R. L. Shoemaker and R. G. Brewer, Phys. Rev. Letters $\underline{28}$, 1430
    (1972); R. G. Brewer and E. L. Hahn, Phys. Rev. $\underline{A8}$, 464
    (1973).

10. J. Keilson and J. E. Storer, Quarterly of Applied Mathematics
    $\underline{10}$, 243 (1952).

11. E. L. Hahn, Phys. Rev. $\underline{80}$, 580 (1950).

12. B. Herzog and E. L. Hahn, Phys. Rev. $\underline{103}$, 148 (1956); J. R.
    Klauder and P. W. Anderson, Phys. Rev. $\underline{125}$, 912 (1962); W.
    B. Mims, Phys. Rev. $\underline{168}$, 370 (1968).

13. H. Y. Carr and E. M. Purcell, Phys. Rev. $\underline{94}$, 630 (1954).

14. A radiative interaction, such as Coherent Raman beats,[9]
    whereby the sample is driven slightly between the pulses
    could account for an intensity dependent dephasing and such
    an effect is observed. It should be noted that the two-pulse
    echo data of Fig. 1 yields Δu = 80 cm/sec and σ = 3900$Å^2$ in
    the absence of any intensity extrapolation.

THEORY OF COHERENT TRANSIENTS[*]

M.O. Scully[+] and R.F. Shea

Department of Physics and Optical Sciences Center
University of Arizona
Tucson, Arizona 85721

INTRODUCTION

In this paper we will be dealing mostly with coherence between levels in an atomic system. An atom is said to be in a coherent superposition of two levels if we can write the state of the system as

$$|\psi\rangle = C_a \exp(-iE_a t/\hbar)|a\rangle + C_b \exp(-iE_b t/\hbar)|b\rangle \tag{1}$$

This state gives rise to a non-zero expectation value for the dipole moment operator

$$p = \langle\psi|e\hat{r}|\psi\rangle = C_a C_b^* \exp[-i(E_a - E_b)t/\hbar]\langle b|e\hat{r}|a\rangle + c.c. \tag{2}$$

The dipole moment oscillates, and, according to Maxwell's Equations, radiates energy at a frequency $(E_a - E_b)/\hbar$. In the following sections we will discuss several cases where a coherent superposition of states is established and the resulting radiation is detected.

Generalizing to the case of several widely separated and non-interacting atoms, we note that we cannot write down a single state vector for this combination of systems. We can, however, in a statistical sense, assign a value to the dipole moment. The procedure is to calculate p for each of the systems and then add up these values with their respective weighting factors. This ensemble average is then

---
* Work supported by the Air Force Office of Scientific Research.
+ Alfred P. Sloan Fellow.

$$p = \sum_i W_i p_i$$

A combination of systems such as this is called a statistical mixture. In contrast when we can write down a single state vector, the system is referred to as a pure case.

The separate treatment necessary to discuss these two cases is awkward and for this reason von Neumann introduced the density matrix to provide a unified treatment of these two kinds of averaging procedures. For a statistical mixture we define the density operator $\hat{\rho}$ as follows

$$\hat{\rho} = \sum_i W_i |\psi_i\rangle\langle\psi_i|$$

The statistical average of the dipole moment operator is then

$$P = \sum_j \langle\psi_j| \left\{ \sum_i W_i |\psi_i\rangle\langle\psi_i| e\hat{r} \right\} |\psi_j\rangle \equiv \text{Tr}(\hat{\rho}e\hat{r}) \tag{3}$$

For a pure case density operator all the weights, save one, are zero and the sum collapses to a single term.

The density operator is more commonly encountered in matrix form. If $|\psi\rangle$ is a linear superposition of two states as in Eq.(1), then we can write the following matrix representation of the density operator in the Hilbert space of these two states

$$\rho = \begin{pmatrix} \rho_{aa} & \rho_{ab} \\ \rho_{ba} & \rho_{bb} \end{pmatrix}$$

$$= \begin{pmatrix} |C_a|^2 & C_a C_b^* \exp[(E_a-E_b)t/\hbar] \\ C_a C_b^* \exp[-i(E_a-E_b)t/\hbar] & |C_b|^2 \end{pmatrix}$$

Atomic coherence is generally associated with the existence of off-diagonal matrix elements in the atomic density matrix.

The remaining portion of this paper is divided into three general sections. The first is a review of some phenomena involving coherence. This section contains discussion of certain aspects of Stark shift spectroscopy. In the latter portions of the paper we explore two areas where coherence manifests itself in somewhat different contents.

ECHOES, SUPERRADIANCE, NUTATIONS AND MUTATIONS

Photon Echo in Gases

As a first example of coherent transient effects, we review
the phenomena of photon echo[1]. The medium we use consists of two-
level atoms with a spread in resonant frequencies (an inhomogeneously broadened medium). A laser pulse of just the right size
(a $\pi/2$ pulse) is used to create a coherent superposition of the
upper and lower atomic states. This superposition gives rise
to a macroscopic dipole moment for the system of atoms.

Since the atoms all have slightly different resonant frequencies, the individual dipoles will slowly get out of phase and
the macroscopic dipole will disappear. If at some later time a
second and larger pulse (a π-pulse) is sent through the system,
the roles of the upper and lower states will be reversed. This
reverses the diffusing process and eventually the individual
dipoles will come back into phase and the polarization will have
built up to what it was at the beginning. The macroscopic dipole
thus created shows up as a pulse emitted from the system, and as
might be expected it is referred to as the echo pulse (see Fig.(1)).

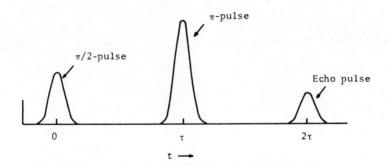

Fig.(1)   Schematic Photon Echo Experiment

The construction of the echo pulse requires that the dipoles
retain some memory of their original phase relations. In realistic
situations there is a finite phase memory time (T_2) which decreases
the amplitude of the echo. As a result photon echo is an important technique for measuring T_2.

Originally photon echo was observed in a solid (ruby)[1] and
some people argued that the technique would not apply to a gas
because the rapid atomic motion would prevent the medium from
returning to its initial state to form a macroscopic polarization.

$$\underline{\hspace{5cm}} \, |a\rangle$$

$$E_a - E_b = \hbar(\omega_a - \omega_b)$$

$$\underline{\hspace{5cm}} \, |b\rangle$$

A $\pi/2$ pulse creates a superposition of states at t=0

$$|\psi_j(t>0)\rangle = (1/\sqrt{2})\{\exp(-i\omega_{a_j} t)|A_j\rangle + \exp(-i\omega_{b_j} t)|b_j\rangle$$

A π pulse at time τ reverses roles of a and b

$$|\psi_j(t+\tau)\rangle = (1/\sqrt{2})\{\exp(-i\omega_{a_j}\tau - i\omega_{b_j} t)|A_j\rangle + \exp(-i\omega_{b_j}\tau$$

$$-i\omega_{a_j} t)|b_j\rangle\}$$

The dipole moment for this system of atoms is sharply peaked at a time 2τ from the initial pulse.

$$P = e\sum_j \langle\psi_j|\hat{r}|\psi_j\rangle = \frac{e}{2}\sum_j \{\langle a_j|\hat{r}|b_j\rangle \exp[i(\omega_{ai} - \omega_{bi})(\tau-t)] + c.c.\}$$

This is referred to as the "photon echo".

They claimed that although a π pulse can "time reverse" the atomic wavefunctions, the center of mass motion of the atoms would not be reversed and that this would destroy the echo. It turns out, however, that photon echo does occur in gases despite the atomic motion, and it is identical to that found in inhomogeneous solids if the $\pi/2$ and π pulses are colinear. If the pulses are not colinear, then the echo intensity is different from that found in a solid (see Ref.2).

Until recently the only examples of photon echo in gases were

with SF_6 and Cs vapor. The work of Brewer and Shoemaker[3] extends
these measurements to other gases and promises to be useful in
studies of collisional relaxation in gases. This new approach
uses a Stark field to pulse the molecules in and out of resonance
with a cw laser. This is, of course, analogous to pulsing the
laser but it is more convenient and allows the experimentalist
to study gases which are not resonant with his laser. After briefly
considering the coherent (superradiant) states first introduced by
Dicke, we will return to a more detailed discussion of these
experiments.

Superradiance

In our discussion of superradiance[4] we will consider a
system of three atoms initially all in their ground states
and described by the ket vector $|---\rangle$. This state interacts with
a quantized radiation field of wave vector $\mathbf{k}$ via the interaction
Hamiltonian

$$V = g \sum_j \exp(i\mathbf{k}\cdot\mathbf{r}_j)\sigma_j{}^\dagger a_k + \text{adjoint}$$

We have considered a resonant situation and have used the rotating
wave approximation to eliminate unnecessary complications. The $a^\dagger$
and a refer to photon creation and annihilation operators while
$\sigma_i{}^\dagger$ raises the ith atom to its upper state and σ_i lowers it. The
various atomic states connected to the radiation field are
generated by applying a perturbation expansion in V to the initial
state

$$\int V |---\rangle \Rightarrow \exp[i\mathbf{k}\cdot\mathbf{r}_1] |+--\rangle + \exp[i\mathbf{k}\cdot\mathbf{r}_2] |-+-\rangle$$
$$+ \exp[i\mathbf{k}\cdot\mathbf{r}_3] |--+\rangle$$

$$\int VV |---\rangle \Rightarrow \exp[i\mathbf{k}\cdot(\mathbf{r}_1+\mathbf{r}_2)] |++-\rangle$$
$$+ \exp[i\mathbf{k}\cdot(\mathbf{r}_1+\mathbf{r}_3)] |+-+\rangle$$
$$+ \exp[i\mathbf{k}\cdot(\mathbf{r}_2+\mathbf{r}_3)] |-++\rangle$$

$$\int\int\int VVV |---\rangle \Rightarrow \exp[i\mathbf{k}\cdot(\mathbf{r}_1+\mathbf{r}_2+\mathbf{r}_3)] |+++\rangle$$

Higher powers of V will not generate states other than these
and the state with which we started.

Although we expect eight different states for a system
of three atoms, we have not been able to generate all of these.
The other states are orthogonal to those above. They may,
however, be easily calculated[5] and in Table I we have listed
all eight states under the approximation that the atoms are
closely spaced compared to a wavelength. This allows us to
factor out the spatial dependence.

We can use these states to calculate the dipole moment for
a system of N atoms. It turns out that the dipole moment is
proportional to N and that, consequently, the intensity of the
field goes as N^2. This is the distinguishing characteristic
of superradiance.

Free Induction Decay and Optical Nutation

Free induction decay (FID) describes the radiation from a
coherently prepared ensemble of atoms. This phenomena, long
familiar in NMR[6], has only recently been observed in the optical
regime[7]. Optical nutation is more familiar[8] but the observations
involving optical FID were such as to link these two phenomena
experimentally.

We consider a gaseous medium and an incident laser beam
resonant or near resonant with the molecules. The motion of
the gas molecules gives a Doppler distribution of resonant
frequencies as shown in Fig.(2).

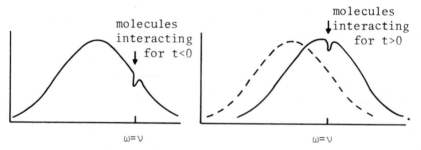

a) Without Stark Field (t<0)   b) With molecules Stark shifted (t>0)

Fig.(2)   Doppler broadened distribution of molecules interacting
          with a laser beam of frequency ν.

Table I. The eight states of the three-atom system are listed in terms of the angular momentum notation as used by Dicke.

	$r = 3/2$	$r = 1/2$	$r = 1/2$
$m = 3/2$	$|+++\rangle$		
$m = 1/2$	$(1/3)^{\frac{1}{2}}(|++-\rangle+|+-+\rangle+|-++\rangle)$	$(1/6)^{\frac{1}{2}}(|++-\rangle+|+-+\rangle-2|-++\rangle)$	$(1/2)^{\frac{1}{2}}(|++-\rangle-|+-+\rangle)$
$m = -1/2$	$(1/3)^{\frac{1}{2}}(|--+\rangle+|-+-\rangle+|+--\rangle)$	$(1/6)^{\frac{1}{2}}(|--+\rangle+|-+-\rangle-2|+--\rangle)$	$(1/2)^{\frac{1}{2}}(|--+\rangle-|-+-\rangle)$
$m = -3/2$	$|---\rangle$		

Since the molecules off resonance do not interact strongly with the laser, only a small velocity group will be important. Within this group the laser establishes a coherent superposition of upper and lower states which, after a time, reaches a steady state limit. A strong Stark field is now applied to the gas shifting the whole Doppler curve with respect to the laser frequency. Two separate transient effects take place. Those molecules initially interacting with the laser are suddenly way off resonance and the coherently prepared superposition decays. This effect is referred to as free induction decay.

Once the Stark field is applied, a new group of molecules suddenly "sees" the laser and exhibits a ringing. This is the optical nutation effect and unless we manage to Stark shift all the molecules away from the laser it will be a part of an experiment involving FID.

Those molecules which for $t \geq 0$ find themselves off resonance (i.e. shifted from a frequency ω to $\omega + \delta\nu$) have an off-diagonal matrix element of the form

$$\rho_{ab}(t,z,\omega+\delta\nu) = (i/2)\left\{\frac{(\wp\mathscr{E}_0/\hbar)[1/T - i(\omega-\nu)]}{(\omega-\nu)^2 + (\wp\mathscr{E}_0/\hbar)^2 + 1/T^2}\right\}$$

$$\cdot \exp[i(kz-\nu t)] \cdot \exp[-i(\omega+\delta\nu-\nu)t - t/T]$$

where $\mathscr{E}_0$ is the amplitude of the laser signal, $\wp$ the dipole matrix element, and T is the decay time of the elements of the density matrix. The exponentials arise from solving the density matrix equations without an electric field. The factor in brackets is just an initial condition representing the steady state p_{ab} established for $t \leq 0$. The resonant denominator is a familiar combination of terms from the natural linewidth (1/T) and power broadening ($\wp\mathscr{E}_0/\hbar$).

The electric field radiated by this ensemble of molecules is found from Maxwell's equations in the slowly varying amplitude and phase approximation

$$[\partial/\partial z + (1/c)(\partial/\partial t)]\mathscr{E} = \alpha'(\hbar/\wp)\int d\omega\sigma(\omega)\,\mathrm{Re}\left\{2i\exp[-i(kz-\nu t)]\right.$$

$$\left.\cdot \rho_{ab}(t,z,\omega+\delta\nu)\right\}$$

Solving for $\mathscr{E}$ we find

$$\mathscr{E} = (\text{amplitude factors}) \cdot \exp\left\{-t/T - t\sqrt{(\wp\mathscr{E}_0/\hbar)^2 + 1/T^2}\right\} \cos(\delta\nu)t$$

We see that the free induction decay signal is a sinusoidal
oscillation at a frequency $\delta\nu$. It decays at a rate given by the
usual $1/T$ plus an additional term. This enhancement comes from
the frequency spread in the phased array of dipoles set up at
$t \leqq 0$ by the laser.

   Since the total electric field is

$$(\mathscr{E}_0 + \mathscr{E})\cos(kz - \nu t)$$

heterodyne detection of this signal will show the FID riding on
top of the laser signal and at a frequency $\delta\nu$. The optical
nutation signal also belongs in the above equation but since this
is essentially Rabi flopping, it oscillates at a frequency $\wp\mathscr{E}_0/\hbar$
instead of $\delta\nu$. This frequency is much slower and appears only as
background. In Fig. (3) we compare the experimental results with
a calculation of FID and optical nutation. The agreement is very
good except for very short times. These differences are due to
the finite rise time of the Stark field ($\sim$40 nsec) which was
not considered in the calculation[9].

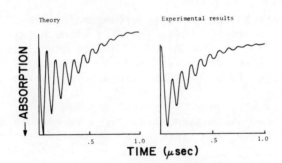

Fig. (3)   The detected signal is the sum of contributions
           from optical nutation and free induction decay.

Two-Photon Superradiance

Two-photon superradiance[10] is seen under circumstances
similar to those discussed in the previous section.   In this case,
however, we begin with a degenerate molecular transition initially
in resonance with the laser.   By means of a Stark field this
degeneracy is removed at t=0.   The levels separated in this
fashion are in a coherent superposition.

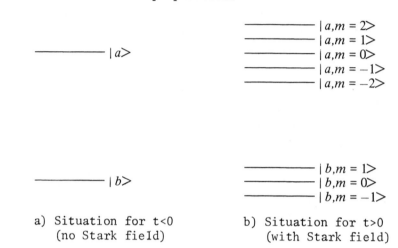

a) Situation for t<0   b) Situation for t>0
   (no Stark field)   (with Stark field)

Fig. (4)   Energy level structures for two-photon superradiance.

When the laser polarization is orthogonal to the Stark
field, the dipole selection rules ($\Delta m = \pm 1$) allow a large number
of transitions.   The two-photon superradiance signal stands out
by virtue of its slower decay rate and its frequency.   This
frequency is shifted from the laser frequency by twice the Stark
splitting.

To find the radiation from this system we calculate the
dipole moment in terms of a trace over the density matrix.

$$p = \mathrm{Tr}\,(\hat{\rho}(t)e\hat{r})$$

$$= \sum_{m} \langle am | \hat{\rho}(t) e\hat{r} | am \rangle + \sum_{m} \langle bm | \hat{\rho}(t) e\hat{r} | bm \rangle$$

By inserting a complete set of states between the operators $\hat{\rho}$ and
$\hat{r}$ and noting that $\hat{r}$ has only off-diagonal matrix elements, we
find that the terms of interest are of the form

$$\langle am|\hat{\rho}(t)|m\pm1\rangle \quad \text{and} \quad \langle bm|\hat{\rho}(t)|am\pm1\rangle$$

To evaluate these we utilize the usual perturbation expansion to first order in V

$$\hat{\rho}(t) = \exp(-iH_0 t/\hbar)\hat{\rho}(0)\exp(iH_0 t/\hbar)$$

$$- (i/\hbar)\exp(-iH_0 t/\hbar)\int_0^t [\exp(iH_0 t'/\hbar)V(t')\exp(-iH_0 t'/\hbar),\hat{\rho}(0)]dt'$$

$$\cdot\exp(iH_0 t/\hbar)$$

We can compute the matrix elements of interest by inserting a complete set of states where necessary. A typical example is given below

$$<a,m|\hat{\rho}(t)|b,m'> = \exp[-i(\omega_{am}-\omega_{bm'})t]<am|\hat{\rho}(0)|bm'>$$

$$-(i/\hbar)\mathcal{E}_0\int_0^t dt'\cos(kz-vt')\exp[-i(\omega_{am}-\omega_{bm'})(t-t')]$$

$$\sum_{m''}\left\{<m''|e\hat{r}|m'><am|\hat{\rho}(0)|am''> \cdot \exp[-i(\omega_{am}-\omega_{am''})t']\right.$$

$$\left. - <m|e\hat{r}|m''><bm''|\hat{\rho}(0)|bm'>\exp[-i(\omega_{bm''}-\omega_{bm'})t']\right\}$$

If $m'=m+1$ then the sum over m'' contains two non-vanishing contributions: $m'' = m\pm1$ and $m+1 = m'' \pm1$. Upon integration this expression yields terms oscillating at $\omega_{bm+1} - \omega_{bm-1}$ and $\omega_{am} - \omega_{am+2}$. These are the terms we are looking for and they represent oscillation at twice the frequency of the Stark shift. The other oscillations are over a wide range of frequencies and will destructively interfere[9].

The appearance of terms like $\langle am|\hat{\rho}(0)|am+2\rangle$ shows that a non-trivial result depends on the existence of these matrix elements. They represent the coherence between the magnetic sublevels at t=0. Had we been more careful and included decay times in our calculation we would have found that the signal decayed as one over the coherence time between these magnetic sublevels.

QUANTUM MECHANICAL COHERENCE IN OTHER CONTEXTS

Generalized Quantum Beats

In previous sections of this paper we have discussed the importance of coherence as a research tool in quantum optics. The final sections will deal with two areas where the quantum theory of coherence can help to provide deeper insights. As a first example we will deal with the phenomena of quantum beats.

Quantum beats have been observed in a number of experiments. They are found, for example, in beam foil spectroscopy[11] when an atom, such as pictured in Fig.(5a), is excited to a coherent superposition of all three levels. The levels $|a\rangle$ and $|b\rangle$ are so close together that their separate spectral lines cannot be resolved. The composite spectral line which is observed, however, is found to have its intensity modulated at a frequency corresponding to the energy difference of the two lower levels.

This phenomena depends crucially upon the coherent excitation of the energy levels. In such a case dipole moments are created between states $|a\rangle$ and $|b\rangle$ and between states $|b\rangle$ and $|c\rangle$. These two dipoles oscillate at slightly different frequencies and the intensity of the electric field is modulated at the frequency difference. The beam foil experimentalist sees his spectral lines formed from a series of light and dark bands.

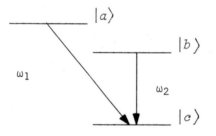
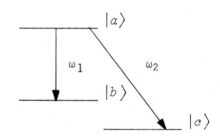

a) Shows quantum beats   b) Does not show quantum beats

Fig.(5)   Quantum beats are observed in one case but not the other.

An analogous problem is the one pictured in Fig.(5b) with two closely spaced lower levels. This figure looks very much like

Initially the atom is in a coherent superposition of states as shown in Fig.(5b). There are no photons present.

$$|\psi(0)\rangle = A|a,0\rangle + B|b,0\rangle + C|c,0\rangle$$

At a later time there exists the possibility that the atom has decayed and that there are photons present.

$$|\psi(t)\rangle = A(t)|a,0\rangle + B_o(t)|b,0\rangle + B_1(t)|b,1\omega_2\rangle$$

$$+ C_o(t)|c,0\rangle + C_1(t)|c,1\omega_1\rangle$$

To look for beats we ask for the probability that there are photons of both frequencies present.

$$\langle\psi(t)|\hat{a}_1^{\dagger}(t)\hat{a}_2(t)|\psi(t)\rangle = \langle 1\omega_1|\hat{a}_1^{\dagger}(t)\hat{a}_2(t)|1\omega_2\rangle\langle c|b\rangle c_1^{*}(t)B_1(t)$$

$$= 0$$

i.e. there are no beats!!

the previous one and from a semiclassical point of view one might expect to find beats here also. A simple minded argument such as we applied above, however, will give the wrong answer for the case of an atom having two closely spaced lower states. The correct result may be readily obtained by using the quantum theory of radiation as outlined above. We find that a "single" atom as shown in Fig.(5b) does not exhibit quantum beats.

Although there are no quantum beats for the single atom just discussed, we do find beats if we consider an ensemble of atoms like the one in Fig.(5b) contained in a small volume. The beats observed under these conditions are referred to as generalized quantum beats and they should be observable in a variety of experiments in quantum optics. These considerations are developed in detail in Ref.(12).

An Example of Coherence in the Quantum Theory of Measurement

For our final discussion of coherence we leave the world
of spectroscopy for a gedanken experiment originally proposed
by Wigner[13,14]. By modifying the familiar Stern-Gerlach experiment we obtain the apparatus pictured in Fig.(6). If the
incoming molecules are polarized in the x-direction, then they
are a coherent superposition of spin up and spin down. The
density matrix for the spins is given by

$$\rho(t=0) = \begin{pmatrix} \tfrac{1}{2} & \tfrac{1}{2} \\ \tfrac{1}{2} & \tfrac{1}{2} \end{pmatrix}$$

By means of the field gradients shown in the diagram,
this initial beam is split into components with spin up and
spin down. The maximum separation of the two beams is taken
to be very large compared to the width of the wavepacket
describing the molecules. At the end of the apparatus these
two beams are brought back together. One of the questions we
would like to ask concerns the polarization of the recombined
beam which emerges from the apparatus. It turns out that the
result depends on the coherence between the two beams.

Couched in the language of density matrices we know that an
x-polarized beam would have the same density matrix it had at
t=0. Since the beams are widely separated in the apparatus,
however, it might appear that they represent two entirely
different atomic beams. If this were true, bringing the two
beams together would generate an incoherent mixture of spin up
and spin down. For such a mixture the density matrix is given
by

$$\rho = \begin{pmatrix} \tfrac{1}{2} & 0 \\ 0 & \tfrac{1}{2} \end{pmatrix}$$

To resolve this question we outline below how one can explicitly
calculate ρ.

We begin with a two-component wavefunction representing
equal contributions of spin up and spin down.

$$\Psi(\mathbf{r},o) = \begin{pmatrix} (1/\sqrt{2})\psi(\mathbf{r},o) \\ (1/\sqrt{2})\psi(\mathbf{r},o) \end{pmatrix}$$

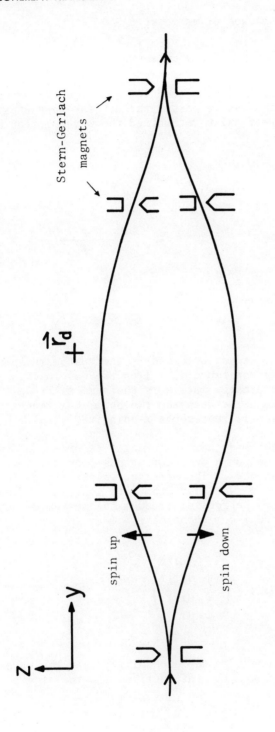

Fig. (6) Modified Stern-Gerlach apparatus used to discuss coherence in the quantum theory of measurement. The initial spin is taken to be a coherent superposition of spin up and spin down.

where $\psi(\mathbf{r},o)$ is a wavepacket of variance σ

$$\psi(\mathbf{r},o) = [\sigma\sqrt{\pi}]^{-3/2}\exp(-|\mathbf{r}|^2/2\sigma^2)\,\exp(i<\mathbf{p}>\cdot\,\mathbf{r}/\hbar)$$

Using standard quantum mechanics (and lots of algebra) we can follow this wavefunction as it passes through the apparatus. When the two beams are finally recombined the wavefunction has the form

$$\Psi(\mathbf{r},t) = \begin{pmatrix} (1/\sqrt{2})\psi(\mathbf{r},t) \\ (1/\sqrt{2})\psi(\mathbf{r},t) \end{pmatrix}$$

with

$$\psi(\mathbf{r},t) = [\sigma\sqrt{\pi}]^{-3/2}(\sigma/\alpha)^{3/2}\exp\left\{-\frac{[\mathbf{r}-(<\mathbf{p}>/m)t]^2}{2\alpha^2}\right\}$$

$$\cdot\exp\left\{\frac{i}{\hbar}<\mathbf{p}>\cdot\,(\mathbf{r}-<\mathbf{p}>t/m)\right\}$$

and

$$\alpha^2 = \sigma^2 + i\hbar t/m$$

The only change that has occurred is the customary spreading of the wavepacket. By constructing ρ from the wavefunction we can now show that the density matrix is the same as it was initially. The coherence between the two beams has been retained and the final result still represents x-polarized spins.

So far we have not made a measurement. In order to discuss measurement "theory" we put a detector at position $\mathbf{r}_d$ along the upper path of our apparatus. The detector is simplified to the point where it becomes a two-level atom with states a and b. To describe these extra degrees of freedom the system-detector wavefunction becomes four-component object.

$$\Psi(\mathbf{r},t) = \begin{pmatrix} \psi_a\uparrow(\mathbf{r},t) \\ \psi_a\downarrow(\mathbf{r},t) \\ \psi_b\uparrow(\mathbf{r},t) \\ \psi_b\downarrow(\mathbf{r},t) \end{pmatrix}$$

Our detector responds to the probability that the wavepacket is found at $\mathbf{r}_d$, and its interaction with the atomic beams is written as

$$V = g\delta(\mathbf{r}-\mathbf{r}_d)(\,|a><b\,|+|b><a\,|\,) \otimes (\,|\uparrow><\uparrow\,|+|\downarrow><\downarrow\,|)$$

Under the assumption that the motion of the molecules is
little effected by the detector we can solve these equations
for the evolution of the wavefunction. If the detector is
initially in its ground state

$$\Psi(\mathbf{r},o) = \begin{pmatrix} 0 \\ 0 \\ \psi_\uparrow(\mathbf{r},o) \\ \psi_\downarrow(\mathbf{r},o) \end{pmatrix}$$

then at the end of the apparatus we find

$$\Psi(\mathbf{r},t) = \begin{pmatrix} -i\,\sin(g_\uparrow)\psi_\uparrow(\mathbf{r},t) \\ -i\,\sin(g_\downarrow)\psi_\downarrow(\mathbf{r},t) \\ \cos(g_\uparrow)\psi_\uparrow(\mathbf{r},t) \\ \cos(g_\downarrow)\psi_\downarrow(\mathbf{r},t) \end{pmatrix}$$

where $\psi(\mathbf{r},t)$ are just the time evolved wavepackets discussed
previously.

We note that the effect of the detector has been to re-
arrange the components of the wavefunction and multiply them
by sine and cosine factors. The couplings $g_\uparrow$ and $g_\downarrow$ are related
to the probability that a wavepacket along the upper or lower
path is found at the position of the detector. The long tail of
the wavepacket is responsible for the finite probability that
molecules along the lower path trigger the detector. This
small but finite overlap of the wavepackets was also the reason
that the two beams retained coherence and emerged as x-polarized.

The density matrix for the system may be constructed from
the wavefunction. Since we are not really interested in the state
of the detector we will trace over these states.

$$\mathrm{Tr}_{\mathrm{Det}}(\rho) = \rho_{\mathrm{spin}} = \begin{pmatrix} |\psi_\uparrow|^2 & \cos(g_\uparrow-g_\downarrow)\psi_\uparrow^*\psi_\downarrow \\ \cos(g_\uparrow-g_\downarrow)\psi_\downarrow^*\psi_\uparrow & |\psi_\downarrow|^2 \end{pmatrix}$$

In the limit that $g_\uparrow \to \pi/2$ and $g_\downarrow \to 0$, this density matrix
becomes

$$\rho_{spin} = \begin{pmatrix} |\psi\uparrow|^2 & 0 \\ 0 & |\psi\downarrow|^2 \end{pmatrix}$$

We see that it is the act of making a measurement that has destroyed the coherence. Though this is an old story, it should be emphasized that it was not necessary to introduce random forces associated with the measurement process to eliminate the off-diagonal elements of ρ. It was the act of tracing (or making a measurement) that changed the density matrix from a pure case to a mixture.

The questions raised by this discussion of measurement "theory" are the kinds of things that have attracted a great deal of interest from physicists in recent years. Though our discussion has been based on a gedanken experiment it is not all that far from the capabilities of present technology and perhaps one day it will be possible to carry out experiments such as the one just discussed. The techniques of coherent spectroscopy would then provide us with valuable input to the never ending discussions concerning the foundations of quantum mechanics.

References

1. I.D. Abella, N.A. Kurnit and S.R. Hartman, Phys.Rev. 141, 391 (1966).
2. M. Scully, M.J. Stephen and D.C. Burnham, Phys.Rev. 171, 213 (1968).
3. R.G. Brewer and R.L. Shoemaker, Phys.Rev.Letters 27, 631 (1971).
4. R.H. Dicke, Phys.Rev. 93, 99 (1954).
5. R. Bonifacio, D.M. Kim and M.O. Scully, Phys.Rev. 187, 441 (1969).
6. E.L. Hahn, Phys.Rev. 77, 297 (1950).
7. R.G. Brewer and R.L. Shoemaker, Phys.Rev.A 6, 2001 (1972).
8. C.L. Tang and B.D. Silverman, in Physics of Quantum Electronics, edited by P.L. Kelley, B. Lax and P.E. Tannenwald (McGraw-Hill, New York, 1966).

9.   F.A. Hopf, R.F. Shea and M.O. Scully, Phys.Rev.A 7, 2105
     (1973).
10. R.L. Shoemaker and R.G. Brewer, Phys.Rev.Letters 28, 1430
     (1972).
11. S. Bashkin, W.S. Bickel, D. Fink and R.K. Wangsness, Phys.
     Rev.Letters 15, 284 (1965).
12. W. Chow, M. Scully and J. Stoner (to be published).
13. E.P. Wigner, Am.J.Phys. 31, 6 (1963).
14. M.O. Scully, R.F. Shea and J.D. McCullen (to be published).

PULSED MICROWAVE ROTATIONAL RELAXATION STUDIES

Herbert M. Pickett[*]

Department of Chemistry, University of California

Berkeley, California 94720

I. INTRODUCTION

In nuclear magnetic resonance, pulse techniques have been developed to a high level of sophistication and are routinely used for
relaxation studies.[1] In the last few years, close analogies to
these pulse techniques have been developed for the microwave and
infrared regions. Dr. Brewer has been active in developing techniques in the infrared, and a number of workers have made contributions in the microwave region. Most of the activity in the microwave region has been centered on forced nutation and double resonance
studies. The first part of this paper will describe a Stark-switched
$\pi/2$ pulse and free induction decay technique for the microwave region. This work was started at Harvard in association with Prof.
E. Bright Wilson and his group, as part of a program directed toward
measurement of rotational relaxation using pulse techniques. The
second part of the paper will describe some recent theoretical results that show how the M dependence of relaxation data can be used
to learn more about intermolecular potentials.

The formalism for the application of magnetic resonance experiments to other areas of spectroscopy was developed a number of
years ago by Feynman, Vernon, and Hellwarth.[2] The connection to
spectroscopic absorptivity and dielectric dispersion has recently
been discussed.[3] Briefly, the analogy between magnetic resonance
and electric dipole spectra in the microwave and optical regions
lies at the fundamental level of the time-dependent solutions for

* Fellow of the Miller Institute for Basic Research, Berkeley, Calif.
Present address: Department of Chemistry, University of Texas,
Austin, Texas 78712

Table I. Generalization of Magnetic Resonance

$$\gamma H_z \longrightarrow \omega_o \qquad\qquad M_z \longrightarrow n_0 - n_1$$

$$\gamma H_y \longrightarrow \mu_{01} \cdot \mathbf{E}/\hbar \qquad M_y + iM_x \longrightarrow \chi' + i\chi'' \quad \underline{or} \quad \rho_{01}$$

the quantum mechanics of absorption. At this level it is recognized that the fundamental purpose of the DC field in magnetic resonance is to create a pair of energy levels, and the Larmor frequency, γH_z, is more generally a transition frequency, ω_o. (See Table I.) Similarly, the gyromagnetic ratio times the RF magnetic field, γH_y, is more generally a dot product of a transition dipole matrix element and the radiation field, divided by $\hbar$. The z component of the magnetization, M_z, is proportional to the difference in population of spin states, and the more general approach considers the difference in populations between the lower and upper states of a transition, $n_0 - n_1$. The quantities M_y and M_x, in a rotating frame, are related to the dispersive and absorptive parts of the susceptibility. In the more general approach, we can use either the complex susceptibility or the off-diagonal element of the density matrix for the transition, since the two are proportional for a two level system.

The generalizations in Table I suggest that the microwave and infrared experiments will be different in several ways from the magnetic resonance experiments. One difference is the way in which inhomogeneities enter the experiment. In magnetic resonance, inhomogeneities in H_z can be a major experimental problem. In the infrared, inhomogeneous broadening occurs because the Doppler shift produces a distribution of values of the effective ω_o. In the microwave region, the Doppler shift also produces inhomogeneities, but the problem is not as serious because the Doppler broadening is often small compared to relaxation broadening. A second difference between magnetic resonance spectra and electric dipole spectra comes in the RF field term. In magnetic resonance the gyromagnetic ratio is often similar for different transitions, while in the infrared and microwave regions μ_{01} will often have widely different sizes. For example, μ_{01} is proportional to M for Q branch transitions. One implication of this second distinction is that some of the pulse techniques used in magnetic resonance are not directly applicable to higher frequency regions. A third difference between magnetic resonance and infrared and microwave spectroscopy is that often one must consider more than simple pairs of interacting states in the higher frequency experiments. This distinction has led to several new effects,[4] and means that much more extensive information about relaxation can be obtained from a single molecular system.

In magnetic resonance it is customary to associate a relaxation
time, T_1, with M_z, and a different relaxation time, T_2, with M_y and
M_x. A similar distinction can be made in the more general case, to
the extent that the two levels involved in a transition are isolated.
Then T_1 is a lifetime for population equilibrium and T_2 is a lifetime
for phase coherence for the two states, as measured by ρ_{01}. It is
the T_2 lifetime which governs the lineshapes of an unsaturated transition, and T_1 enters only when there is saturation. When more than
two levels are involved in the relaxation, it is useful to have an
equation for the time dependence of relaxation which retains the
distinction between T_1 and T_2-like processes. In magnetic resonance,
the Redfield equation is used where many levels are involved,[3]

$$-\frac{d}{dt}\,\rho_{jj'} = \frac{i}{h}\left[H,\rho\right]_{jj'} + \sum_{k,k'} R_{jj';kk'}\,\rho_{kk'}, \quad (1)$$

in which H is the Hamiltonian in the absence of collisions, and the
four-dimensional matrix R governs the relaxation. A similar equation
can be used for infrared and microwave relaxation. The elements of
R coupling diagonal elements of the density matrix, i.e. the populations of the states, form a submatrix of R which is equivalent
to Gordon's Π matrix.[6] The elements of R coupling the off-diagonal
elements of ρ together correspond to the various T_2 relaxation times
possible. There are also elements of R coupling off-diagonal elements
of ρ to diagonal elements, but these can be shown to be zero, except
in special cases. For infrared experiments in which only a particular velocity group is excited because of the Doppler shift, Eq.(1)
must be modified to include the effects of translation. The theory
of the experiment will be even more complicated in this case, but
the information could be very useful.

II. THE EXPERIMENTS

It is important to be able to design an experiment to measure
only selected elements of R, with as few assumptions as possible
about the nature of the undetermined elements, because in a system
of N levels, the number of elements of R goes as N^4. The T_2-like
elements of R can be measured selectively using unsaturated linewidths, but there are a number of difficulties at lower pressures.
One major problem is that one must work at very low incident radiation powers to avoid saturation. In addition, there may be
systematic errors due to the presence of nutation effects accompanying
modulation. A $\pi/2$ pulse and free induction decay is an alternative
way to measure the same quantities in the time domain at relatively
high incident power levels.

Our free induction decay experiment is similar in some ways
to the method for producing free induction decays used by Brewer
and Shoemaker in the infrared,[7] in which the radiation is effectively
pulsed by sudden changes in the resonant frequency of a transition

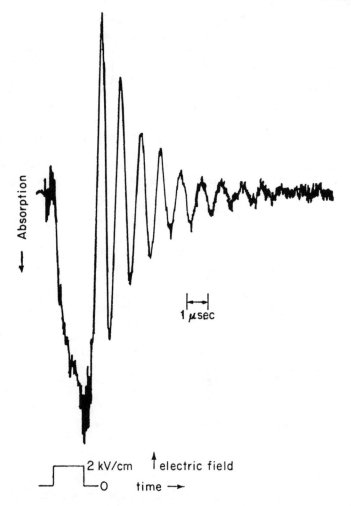

Figure 1. Stark-switched $\pi/2$ pulse and free induction decay in OCS at a pressure of 0.02 Torr for the J=3-2, M=0 transition. The frequency of the microwave source is in resonance with the transition while the electric field is present.

using the Stark effect. In the microwave experiments, the klystron frequency is resonant with a rotational transition in the presence of a large electric field, but is far from resonance with any transition in the absence of the field. Starting at zero field, we apply a voltage pulse to plates in the sample cell such that the the molecules suddenly come into resonance with the microwave radiation. The duration of the pulse is adjusted so that $\mu_{01} E_o t / \hbar = \pi/2$, in which μ_{01} is the transition dipole, E_o is the peak field strength of the microwave radiation, and $\underline{t}$ is the pulse length.

When the pulse is turned off, the molecules emit in a coherent way
at their resonant frequency, which is now the resonant frequency
at zero field. The microwave incident signal is still at a frequency
corresponding to the resonant frequency during the pulse, and the
two signals undergo temporal interference to give a heterodyne beat.

For molecules with a second-order Stark effect, this kind of
experiment requires a pulse generator which is capable of producing
up to 1400 V with a 250 nsec rise time into the capacitive load of
the cell. The generator must be capable of delivering instantaneous
currents of over 25 Amps to obtain the required risetime and final
voltage. A transistorized pulse generator with these specifications
was constructed, and the first experiment observed with it is shown
in Figure 1. The experiment was performed with a field of 2KV/cm
parallel to the microwave polarization, using the M=0 Stark lobe of
the OCS J=3-2 transition. At these fields, the transition lies
approximately 1 MHz below the zero-field resonance frequency of
36 488.811 MHz. The frequency of the microwave source was phase-
locked to be in resonance with the M=0 transition in the presence
of the field, and the pulse duration was adjusted to the $\pi/2$ condition by switching off the pulse when the absorption reached a
maximum. The decay signal at the crystal detector, after averaging
with a box car integrator, was an exponentially-damped cosine function. At a pressure of approximately 0.02 Torr, the time constant
of the exponential is 1.8 μsec. The frequency of the cosine is
approximately 1 MHz (corresponding to the difference between the
frequency of the microwave source and the zero-field resonant frequency of the transition). It can be shown that the free induction
decay is the Fourier transform of a conventional lineshape which
has been recorded in the limit of low power and low modulation
frequency. The time-domain decay in this experiment is nonetheless
observed under conditions of relatively high power (10mW) and is
modulated near 1 MHz. The resulting improved conversion efficiency
and rejection of 1/f noise at the crystal detector can yield important gains in signal to noise ratios in many situations. The lineshape obtained from this experiment, given the uncertainties in
this first measurement, is in agreement with what would be expected
from linewidth measurements,[8] and more careful measurements of free
induction decays are currently under way at Harvard. The measured
decay time constant is a collisional T_2 lifetime, since effects of
Doppler broadening and wall collisions are small.

In the future, pulse methods should also be very useful for
measurements of T_1-like relaxation times. For example, with a π
pulse one could invert the populations of a pair of levels in a time
short compared with relaxation times, and then follow the effect of
this change in population on different levels of the molecule. This
could be an improvement over present methods of four-level double
resonance[9] in which a pump transition is saturated for comparatively
long times, thereby producing changes in the population of many of

the levels of the system. The changes in population after the π pulse can be monitored with very low power radiation or, alternatively, by $\pi/2$ pulses. With the use of low power radiation, the absorption would not be strictly proportional to populations, but the changes in population could be extracted once the T_2 for the pair of probe levels is known.[10] A $\pi/2$ pulse could be used because the initial amplitude of the free induction decay is directly proportional to the population difference of the levels probed.

A number of workers in other laboratories are looking at time-domain rotational relaxation experiments involving forced nutation or double resonance. Forced nutation coherence effects are, in principal, observable on any Stark-modulated microwave spectrometer by connecting a box car integrator and a suitable wide band amplifier up to the detector. The effect was observed for the first time about five years ago,[11] and since then a number of reports of similar experiments have appeared.[12,13,14] More measurements of nutation are underway at a number of laboratories, but detailed analysis of the results require careful correction for non-uniformity of the power in the cell, and for the M dependence of the nutation. In addition, there is the problem that the experiment measures complicated mixtures of T_1 and T_2-like relaxation times, particularly if any M dependence of the relaxation is to be explicitly included. Nonetheless, important information can be gained, and the various relaxation times can be extracted, if the pressure variable is exploited fully to separate effects of multiple collisions. Another kind of time-domain experiment in active use in at least two laboratories, involves time-domain measurements of recovery from saturation following three-level double resonance. In Prof. Gwinn's group at Berkeley the pump radiation comes from a microwave klystron,[13] while the experiments at MIT have used an infrared laser.[14] As with the nutation experiments, the analysis in terms of relaxation rates can be complex, but extraction of useful relaxation information should be helped by full use of the pressure dependence of the waveforms.

III. Theory

The observation of these microwave and infrared coherence effects has deepened our understanding about what happens to molecular states when light is absorbed. The effects should also provide us with a variety of very powerful methods with which to study relaxation. The actual relaxation rates obtained in these experiments may be of some use in themselves, but generally speaking, their major usefulness lies in what these rates can tell about collisions and intermolecular forces. It is not presently possible to convert the macroscopic relaxation data directly to microscopic collision quantities because of the extensive statistical averaging that must be done in the steps going from the molecular level to the bulk gas properties. The information about collisions must

come in a more indirect way. First, a theoretician must choose a
model for the intermolecular potential, and a set of approximations.
Then the relaxation rates must be computed. The faithfulness and
accuracy of the model will be determined by the agreement between
calculation and experiment, and by the sensitivity of the calculated
rates to changes in the potential. With the greater variety of data
available from pulse techniques, it may be more difficult to find
models for the collisional potential which agree with all the experimental data. On the other hand, a model of the collisions which
passes the harder tests is likely to be a more accurate model.

If the results of the pulse techniques and other relaxation
experiments are to be compared with theory, the results will have
to be put in a form which is well-defined for a theoretical calculation and which maintains certain experimental distinctions.
Until recently, there was really no reason to worry about differences
between T_1 and T_2-like rotational relaxation rates, because the
only measurements of relaxation were linewidths. The introduction
of double resonance methods and pulse techniques has made the
distinction between the kinds of rates seem much more important.
It must also be recognized that there may be M dependence of the
relaxation rates, and the particular average of M states found in
each experiment should be identified. To maintain these distinctions,
the form of the Redfield equation, Eq.(1), seems useful. The
equation clearly distinguishes between measurements of T_1 and T_2,
and it is applicable to a multilevel system.

Before Eq.(1) is used for the analysis of experimental data,
it is important to determine whether the equation is appropriate for
gas phase collisions. In the past year, the relation between the R
supermatrix and the S matrix of scattering theory has been developed,[16]
and the equation does appear appropriate for low density gases when
the perturbation of ρ from equilibrium is small. (The S matrix is
a single-collision transformation matrix which operates on the wave
vector of the colliding molecules just before the collision to
produce the wave vector just after the collision.) For experiments
in the microwave region, limitations on the use of Eq.(1) are easily
met because the energy of a microwave photon is much smaller than
kT, and the pressure is generally low.

The definition of R in terms of S, in this theory, leads to
a number of useful properties for R. It can be shown that there
is conservation of particles and detailed balance at equilibrium.
In addition, many of the elements of R connecting diagonal elements
of ρ to off-diagonal elements are zero. By looking at the elements
of S that contribute to an approximate T_1 and T_2 for a transition,
is is possible to see that the two relaxation times should have
values within an order of magnitude of each other. The value of
the T_1 rate is roughly an arithmetical mean of certain S matrix
elements, while the T_2 rate is roughly a geometric mean of the
same elements.

The definition of R in terms of S can also be used to derive
a relation for the M dependence of R. This derivation uses the
requirement that S should be rotationally invariant if laboratory
orientations are not imposed on the molecule by external fields.
In the language of irreducable spherical tensors, the rotational
invariance is equivalent to a requirement that the S matrix be a
tensor of order zero. Total angular momentum is therefore conserved,
and the S matrix can be expanded as an infinite sum of scalar
products of three tensor operators. One tensor in this product
depends on the translational coordinates, and the other two tensors
depend on the internal degrees of freedom of the two molecules in
the collision. Use of the Wigner-Eckart theorem for each of these
tensors in the sum of products, followed by some tensor algebra, then
yields the following expression for the M dependence of the R matrix,

$$\langle J_f, M_f, J_f', M_f' \mid R \mid J_i, M_i, J_i', M_i' \rangle = \sum_L \langle J_f, J_f' \parallel R^L \parallel J_i, J_i' \rangle$$

$$\times (-)^{J_f - M_f - J_f' + M_f'} \begin{pmatrix} J_f & L & J_i \\ -M_f & P & M_i \end{pmatrix} \begin{pmatrix} J_f' & L & J_i' \\ -M_f' & P & M_i' \end{pmatrix}, (2)$$

in which the quantum numbers for internal degrees of freedom, other
than J and M, have been suppressed. The sum over L in Eq.(2) is
finite, because of the 3j symbols, and it is possible to invert Eq.(2)
to give a relation for the elements of R^L in terms of the M dependence of R. The elements of R^L contain contributions from only those
parts of the S matrix which transform like a tensor of order L in
the space of the rotational coordinates of the molecule under study.
Because of the close relation between the S matrix and the intermolecular potential, the various reduced matrix elements R^L contain
information about the relative importance of different parts of the
potential. For example, the dipolar part of the potential will affect
only $R^{(1)}$ in the first order Born approximation, while in second
order only $R^{(0)}, R^{(1)}$, and $R^{(2)}$ will be affected. Therefore, when
the relaxation is split up into the contribtions from the various
reduced matrix elements in Eq.(2), it will be possible to get a
quantitative idea of the importance of various parts of the potential
from the experimental data alone. Eq.(2) also gives some insight
into what parts of the potential are measured by each experiment.
For example, because of the limits on L implied by the 3j coefficients,
the T_2 for a J=1-0 transition measures only $R^{(0)}$. To learn about
first-order contributions from the dipole part of the potential, i.e.
$R^{(1)}$, one would have to look at the T_2 for a transition which did
not involve the J=0 state. Thus, Eq.(2) can be used not only to
tell about the importance of parts of the potential, but it can
also be used to suggest which experiments will be sensitive to a
given part of the potential.

IV. CONCLUSION

We have seen that the analogy between magnetic resonance and infrared and microwave spectroscopy can suggest a variety of novel ways of measuring relaxation. One of these new methods for the microwave region, which has been described, uses Stark switching to obtain a $\pi/2$ pulse and free induction decay. On the theoretical side, an examination of the relation of macroscopic relaxation rates to macroscopic collision potentials has shown that the new pulse methods should provide important information about the potentials, particularly if the M dependence of the rates is determined.

ACKNOWLEDGEMENTS

The author would like to thank E. Bright Wilson, Herbert L. Strauss, and the Miller Foundation for their encouragement, financial support, and hospitality. The experimental work described here was performed with the very able assistance of Mr. Steven Coy. The pulse generator was designed with help from the staff of the Harvard Electronics Design Center. The author would also like to acknowledge stimulating discussions with J. Cohen, R. Schwendeman, R. Brewer, W. Gwinn, J. Steinfeld, and W. Miller.

REFERENCES

1. T. C. Farar and E. D. Becker, Pulse and Fourier Transform NMR, (Academic Press, New York, 1971).
2. R. P. Feynman, F. L. Vernon, Jr., and R. W. Hellwarth, J. Appl. Phys., 28, 49 (1957).
3. R. H. Schwendeman and H. M. Pickett, J. Chem. Phys., 57, 3511 (1972).
4. R. L. Shoemaker and R. G. Brewer, Phys.Rev.Lett., 28, 1430 (1972).
5. A. G. Redfield, Adv. Mag. Res., 1, 1 (1965).
6. R. G. Gordon, J. Chem. Phys., 46, 4399 (1967).
7. R. G. Brewer and R. L. Shoemaker, Phys. Rev. A, 6, 2001 (1972).
8. C. O. Britt and J. E. Boggs, J. Chem. Phys., 45, 3877 (1966).
9. a. P. W. Daly and T. Oka, J. Chem. Phys., 53, 3272 (1970), and references therein; b. R. G. Gordon, P. E. Larson, C. H. Thomas, and E. B. Wilson, J. Chem. Phys., 50, 1383 (1969).
10. S. R. Brown, J. Chem. Phys., 56, 1000 (1972).
11. H. W. Harrington, Symp. on Molec. Specty., Columbus, Ohio, 1968.
12. R. H. Schwendeman and A. H. Brittain, Symp. on Molec. Specty., Columbus, Ohio, 1970.
13. E. B. Macke and P. Glorieux, Chem. Phys. Lett., 14, 85 (1972).
14. J. M. Levy, J. H. S. Wang, S. G. Kukolich, and J. I. Steinfeld, Phys. Rev. Lett., 29, 395(1972).
15. S. R. Brown, Ph. D. thesis, University of Calif., Berkeley, June 1972.
16. H. M. Pickett, to be published.

OFF-RESONANCE PULSE PROPAGATION IN A TWO-LEVEL SYSTEM--DISTANCE DEPENDENCE OF THE CARRIER FREQUENCY*

E.L. Hahn and J.C. Diels

Department of Physics, University of California

Berkeley, California 94720

Self-induced transparency[1] (SIT) is a manifestation of distortionless pulse propagation of a traveling wave in the form of a pulse propagating through an inhomogeneously broadened two-quantum system. Although the carrier frequency in the pulse is at or near resonance with the quantum oscillators of the medium (assumed to be undamped), no pulse losses occur because the non-linear process of induced absorption during the first half portion of the pulse is balanced by induced emission during the second half of the pulse. This particular non-linear process neutralizes the pulse spreading and distortion which would normally be expected for a weak pulse, as predicted by linear optical dispersion theory.

The analysis of SIT has been characterized by the absence of phase modulation of the carrier wave, where it has been pointed out[1] that the assumption of a constant carrier frequency is self-consistent with conditions imposed by the slowly varying envelope approximation. The condition of constant carrier frequency applies for a 2π hyperbolic secant (2π h.s.) pulse propagating on or off-resonance, or for a non 2π h.s. pulse applied only at exact resonance of a symmetrically inhomogeneously broadened line. We present a brief summary[2] and some novel viewpoints concerning phase modulation during propagation which will occur when a pulse is applied off-resonance. The result is that frequency pushing or pulling of the carrier frequency will occur with respect to the resonance absorption (or emission) line, depending upon particular initial conditions. The model is extended to some aspects of continuous wave radiation applied far off-resonance.

*Supported by the National Science Foundation.

Analysis according to linear dispersion theory shows that a propagating pulse does exhibit some degree of frequency modulation (chirp) as the pulse spreads with increasing propagation distance. However, the pulse envelope distorts in such a way that the average of the chirp over the pulse envelope intensity gives net zero average frequency change. This is not the case when non-linear exchange of energy occurs between radiation and medium. Although net frequency shifts may be unobservably small for a finite propagation distance, the shift becomes cumulative with increasing propagation distance. Usually a linear analysis of weak light propagation suffices for typical laboratory distances, but considerations of non-linearity may be necessary in those cases where distances become enormous, for example, through interstellar media, where phase modulation effects may build up.

In order to formulate the problem semi-classically, we summarize a few necessary relationships. The distance dependence of the average carrier frequency

$$\omega_{av} = \omega + \langle \dot{\phi}(z) \rangle$$

is formulated for a circularly polarized plane wave

$$E(z,t) = \mathcal{E}(z,t)e^{i[\omega t - kz + \phi(z,t)]} + cc, \tag{1}$$

propagating in the z direction. The average frequency shift $\langle \dot{\phi}(z) \rangle$ toward or away from an applied carrier frequency ω is given by the first moment of deviation

$$\Omega = \langle \dot{\phi} \rangle = \frac{\int_{-\infty}^{\infty} \dot{\phi}(z,t)\mathcal{E}^2(z,t)dt}{\int_{-\infty}^{\infty} \mathcal{E}^2(z,t)dt} = \frac{\int_{-\infty}^{\infty} \Omega|\tilde{\mathcal{E}}(z,\Omega)|^2 d\Omega}{\int_{-\infty}^{\infty} |\tilde{\mathcal{E}}(z,\Omega)|^2 d\Omega}, \tag{2}$$

where

$$\tilde{\mathcal{E}}(z,\Omega) = \int_{-\infty}^{\infty} \mathcal{E}(z,t)e^{i\Omega t + i\phi(z,t)}dt.$$

The second moment of deviation is

$$\langle \Omega^2 \rangle = \frac{\int_{-\infty}^{\infty} \Omega^2|\tilde{\mathcal{E}}(z,\Omega)|^2 d\Omega}{\int_{-\infty}^{\infty} |\tilde{\mathcal{E}}(z,\Omega)|^2 d\Omega} = \langle \dot{\phi}^2(z) \rangle + \frac{\int_{-\infty}^{\infty} \dot{\mathcal{E}}^2(z,t)dt}{\int_{-\infty}^{\infty} \mathcal{E}^2(z,t)dt}, \tag{3}$$

where

$$\langle \dot{\phi}^2 \rangle = \frac{\int_{-\infty}^{\infty} \dot{\phi}^2(z,t)\mathcal{E}^2(z,t)dt}{\int_{-\infty}^{\infty} \mathcal{E}^2(z,t)dt}. \tag{4}$$

The quantity

$$\frac{\int_{-\infty}^{\infty} \dot{\mathcal{E}}^2(z,t)dt}{\int_{-\infty}^{\infty} \mathcal{E}^2(z,t)dt} = \frac{1}{\tau^2} \tag{5}$$

is a measure of the reciprocal mean square pulse width for a bell-shaped pulse envelope.

The above definitions for $\langle \dot{\phi} \rangle$ and $\langle \dot{\phi}^2 \rangle$ can be connected with the semi-classical Bloch-Maxwell equations:

$$\left.\begin{array}{l}
\dfrac{\partial u}{\partial t} = (\Delta\omega - \dot{\phi})v - \dfrac{u}{T_2} \; ; \\[3mm]
\dfrac{\partial v}{\partial t} = -(\Delta\omega - \dot{\phi})u - \dfrac{\kappa^2}{\omega_o}\mathcal{E}W - \dfrac{v}{T_2} \; ; \\[3mm]
\dfrac{\partial W}{\partial t} = \omega_o \mathcal{E}v;
\end{array}\right\} \tag{6}$$

$$\frac{\partial \mathcal{E}}{\partial z} + \frac{\eta}{c}\frac{\partial \mathcal{E}}{\partial t} = -\frac{2\pi(\omega + 2\Delta\omega_o)}{\eta c}\int_{-\infty}^{\infty} vg(\delta)d\delta; \tag{7}$$

$$\mathcal{E}\frac{\partial \phi}{\partial z} + \frac{\eta}{c}\mathcal{E}\frac{\partial \phi}{\partial t} = \frac{-2\pi(\omega + 2\Delta\omega_o)}{\eta c}\int_{-\infty}^{\infty} ug(\delta)d\delta. \tag{8}$$

The usual definitions of u and v here are functions of z and t. The off-resonance frequency is initially

$$\Delta\omega = \Delta\omega_o + \delta,$$

where

$$\Delta\omega = \omega_o - \omega,$$

and ω_o is the average resonance frequency of the two-level system. The inhomogeneous spectral deviation from ω_o is δ, and $g(\delta)$ is defined as the normalized spectral distribution function. The remaining parameters including the homogeneous damping time T_2 are further defined in reference 1. The energy expectation value of the number density N of the two-level system is given by W. At equilibrium, prior to the appearance of the pulse, the initial energy is

$$W_o(\delta) = \pm \frac{N}{2}\hbar\omega_o g(\delta)d\delta,$$

with a minus sign applicable for an absorber and a plus sign for an emitter. For an emitter, a pumping mechanism is assumed to maintain $W_o = +|W_o|$ for all z when $\mathcal{E} = 0$. When induced emission occurs ($\&\neq 0$), the pumping mechanism is assumed to be too slow, or virtually absent, to compete with the driving radiation. After subsidence of the pulse at time t = + ∞, we define

$$\Delta W(\delta) = W(z,\delta) - W_o(\delta)$$

as the energy absorbed by the two-level system at distance z.

Upon combining Eqs. (6),(7), and (8) with (2),(4), and (5), the result for the first moment distance deriviative is

$$\frac{d\langle\dot\phi\rangle}{dz} = - \frac{(\omega + 2\Delta\omega_o)}{\mathcal{J}(z)\omega_o} \int_{-\infty}^{\infty} \Delta W(\delta,z)(\Delta\omega - \langle\dot\phi\rangle)g(\delta)d\delta + \frac{2\langle k'\rangle}{T_2}, \quad (9)$$

where

$$\mathcal{J}(z) = \frac{nc}{4\pi}\int_{-\infty}^{\infty}\&^2(z,t)dt \quad (10)$$

is the pulse energy, and

$$\langle k'\rangle = -\left(\int_{-\infty}^{\infty}\frac{\&^2\frac{\partial\phi}{\partial z} dt}{\int_{-\infty}^{\infty}\&^2 dt}\right) = (\frac{1}{2\mathcal{J}}\int_{-\infty}^{\infty}\int_{-\infty}^{\infty}u\&g(\delta)d\delta dt)(\omega+2\Delta\omega_o) \quad (11)$$

is the average contribution of the propagation vector because of the two-level system. The expression for $d\langle\dot\phi^2\rangle/dz$, including a scattering loss term $-\sigma\varepsilon/2$ added to the right side of (7), gives

$$\frac{d\langle\dot\phi^2\rangle}{dz} = \frac{1}{\mathcal{J}}\left\{\int_{-\infty}^{\infty}\int_{-\infty}^{\infty}dtd\delta g(\delta)\left[\omega v\&(\langle\dot\phi^2\rangle + \frac{1}{T_2}Z - \Delta\omega^2) + \frac{\omega}{\omega_o}\kappa^2\&^2\left(\frac{W}{T_2} - \frac{\dot W}{2}\right)\right.\right.$$

$$+ \frac{2\omega\delta\&u}{T_2}\left.\right\} +4\left(\frac{\Delta\omega_o}{T_2}\right)\left(\frac{\langle k'\rangle\omega}{\omega+2\Delta\omega_o}\right) - \frac{1}{\mathcal{J}}\left[\frac{d}{dz}(\mathcal{J}/\tau^2) + \frac{\sigma\mathcal{J}}{\tau^2}\right] \quad (12)$$

where

$$z' = \left(\frac{\omega + 2\Delta\omega_o}{\omega}\right)z.$$

The above relations are valid in the approximation that

$$\left(\frac{\Delta\omega}{\omega}\right)^2 << 1 \text{ and } \frac{\dot\phi}{2\omega} << 1. \quad (13)$$

Equation (9) expresses two frequency alteration effects on the pulse carrier. For no initial phase modulation ($\langle\dot\phi\rangle = 0$), near z = 0 the first term on the right side of (9) contributes to frequency pushing, and the second term contributes to frequency pulling with respect to an absorbing line resonant at ω_o. For an emitter the converse takes place.

ENERGY BALANCE CONSIDERATIONS

One expects that energy from the radiation must contribute both to excitation of the atoms and to frequency changes in the forward scattered plane wave. This must take place at the expense of the light intensity $\mathcal{E}^2$. Consider an average of $\boldsymbol{\eta}$ light quanta in the pulse, where each quantum has mean energy $\hbar[\omega + \langle \dot{\phi} \rangle]$. The pulse energy is expressed by

$$\mathcal{T} = \boldsymbol{\eta}\,\hbar[\omega + \langle \dot{\phi} \rangle]. \tag{14}$$

For mean frequency shifts to occur, we must have

$$\frac{d\mathcal{T}}{dz} = \hbar[\omega + \langle \dot{\phi} \rangle]\,\frac{\partial \boldsymbol{\eta}}{\partial z} + \hbar\boldsymbol{\eta}\frac{\partial \langle \dot{\phi} \rangle}{\partial z} - \hbar\boldsymbol{\eta}\langle\frac{d(\phi/T_2)}{dz}\rangle. \tag{15}$$

The first two terms of (15) express energy conservation in the limit of no random scattering ($1/T_2 = 0$). This would apply to coherent forward plane wave scattering expressed by the first term on the right side of (9). Normally radiation is absorbed at the driving frequency $\omega + \langle \dot{\phi} \rangle$, and scattered at that same frequency if no losses occur. In the non-linear rearrangement process involving both scattering and losses, the derivatives $\partial \boldsymbol{\eta}/\partial z$ and $\partial \langle \dot{\phi} \rangle/\partial z$ adjust to give energy balance. The second term on the right side of (15) accounts for frequency shifts ϕ/T_2 associated with random phase and momentum scattering.

Our problem is taken in the retarded time frame $t' = t-\eta(z/c)$, so that the time derivatives on the left side of (7) and (8) are eliminated. Let $T_2 = \infty$ and consider the sharp line case ($\delta = 0$). Combination of Eq. (7) with the $\dot{W}$ relation in (6) gives

$$\frac{d\mathcal{T}}{dz} = -\left(1 + \frac{\Delta\omega_o}{\omega_o}\right)\Delta W. \tag{16}$$

We note that Eq. (16) is usually seen in the form on-resonance where $\Delta\omega_o = 0$, $\Delta W = |\alpha|\mathcal{T}$, and α is the absorption coefficient. The ordinary decay law $\mathcal{T} = \mathcal{T}_o \exp(-|\alpha|z)$ results. Whether or not resonance is obtained, ΔW only relates to the rate at which quanta are absorbed from the field. Therefore

$$\hbar[\omega + \langle \dot{\phi} \rangle]\,\frac{\partial \boldsymbol{\eta}}{\partial z} = -\Delta W. \tag{17}$$

If ΔW peaks about ω_o, and $\omega + \langle \dot{\phi} \rangle$ is off-resonance, then the quantum frequency change is obtained from the $\Delta\omega_o(\Delta W/\omega_o)$ term in (16), set equal to $\hbar\boldsymbol{\eta}(\partial\langle\dot\phi\rangle/\partial z)$ in (15):

$$\frac{\partial\langle\dot\phi\rangle}{\partial z} = -\left(\frac{\omega + \langle\dot\phi\rangle}{\omega_o}\right)\left(\frac{\Delta W\Delta\omega_o}{\boldsymbol{\mathscr{T}}}\right) \tag{18}$$

The term in T_2, on the right side of (15), would also appear in (18), if included. The result is essentially that of Eq. (9), derived carefully from the semi-classical equations, where we see that the $\langle\dot\phi\rangle$ and $2\Delta\omega_o$ terms may be dropped when compared to ω alone. But to the term $\Delta\omega_o = \omega_o - \omega$, $\langle\dot\phi\rangle$ should be added (as $-\langle\dot\phi\rangle$) because $\langle\dot\phi\rangle$ may be comparable to $\Delta\omega_o$. Therefore $\Delta\omega \rightarrow \Delta\omega_o - \langle\dot\phi\rangle$ in (18) above.

In making this phenomenological approach according to (16), it is important to retain the next order terms (keeping $2\Delta\omega_o/\omega_o$) usually dropped in the slow wave approximation. These come from the factors $\omega + 2\Delta\omega_o$ in (7) and (8). Physically they signify that to some extent the radiation due to the much slower rates of change of $\dot u$ and $\dot v$ must be taken into account to provide the necessary asymmetry in the way $\boldsymbol{\mathscr{T}}$ changes with changes in $\langle\dot\phi\rangle$ if (15) is to be applied. This can be seen in the sharp line case, letting $\langle\dot\phi\rangle = 0$ at $z = 0$. The ratio of Eq. (16) to Eq. (18) gives

$$\frac{\partial\boldsymbol{\mathscr{T}}}{\partial\langle\dot\phi\rangle} \approx \left(\frac{\omega_o + \Delta\omega_o}{\omega}\right)\left(\frac{\boldsymbol{\mathscr{T}}}{\Delta\omega_o}\right) \quad , \tag{19}$$

valid for $\Delta\omega_o \neq 0$.

For $\Delta\omega_o$ positive below resonance, Eq. (19) states the $\partial\boldsymbol{\mathscr{T}}/\partial\langle\dot\phi\rangle$ is positive for frequency pushing (absorber) or pulling (emitter). If the quantum energy $\hbar[\omega + \langle\dot\phi\rangle]$ reduces, the pulse field intensity $\mathscr{E}^2$ increases (if τ does not change appreciably) at the expense of reduction in carrier frequency, other variables being held constant. Conversely for $\Delta\omega_o$ negative, above resonance, $\mathscr{E}^2$ would decrease.

WEAK PULSE BEHAVIOR

For weak pulses the condition $\Delta W \ll W_o$ permits an analytic expression for $d\langle\dot\phi\rangle/dz$. Assume a Gaussian inhomogeneously broadened two-level spectrum, $T_2 = \infty$, and a pulse which preserves a Gaussian shape during propagation. In the approximation $\Delta\omega_o/\omega \ll 1$ and $\langle\dot\phi\rangle \ll \Delta\omega_o$, one can integrate Eq. (9) on the right side and obtain the approximate result

$$\frac{d\langle\dot\phi\rangle}{dz} \approx 4\pi^{3/2}\left(\frac{\kappa^2 W_o}{\eta c}\right)[\Delta\omega_o - \langle\dot\phi\rangle]\tau s^3 e^{-[\Delta\omega_o - \langle\dot\phi\rangle]^2\tau^2 s^2} \quad , \tag{20}$$

where

$$s = \frac{T_2{}^*}{\sqrt{T_2{}^{2*} + \tau^2}}.$$

The width of the inhomogeneous Gaussian line spectrum corresponds to $1/T_2{}^*$. Eq. (20) indicates frequency pushing for an absorber and frequency pulling for an emitter. Experimental results with ruby laser pulses, to be reported in a later paper, show, for input pulses above or below the π area threshold for onset of SIT, that the output pulse is strongly phase modulated according to (20). The pulse exhibits some delay, less than that for a 2π h.s. SIT pulse but tends to show asymmetric shape. Its appearance may easily be mistaken for a real SIT pulse: the pulse could be appropriately labeled pseudo self-induced transparency.

QUASI-CONTINUOUS WAVE LIMIT; HOMOGENEOUS BROADENING

For conditions giving Eq. (20), the frequency change disappears for long pulse widths such that $\Delta\omega_0 \tau \gg 1$ or $\tau \gg T_2{}^*$. The pulse Fourier components then essentially disappear at the frequency ω_0 about which the line spectrum peaks. In this situation for a weak pulse, the carrier wave may be considered virtually continuous for τ very large, and the two-level system responds adiabatically to the radiation.

In the absence of inhomogeneous broadening ($\delta = 0$), the only mechanism which remains to produce any further phase modulation propagation effects is that due to homogeneous broadening, associated with the parameter $1/T_2$. In this case, it is the amplitude of the line Fourier spectrum at the frequency $\omega + \langle \dot\phi \rangle$ of the carrier which is important. Since the wave is now virtually continuous, with an amplitude changing very slowly due to propagation loss (for an absorber) or gain (for an emitter), the steady state solutions to Bloch's equations (6) may be used, where $\dot u = \dot v = 0$, $W_0 \approx W$, and W is finite but small. These solutions are coupled with Eqs.(9) and (12) in the far off-resonance limit $\Delta\omega_0 T_2 \gg 1$, with the following restrictions:

$$\Delta\omega_0 \ll \omega; \quad \langle \dot\phi \rangle \ll \Delta\omega_0; \quad \kappa^2 \mathcal{E}^2 \ll \Delta\omega_0{}^2 \tau^2; \tau \gg T_2; \quad \alpha \mathcal{T}/W_0 \ll 1;$$

$$\dot u = \dot v = 0; \quad W \sim W_0; \quad \text{and} \quad \dot W \neq 0.$$

The time interval for measurement of $\langle \dot\phi \rangle$ and q needs only to be long, and not beginning and ending with $\mathcal{E} = 0$. Including terms up to order $1/(\Delta\omega_0)^3$, the distance derivative of the first moment is given by

$$\frac{d\langle \dot{\phi} \rangle}{dz} = \frac{\omega}{\omega_o} \frac{\alpha}{\Delta\omega_o} \left[2q - \frac{\kappa^2 \langle \mathscr{E} \rangle^2}{2} \right] + \dots, \tag{21}$$

where the absorption coefficient at frequency ω is

$$\alpha = \frac{4\pi W_o \kappa^2}{\eta c \Delta\omega_o^2 T_2} \quad ; \quad q = \langle \dot{\phi}^2 \rangle - \langle \dot{\phi} \rangle^2 ;$$

and

$$\langle \mathscr{E}^2 \rangle = \int_{-\infty}^{\infty} \mathscr{E}^4 dt / \int_{-\infty}^{\infty} \mathscr{E}^2 dt.$$

Eq. (21) expresses either frequency pushing or pulling from the point of view of non-linear changes in the classical dispersion $k' \sim \partial\phi/\partial z = \eta'$. Suppose ω is applied below resonance ($\Delta\omega_o = +|\Delta\omega_o|$) for an absorber ($\alpha = -|\alpha|$). If $\kappa^2 \langle \mathscr{E}^2 \rangle > q$ is the initial condition in (21), dispersion saturation is dominant and the effective **refractive** index η' decreases during the presence of $\mathscr{E}$, and the average value of $\partial\phi/\partial z$ decreases with time for initial progation distance z. A corresponding increase in wave velocity

$$V_w = \frac{\omega + \dot{\phi}}{k - \frac{\partial\phi}{\partial z}}$$

occurs. As a function of increasing z, the field oscillations compress in time, $\omega + \langle \dot{\phi} \rangle$ increases, and therefore $d\langle \dot{\phi} \rangle/dz$ is positive. Conversely, $\omega + \langle \dot{\phi} \rangle$ decreases for $\omega_o < \omega$, and $d\langle \dot{\phi} \rangle/dz$ is negative. Hence, frequency pulling toward the resonance at ω_o is associated with dispersion saturation for an absorber; and conversely, frequency pushing occurs for an emitter ($\alpha = +|\alpha|$).

In (21) the predominance of chirping or mean square frequency modulation occurs for $q > \kappa^2 \langle \mathscr{E}^2 \rangle$. All the signs of $d\langle \dot{\phi} \rangle/dz$ which apply in the case of dispersion saturation now reverse for the initial propagation. In the case of the absorber, strong chirping implies a significant variation of $\omega + \dot{\phi}$ about the average value $\omega + \langle \dot{\phi} \rangle$. For $\omega_o > \omega$, the refractive index increases more for $\omega + \dot{\phi} > \omega + \langle \dot{\phi} \rangle$ than for $\omega + \dot{\phi} < \omega + \langle \dot{\phi} \rangle$ because

$$\frac{d^2\eta' (\omega + \langle \dot{\phi} \rangle)}{d\omega_{av}^2} > 0;$$

and the converse effect occurs for $\omega_o < \omega$. The net effect is an increase in η' for $\omega_o > \omega$ and a net decrease in η' for $\omega_o < \omega$, which in turn signify corresponding decreases and increases in wave velocity. The carrier frequency $\omega + \langle \dot{\phi} \rangle$ then pushes away from resonance. This opposes the pulling effect due to saturation. Both pushing and pulling effects are therefore embodied in Eq. (21).

A general solution for $\langle \dot{\phi}(z) \rangle$ from Eq. (21) is difficult to set forth here because of the non-linearity involved and because of the perturbation expansion imposed. To first order one obtains approximately

$$\frac{dq}{dz} \cong 0. \tag{22}$$

This is a consequence of the adiabatic requirement that q can never be permitted to be so large that $\langle \dot{\phi} \rangle$ is not a sufficiently unique value. If so, it would violate the assumption of slowly varying Bloch solutions.

For an absorber, appreciable changes in $\langle \dot{\phi} \rangle$ and q would be difficult to observe because of radiation attenuation for distances $z \gg |\alpha|^{-1}$. For an emitter, consider the case where initially

$$\kappa^2 \langle \mathcal{E}^2 \rangle /2 \gg 2q \tag{23}$$

and assume $\langle \mathcal{E}^2 \rangle = \langle \mathcal{E}_o^2 \rangle e^{+|\alpha|z}$. Eq. (21) then predicts initial frequency pushing. However, higher order terms dependent upon $1/(\Delta\omega_o)^n$, where $n > 3$, show that q will begin to increase so that the inequality given by (23) will head toward a reversal with increasing $|\alpha|z$. Thus, it appears that frequency pulling eventually takes over for an emitter. This turnabout is not easily analyzed here, however. Although the initial conditions are different, and pertain to pulses, changes in frequency as function of distance of this nature have been shown to occur in earlier computer calculations.[2]

In the course of experiments which yield self-induced transparency in EPR microwave transmission at X-band frequencies, spectrum analyzer measurements of $\langle \dot{\phi} \rangle$ and q have been made[3] for a pulse off-resonance propagating through a free radical two-level medium in a waveguide. The off-resonance distance dependence of pulse spectrum shifts are found to be in quantitative accord with Eq. (20) for inhomogeneous broadening. For homogeneous broadening Eq. (21) is shown to apply qualitatively from preliminary measurements.

A further study of our model is being extended in an attempt to account for certain interstellar emission line observations, taking into account Doppler broadening and multi-level transitions of the medium.

REFERENCES

[1]S.L. McCall and E.L. Hahn, Phys. Rev. 183, 457 (1969).
[2]J.C. Diels and E.L. Hahn, Phys. Rev. A 8, 1084 (1973).
[3]S.B. Grossman and E.L. Hahn, Bull.Am. Phys. Soc, Berkeley Meeting, December 1973.

OBSERVATION OF SLOW VELOCITIES AND SELF-STEEPENING OF OPTICAL

PULSES DESCRIBED BY ADIABATIC FOLLOWING

D. Grischkowsky*, Eric Courtens**, and J. A. Armstrong*

IBM Thomas J. Watson Research Center

Yorktown Heights, New York 10598

ABSTRACT

The observation and the theoretical explanation of phenomena which accompany near-resonant pulse propagation in atomic vapors are described. These phenomena bridge the gap between linear dispersion theory, which applies in the far wings of lines, and strictly resonant effects characteristic of self-induced transparency. The results indicate that very low propagation velocities and strong pulse reshaping are not unique to transparency situations.

Short pulses of narrow-line, low-intensity, dye laser light nearly resonant with the Zeeman split $^2P_{1/2}$ resonance line (7948 Å) of rubidium were observed to propagate through the dilute vapor as slowly as c/14 [1]. Low-intensity pulses were essentially undistorted by passage through the vapor. High-intensity pulses not only propagated slowly but were self-steepened in the process [2]. Their risetimes changed from typically 4 nsec to less than 1 nsec and complicated envelope shapes developed.

All observations are very well accounted for by the adiabatic following model in which the pseudomoments of the atoms remain closely aligned along the effective field of the laser light. At low intensity, the Rabi precession frequency is very small compared

* Work of this author was partially supported by ONR under Contract No. N00014-70-C-0187.

** Permanent address: IBM Zurich Research Laboratory, 8803 Rüschlikon, Switzerland.

to the frequency difference between the light and the resonance
line. In this limit, the approximation is equivalent to linear
dispersion theory. The low-level pulse velocity is $v_g = \partial\omega/\partial k$, in
good agreement with experiment. At high intensity, adiabatic
following predicts a nonlinear dielectric response demonstrated
earlier in self-focusing [3] and self-defocusing [4] experiments.
This leads to two self-steepening mechanisms, which have familiar
analogs in other nonlinear optics media: i) an intensity
dependence of the pulse velocity, ii) self-phase modulation
together with strong group velocity dispersion. The numerical
integration of the resulting equations gives excellent quantitative
agreement with observations. Further, it indicates optical shock
formation on the leading edge of the pulses.

INTRODUCTION

The present work describes phenomena that accompany the
propagation of nearly resonant optical pulses in dilute vapors.
By nearly resonant it is meant that the frequency difference
between the light and the resonance line is very small compared to
the resonant frequency, but the frequency difference is larger than
a few Doppler widths. The phenomena can be divided in two
categories. Firstly, the pulse propagation velocity can become
much smaller than c. For low-intensity pulses this is the most
prominent effect. Quantitative measurements are in accordance with
the Sommerfeld-Brillouin group velocity theory. Secondly, more
intense pulses become strongly distorted in this process. This
phenomenon, which can be described as self-steepening, bears a
resemblance to phenomena predicted to occur, but never directly
observed, in other nonlinear optical media. There is indication
that the present steepening leads to optical shock formation at the
front of the pulses. By optical shock we mean an extremely sharp
front propagating with little reshaping due to the compensating
influence of the steepening mechanisms and of some other dispersive
mechanism.

Accurate measurements of group velocity in dilute media are
surprisingly scarce. Most previous measurements were obtained by
placing the material under study within mode-locked laser cavities
and measuring the change in the pulse repetition rate [5-8]. In
particular, using this technique Casperson and Yariv [8] observed
slow pulse velocities of c/2.5. On the other hand, at resonance
the coherent phenomenon of self-induced transparency [9] can
produce extremely slow pulse velocities for pulses with an area of
the order of 2π. In this case pulse velocities of c/1000 have been
measured [10]; these velocities are not describable by the usual
group velocity theory.

The possibility of observing self-steepening of intense optical pulses and the formation of optical shock waves has been extensively discussed in the literature [1,2,11-18]. An early problem considered in detail was that of pulse propagation in a medium with an intensity dependent index of refraction and no frequency dispersion [12-13]. For this case the pulse reshaping is entirely caused by the intensity dependence of the pulse velocity v_p. The pulse propagates as a "simple wave", i.e. a wave described by $I(t,z) = I(t-z/v_p)$, until the formation of the optical shock, even though the instantaneous frequency may change substantially due to self-phase modulation. The propagation distances and input powers required for significant self-steepening and for the formation of the optical shock under these conditions have remained experimentally unattainable. By including linear dispersion in the analysis, self-steepening was predicted to occur in much shorter propagation distances [11,14,17,18]. The main cause is self-phase modulation which sweeps the frequency of the input pulse. Consequently, the pulse reshapes due to the linear dispersion of its velocity. Even though the experimental parameters required for observation of self-steepening in this case are much more favorable, direct experimental verification of this effect has not been obtained.

Compared to the nonlinearity and dispersion of the pulse velocity in the Kerr liquids usually discussed, the nonlinearity and dispersion of the pulse velocity for near-resonant light in alkali vapors is enormous [1-4,15,17]. In particular the nonlinearity of the pulse velocity is easily 10^9 times larger in the vapors. Thus, by using these vapors, experiments demonstrating steepening become possible. Some of the earlier theoretical work concerning near-resonant propagation predicted simple wave propagation with a nonlinear pulse velocity [1,15]. This work was oversimplified in that the change in the instantaneous frequency caused by self-phase modulation was neglected. This was a serious omission, as the pulse velocity is exceptionally dispersive. Other theoretical work [17] which discussed near-resonant pulse propagation considered the effect of self-phase modulation and dispersive pulse velocity on pulse reshaping, but neglected to consider the nonlinear pulse velocity.

In contrast to self-induced transparency, our low intensity results remained the same when the pulse width was changed from 3 nsec to 18 nsec, when the intensity was changed by 400, and when the input pulse areas were changed from 0.5π to 50π. However, as in self-induced transparency, most of the energy in the propagating pulse is contained by the atoms as coherent excitation, the spatial extent of the pulse in the resonant material is reduced by the factor v_p/c, and the absorption is due to the relaxation processes which determine the homogeneous linewidth of the resonance line.

The present observations confirmed that "the <u>qualitative</u>
observation of a pulse delay beyond the input pulse tail is <u>not</u>
sufficient evidence of (self-induced) transparency" as first
pointed out by Courtens and Szöke [19].

Adiabatic following describes the propagation of near-resonant
light rather accurately. This model bridges the gap between the
region where linear dispersion theory is applicable and the region
where the entire set of Bloch equations must be coupled to the wave
equation. The approximation provides an immediate solution to the
point response problem in that the pseudomoments are closely
aligned along the effective field of the incident light. The near-
resonant atomic response is thus expressed as a nonlinear electric
susceptibility. This is an important simplification, which allows
for comparison with other nonlinear systems and for convenient
numerical integration of the coupled Maxwell-Bloch equations. The
adiabatic following approximation applies when two conditions are
satisfied: (1) the direction of the effective field $\vec{\mathcal{E}}_e$ (in the
rotating frame in which the circularly polarized electric field $\vec{E}$
of the incident light appears stationary) should change slowly
compared to the precession frequency of the pseudomoment $\vec{p}$ about
$\vec{\mathcal{E}}_e$, (2) the pulsewidth should be short compared to T_1 and T_2 of
the atomic system. A low-intensity pulse is one for which the
angle between the effective field vector $\vec{\mathcal{E}}_e$ and its original
position along the Z axis of the rotating frame (Fig. 1) never
becomes large. Under this condition the adiabatic following
results are equivalent to those of linear dispersion theory. At
higher intensities self-steepening is obtained. Adiabatic
following predicts two pulse reshaping mechanisms: i) an intensity
dependent pulse velocity, ii) self-phase modulation together with
strong group velocity dispersion. Numerical integration gives
quantitative agreement with observations and indicates optical
shock formation on the leading edge.

The Theory of Nonlinear Wave Propagation
in the Adiabatic Following Approximation

Before turning to the actual experimental results it is
helpful to develop the theory of wave propagation within the
adiabatic following approximation. The electric field is described
by its slowly-varying envelope E and its slowly-varying phase ϕ,
the instantaneous angular frequency being $\omega' = \omega + \partial\phi/\partial t$. It is
convenient to use, instead of E, the instantaneous Rabi precession
frequency $\mathcal{E} = \sqrt{2}\, p_{12}\, E/\hbar$, where p_{12} is the absolute value of the
electric dipole matrix element for the σ transition of interest
($p_{12} = 6.16 \times 10^{-18}$ esu for the $P_{1/2}$ line of Rb). The atoms are
described by the in-phase and out-of-phase components of the
polarization, u and v, respectively, and by the atomic energy

ADIABATIC FOLLOWING

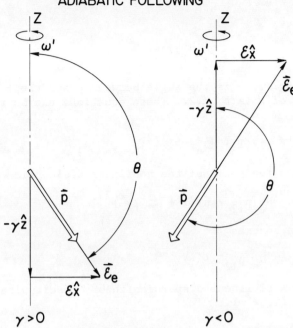

Figure 1.   Adiabatic following of the effective field $\vec{\mathcal{E}}_e$ (in the rotating frame in which $\vec{E}$ appears stationary) by the pseudomoment $\vec{p}$, when the instantaneous angular frequency ω' of the laser light is either less than or greater than the atomic resonant frequency ω_0. $\mathcal{E} = \sqrt{2} \, p_{12} \, E/\hbar$ is the angular Rabi precession frequency; E is the electric field of the input circularly polarized light; p_{12} is the absolute value of the matrix element of the electric dipole moment for the σ transition; $|\vec{p}| = p_{12}/\sqrt{2}$; $\gamma = \omega_0 - \omega'$.

density W. In a frame (X,Y,Z) rotating at the instantaneous field frequency ω', $\mathcal{E}$ is along the X axis; u and v are N_e times the X and Y components of the pseudomoment $\vec{p}$, respectively, where N_e is the effective number density. The effective field $\vec{\mathcal{E}}_e$ is defined in angular frequency units so that the precession equation for $\vec{p}$ reduces to $\partial \vec{p}/\partial t = \vec{\mathcal{E}}_e \times \vec{p}$. As shown in Fig. 1 the components (X,Y,Z) of $\vec{\mathcal{E}}_e$ are $(\mathcal{E}, 0, -\gamma)$, respectively; $\gamma = \omega_0 - \omega'$ is the instantaneous angular frequency offset, where ω_0 is the resonant frequency. The adiabatic following approximation states that $\vec{p}$ remains closely aligned with $\vec{\mathcal{E}}_e$ provided the direction of $\vec{\mathcal{E}}_e$ changes sufficiently slowly, as described in the introduction. The in-phase component u of the macroscopic polarization $\vec{P}$ of the Rb vapor is then, from Fig. 1:

$$u = N_e |\vec{p}| \sin\theta \tag{1}$$

with

$$\sin\theta = \mathcal{E} / [\gamma(1 + \mathcal{E}^2/\gamma^2)^{1/2}] \tag{2}$$

As shown in Fig. 1, θ is the angle between $\vec{p}$ and the + Z direction. The magnitude of $\vec{p}$ is $p_{12}/\sqrt{2}$. These equations can be rewritten as [1-4,15,20]:

$$u = N_e |\vec{p}| \mathcal{E} / [\gamma(1 + \mathcal{E}^2/\gamma^2)^{1/2}] \tag{3}$$

Using Eq. 3 the real part of the nonlinear dielectric constant is

$$\varepsilon = 1 + 4\pi N_e p_{12}^2 / [\hbar\gamma(1 + \mathcal{E}^2/\gamma^2)^{1/2} \tag{4}$$

For $\mathcal{E}^2 \ll \gamma^2$, ε becomes

$$\varepsilon_o = 1 + 4\pi N_e p_{12}^2 / (\hbar\gamma) \tag{5}$$

which is the usual linear dispersion theory result with the neglect of relaxation ($T_2 = \infty$).

The adiabatic following condition implies that the out-of-phase component of the polarization v is much smaller than the in-phase component u. In the next order of approximation, u is used to calculate v from the Bloch equation

$$v = (\partial u/\partial t - u/T_2)/\gamma \tag{6}$$

Here $1/T_2 = 1/2\tau + 1/T_c$, where τ is the radiative lifetime of the excited state (28 nsec for the $^2P_{1/2}$ line) and T_c is the time between dephasing collisions [21].

These results are introduced in the coupled, reduced wave equations.

$$\frac{\partial E}{\partial z} + \frac{1}{c} \frac{\partial E}{\partial t} = \frac{-2\pi\omega v}{c} \tag{7a}$$

$$\frac{\partial \phi}{\partial z} + \frac{1}{c} \frac{\partial \phi}{\partial t} = \frac{-2\pi\omega u}{cE} \tag{7b}$$

These equations are written for $n_o = \sqrt{\varepsilon_o} = 1$ since $(n_o-1) < 10^{-4}$ in the present experimental conditions. Equation (7a), and the partial time derivative of Eq. (7b), are rewritten in the form

$$\frac{\partial \mathcal{E}}{\partial z} + \frac{1}{v_p} \frac{\partial \mathcal{E}}{\partial t} = (\frac{1}{v_p} - \frac{1}{c}) \frac{\mathcal{E}}{\gamma} \frac{\partial \gamma}{\partial t} - \frac{\alpha}{2} \mathcal{E} \tag{8a}$$

$$\frac{\partial \gamma}{\partial z} + \frac{1}{v_p} \frac{\partial \gamma}{\partial t} = - (\frac{1}{v_p} - \frac{1}{c}) \frac{\mathcal{E}}{\gamma} \frac{\partial \mathcal{E}}{\partial t} \tag{8b}$$

where the nonlinear pulse velocity v_p and the nonlinear absorption coefficient α are given by

$$\frac{1}{v_p} = \frac{1}{c}\{1 + 2\pi\omega N_e p_{12}^2 / [\hbar\gamma^2 (1 + \mathcal{E}^2/\gamma^2)^{3/2}] \} \tag{9a}$$

$$\alpha = \frac{2}{T_2} (\frac{1}{v_p} - \frac{1}{c}) (1 + \frac{\mathcal{E}^2}{\gamma^2}) \tag{9b}$$

Low intensity light is defined by $\mathcal{E}^2 \ll \gamma^2$. In this case Eqs. (8) reduce approximately to

$$\frac{\partial \mathcal{E}}{\partial z} + \frac{1}{v_g} \frac{\partial \mathcal{E}}{\partial t} = (\frac{1}{v_g} - \frac{1}{c}) \frac{\mathcal{E}}{\gamma} \frac{\partial \gamma}{\partial t} - \frac{\alpha_o \mathcal{E}}{2} \tag{10a}$$

$$\frac{\partial \gamma}{\partial z} + \frac{1}{v_g} \frac{\partial \gamma}{\partial t} = 0 \tag{10b}$$

where v_g is the usual group velocity and α_o is the low-level absorption coefficient given by

$$\frac{1}{v_g} = \frac{1}{c}\{1 + 2\pi\omega N_e p_{12}^2 / (\hbar\gamma^2)\} \tag{11a}$$

$$\alpha_o = \frac{2}{T_2} (\frac{1}{v_g} - \frac{1}{c}) \tag{11b}$$

The first term on the right hand side of Eq. (10a) is caused by linear dispersion and could have been obtained by second differentiation in ω of the dielectric constant. In the absence of frequency modulation ($\partial\gamma/\partial t = 0$) the pulse envelope propagates as an attenuated simple wave

$$\mathcal{E}(t,z) = \mathcal{E}_o(t-z/v_g) \exp (-\alpha_o z/2) \tag{12a}$$

In the presence of initial frequency modulation Eqs. (10) predict reshaping by dispersion. This is due both to the first term on the right hand side of Eq. (10a) and to the γ dependence of v_g. The frequency, however, does propagate as a simple wave:

$$\gamma(z,t) = \gamma(t - z/v_g) \tag{12b}$$

Equation (11a) for v_g can be recast in the form

$$\frac{1}{v_g} = \frac{1}{c} \{1 + \omega \frac{\partial n_o}{\partial \omega}\} \tag{13}$$

provided $n_o = \sqrt{\varepsilon_o} \overset{\sim}{=} 1$. In this limit this result is the same as that obtained from linear dispersion theory [22-25]. This is not surprising, since the condition $\mathcal{E}^2 << \gamma^2$ means that the angle between p and the Z axis remains very small and that saturation can be neglected. In this case the equation of motion reduces to the equation of a damped harmonic oscillator. In the adiabatic following approximation the oscillator is excited at the driving frequency ω' without transient oscillations at the resonant frequency ω_o.

For unmodulated input pulses with $\gamma = \delta$ = constant, Eqs. 8, valid for arbitrary intensity, are conveniently rewritten in terms of a coupling strength $S = 2\pi N_e \omega p_{12}^2/(\hbar \delta^2)$, a normalized distance $\zeta = z/c$ and a local time $\tau = t - z/v_g$. The low-level group velocity is $v_g = c/(1 + S)$ provided $|\delta| << \omega$. The equations are:

$$\frac{\partial \mathcal{E}}{\partial \zeta} = M \frac{\partial \mathcal{E}}{\partial \tau} + N \mathcal{E} \frac{\partial \gamma}{\partial \tau} - \frac{c\alpha \mathcal{E}}{2} \tag{14a}$$

$$\frac{\partial \gamma}{\partial \zeta} = M \frac{\partial \gamma}{\partial \tau} - N \mathcal{E} \frac{\partial \mathcal{E}}{\partial \tau} \tag{14b}$$

where $M = S[1-(\delta/\gamma)^2(1+\mathcal{E}^2/\gamma^2)^{-3/2}]$ and $N = S \delta^2 \gamma^{-3}(1+\mathcal{E}^2/\gamma^2)^{-3/2}$. The terms in M account for the intensity and frequency dependences of the pulse velocity. These constitute one steepening mechanism. The term in N in Eq. (14a) describes nonlinear dispersion. For small $\mathcal{E}$ it becomes identical to the usual result of linear dispersion theory which can be obtained by second differentiation in ω of the dielectric constant ε_o. The term in N in Eq. (14b) describes self-phase modulation. The combination of dispersion and self-phase modulation is another steepening mechanism. Under present experimental conditions both mechanisms contribute to the final pulse shape, the second being the stronger. The self-steepening is identical on both sides of the resonance line since Eqs. (14) are unchanged when γ changes signs.

In order to compare theory and experiment it is necessary to account for the ten hyperfine components (ω_{oi}, i = 1 to 10) of the Rb line, for the diffraction of the spherical wave, and for the absorption caused by T_2. One defines ten coupling strengths S_i, ten offsets $\gamma_i = \Gamma + \delta_i^2$, where $\Gamma = -\partial\phi/\partial t$, and the group velocity $v_g = c/(1 + \sum_i^i S_i)$. Equations (14) are replaced by:

$$\frac{\partial \mathcal{E}}{\partial \zeta} = m \frac{\partial \mathcal{E}}{\partial \tau} + n \, \mathcal{E} \, \frac{\partial \Gamma}{\partial \tau} - \frac{\mathcal{E}}{(\zeta_o + \zeta)} - \frac{c\alpha \mathcal{E}}{2} \tag{15a}$$

$$\frac{\partial \Gamma}{\partial \zeta} = m \frac{\partial \Gamma}{\partial \tau} - n \, \mathcal{E} \, \frac{\partial \mathcal{E}}{\partial \tau} \tag{15b}$$

where m and n are sums over ten terms similar to M and N, ζ_o is the time of travel from the center of the spherical wave to the cell entrance at $\zeta = 0$, and $\alpha = 2(c \, T_2)^{-1} \sum_i S_i \, \delta_i^2 \, \gamma_i^{-2} (1 + \mathcal{E}^2/\gamma_i^2)^{-1/2}$. The term containing ζ_o in Eq. (15a) can be derived from the three dimensional scalar wave equation as the major diffraction term for a spherical wave. Its present use is of course approximate; in particular defocusing effects are not taken into account.

Experimental Arrangement

The experimental arrangement shown in Fig. 2 is similar to that of Ref. 4 which describes the ruby laser pumped dye laser

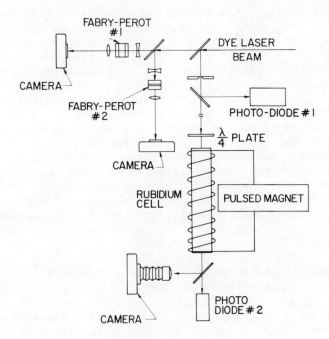

Figure 2.   Schematic diagram of the experiment.

and the Rb cell. Two Fabry-Perot interferograms were taken of
each pulse. Both Fabry-Perot (F.P.) interferometers had a finesse
of about 100. F.P.#1 had a spectral range of 0.33 cm^{-1} and F.P.#2
a spectral range of 2.5 cm^{-1}. F.P.#1 monitored the linewidth $\delta\omega$
(Fig. 3a), while F.P.#2 measured the frequency difference between
the dye laser output and the $^2P_{1/2}$ line of a Rb spectral lamp.
The main beam went through a 1 mm dia. aperture, 525 cm before the
entrance window of the Rb cell. Consequently, the intensity
profile of the beam at the Rb cell was the Fraunhofer diffraction
pattern of the aperture. The input pulses shown in Figs. 3b and 5
were monitored with the ITT biplanar photodiode #1 connected to a
Tektronix 519 oscilloscope. The main beam went through a quarter-
wave plate, which determined the ratio of σ^+ to σ^- light. After
passing through attenuating filters, the pulse propagated through
the 100 cm Rb cell while an adjustable pulsed magnetic field of
typically 8 kG was applied. A near-field photograph was taken of
the beam at the exit window of the cell to check for self-
defocusing; the carrier frequency ω of the pulse was always on the
low-frequency side of the line in order to avoid problems caused by
self-focusing [3]. The output pulses of Figs. 3c and 5 were
observed with the ITT biplanar photodiode #2 connected to a
Tektronix 7904 oscilloscope externally triggered by the 519. The
photodiodes were calibrated against a TRG thermopile. This set-up
was adequate to measure the slow pulse velocities of the low-
intensity pulses.

At high intensities, self-defocusing was evident with the
arrangement of Fig. 2, which was therefore modified as follows. An
f = - 103 cm lens was placed 525 cm from the 1 mm aperture; 20 cm
beyond the lens the beam entered the cell. The beam profile at the
lens was the Fraunhofer diffraction pattern of the aperture. The
negative lens changed the radius of the spherical wave from 525 cm
to 86 cm at the lens position, and this strongly suppressed self-
defocusing effects. At the same time an additional Fabry-Perot
interferometer was added to the system. The interferometer with a
resolution of 0.01 cm^{-1} monitored the spectrum of the output pulse
in order to check for self-phase modulation. Finally, in order to
be able to compare the experimental results with the spherical wave
theory described above, the output window of the cell was imaged
with unit magnification on a 3 mm aperture placed in front of the
photodiode #2, so that only the central portion of the beam was
recorded.

Experimental Results and Comparison with Theory:
Low-Intensity Case

Figure 3 shows a single pulse of low-intensity linearly-
polarized dye laser light (b) which has been split (c) into two

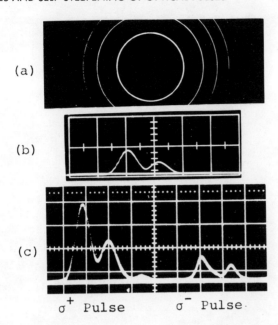

(a)

(b)

(c)

σ^+ Pulse σ^- Pulse

Figure 3.   (a) Fabry-Perot interferogram of the dye laser pulse
shown in (b) and (c). The interferometer had a finesse
of ∿ 100 and a spectral range of 0.33 cm^{-1}. The
measured linewidth $\delta\omega/2\pi c$ was less than 0.005 cm^{-1}.
(b)   The low-intensity linearly polarized input pulse to
the Rb cell. The pulse was observed with an ITT
biplanar photodiode and a Tektronix 519 oscilloscope.
The sweep speed was 5 nsec per division and the peak
power corresponded to 7 watts entering the Rb cell.
(c) The resulting two circularly polarized σ^+ and σ^-
output pulses from the 100 cm Rb cell. The pulses were
observed with an ITT biplanar photodiode and a Tektronix
7904 oscilloscope. The sweep speed was 5 nsec per
division, and the measured pulse separation was 26 nsec,
corresponding to a pulse velocity of c/9 for the σ^-
pulse. The Rb cell temperature was 140°C, and the cell
was in a magnetic field of 8.5 kG. The frequency of the
laser light was 0.25 cm^{-1} below the center frequency of
the σ^- hyperfine components of the Zeeman split $^2P_{1/2}$
resonance line of Rb and 1.30 cm^{-1} below the center
of the σ^+ components.

pulses of σ^+ and σ^- circularly polarized light by passage through
the Rb vapor cell. The σ^+ pulse has not been much delayed or
distorted by passage through the vapor. The σ^- pulse has been
strongly delayed and slightly reshaped. It has also been
attenuated by the factor $\exp(-\alpha_o L)$, where α_o is the small-signal

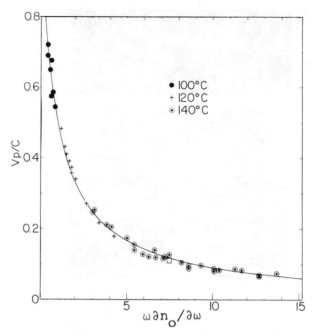

Figure 4.   Comparison with Eq. (13) of the measured pulse
             velocities of the low-intensity σ^- pulses through the Rb
             vapor. Data were taken with the Rb vapor cell
             temperature at 100°C, 120°C, and 140°C. The σ^- pulse
             of Fig. 3 is indicated by the open square.

absorption coefficient on the Lorentzian wing of the line and L_+ is
the cell length. With a circularly polarized input, only the σ^+ or
the σ^- pulse is seen. Switching from one polarization to the other
is accomplished by rotating the quarter-wave plate of Fig. 2 by 90
degrees. Knowing the transit time of the σ_- pulse ($\sim$ 4.2 nsec),
the 26 nsec separation between the σ^+ and σ^- pulses seen in Fig. 3c
indicates a σ^- pulse velocity v_g = c/9. The absorption
coefficient of the σ^- pulse is determined from the relative heights
of the σ^+ and σ^- pulses, as α_o = 1.2 x 10^{-2} cm^{-1}.

     These measurements have been performed as a function of cell
temperature, thereby adjusting the effective number density N_e, and
as a function of frequency offset δ. All the measurements can be
presented in a single diagram (Fig. 4) which is a representation of
Eq. (13). The parameters entering $\omega \partial n_o / \partial \omega = \sum_i S_i$ are all
experimentally measured except for N_e which at each temperature was

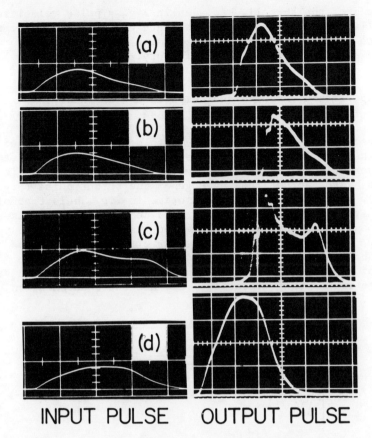

INPUT PULSE   OUTPUT PULSE

Figure 5. High intensity σ^- input pulses to the 100 cm Rb cell at
120°C and the resulting output pulses. The detectors
were ITT biplanar photodiodes; the input was observed
with a Tektronix 519 oscilloscope, which triggered a
Tektronix 7904 oscilloscope monitoring the output.
(a) $(\nu_o-\nu)/c = 0.24$ cm^{-1}, where ν_o is the center
frequency of the Rb line (7948 Å) and ν is the input
frequency. (b) $(\nu_o-\nu)/c = 0.20$ cm^{-1}. (c) $(\nu_o-\nu)/c =$
0.23 cm^{-1}. (d) $(\nu_o-\nu)/c = 0.78$ cm^{-1}.

determined from an adjusted equation of state giving the best
agreement with the data. This procedure gave for the total number
density $N = 2 N_e$, $N = 0.55$ x 10^{13}/cm^3 at 100°C, $N = 1.8$ x 10^{13}/cm^3
at 120°C and $N = 5.5$ x 10^{13}/cm^3 at 140°C. It also gave $N = 2.3$ x
10^{13}/cm^3 at 124°C, which agrees well with the value $N=2$ x 10^{13}/cm^3
used in Ref. 4. For the data of Fig. 4, $(\omega_o-\omega)$ was known to about
± 0.02 cm^{-1}; the pulse delays were measured to approximately ± 1
nsec for large delays and ± 0.5 nsec for small delays. A large
part of the scatter of the data was due to the ± 1°C temperature

fluctuations of the Rb cell. As seen in Fig. 4 pulse velocities as
slow as c/14 have been observed. The data point indicated by the
open square is the σ^- pulse of Fig. 3. The agreement between
theory and experiment is very satisfactory.

Experimental Results and Comparison with Theory: High Intensity Case

    Intense pulses of σ^- light were both delayed and reshaped as
shown in the examples of Fig. 5. We believe this is the first
direct observation of self-steepening. The self-steepening is
manifested by the abrupt leading edge of the output pulse (Figs.
5a and 5b). The observed 1 nsec risetime in Fig. 5b is that of the
detection system. More intense, or more nearly resonant pulses
develop complicated envelopes with abrupt rises and multiple peaks
(Fig. 5c). The pulse of Fig. 5d was far off resonance, shows
little reshaping or attenuation, and propagated with a group
velocity of 0.8 c. Comparing the pulses of Figs. 5a–5c with Fig.
5d, one determines their velocity $\lesssim$ c/4 and their attenuation $\lesssim$ 30%.

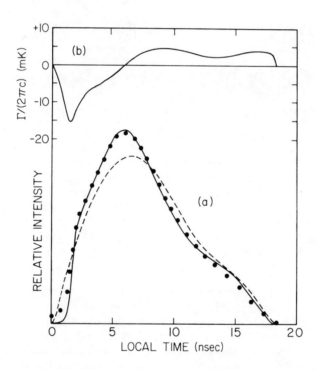

Figure 6.   The output pulse (dots) of Fig. 5a compared with the
             calculated pulse (solid line) and the normalized input
             (dashed line).   (b) Calculated self-phase modulation
             $\Gamma = - \partial\phi/\partial t$ in mK (10^{-3} cm^{-1}).

Figure 6 shows the result of applying Eqs. (15) to the input pulse of Fig. 5a. For comparison to the output (solid line) the input pulse (dashed line) has been drawn as if it had gone through the cell at the low-level group velocity subject only to linear absorption and diffraction. With this normalization, and except for the intensity dependence of α, the energies of the input and output pulses are equal. In particular the output peak is higher than the input due to the time compression, in good agreement with the calibration deduced from Fig. 5d. The calculated output agrees remarkably well with the measured points. The integration indicates that the leading edge of the pulse was still steepening at the end of the cell. Figure 6b shows that the predicted self-phase modulation is proportional to the time derivative of $\mathcal{E}^2$ in conformity with Eq. (15b). Its extent is of the order of the resolution of the output interferometer, and indeed broadening could not be clearly observed. All parameters of the calculation were experimentally measured. The input power was adjusted to 170 W, while the measured value was 200 W; this is within the accuracy of the calibration but might also indicate some defocusing. The adjusted power corresponds to a peak input intensity of 500 W/cm^2, a pulse area [9] $\sim 40\pi$, and a maximum $\mathcal{E} \sim 0.06$ cm^{-1}. The excellent agreement between experiment and theory for this and other similar pulses shows that the adiabatic following approximation is valid during reshaping.

In Fig. 5b the measured 1 nsec output risetime is that of the oscilloscope. Some spectral broadening mostly to the high frequency side, as predicted, was observed. Numerical integration of this pulse produced a steep front 30 cm before the output window; the integration of Eqs. (15) could not be continued beyond this point. The fact that a risetime limited leading edge was still obtained at the output strongly suggests that a shock traveled the remaining 30 cm. By shock we mean an extremely sharp front propagating with little reshaping, due to a dynamic balance between the steepening mechanisms of Eqs. (15) and a smoothing mechanism, not taken into account in the equations and operating only during extremely rapid variations of the field.

In Fig. 5c a complicated output envelope is obtained. In such cases there is usually insufficient detailed knowledge about the input pulse to be able to fit meaningfully theory and experiment. The interaction is so strong that the reshaping predicted by Eqs. (15) becomes extremely sensitive to the exact shape of the input. This instability is typical of the type of partial differential equations being solved. It turns out that the discriminant of the system of Eqs. (15) is usually negative, meaning that the system is elliptic. However, as usual in propagation problems, the boundary conditions are hyperbolic. The problem is therefore "improperly posed" in the mathematical sense, and it is well known that this

leads to singular solutions [16,26]. This instability is well
illustrated by the pulse of Fig. 5c. Other pulses, with apparently
similar input envelopes, evolved to quite different output shapes
when the interaction is as strong. One common feature however is
the sharp leading edge.

The authors would like to acknowledge useful discussions with
P. D. Gerber, R. Landauer, and Michael M. T. Loy. The skilled
technical assistance of R. J. Bennett was essential to this work.

REFERENCES

1. D. Grischkowsky, Phys. Rev. A (to be published).
2. D. Grischkowsky, Eric Courtens, and J. A. Armstrong (to be
   published).
3. D. Grischkowsky, Phys. Rev. Lett. 24, 866 (1970).
4. D. Grischkowsky and J. A. Armstrong, Phys. Rev. A 6, 1566
   (1972).
5. J. A. Carruthers and T. Bieber, J. Appl. Phys. 40, 426 (1969).
6. A. Frova, M. A. Duguay, C. G. B. Garrett and S. L. McCall, J.
   Appl. Phys. 40, 3969 (1969).
7. F. R. Faxvog, C. N. Y. Chow, T. Bieber and J. A. Carruthers,
   Appl. Phys. Lett. 17, 192 (1970).
8. L. Casperson and A. Yariv, Phys. Rev. Lett. 26, 293 (1971).
9. S. L. McCall and E. L. Hahn, Phys. Rev. 183, 457 (1969).
10. R. E. Slusher and H. M. Gibbs, Phys. Rev. A 5, 1634 (1972).
11. L. A. Ostrovskii, Zh. Eksp. Teor. Fiz. 51, 1189 (1966) [Sov.
    Phys. JETP 24, 797 (1967)].
12. R. J. Joenk and R. Landauer, Phys. Lett. 24A, 228 (1967).
13. F. DeMartini, C. H. Townes, T. K. Gustafson and P. L. Kelley,
    Phys. Rev. 164, 312 (1967).
14. T. K. Gustafson, J. P. Taran, H. A. Haus, J. R. Lifsitz and
    P. L. Kelley, Phys. Rev. 177, 306 (1969).
15. V. M. Arutyunyan, N. N. Badalyan, V. A. Iradyan and
    M. E. Movsesyan, Zh. Eksp. Teor. Fiz. 58, 37 (1970) [Sov.
    Phys. JETP 31, 22 (1970)].
16. B. B. Kadomtsev and V. I. Karpman, Usp. Fiz. Nauk 103, 193
    (1971) [Sov. Phys. USPEKHI 14, 40 (1971)].
17. B. Ya. Zel'dovich and I. I. Sobel'man, ZhETF Pis. Red. 13, 182
    (1971) [JETP Lett. 13, 129 (1971)].
18. F. Shimizu, IEEE J. Quantum Electron 8, 851 (1972).
19. E. Courtens and A. Szöke, Phys. Lett. 28A, 226 (1968).
20. I. I. Rabi, N. F. Ramsey and J. Schwinger, Rev. Mod. Phys. 26,
    167 (1954).
21. Ch'en Shang-Yi, Phys. Rev. 58, 884 (1940).
22. L. Brillouin, "Wave Propagation and Group Velocity," (Academic
    Press, New York, 1960), p. 119.

23. E. O. Schulz-Dubois, Proc. IEEE $\underline{57}$, 1748 (1969).
24. C. G. B. Garrett and D. E. McCumber, Phys. Rev. A $\underline{1}$, 305 (1970).
25. M. D. Crisp, Phys. Rev. A $\underline{4}$, 2104 (1971).
26. M. J. Lighthill, Proc. Roy. Soc. $\underline{A299}$, 28 (1967).

SPECTROSCOPY III

SATURATION SPECTROSCOPY OF ATOMS[*]

T. W. Hänsch

Department of Physics, Stanford University

Stanford, California 94305

Laser saturation spectroscopy provides a powerful way to eliminate Doppler broadening of spectral lines of absorbing gases. This is accomplished by studying the nonlinear interaction of the absorber with two monochromatic laser waves which are in general traveling in opposite directions. At least one of these waves has to be sufficiently strong to produce a spectral hole burning, i.e. a velocity selective partial saturation of the absorbing transition. Hole burning in Doppler-broadened lines was first discussed by Bennett[1] in an early study of gas laser modes. One of its most conspicuous manifestations is the well-known Lamb dip in the power versus frequency characteristic of a single mode gas laser, which was theoretically predicted by Lamb[2] and first observed by Javan and co-workers.[3] Lee and Skolnick[4] were the first to place a gaseous absorber inside the cavity of a gas laser and to observe the saturation resonances as "inverted Lamb dips." This phenomenon has subsequently been used in many laboratories to stabilize the frequency of gas lasers or to spectroscopically study molecular absorption lines in coincidence with existing gas laser lines in unprecedented resolution.[5-7] A very impressive example has been reported by Hall and Bordé at this conference.

Some two years ago we demonstrated at Stanford[8] that it is possible, with organic dye lasers, to generate widely tunable coherent radiation with a bandwidth of only a few MHz, i.e. a small fraction of typical Doppler line width in the visible. Such a dye

[*]Work supported by the National Science Foundation under Grant GP-28415 and the U.S. Office of Naval Research, Contract ONR-71.

laser and a new, convenient and sensitive method of saturation
spectroscopy enabled us to study the yellow Na D lines in saturated
absorption and to optically resolve their hyperfine splitting.[9]

This first application of saturation spectroscopy to atomic
resonance lines made it clear that the advent of stable,
monochromatic, tunable lasers is turning the new technique into a
general, widely applicable spectroscopic method, which no longer
depends on accidental coincidences of spectral lines. One may now
ask to what extent laser saturation spectroscopy can supplant or
complement the older "classic" techniques of atomic spectroscopy
without Doppler broadening, such as level crossing or double
resonance spectroscopy. It appears not very likely that one can
surpass the accuracy of e.g. the radiofrequency double resonance
method in measurements of the fine and hyperfine splitting of atomic
energy levels, since saturation spectroscopy in the optical region
requires a much more formidable relative resolution. And, of
course, the potential of the classic techniques has also been
greatly enhanced by the development of tunable lasers. They are now
becoming applicable to excited states, to ions, molecules, and rare
atoms, for which strong resonance lamps have not been available.
There are, however, many situations, where the conventional methods
are not applicable, and where saturation spectroscopy can become a
very valuable complement, in particular if closely spaced lines
without any shared common level have to be resolved, such as isotope
shifts in isotopic mixtures, or hyperfine components in certain
molecular absorption lines.[10] And laser saturation spectroscopy is
certainly especially valuable wherever a precise wavelength
measurement of an optical transition is required. The only
alternative for some of these investigations would be the
elimination of Doppler broadening by transverse observation of a
well collimated atomic or molecular beam. Very impressive
resolutions have recently been reported by several workers[11,12] who
excited resonance fluorescence from such beams with a monochromatic
laser in a crossed beam arrangement. Atomic beams, however, are
sometimes difficult or impossible to prepare, in particular for
short living excited species, and the observation of a gas sample in
a glass bottle via saturation spectroscopy has certainly the
advantage of attractive experimental simplicity.

In our experiments at Stanford we have been using a particularly
convenient and conceptually simple method of saturation spectroscopy:
The absorbing gas sample is contained in a glass cell outside the
laser resonator. A beam splitter divides the laser output into a
weak probe beam and a stronger saturation beam which are sent in
nearly opposite directions through the absorber. If the laser is
tuned to an atomic resonance frequency, both light waves are
interacting resonantly with the same atoms, those with practically
zero axial velocity, and the saturating beam can bleach a path for

the probe. The saturation signal can be observed with high
sensitivity by periodically blocking the saturating beam with a
chopper and detecting the resulting modulation of the probe with a
phase sensitive amplifier. The detection of saturation signals
via the intensity modulation of a weak probe beam has first been
reported by Hänsch and Toschek[13] in a study of coupled gas laser
transtions. The present method has been used before with gas lasers
by Smith and Hänsch[14] and Hänsch et al.[10] and, independently, by
Bordé.[15]

This method has several advantages over the more often used
inverted Lamb dip technique with intracavity absorber. The
saturation spectra are easier to interpret because one does not have
to worry about nonlinear interactions between absorber and laser
medium and one does not have to take any resonator properties into
account. The modulation method eliminates moreover any slowly
varying background and provides a well defined zero line in the
saturation spectrum. In the limit of low laser intensities and a
large Doppler width, compared to the natural linewidth, it is
possible to give a quantitatively correct description of the
saturation signals simply in terms of the velocity selective
population changes of the absorbing levels.[10] At higher intensities
it is necessary to take atomic coherence effects into account, which
are associated with the off-diagonal elements of the density matrix.
But as long as the probe beam is weak and only the saturating beam
is allowed to be strong, it is still possible to calculate the line
shape in the saturation spectrum analytically.[16,17] A strong
standing wave field, as in the Lamb dip method, on the other hand,
requires complicated numerical computer calculations.[18,19]

Various other variations of laser saturation spectroscopy have
been reported. Particuarly noteworthy is the possibility to observe
the saturation resonances in the fluorescent sidelight rather than
in absorption, as demonstrated by Freed and Javan.[20] Very dilute
samples of negligible absorption can be studied in this way, in
particular if the sensitivity is increased by the method of
intermodulated fluorescence, as described by Sorem and Schawlow[21]
and analyzed by Shimoda.[22]

The laser used in our experiments is a pulsed tunable dye
laser, repetitively pumped by a nitrogen laser.[8] Wavelength
selection is achieved by an echelette grating in Littrow mount
together with a beam expanding telescope and a tilted Fabry-Perot
etalon. An additional external confocal interferometer reduces the
bandwidth to less than 10 MHz and stretches the pulse length from
about 5 nsec to some 30 nsec. Typical output powers are in the
order of watts, but kilowatt pulses have been produced with an
additional dye laser amplifier. A partly saturated amplifier stage
can also reduce the large random amplitude fluctuations which are

observed if the radiation from the quasi-superradiant dye laser is
sent through the external filter.[23]

The past two years have brought very impressive progress in the
development of narrowband continuous dye lasers[24-26] and their
tuning range has been extended over a large fraction of the visible
spectrum. These lasers will certainly become important tools for
saturation spectroscopy, because they can provide a better signal
to noise ratio than a fluctuating pulsed laser. And cw operation
eliminates the possibility of uncontrolled transient phenomena in
the sample.

The nitrogen-laser pumped pulsed dye laser remains nonetheless
a very attractive instrument for saturation spectroscopy. It has a
wide conveniently accessible tuning range, which can easily be
extended into the ultraviolet by frequency doubling. The high
available peak powers make it possible to saturate transitions of
low oscillator strength. And the short pulse duration permits novel
experiments with time resolution, which can be partcularly interesting
for the investigation of collision and relaxation processes.

The laser pulse shape after the external filter exhibits a fast
rise time and a slower exponential decay, which is rather well
suited for a theoretical analysis of the atomic response, and it is
relatively easy to confirm, up to third order perturbation theory,
that the pulsed operation does not introduce any shifts of the line
centers in the saturation spectrum.

The problem of noise due to random laser amplitude fluctuations
can be reduced by proper signal processing. A particularly helpful
step is the use of two probe beams, both passing through the
absorber, but only one crossing the bleached region. It is then
possible to use a sensitive differential detection scheme.[9] We are
also exploring the possibilities of on-line digital data processing
with a small computer, which should permit one to normalize the
observations to constant laser intensity or to sort them according
to pulse intensity and to analyze them for a possible intensity
dependence of the line positions.

Some of the new possibilities opened by pulsed laser saturation
spectroscopy have been illustrated by the first experiments with the
Na D lines.[9] Measurements with an optically delayed probe pulse
revealed that a velocity selective optical pumping and fluorescence
cycle leads to a remanent hole burning in the velocity distributions
of the two stable hyperfine components of the Na ground state, which
can be observed many radiative life times after the bleaching. By
observing the saturation resonances in the presence of a buffer gas
as a function of the probe delay one can study the return of the
perturbed atomic velocity distribution to thermodynamic equilibrium

due to gas kinetic collisions in unprecedented detail. Measurements
of the power broadening for different probe delays revealed moreover
a contribution to the line width due to optical nutations of the Na
atoms in the strong sturating field.[27]

   The great potential of laser saturation spectroscopy, as
compared with conventional methods of high resolution spectroscopy,
is perhaps demonstrated even better by our subsequent study of the
red Balmer line H_α of atomic hydrogen.[28]   Using a simple Wood
discharge tube at room temperature we have, for the first time, been
able to resolve single fine structure components of H_α and to
observe the Lamb shift directly in the optical absorption spectrum.
The use of a pulsed dye laser made it possible to observe the
saturated absorption in the afterglow of the gas discharge and thus
to minimize Stark shifts and Stark broadening.

   By reducing the finite crossing angle between saturating beam
and probe to a few mrad and by further reducing the laser bandwidth
we have in the meantime been able to optically resolve even the 177
MHz hyperfine splitting of the $2S_{1/2}$ state in the $3P_{3/2} - 2S_{1/2}$
component of H_α. And measurements with a time delayed probe
permitted the direct observation of the velocity redistribution of
the short living 2P atoms due to resonance trapping of the emitted
ultraviolet Lyman alpha radiation.

   We are presently working at an attempt to utilize the optical
resolution of H_α for a new precision measurement of the wavelength
of one of its components which should yield a considerably improved
value of the Rydberg constant, one of the cornerstones in the
evaluation of the fundamental constants.

   Hydrogen has always attracted the attention of atomic
spectroscopists because its simplicity permits detailed and accurate
comparison with theoretical models. And much of the progress in
atomic theory, from Bohr and Sommerfeld to Dirac and modern quantum
electrodynamics is based on increasingly refined spectroscopic
studies of the hydrogen atom.[29] It is therefore not surprising that
countless efforts have been undertaken in many laboratories to study
the visible lines of the hydrogen spectrum with ever improving
resolution. Unfortunately, Doppler-broadening is particularly large
for the light hydrogen, almost 6000 MHz for the red Balmer line at
room temperature, and masks the important details of its fine
structure. Our present accurate knowledge of this structure is
almost entirely based on RF and level crossing spectroscopy. The
optical spectra always remained unresolved blends of fine structure
components, even if Doppler broadening was reduced by operation of
the gas discharge at cryogenic temperatures and by use of the
heavier isotopes deuterium and tritium.

    All previous attempts to determine the Rydberg constant from
the wavelength of hydrogen Balmer lines had therefore to rely on
cumbersome deconvolution procedures, and serious doubts have been
expressed[30],[31] in the accuracy of the presently accepted value of
the Rydberg constant.[32]   These doubts can be only partly alleviated
by three very recent new measurements of the Rydberg constant by
Masui,[33] and Kibble et al.,[34] and Kessler[35] using conventional high
resolution spectroscopy of gas discharges. Taylor[31] has emphasized
the importance of a new precision measurement of the Rydberg by
pointing out that a possible error of a few parts in 10^7 would
severely limit the interpretation of several precision experiments
in physics. The most critical examples are a determination of the
fine structure constant from e/h as measured via the Josephson
effect, and the evaluation of the proton and muon polarizability
from measurements of the hydrogen and muonium hyperfine structure.

    A prerequisition for any precise wavelength measurement is an
accurate wavelength standard. In our experiment we are using an
iodine stabilized He-Ne laser whose wavelength has been compared
with the present krypton standard by Schweitzer et al.[36] at NBS. A
pressure-tuned Fabry-Perot interferometer with plane, silvered
plates with an invar spacer of 30 - 65 mm length is kept in exact
resonance with this laser so that its optical spacing is a known
half-integer multiple of the standard wavelength. The absolute
spacing is determined within one order by comparing the fringes of
several known spectral lines provided by Cd and Hg vapor lamps.
Part of the dye laser light is sent through this interferometer,
while the saturation spectrum is recorded, and the resulting
transmission maxima are recorded simultaneously as frequency
markers. The effect of nonlinearities in the dye laser tuning
mechanism can be minimized by chosing a proper order of the interferometer so that the marker is recorded practically simultaneously
with the fine structure component of interest.

    Extensive preliminary measurements[37],[38] confirmed that the
known fine structure intervals can be reproduced in the saturation
within a few MHz. The line separations do not change by more than
10 MHz, if gas pressure and current are varied over the range
easily accessible in the Wood discharge tube.

    The absolute wavelength measurement has meanwhile made
considerable progress, thanks to the efforts of Munir Nayfeh and
Miss Siu Au Lee in our laboratory, but the data acquired so far
seem to be subject to a serious systematic error, and further work
is required. The present measurements, evaluated with the help of
the calculations by Garcia and Mack,[39] yield a Rydberg value in
agreement with previous measurements only for the deuterium isotope.
The value obtained from light hydrogen is almost 0.7 ppm too low.
The statistical errors of these measurements are very encouraging,

however, and indicate that it will be possible, using laser
saturation spectroscopy, to obtain a new value for the Rydberg,
which will be accurate within 2.10^{-8}.

If, against all expectations, the observed discrepancy in our
Rydberg measurement should persist, it would be extremely desirable
to attempt precise wavelength measurements of the higher members of
the hydrogen Balmer series, and to extend the saturation measurements
to the tritium isotope and perhaps to heavier hydrogen-like ions, to
reveal any possible systematic dependence on the nuclear parameters.

And there certainly remain other interesting problems to be
studied with the new method of saturation spectroscopy with
tunable dye lasers, even if one restricts one's attention to the
hydrogen atom. Just one more example is the theoretically predicted
dependence of the Lamb shift on the radiation intensity.[40]

REFERENCES

[1] W. R. Bennett, Jr., Phys. Rev. 126, 580 (1962).

[2] W. E. Lamb, Jr., Phys. Rev. A134, 1429 (1964).

[3] A. Szöke and A. Javan, Phys. Rev. Letters 10, 521 (1963).

[4] P. H. Lee and M. L. Skolnick, Appl. Phys. Letters 10, 3641 (1971).

[5] K. Shimoda and T. Shimizu, *Progress in Quantum Electronics, Vol 2, Part 2*, J. H. Sanders and S. Stenholm, eds., Pergamon Press, Oxford, 1972.

[6] R. G. Brewer, Science 178, 247 (1972).

[7] J. L. Hall, *Atomic Physics III*, S. J. Smith and G. K. Walters, eds., Plenum Press, New York (1973), pp. 615.

[8] T. W. Hänsch, Appl. Opt. 11, 895 (1972).

[9] T. W. Hänsch, I. S. Shahin, and A. L. Schawlow, Phys. Rev. Letters 27, 707 (1971).

[10] T. W. Hänsch, M. D. Levenson, and A. L. Schawlow, Phys. Rev. Letters 26, 946 (1971).

[11] W. Hartig and H. Walther, Appl. Phys. 1, 171 (1973).

[12] S. Ezekiel, this Volume.

[13] T. W. Hänsch and P. Toschek, IEEE J. Quant. Electr. QE-4, 467 (1968).

[14] P. W. Smith and T. W. Hänsch, Phys. Rev. Letters 26, 946 (1971).

[15] C. Bordé, C. R. Acad. Sci. Paris, 271, 371 (1970).

[16] E. V. Baklanov and V. P. Chebotaev, Soc. Phys. JETP 33, 300 (1971).

[17] S. Haroche and F. Hartmann, Phys. Rev. A6, 1280 (1972).

[18] S. Stenholm and W. E. Lamb, Jr., Phys. Rev. 181, 618 (1969).

[19] B. J. Feldmann and M. S. Feld, Phys. Rev. A1, 1375 (1970).

[20] C. Freed and A. Javan, Appl. Phys. Letters 17, 53 (1970).

[21] M. S. Sorem and A. L. Schawlow, Opt. Comm. 5, 148 (1972).

[22] K. Shimoda, Appl. Phys. 1, 77 (1973).

[23] S. M. Curry, R. Cubeddu, and T. W. Hänsch, Appl. Phys. 1, 153 (1973).

[24] H. Walther, this Volume.

[25] R. L. Barger, M. S. Sorem, and J. L. Hall, Appl. Phys. Letters 22, 573 (1973).

[26] H. W. Schroder, H. Welling, and B. Wellegehausen, Appl. Phys. (1973), to be published.

[27] T. W. Hänsch and I. S. Shahin, Opt. Comm. (1973), to be published.

[28] T. W. Hänsch, I. S. Shahin, and A. L. Schawlow, Nature 235, 63 (1972).

[29] G. W. Series, *Spectrum of Atomic Hydrogen*, Oxford University Press, Oxford, 1957.

[30] G. W. Series, in *Proceedings of the International Conference on Precision Measurements and Fundamental Constants*, Nat'l Bur. Std. (U.S.), Spec. Publ. No. 343 (U.S. GPO, Washington, D.C., 1971), pp. 63.

[31] B. N. Taylor, private communication.

[32] B. N. Taylor, W. H. Parker, and D. N. Langenberg, Rev. Mod. Phys. 41, 375 (1969).

[33] T. Masui, in *Proceedings of the International Conference on Precision Measurements and Fundamental Constants*, Nat'l Bur. Std. (U.S.), Spec. Publ. No. 343 (U.S. GPO, Washington, D.C., 1971), pp. 83.

[34] B. P. Kibble, W. R. C. Rowley, R. E. Shawyer, and G. W. Series, J. Phys. B (1973), to be published.

[35] E. G. Kessler, Jr., Phys. Rev. A7, 408 (1973).

[36] W. G. Schweitzer, Jr., E. G. Kessler, Jr., R. D. Deslattes, H. P. Layer, and R. D. Whetstone, Appl. Opt., to be published.

[37] I. S. Shahin, Thesis, Stanford University (1972), M. L. Report No. 2099.

[38] I. S. Shahin, T. W. Hänsch, M. Nayfeh, and A. L. Schawlow, Phys. Rev., to be submitted.

[39] J. D. Garcia and J. E. Mack, J. Opt. Soc. Am. 55, 654 (1965).

[40] C. R. Stroud, Phys. Rev. A3, 1044 (1971).

MOLECULAR BEAM SPECTROSCOPY WITH ARGON AND DYE LASERS

Shaoul Ezekiel

Department of Aeronautics and Astronautics
and Research Laboratory of Electronics
Massachusetts Institute of Technology
Cambridge, Massachusetts   02139

The advent of single frequency tuneable lasers, e.g. dye lasers, will clearly open up the field of high resolution absorption spectroscopy in the visible region of the spectrum. However a tuneable narrow spectral width laser is not the only prerequisite for high resolution spectroscopy. Since doppler broadening is typically several hundred megahertz in the visible region, techniques for reducing doppler as well as collisional broadening must be employed. The method of saturated absorption[1-9] is a good and convenient way of eliminating doppler broadening. Another method is to use molecular beam techniques[10] which can eliminate both doppler and collisional broadening as well as collisional shift.

A typical molecular beam set-up would involve a collimated tuneable laser exciting a beam of atoms or molecules at right angles. Collisional broadening is eliminated because of the collisionless flow of molecules in the beam, and doppler broadening is determined by the angular width of the molecular beam which is at the disposal of the experimenter. The measurement of the interaction between laser and molecular beam depends on the transition that is excited. Because of the low density of molecules in the beam, absorption out of the laser beam is not generally feasible to detect so that other methods have to be used. For example, if the excited state decays radiatively and the lifetime is shorter than about 10^{-5} seconds, then it is very

convenient to detect the re-radiated light or fluorescence at the interaction region. For long lived excited
states, one may have to resort, if applicable, to
magnetic or electric resonance techniques, often used
in microwave spectroscopy, to detect any change in the
magnetic or electric substate due to laser excitation.

To illustrate the use of laser-molecular beam
techniques for high resolution spectroscopy we have been
conducting experiments[11] using a 5145Å argon ion laser
and a molecular beam of I_2. The experimental arrangement
is shown in Fig. 1. A single frequency tuneable 5145Å
argon laser excites an I_2 molecular beam at right angles.
The laser induced fluorescence is collected by a lens
and focused onto a photomultiplier.

Single frequency laser operation is achieved by the
use of a Fox-Smith mode selector[12]. A feedback loop with
a bandwidth of 1 kHz holds the laser frequency at the
resonance frequency of the mode selector[13]. The laser
frequency may be scanned smoothly across the gain curve
by linearly changing the length of the intracavity mode
selector.

The molecular beam apparatus is a high vacuum
chamber into which I_2 molecules, at room temperature,
effuse through a narrow primary slit. The geometrical
width of the I_2 beam is restricted to less than 0.5
milliradians by a secondary slit farther down the
apparatus.

The laser induced fluorescence is collected by a
lens that is mounted external to the beam apparatus,
and is then focused onto a photomultiplier with an S-20
photocathode.

The technique used to measure the fluorescence is to
chop the laser (or molecular) beam at 1 KHz and to
synchronously detect the fluorescence incident on the
photomultiplier with a lock-in amplifier. Details of
the experimental procedure have been published elsewhere[13]. The $I_2{}^{127}$ transitions excited, shown in
Fig. 2, are the hyperfine components of the P(13) (43-0)
and the R(15) (43-0) lines between the $^1\Sigma_g^+$(X) and
$^3\Pi_{ou}^+$(B) electronic states which fall within the 5145Å
laser bandwidth.

The measured spectral width in Fig. 2 is typically
650 kHz (FWHM) i.e. a resolution of 1 x 10^{-9}. Higher

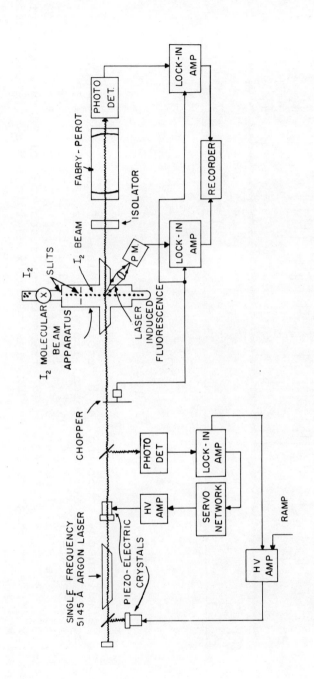

Fig. 1. Schematic diagram of apparatus for Molecular Beam Spectroscopy.

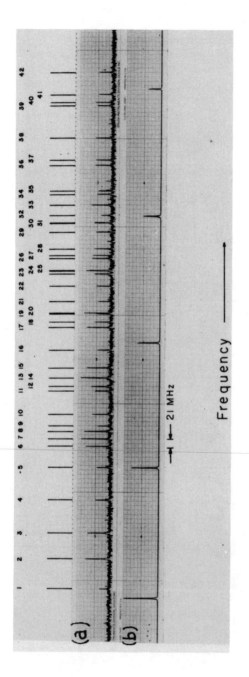

Fig. 2. Simultaneous recording, as a function of laser frequency, of (a) hyperfine structure of I2 molecular beam, typical linewidth 650 kHz (FWHM) (τ = 10 m sec). The P(13) lines are 1, 2, 3, 4, 6, 8, 9, 11, 12, 13, 15, 16, 18, 19, 22, 23, 27, 30, 31, 32, 35. The R(15) lines are 5, 7, 10, 14, 17, 20, 21, 24, 25, 26, 28, 29, 33, 34, 36, 37, 38, 39, 40, 41, 42. (b) transmission resonances of Fabry-Perot interferometer (free spectral range 300 MHz, instrument linewidth 1 MHz, frequency scan rate 10 MHz/sec).

resolution has also been obtained[14] as shown in Fig. 3 where the linewidth is 300 kHz (FWHM) i.e. a resolution of 5×10^{-10}. The natural width of the I_2 lines is inferred to be about 100 kHz from lifetime measurements in the beam[10]. Broadening mechanisms such as small angle collisions in the beam, laser jitter, nonorthogonality between laser and molecular beam are probably responsible for the observed residual width and this is being investigated.

In the absence of the above broadening mechanisms, the linewidth in an orthogonally excited molecular beam should be the natural width modified by a doppler broadening due to the geometric angle of the beam which must take into consideration the beam profile and the velocity distribution of the molecules[15]. Transit time broadening must be carefully examined when designing the interaction region particularly when the spontaneous lifetime becomes comparable to the transit time.

The observed spectrum, Fig. 2, is due to nuclear electric quadrupole and spin rotation interactions in the I_2 molecule on the $P(13)$ and $R(15)$ lines. The spectrum has been fitted[16] to obtain a quadruple coupling strength difference of $\Delta eQq = 1906 \pm 2$ MHz and a spin rotation interaction strength difference of $\Delta C_I = 181 \pm 7$ kHz between the upper and lower levels of the $P(13)$ transition. For the $R(15)$ transition, $\Delta eQq = 1905 \pm 2$ MHz and $\Delta C_I = 167 \pm 5$ kHz have been obtained. The standard deviation between the measured and calculated relative transition frequencies was about 1.3 MHz.

With a stronger and geometrically broader beam, other weaker transitions have been uncovered. Several of these weak lines are due to $\Delta F = 0$ transitions[16, 17, 18] while the strong lines in Fig. 2 are due to $\Delta F = \Delta J$ transitions. Fig. 4 shows the derivative of the I_2 spectrum with the weak $\Delta F = 0$ lines clearly visible dispersed among the strong $\Delta F = \Delta J$ lines shown in Fig. 2.

Because of the limited tuneability of gas lasers, it is only possible to explore coincidences between existing laser lines and molecular transitions. Tuneable cw dye lasers are much more useful for spectroscopy[19-23]. However, cw dye lasers are not without problems

Recently we have been investigating the use of a cw dye laser to excite narrow resonances, for example,

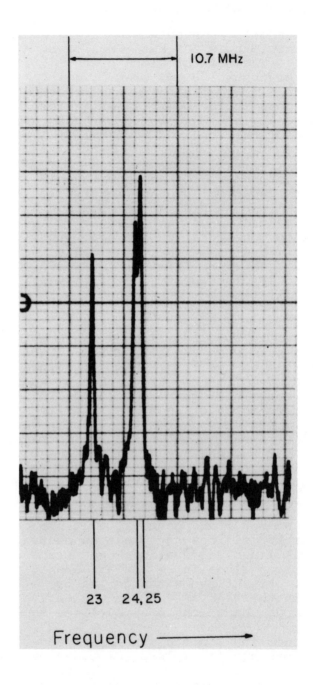

Fig. 3. I$_2$ hyperfine structure with a resolution of 5 x 10^{-10} or 300 kHz (FWHM), (τ = 10 m sec, frequency scan rate 5 MHz/sec).

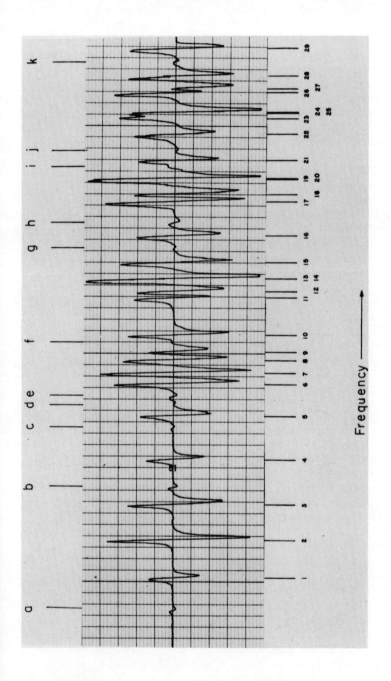

Fig. 4. Derivative of I_2 hyperfine structure showing weaker lines ($\Delta F = 0$) which are not clearly observable in Fig. 2 (a) ($\tau = 100$ m sec, frequency scan rate 10 MHz/ sec). $\Delta F = 0$ transitions are labeled a, b, c, d, e, f, etc.

in a molecular beam of I_2. We used a single frequency
jet stream cw dye laser[24] that can be smoothly tuned
over at least 1.2 GHz.

The basic dye laser cavity is similar to the
astigmatically compensated design of Dienes, Ippen and
Shank[25] as shown in Fig. 5. The dye, rhodamine 6G
dissolved in ethylene glycol, flows out of a nozzle
that is designed to give a stream of optical quality.[26]
Single frequency operation is achieved with the aid of
3 dispersion prisms two etalons and a Fox-Smith mode
selector. Details of the design of the jet stream dye
laser is published elsewhere[24]. The dominant laser
jitter was found to be $\pm$ 1 MHz at a rate of approx. 100 Hz.
The source of the jitter is being investigated. There
was no evidence of high frequency jitter. Fluctuation
in the power of the single frequency laser was less
than 1% and was similar in character to the noise in the
argon laser pump. No attempt was made to stabilize
the dye laser power in the present experiments. Continuous tuning was achieved by locking the laser
frequency to the resonance frequency of the mode selector
by means of low bandwidth feedback loop and then linearly
changing the length of the mode selector[24].

High resolution absorption spectroscopy around
5900Å was performed with the dye laser by exciting a
molecular beam of I_2 at right angles as described above.
I_2 has numerous absorption bands that fall within the
gain bandwidth of rhodamine 6G. Figure 6 shows several
of the hyperfine structure lines of I_2^{127} around 5900Å.
The widths of the narrowest lines are typically 3 MHz
(i.e. a resolution of 5×10^{-9}) and are evidently limited
by laser jitter. The observed widths, after allowing
for laser jitter of $\pm$ 1 MHz and other broadening
mechanisms in our set-up, are in reasonable agreement
with the natural width of less than 1 MHz predicted from
recent lifetime measurements[27] in I_2 around 5900 Å.

In the above experiments, the laser frequency is
determined by the mode selector. Even though the mode
selector is made of invar, it is still subject to thermal
and acoustic effects. For high resolution measurements,
particularly where the tuning must be slow because of
signal to noise limitations, the stability of the mode
selector is not adequate and long term stabilization of
the laser frequency is needed. In this way, a second
laser that is locked to a frequency offset with respect
to the stabilized laser may be used as the tuneable

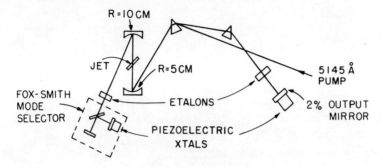

Fig. 5. Single frequency dye laser cavity.

source by simply varying the frequency offset[28].
Another method, is to externally shift the frequency of
the stabilized laser by means of, say, an acousto-optic
frequency shifter so that only one laser is needed[29].
The tuning is achieved by varying the oscillator
frequency driving the acoustic cell.

The I_2 resonances in Fig. 2 are ideal reference
frequencies for long term laser frequency stabilization[30].
We have stabilized the 5145Å argon laser to one of the
lines in Fig. 2. The experimental arrangement is shown
in Fig. 7.

For the purpose of frequency stabilization, a
derivative of the I_2 transitions is needed. This is
achieved by externally modulating the laser frequency
by means of a vibrating mirror and synchronously
detecting the fluorescence in lock-in amplifier A, in
Fig. 7. The derivative of the first three lines (on the
left side of Fig. 2) is shown in Fig. 8(a). By means
of a feedback loop, the resonance frequency of the Fox-
Smith mode selector (and, in turn, the laser frequency)
is locked to the center of a selected I_2 line i.e. the
zero of the derivative of the line[13]. The transition
used in the experiments discussed here is the second from
the left in Fig. 2, and Fig. 4, where it is clear that no
other line is near by, with the present resolution of
5×10^{-10}.

The frequency drift of the stabilized laser is
monitored by examining the fluorescence signal associated

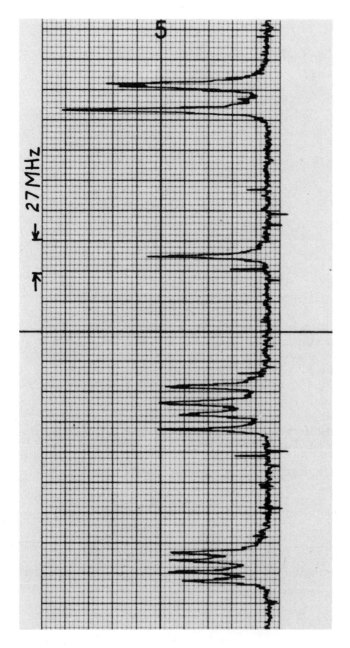

Fig. 6. Hyperfine structure of I_2 molecular beam excited by dye laser operating near 5900 Å [τ = 10 msec scan rate 20 ^{MHz}/sec].

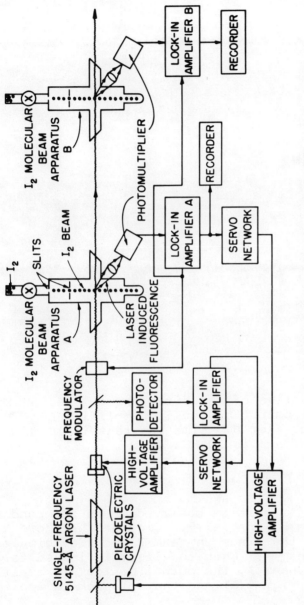

Fig. 7. The experimental set-up for Molecular Beam Stabilized Laser

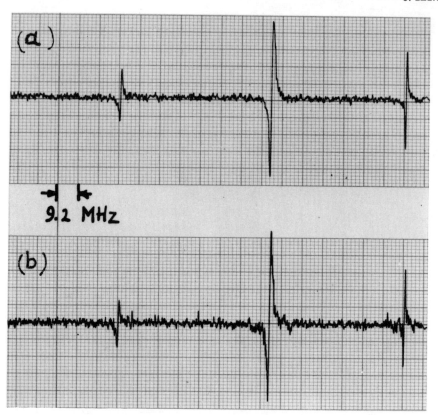

9.2 MHz

Fig. 8. Derivative of the first three lines on the left
in Fig. 2; (a) output of lock-in amplifier A; (b) output
of lock-in amplifier B. Time constant is 10 m sec.
Frequency scan rate is 10 MHz/sec.

with the same transition observed in an independent I_2
beam apparatus B, in Fig. 7. Figure 8(b) shows the
derivative of the lines in beam B recorded simultaneously
as those in beam A. When the laser is locked to an I_2
transition in beam A, the output of lock-in amplifier
A is shown in Fig. 9(A) and that of lock-in amplifier B
in Fig. 9(B).

        The left side of Fig. 9 (section a) is the derivative
of the selected I_2 line as a function of time when the
laser frequency is tuned manually back and forth over
the line prior to lock. After the laser is locked the

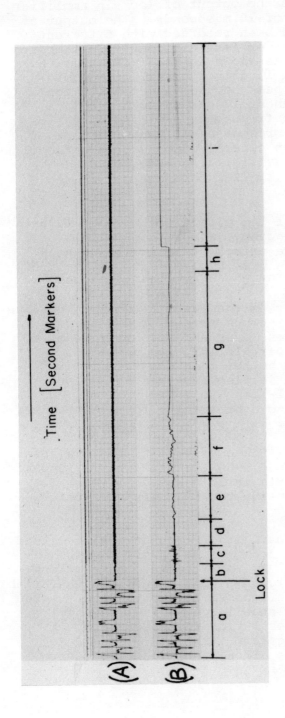

Fig. 9. Stabilized Laser drift data. (A) Output of lock-in amplifier A before and after lock. (B) Output of lock-in amplifier B: sec(a) derivative signal when manually tuning laser over I2 line (τ = 10 m sec, recorder sensitivity 40 kHz/box): secs(b) to (i) are lock signals with different time constant τ and recorder sensitivity RS: sec(b) τ = 10 m sec; RS = 40 kHz/box; time scale 1 sec/box: sec(c) τ = 10 m sec; RS = 40 kHz/box; time scale 5 sec/box: sec(d) τ = 1 sec; RS = 40 kHz/box: sec(e) τ = 1 sec; RS = 2 kHz/box: sec(f) τ = 1 sec; RS = 1 kHz/box: sec(g) τ = 10 sec; RS = 1 kHz/box: sec(h) τ = 100 sec; RS = 1 kHz/box: sec(i) τ = 100 sec; RS = 200 Hz/box.

rest of Fig. 9(A) shows the output of lock-in amplifier
A with a time constant of 10 m seconds. The output of
lock-in amplifier B has been recorded with different
time constants and recorder sensitivities to obtain a
measure of laser drift as shown in Fig. 9(B) sections
(b) to (i). In section i, only a small portion of
which is included, the peak to peak laser drift for a
run of 20 minutes was about 100 Hz or $\Delta\nu/\nu \sim 10^{-13}$. In
terms of RMS drift, the frequency stability is a few
parts in 10^{14}. In the present experiments, actual
laser drift has not been clearly detected since the drift
data in Fig. 9(B) is very similar to that obtained by
integrating the photomultiplier dark current in the
absence of fluorescence.

An investigation of the effect of external magnetic
fields on the laser frequency gave an upper limit of
$\Delta\nu/\nu < 10^{-14}$/gauss. The effect of intensity changes
in the laser exciting I_2 beam B gave an upper limit of
$\Delta\nu/\nu < 4 \times 10^{-14}$ for a 10% change in laser intensity.

The reproducibility of the laser frequency was also
examined. With a rather crude method of determining
the orthogonality between laser and molecular beam, a
reproducibility of one part in 10^{11} was achieved. With
care, a reproducibility better than one part in 10^{13}
should be possible. It is also feasible to reflect the
laser back through the I_2 beam to convert shifts due
to non-orthogonality into a broadening to obtain better
reproducibility[10, 13].

In a similar way, we have stabilized the dye laser
to an I_2 transition around 5900Å. A long term frequency
drift of less than 3 parts in 10^{11} was achieved. The
drift was again monitored by use of an independent I_2
beam[31].

We are at present using the frequency shift method
described above to obtain a precise measurement of the
I_2 natural width as well as the frequency separations
in the I_2 spectrum in Fig. 2. With commercially available
acousto-optic cells, it is possible to tune about $\pm$ 200
MHz with respect to the stabilized laser frequency.
The ultimate limit on resolution in this type of
spectroscopy is the long term stability of the laser
which in the case of the argon laser described above,
is several parts in 10^{14}.

Finally, laser molecular beam techniques may be used as a diagnostic tool in molecular beam experiments. The velocity distribution in a molecular beam may be observed in a continuous way by using a narrow angle beam that is excited non-orthogonally by the laser. In addition, long term continuous experiments on a single velocity class of molecules may be performed. Here, a laser is locked to a resonance observed in an orthogonal molecular beam and the stabilized laser frequency is externally shifted by means of, say, an acousto-optic shifter, as mentioned above, so as to excite a selected velocity-class in a second molecular beam that is non-orthogonal with the laser beam.

The use of laser-molecular beam techniques for high resolution spectroscopy has been clearly demonstrated. Molecular beam spectroscopy is not limited to ground state transitions. Beams of metastable atoms and molecules can also be generated[32]. For excited state spectroscopy, the lower level has to be populated at the interaction region. Direct or indirect optical excitation of the lower level introduces only small recoil shift and is preferred over electron impact excitation.

It should be noted that transitions observed in a molecular beam are free from any background doppler profile that is normally encountered in saturated absorption spectroscopy. This means that the line center and the lineshape of the transitions may be determined with greater precision. Moreover, crossover transitions[17,33] are not observed in molecular beam spectroscopy.

The work described here would not have been possible without the able contributions of my graduate students Lloyd Hackel, Douglas Youmans, Robert Grove, Fred Wu and, earlier, Tim Ryan and Ken Allen. Helpful discussions with Prof. R. Weiss are very much appreciated. The argon laser-iodine work was supported by the Air Force Office of Scientific Research. The dye laser development was supported by the Joint Services Electronics Program.

REFERENCES

1. P. H. Lee and M. L. Skolnick, App. Phys. Lett. 10, 303, (1967).

2. R. L. Barger, and J. L. Hall, Phys. Rev. Letters, 22, 4, (1969).

3. G. R. Hanes and K. M. Baird, Metrologia, 5, 32, (1969).

4. G. R. Hanes and C. E. Dahlstrom, App. Phys. Letters, 14, 362, (1969).

5. P. Rabinowitz, R. Keller, and J. T. Latourrette, App. Phys. Letters, 14, 376, (1969).

6. F. Shimizu, Appl. Phys. Letters, 14, 378, (1969).

7. T. W. Hänsch, M. D. Levenson, and A. L. Schawlow, Phys. Rev. Letters, 26, 946 (1971).

8. C. Freed, A. Javan, Appl. Phys. Letters, 17, 53, (1970).

9. M. S. Sorem and A. L. Schawlow, Optics Communications, 5, 148, (1972).

10. S. Ezekiel and R. Weiss, Phys. Rev. Letters, 20, 91, (1968).

11. D. G. Youmans, L. A. Hackel and S. Ezekiel, J. Appl. Phys. 44, 2319, (1973).

12. P. W. Smith, IEEE, J. Quant. Elect. QE-1, 343, (1965).

13. T. J. Ryan, D. G. Youmans, L. A. Hackel, and S. Ezekiel Appl. Phys. Letters, 21, 320, (1972).

14. L. A. Hackel, D. G. Youmans, S. Ezekiel, Proceedings IV International Symposium on Molecular Beams, Cannes, France, (1973).

15. L. A. Hackel and S. Ezekiel (to be published).

16. D. J. Ruben, S. G. Kukolich, L. A. Hackel, D. G. Youmans and S. Ezekiel, Chem. Phys. Letters, (to be published).

17. M. S. Sorem, T. W. Hansch and A. L. Schawlow, Chem. Phys. Letters 17, 300 (1972).

18. D. J. Ruben, S. G. Kukolich, L. A. Hackel, D. G. Youmans and S. Ezekiel (to be published).

19. F. Schuda, M. Hercher and C. R. Stroud, Jr., Appl. Phys. Letters, 22, 360, (1973).

20. W. Hartig and H. Walther, Appl. Phys. 1, 171, (1973).

21. R. L. Barger and M. S. Sorem, J. L. Hall, Appl. Phys. Letters, 22, 573, (1973).

22. H. T. Duong, P. Jacquinot, S. Liberman, J. L. Picqué, J. Pinard and J. L. Vialle, Optics Communications, 7, 371, (1973).

23. W. Lange, J. Luther, B. Nottbeck and H. W. Schröder, to be published in Optics Communications.

24. R. E. Grove, F. Y. Wu, L. A. Hackel, D. G. Youmans and S. Ezekiel, Appl. Phys. Letters, 15 October (1973).

25. A. Dienes, E. P. Ippen and C. V. Shank, Appl. Phys. Letters, 19, 258, (1971).

26. P. K. Runge and R. Rosenberg, IEEE, J. Quant. Elect. QE-8, 910, (1972).

27. G. A. Capelle, H. P. Broida, J. Chem, Phys. 58, 4212, (1973).

28. R. L. Barger and J. L. Hall, Appl. Phys. Letters 22, 196 (1973).

29. L. A. Hackel, D. G. Youmans, and S. Ezekiel (to be published).

30. L. A. Hackel, D. G. Youmans, and S. Ezekiel, Proc. 27th Annual Symposium on Frequency Control, U. S. Army Electronics Command, Fort Monmouth, N. J., (1973).

31. R. E. Grove, F. Wu, and S. Ezekiel (to be published).

32. S. Ezekiel, Research Lab. of Electronics Quarterly Progress Report, January 15, 1967, p. 70-73.

33. J. L. Hall, and C. Bordé, Phys. Rev. Letters 30, 1101 (1973); H. R. Schlossberg and A. Javan, Phys. Rev. 150, 267, (1966).

SELF-INDUCED EMISSION IN OPTICALLY PUMPED HF GAS:

THE RISE AND FALL OF THE SUPERRADIANT STATE*

I. P. Herman,[+] J. C. MacGillivray, N. Skribanowitz,[#] and M. S. Feld[≠]

Department of Physics, Massachusetts Institute of Technology

Cambridge, Massachusetts 02139

I. INTRODUCTION

In 1954 R. H. Dicke pointed out that the spontaneous emission rate of an assembly of atoms would be much greater than that of an isolated atom.[1] This effect, called superradiance, is due to cooperation of the atoms coupled via the common radiation field. In his treatment Dicke distinguished two regimes, characterized by whether the atoms are confined to a region small or large compared to the wavelength of the emitted radiation. In the former case the theoretical formulation is straightforward, and the predictions have been confirmed in the microwave region.[2] For extended samples, as occur in the optical range, the theory is more complex, since propagation effects must be taken into account. Several theoretical treatments have been given,[3-8] but as yet there is no general agreement as to the details of the radiation process. Experimentally, a number of coherent optical effects closely related to the concepts used to treat superradiance had been observed.[9-15] In a recent paper[16], the first

---

*Work supported in part by National Science Foundation and Research Corporation.

[+] Hertz Foundation predoctoral fellow.

[#] Present Address: Physik Department, Technische Universität, Munich, Germany.

[≠] Alfred P. Sloan Research Fellow.

observation of superradiant pulse evolution was reported in far-infrared transitions of optically pumped HF gas, and an analysis was given. The present paper is a continuation and elaboration of that work.

In the experiments a long sample cell of low pressure HF gas is pumped by a short intense pulse from an HF laser operating on a single R or P branch transition to the vibrational ground state (Fig. 1). This produces a nearly complete population inversion between two adjacent rotational levels in the first excited vibrational state, corresponding to a transition in the 50-250 μm range.[17] The infrared radiation from this transition is studied as a function of time. There are no mirrors, and care is taken to minimize feedback.[18]

The observed output is shown in Fig. 2(c). After a considerable delay (~1-2 μsec) with respect to the ~100 nsec pump pulse [Fig. 2(a)], the radiation is emitted in a series of intense, short (~100 nsec) bursts of diminishing size ("ringing"); the process is completed within a few μsec. The radiation pattern is highly directional; almost all of the radiation is emitted into a very small angle along the axis of the pump beam.

If the radiation emitted by this system were incoherent spontaneous emission, then it would have a long exponential decay (the radiative lifetime of these transitions is in the 1-10 second range), and the radiation would be emitted isotropically [Fig. 2(b)]. Furthermore, the observed peak intensity is ten orders of magnitude greater than that expected for incoherent radiation. It is therefore clear that the process observed is not incoherent spontaneous emission. It is also clear that the output signal is not "amplified spontaneous emission"[19] in the usual sense, since the transit time through the cell is over one hundred times

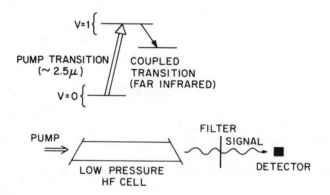

Figure 1.   HF level scheme and schematic of experimental set-up.

shorter than the pulse evolution time.

The observed radiation is also distinct from that of an "ordinary" laser system, even one which can oscillate without mirrors, such as a saturated molecular nitrogen laser (3300 Å). [20] In the nitrogen system the peak output intensity is directly proportional to the total population difference between the levels of the laser transition. Therefore, when the length or pressure is increased, the peak intensity increases proportionally. In contrast, the peak intensity of our output

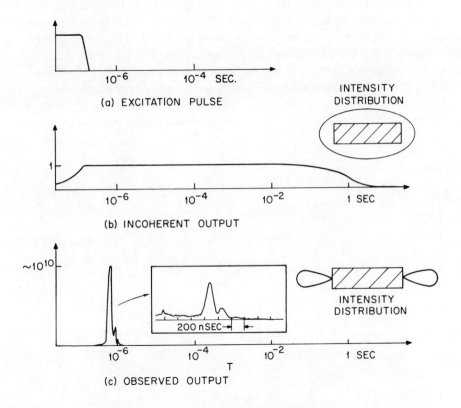

Figure 2.   Comparison of observed output and incoherent spontaneous emission. Time is plotted on a logarithmic scale.   (a) Pump laser pulse.   (b) Output expected from incoherent spontaneous emission, exhibiting exponential decay and an isotropic radiation pattern.   (c) Observed output, exhibiting ringing, a highly directional radiation pattern, and a peak intensity of $\sim 10^{10}$ times that of (b). The inset shows the time evolution of the same pulse with a linear time scale.

pulse is proportional to the square of the pressure. The experimental
data is shown in Fig. 3.

The observed output signal is therefore distinct from both incoherent emission and normal laser radiation. The N^2 dependence and the directionality of the observed radiation agree with the predictions of Dicke[1,21] for the behavior of a superradiant system. The present analysis of a superradiant system is based on the semiclassical formalism [22] which as shown below predicts the same N^2 dependence and directionality as Dicke's analysis. In addition, it makes possible a detailed description of the time evolution of the system and allows one to establish the specific conditions under which superradiance can occur in an extended sample. Furthermore, in the special case of "limited superradiance" described below, the results of the present analysis reduce to those obtained in previous theoretical treatments of superradiance.[7,8] The interpretation of the present experiment as the observation of Dicke superradiance is based on the connections of the present theory with previous treatments and the agreement of this theory with our experiments.

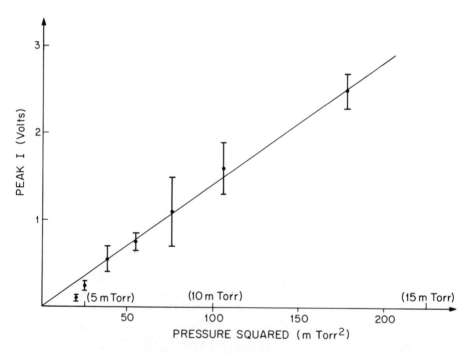

Figure 3.   Peak intensity of superradiant pulse at 84 μm (J=3→2),
             pumped by the P_1(4) laser line, as a function of the square of
             the HF pressure in the sample cell.

The remainder of the paper is divided into the following sections:

II. DESCRIPTION OF SYSTEM

The experimental arrangement consisted of an HF pump laser, a sample cell and the detection system (Fig. 1). The pump laser produced $R_1(J)$ and $P_1(J)$ pulses ($\sim 2.5\mu$m) with peak powers of a few kW/cm^2. (The notation is defined in Table I below.) The early experiments used a helical pin laser, described previously,[17,23] which produced pulses of 200-400 nsec duration. In recent experiments, a double-discharge-initiated laser of the type described by Wenzel and Arnold was used.[24,25] The new laser provided much more reproducible pulses and could produce pulses as short as 50-100 nsec.

The stainless steel or Monel cells containing the room tempera-

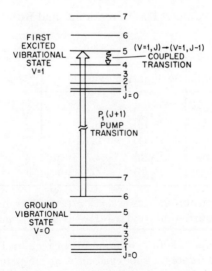

Figure 4.   HF energy levels and transitions. The superradiant output pulse occurs at the coupled transition.

ture HF gas ranged in length from 30 to 100 cm, with inner diameters
between 12 and 28 mm. They had silicon Brewster windows coated on the
inside with a thin layer of Halocarbon stopcock grease to prevent corrosion. The HF gas was purified by freezing and distillation. The sample
cell HF pressures, typically in the ~1-20 mTorr region, were measured
with a Hastings vacuum gauge calibrated for HF. To observe the output
pulses a helium cooled In-Ge detector was used, followed by a fast pre-
amplifier and a pulse amplifier. The overall rise time of the system was
below 10 nsec, giving ample resolution of the pulse shapes. The wavelengths of the output pulses were determined with the aid of a monochromator.

At room temperature the v=1 vibrational level is ~ 20 kT above
the ground state, so that all of its rotational sublevels are virtually un-
populated. Therefore, by optically pumping the sample cell with one of
the vibrational lines in the v=0 to v=1 band of the pump laser, a complete
population inversion between two adjacent rotational levels in the v=1
vibration level is produced (Fig. 4). The corresponding transitions (at
which superradiant output radiation occurs) fall in the 50-250 μm region.
Table I gives the wavelengths and notation of the pump and superradiant
transitions. All of these lines have been observed previously by optical
pumping at much higher pressures.[17]

The sample cell was optically pumped in a single pass, leading to

TABLE I. Optically Pumped HF Vibrational and Rotational Transitions

Pump Transitions [a]				Superradiant Transition		
Designation	Wavelength	Designation	Wavelength	$J_{upper} \rightarrow$	J_{lower}	Wavelength
$R_1(0)$	2.50 μm	$P_1(2)$	2.58 μm	1	→ 0	252.7 μm
$R_1(1)$	2.48 μm	$P_1(3)$	2.61 μm	2	→ 1	126.4 μm
$R_1(2)$	2.45 μm	$P_1(4)$	2.64 μm	3	→ 2	84.4 μm
$R_1(3)$	2.43 μm	$P_1(5)$	2.67 μm	4	→ 3	63.4 μm
$R_1(4)$	2.41 μm	$P_1(6)$	2.71 μm	5	→ 4	50.8 μm

$R_1(J)$ signifies the (v=1, J+1) → (v=0, J) transition.
$P_1(J)$ signifies the (v=1, J-1) → (v=0, J) transition.

[a] The system can be pumped by either an R branch or P branch
transition in order to obtain superradiance at the indicated
transition.

superradiant output pulses with peak intensities in the 100 $\mu W/cm^2$ range and widths from 20 to several hundred nsec, depending on the gas pressure and pump laser power. This corresponds to pulse areas (Sec. IV) of order unity. At the low HF pressures used the gas is heavily saturated by the pump radiation and consequently absorbs only a small fraction of the incident energy. Nevertheless, the gains at the rotational transition are sizeable and can be in excess of 20% per centimeter.[18]

An oscilloscope trace of the output is shown in the inset of Fig. 2(c). At pressures below 5 mTorr the superradiant pulses were delayed by 500 to 2000 nsec past the beginning of the pump pulse. Decreasing the pressure increased the delays, broadened the pulses, and decreased their magnitude [Fig. 10(a), (b) below]. Above 10 mTorr, the pulses were single spikes 40 nsec in width which occurred during the pump pulse. Furthermore, decreasing the pump power increased the pulse delay and width and decreased its amplitude (Fig. 5). The pulses often exhibited ringing with as many as four lobes. Similar pulses were also seen in the backward direction, i.e. propagating antiparallel to the travelling-wave pump radiation.

III. PHYSICAL PRINCIPLES

In his original treatment, Dicke[1] considered the radiative decay of an assembly of molecules. For a point sample, he described the entire collection of molecules as a single quantum mechanical system

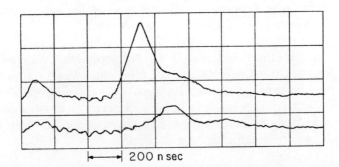

200 n sec

Figure 5.   Oscilloscope traces of superradiant pulses at 84 μm (J=3→2), pumped by the $R_1(2)$ laser line. The pump field intensity in the bottom trace is approximately 40% less than in the top trace.

characterized by quantum numbers r (the cooperation number) and m (half the number difference between the upper and lower levels), which can have integral (or half-integral) values $0 \leq r \leq N/2$ and $|m| \leq r$, where N is the total number of molecules. The spontaneous emission intensity from the state $|r, m\rangle$ to $|r, m-1\rangle$ is then given by

$$I = I_0 (r+m)(r-m+1), \tag{1}$$

where I_0 is the intensity radiated by a single molecule. He then showed that a totally inverted system ($r = m = N/2$) will evolve into a "superradiant" state ($r = N/2$, $m \sim 0$), whose intensity is $I \sim I_0 (N^2/4)$. This is larger by a factor of $N/4$ than the intensity $I = N I_0$ of N incoherent atoms.

For a geometrically large sample, Dicke showed that an array of quantum mechanical dipoles phased along an axis in the direction $\vec{k}$ could also produce superradiant emission. If the dipoles are linearly polarized and perpendicular to $\vec{k}$, then the intensity emitted in the direction $\vec{k}'$ is[1]

$$I(\vec{k}') = I_0(\vec{k}') \frac{N^2}{4} | [e^{i(\vec{k} - \vec{k}') \cdot \vec{r}}]_{av} |^2 \tag{2}$$
$$+ \text{(terms proportional to N)}.$$

Here $I_0(\vec{k}')$ is the radiation intensity of a single molecular dipole in the direction $\vec{k}'$, and $[\ ]_{av}$ denotes an average over the entire volume. When $|\vec{k} - \vec{k}'|$ is small, the average is ~ 1 and $I(\vec{k}')$ is proportional to N^2. Otherwise the average is $\ll 1$ and the radiation rate is greatly reduced. Therefore, most of the radiation is emitted into a small solid angle along the axis. In our case, this solid angle $\sim \lambda^2/A$, where λ is the wavelength of the transition and A is the cross-sectional area of the sample, and the total radiation is enhanced by a factor $\sim N \lambda^2/A \sim 10^{10}$.

The characteristic directionality and N^2 dependence of superradiant emission in an extended sample are not inherently quantum mechanical, and can be understood in terms of a simple classical model. Consider the radiation emitted by a linearly polarized array of dipoles enclosed in a volume $V = AL$, whose macroscopic polarization density is in the form of a plane wave travelling in the z direction:

$$\vec{P} = n\mu\hat{x}e^{i(\omega t - kz)}, \tag{3}$$

where $n = N/V$ is the number density of molecules, μ is the dipole moment of a single radiator, k is the wavenumber of the transition of frequency ω, and $\hat{x}$ is a unit vector describing the polarization direction of the wave.

The total far-field intensity distribution from such a system in the direction $\vec{k}'$, obtained by summing the electric field contributions from individual dipoles,[26] is

$$I(\vec{k}') = I_0\,(\vec{k}')\frac{N^2}{4}\,|\,[e^{i(\vec{k}-\vec{k}')\,\cdot\,\vec{r}}\,]_{av}|^2,\qquad(4)$$

where $\vec{k}=k\hat{z}$ and other notation is as in Eq. (2). As discussed previously, the $[\ ]_{av}$ factor is sharply peaked in the forward direction, and coherent radiation proportional to N^2 is emitted into a small solid angle along the z axis. In other directions, $[\ ]_{av}\sim N^{-1}$, and the emitted radiation is proportional to N. The radiation pattern is similar to that of an "endfire" antenna array.[21] Except for terms of order N, Eq. (4) is the same as Dicke's result for an extended superradiant system [Eq. (2)]. The N^2 dependence and highly directional radiation pattern predicted by these equations are observed in our experiments.

As can be seen from this discussion, a superradiant state in an extended medium can be visualized as one in which a macroscopic polarization as in Eq. (3) is established over a region of space. From a classical point of view this polarization is equivalent to a phased array of dipoles. In a quantum mechanical picture these dipoles can be represented as coherent mixtures of the stationary states of the molecules.[21]

For a system in which the Fresnel number $2A/\lambda L\gg 1$ ("disk"), the solid angle within which the radiation adds coherently is[7]

$$f = \frac{\Delta\Omega}{4\pi} = \frac{\lambda^2}{4\pi A}.\qquad(5)$$

In our HF system, in which the Fresnel number is approximately 1, the correct formula[7] for f gives a value only slightly different from that of Eq. (5), and $f\sim 2\times10^{-5}$. Taking advantage of the sharply peaked nature of $[\ ]_{av}$, Eq. (4) may be integrated over all solid angles to obtain

$$I = I_0\frac{N^2}{4}\frac{3\lambda^2}{8\pi A},\qquad(6)$$

where

$$I_0 = \frac{4}{3}\frac{|\mu|^2\omega^4}{c^3} = \frac{\hbar\omega}{T_{sp}}\qquad(7)$$

is the average power radiated by an isolated molecular dipole and T_{sp} is the corresponding lifetime. In other words, all of the molecules in the sample radiate in phase over the small solid angle f, giving rise to an enhancement factor $\sim Nf$.

Since the system radiates into such a small solid angle, the total electric field can be approximated by a plane wave. The electric field is driven by the polarization, Eq. (3), according to the wave equation,

$$\nabla^2 \vec{E} - \frac{1}{c^2}\frac{\partial^2 \vec{E}}{\partial t^2} = \frac{4\pi}{c^2}\frac{\partial^2 \vec{P}}{\partial t^2},$$

(8)

which has the solution

$$\vec{E} = \vec{E}_o e^{i(\omega t - kz)},$$

(9)

$$\vec{E}_o = -2\pi i k z n \mu \hat{x}.$$

(10)

The average power radiated is then given by

$$I = \frac{1}{2}\int_o^L \vec{P} \cdot \vec{E}^* \, A \, dz = N^2 I_o \frac{3}{4}\left(\frac{\lambda^2}{8\pi A}\right),$$

(11)

which agrees with Eq. (6). The power radiated by the macroscopic polarization is $[(3/4)(n\lambda^2 L/8\pi)]$ times the radiation of N isolated dipoles. This enhancement is caused by the phase coherence among dipoles over an extended region of space, leading to constructive interference in the forward direction. Combining Eqs. (7) and (11), the radiated power I can be written in the form

$$I = N\frac{3}{4}\frac{\hbar\omega}{T_R},$$

(12)

where

$$T_R = T_{sp}\,(8\pi/n\lambda^2 L).$$

(13)

Thus, when the sample radiates as a collective system, the radiative enhancement factor changes the lifetime from T_{sp} to $\sim T_R$. T_R can therefore be interpreted as a characteristic radiation damping time of the collective system. For the HF system, $T_{sp} \sim 1$ sec and $n \sim 10^{11}\text{--}10^{12}/cm^3$, so $T_R \sim 10^{-8}$ sec. In order for the collective mode to dominate it is necessary that $T_R \ll T_{sp}$. This requires that $n\lambda^2 L \gg 1$, i.e. there must be a large number of molecules in a "diffraction volume" $\lambda^2 L = (\lambda^2/A)AL$. This requirement also insures that there are many molecules in a cylinder of cross section λ^2 and length L, so that a coherent wavefront can be properly reconstructed.

Consider next the process by which a macroscopic polarization builds up in a system which was initially totally inverted (the two adjacent rotational levels of interest in the HF v=1 level). In the Bloch formalism[27] this initial state corresponds to a vector pointing straight up, in analogy to a rigid pendulum balanced exactly on end. Just as the pendulum is unstable to small fluctuations, so the excited molecular system is unstable to a small perturbing field. The radiation process can therefore be initiated by spontaneous emission from one of the excited molecules or by background thermal radiation. This provides a weak propagating electric field which induces a small macroscopic polarization in the medium. This polarization acts as a source to create an additional electric field in the medium which, in turn, produces more polarization. This regenerative process gives rise to a growing electric field and an increasing polarization throughout the medium (Fig. 6). Therefore, a

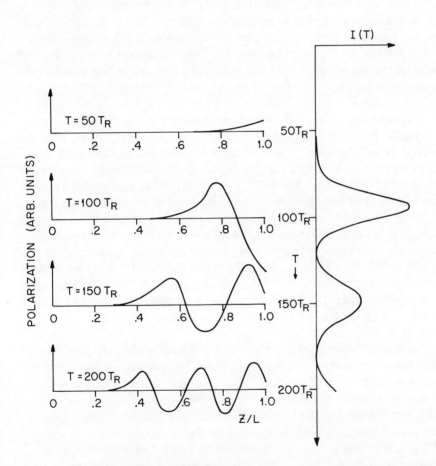

Figure 6.   Sketch of the buildup of polarization in the medium, showing the polarization as a function of z at $T = 50T_R$, $100\ T_R$, $150\ T_R$, and $200\ T_R$. The corresponding output intensity pattern is shown at the right.

superradiant state slowly evolves over a sizeable portion of the sample cell. In the Bloch formalism, this superradiant state corresponds to the Bloch vector pointing sideways, at which time radiation is emitted at a greatly enhanced rate. This process leads to a rapid de-excitation of the medium, after which essentially all of the population is in the lower level [Fig. 9(c), below], i.e. the Bloch vector points downwards.

In Fig. 6, the polarization envelope, $\mathcal{P}$ (z, T), is plotted throughout the sample at several instants of time. (The rapidly varying oscillations at the carrier frequency are not shown.) Notice that $\mathcal{P}$ varies slowly in space and time throughout the medium, giving rise to several regions of locally uniform polarization. The nonuniformity of $\mathcal{P}$ over the entire sample is due to propagation effects in a high gain medium (see below), a crucial point which was not appreciated in some earlier work. The ringing of the output pulse (Fig. 6) is a direct consequence of the spatial variation of $\mathcal{P}$. This feature also determines the width of the output pulses (Sec. VI).

As mentioned in Sec. II, superradiant pulses are observed together in both the forward and backward directions with respect to the pump beam. This is so because the transit time through the sample cell is negligible, so that retardation effects in the excitation process are unimportant. It should be noted that the forward and backward waves are small during most of the pulse evolution time and do not interact appreciably. Even when they become large, competition between the waves is still negligible because they build up in different ends of the medium. The waves can only interact when they grow large in the same region of the medium (~ 150 T_R, Fig. 6), by which time the main pulse has already been emitted. Accordingly, it is an excellent approximation to assume that the oppositely travelling waves evolve independently. Henceforth, we will only deal with the forward travelling wave.

The time evolution of the radiation emitted by the initially inverted system depends on many factors including broadening, diffraction loss and level degeneracy. However, as shown in Sec. V, in a high gain system the major features of the output radiation pulse are determined by the single parameter T_R, and a normalized curve can be drawn which gives the output intensity I(T) multiplied by T_R^2 as a function of time in units of T_R. This curve (Fig. 7) consists of a burst of radiation with ringing preceeded by a delay. This curve's scaling properties show that when T_R is halved, the system radiates twice as fast and the peak intensity is quadrupled. This curve exhibits the N^2 intensity dependence which is characteristic of superradiant emission, since the peak intensity is proportional to T_R^{-2} and T_R is proportional to N^{-1}. Experimental data exhibiting these scaling properties is shown in Fig. 5.

Note from Fig. 7 that the delay time T_D from the pump pulse to the peak of the first lobe of emitted radiation is much greater than T_R. This may be understood by considering a pendulum which is initially balanced exactly on end. The time required for the pendulum to fall following a small perturbation can be much longer than the oscillation period. In an analogous manner, the time necessary for the completely inverted system to develop a macroscopic polarization can be much longer than the collective radiation time T_R. This point will be discussed in Sec. VI.

IV. THEORETICAL CONSIDERATIONS

In numerous treatments of superradiance, the radiation field is quantized, and the molecular system is described in terms of collective Dicke states. As pointed out by Arecchi et al.,[22]

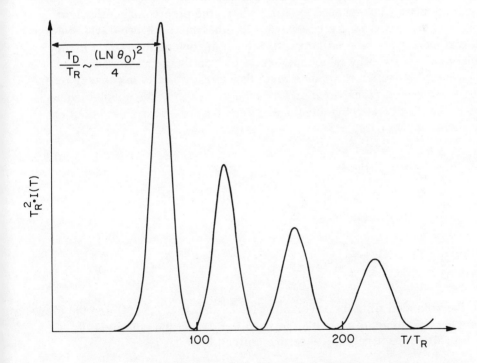

Figure 7.   Normalized output curve. This curve is the output response to a small rectangular input pulse of area θ_0 in a nondegenerate system where $T_2 = T_2^* = \infty$ and $\kappa L = 0$. The time scales as T_R and the intensity scales as T_R^{-2}.

these states can be used to construct a new set of states, Bloch states, which also describe superradiant ensembles. These new states can be treated by means of the semiclassical formalism, in which the molecular system is quantized but the e. m. field is treated classically. The semi-classical approach does not hold during the initial stages of the evolution of the superradiant state, when the number of photons in the diffraction mode is small. As shown below, in the first time interval $\sim T_R$ after excitation, the number of photons emitted into the diffraction mode is greater than one, and from then on the semiclassical treatment is justi-fied. The pulse evolution time in our experiments is $\sim 100 T_R$ so that the details of the fluctuations in the initial stage are not very important in determining the overall behavior. Furthermore, the numerical results show that the superradiant output pulse is insensitive to the specific form of the source fluctuations (see Sec. V).

Adopting the semiclassical formalism, we consider a gas of two-level molecules interacting with an electromagnetic wave. As in the dis-cussion following Eq. (4), we confine our attention to a system of large Fresnel number (disk-shaped system) where the electromagnetic field can be approximated by a plane wave. In this limit, the two-level system is described by the coupled Maxwell-Schroedinger equations in a form which has been given by many authors. [28, 29] In the notation of Icsevgi and Lamb, [28] extended to include level degeneracy, as is necessary in treating rotational transitions of a molecular system, the coupled equa-tions in the slowly-varying envelope approximation, written in complex form, are

$$\frac{\partial \mathcal{E}}{\partial z} = -\kappa \mathcal{E} + 2\pi k < \mathbf{P} >_{v, M} \quad , \tag{14}$$

$$\frac{\partial \mathbf{P}}{\partial T} = -(\gamma - ikv)\mathbf{P} + \frac{(\mu_z)^2_M}{\hbar} \mathcal{E}n + \Lambda_P \quad , \tag{15}$$

$$\frac{\partial n}{\partial T} = \Lambda - \gamma n - (1/\hbar) \operatorname{Re}(\mathcal{E}^* \mathbf{P}). \tag{16}$$

Here the retarded time $T = t - z/c$, and $\mathcal{E}(z, T)$ and $\mathbf{P}(z, T, v, M)$ are the slow-ly varying envelopes of the electric field $\vec{E}(r, t)$ and the polarization $\vec{P}(r, t, v, M)$, respectively. These are defined by

$$\vec{E}(r, t) = \hat{x} E(z, t) = \hat{x} \operatorname{Re}[\mathcal{E}(z, T)e^{i(\omega t - kz)}] \quad , \tag{17}$$

$$\vec{P}(r, t, v, M) = \hat{x} P(z, t, v, M) = \hat{x} \operatorname{Re}[\mathbf{P}(z, T, v, M)e^{i(\omega t - kz)}] \tag{18}$$

at position z and time t.   Note that $\mathcal{E}$ and $\mathcal{P}$ are complex, and that $\mathcal{E}(z, T)$ = $|\mathcal{E}(z, T)|$ exp$(i\varphi)$ where $\varphi(z, T)$ is a slowly varying real phase.   In these equations, $n(z, T, v, M) = n_2 - n_1$, with n_2 and n_1 the number densities of molecules in the upper and lower levels of the superradiant transition, respectively, κ is a linear loss term to take account of diffraction, and $\gamma = T_2^{-1}$ is the homogeneous decay rate.   For simplicity, the decay rates of n_1, n_2, and $\mathcal{P}$ are all assumed to be equal. $\Lambda = \lambda(z, T, M)W(v)$ is a source term describing the rate of production of n, due to optical pumping in our case, where $W(v)$ is defined below. $\Lambda_P(z, T, v, M)$ is a polarization source term describing the rate of production of $\mathcal{P}$ due to spontaneous emission (the exact form will be discussed below).   The symbol $< >_{v, M}$ denotes an average over the velocity distribution and over degenerate M states of the rotational levels:

$$<f(v, M)>_{v, M} \equiv \int_{-\infty}^{\infty} dv \; \sum_{M=-J}^{J} (2J+1)^{-1} f(v, M)W(v). \qquad (19)$$

Here $W(v)$ is the distribution of molecular velocities, normalized to unity.

The number of molecules produced in the upper state and their velocity distribution depend on the pump field intensity.   For an intense laser the pump transition is saturated and appreciable power broadening occurs.   In the experiments this broadening is such that the gain bandwidth at the rotational transition is far larger than its homogeneous width (~ 25 kHz/mTorr), yet still less than the Doppler width of the rotational transition at room temperature (~ 10 MHz).   In this regime, the excitation process produces excited molecules in the $v = 1$ state (the upper level of the two-level system) with a Lorentzian velocity distribution $W(v)$ whose width u_1 depends on the intensity of the pump pulse:[30, 31]

$$W(v) = \frac{u_1/\pi}{v^2 + u_1^2} \quad , \qquad (20)$$

where

$$u_1 \approx \mu_p \mathcal{E}_p / \hbar k_p. \qquad (21)$$

Here k_p is the wavenumber, and μ_p the matrix element of the pump transition, and $\mathcal{E}_p$ is the amplitude of the pump field.   (In our case $k_p \sim 2\pi/2.5 \mu$m, $\mu_p \sim 10^{-19}$c.g.s. units, and for pump power I=1 kW/cm^2, $\mathcal{E}_p \sim$ 3 e.s.u., so that $u_1 \sim 1.2 \times 10^4$cm/sec.) Note that u_1 can be a sizeable fraction of the thermal velocity spread of the molecules ($u \sim 5 \times 10^4$cm/sec for HF).[30]   The number density of excited molecules is then:

$$n_0 = \frac{1}{2}(\frac{\mu_p \mathcal{E}_p / \hbar}{k_p u \; \sqrt{\pi}}) n_T, \qquad (22)$$

where n_T is the number density of molecules in the $v = 0$ rotational level which is being pumped.   The gain bandwidth Δ of the superradiant transition is $\Delta = ku_1$, so that the total gain αL is

$$\alpha L = 2\pi k \frac{|\mu_z|^2}{\hbar} \frac{n_o}{\Delta} L. \tag{23}$$

Therefore αL is independent of the pump field intensity I.   On the other hand, T_R^{-1} is proportional to n_o (Eq. 13) and therefore to I. Accordingly, in the experiments the gain αL is chosen by fixing the sample cell pressure, and T_R can then be adjusted independently by varying the pump intensity.

For a general system which is both homogeneously and inhomogeneously broadened, the initial total field gain αL is

$$\alpha L = 2\pi k \frac{|\mu_z|^2}{\hbar} n_o L T_2', \tag{24}$$

where n_o is the initial inversion density.   The characteristic broadening time T_2' depends on both the collision time T_2 and the inverse Doppler width T_2^* .   In the Doppler-broadened limit, which applies in the HF system, $T_2' \approx T_2^*$.   Note that in our case T_2^* is an effective Doppler width, determined by saturation broadening as described above.   Combining Eqs. (7), (13), and (24) gives the relationship[32]

$$\alpha L = T_2'/T_R. \tag{25}$$

This shows that for a system with fixed T_2' the gain and the characteristic radiation time T_R are related and cannot be specified independently. As will be seen in Section VI this relationship has important consequences.

Although the propagation and evolution of the electric field in the medium does not follow simple exponential gain, the time integral of the electric field obeys an exponential law, the area theorem.[10] This states that for a nondegenerate Doppler-broadened collisionless system subjected to an incident field $\mathcal{E}(z = 0, t)$ of constant phase, the pulse area $\theta(z)$ obeys the equation

$$\tan [\theta(z)/2] = \tan (\theta_o/2) e^{\alpha z}, \tag{26}$$

where

$$\theta(z) = (\mu_z/\hbar) \int_{-\infty}^{\infty} \mathcal{E}(z, t)dt, \tag{27}$$

and $\theta_o = \theta(z = 0)$.   Equation (26) shows that in a high gain medium the area of a pulse of small initial area begins to grow exponentially,

$$\theta(z) = \theta_o e^{\alpha z}, \tag{28}$$

but then evolves towards π. In this process the field envelope may develop positive and negative lobes (ringing) whose contributions to the area nearly cancel one another. Accordingly, in a high gain system the pulse energy continues to grow even though the area remains constant.

The presence of level degeneracy does not invalidate this conclusion, although level degeneracy can inhibit pulse propagation in an absorber. The considerations of Rhodes, Szöke, and Javan[33] applied to an amplifier show that level degeneracy does not prevent pulse formation, and that pulses of increasing energy and area near π can evolve.

As discussed earlier, the pulse evolution process is initiated by spontaneous emission from the excited molecules and background thermal radiation. In our model, spontaneous emission is simulated by a randomly phased polarization source term Λ_P which is distributed throughout the medium. Its amplitude may be determined by requiring that Eqs. (14) and (15) give the correct intensity at thermal equilibrium. A calculation[30,34] shows that

$$\Lambda_P = 2\left[\frac{\Delta T}{T_2} \frac{\Delta z}{L} n_2 A L\right]^{1/2} (\frac{\mu_z\sqrt{2}}{A}) \sum_j \sum_m \delta(z-z_j)\delta(T-T_m)e^{i\Phi_{jm}}, \tag{29}$$

where z_j and T_m are random space and time points, respectively, $\Delta z \Delta T$ is the average area associated with one spontaneous emission event in the space-time grid, and Φ_{jm} is a random phase which is different for each pair (j, m). (In the computer analysis z_j and T_m are each chosen at equal intervals Δz and ΔT, respectively.) Note that Λ_P is proportional to μ_z and depends on the total number of excited molecules $n_2 A \Delta z \Delta T$ associated with one point in the space-time grid. The square root dependence occurs because the radiation from independent spontaneous emission events adds incoherently.

In the computer model, background thermal radiation is simulated by an input field of randomly fluctuating phase, whose intensity depends upon the bandwidth and the solid angle of the input radiation which interacts with the system. For a high gain system,[30]

$$\frac{|\mathcal{E}(z=0, T)|^2}{8\pi} = \left[\frac{\hbar\omega}{e^{\hbar\omega/kT}-1}\right]\left[\frac{4\pi(\omega/2\pi)^2}{c^3}\right]\left[\frac{\lambda^2}{4\pi A}\right]\left[\frac{ku_1}{\sqrt{2\pi\alpha L}}\right], \tag{30}$$

$$cA|\mathcal{E}(z=0, T)|^2/8\pi = [(\hbar\omega)/(e^{\hbar\omega/kT}-1)] (ku_1)/(\sqrt{2\pi\alpha L}). \tag{31}$$

The factors on the right side of Eq. (30) are, respectively, the background thermal radiation energy per mode, the number of modes per unit volume per frequency interval, the solid angle factor f, and the gain-dependent effective bandwidth. Using the value of $\mathcal{E}(z=0, T)$ of Eq. (30) as an input boundary condition, and using Eq. (29) for Λ_P, Eqs. (14)–(16) can be numerically integrated to obtain the output radiation intensity. An example of the computer results using this approach is given in Fig. 8(a) below.

Alternatively, in describing emission in our experiment it is possible to replace the distributed polarization source by an "equivalent" input electric field. This approximation is valid because spontaneous emission near the input face is most important in initiating the radiation process in a high gain system, and because the transit time is negligible.[35] The latter approach has the advantage that when phase fluctuations are neglected, it allows us to use the results of the area theorem, which is derived under the assumption of an external field incident upon a sample with no distributed source. The amplitude of this equivalent field is found by requiring that near thermal equilibrium the output radiation produced by it agrees with the predictions of Einstein's equation for intensity gain. This electric field is then added incoherently to the incident field due to background thermal radiation. For a high gain system the total equivalent input field is then given by[30]

$$\frac{cA}{8\pi} \, \mathcal{E}_{eff}^2 \, (z=0, T) = \frac{\hbar\omega}{T_R} \, (2\alpha L)^{-3/2}[1 + (e^{\hbar\omega/kT}-1)^{-1}], \qquad (32)$$

where the terms in brackets are the contributions of spontaneous emission and background thermal radiation, respectively. The output pulse shapes have been calculated by computer using both the distributed source model and the equivalent input field model. The results for a particular set of initial conditions are shown in Fig. 8. As can be seen from the figure, the two approaches give almost identical output pulses. In this paper extensive use is made of the equivalent input field approach in fitting our experimental data. The data is now being analyzed in detail using the distributed polarization source model. The results will be presented elsewhere,[30] but preliminary analysis shows that the distributed nature of Λ_P has virtually no effect upon the predicted outputs which fit our experimental data. This finding strongly supports the validity of the equivalent field approach and confirms the interpretation and analysis reported earlier.[16]

The effective input field can only influence the evolution of the superradiant state for a short time after the system is pumped into exci-

tation.   The situation is analogous to the falling of a rigid pendulum initially balanced on end: A small perturbation can initiate the motion, but similar perturbations have almost no effect once the motion is under way.
Under the influence of $\mathcal{E}_{eff}$ a macroscopic polarization builds up in the
medium.   As the field produced by this polarization grows, the importance of $\mathcal{E}_{eff}$ diminishes.   The output field more than doubles in a time
T_R after excitation (Sec. IV), and the effective input field can be set
equal to zero after this time.   This has been verified by computer calculations.   The effective input pulse, therefore, has an effective area

$$(\theta_o) \sim \frac{\mu \mathcal{E}_{eff}}{\hbar} T_R. \tag{33}$$

Using Eqs. (7), (13) and (32), this becomes

$$(\theta_o)^2 \approx 2 N_2^{-1} (\alpha L)^{-3/2}[1 + (e^{\hbar \omega/kT} - 1)^{-1}]. \tag{34}$$

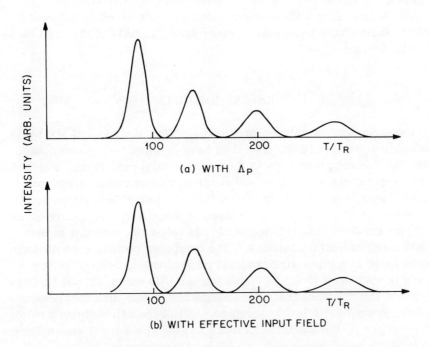

(a) WITH Λ_P

(b) WITH EFFECTIVE INPUT FIELD

Figure 8.   Comparison of the output intensity predicted by (a) the distributed source model and (b) the equivalent input field model for a typical set of input parameters.   The two curves
are essentially identical.

For our system, $\theta_0 \sim 10^{-7}$, and in order for the output to evolve to a pulse of area $\sim \pi$, αL [Eq. (23)] must be $\gtrsim 20$. This is readily achieved in the experiments.

Consider next the influence of phase fluctuations of the source terms on the pulse evolution process. As seen above, coherent radiation is emitted into a [single] diffraction mode down the pump excitation axis of the sample. The average number of photons ν emitted into this mode in a time t is

$$\nu = \frac{N}{T_{sp}} \frac{\lambda^2}{4\pi A} t \tag{35}$$

and therefore, the first photon is emitted in a time $\sim T_R$. In the far infrared, background thermal radiation also contributes at a comparable rate [see Eq. (32)]. Since the input field is only important for a time $\sim T_R$ after excitation, the spontaneous emission of only about one photon affects the output, and phase fluctuations associated with subsequent spontaneous emission can be neglected. Since the observed output is similar to that of an equivalent system with no distributed sources which is subjected to a short constant-phase input pulse, many useful results of the area theorem can be applied.

V. NUMERICAL RESULTS

Using the analysis of the previous section, the coupled Maxwell-Schroedinger equations [Eqs. (14)-(16)] have been solved by computer. In fitting the theoretical curves to the experimental data (Figs. 9 and 10), all of the parameters are expressed in terms of three quantities: the estimated linear loss, κ, due to diffraction; the pump field intensity, I; and the sample cell pressure, p. Values of these parameters are given in the figure captions. Level degeneracy is taken into account in both pump and superradiant transitions. The computer results show that the output intensity at a given superradiant transition depends on the pump transition branch used (P or R). This agrees with experimental findings and is due to the different matrix elements and widely different populations of the ground state levels selected. All of the calculations assume that at pressure p, the collision time T_2 is given by $T_2 \cdot p = 7$ sec$\cdot$mTorr. As can be seen, good agreement is obtained throughout.

The numerical results show that the output is insensitive to the exact form of the input field. Delta function, step function, Gaussian pulses, and pulse trains of intensity consistent with the magnitude of the input radiation field all give output pulses of about the same shape and

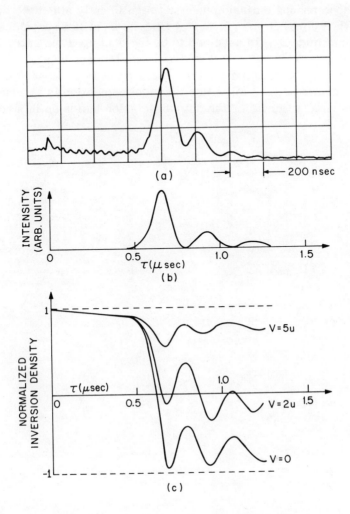

Figure 9. (a) Oscilloscope trace of superradiant pulse at 84 μm (J=3→2),
pumped by the R_1(2) laser line. The small peak on the scope
trace at T=0 is the 2.5 μm pump pulse, highly attenuated.
(b) Theoretical fit to (a). The parameters are I = 1 kW/cm^2,
p = 1.3 mTorr and κL = 2.5 for L = 100 cm   The corresponding
values of T_2^* and T_R are 330 and 6.1 nsec, respectively. Note
that $\alpha L = T_2^*/T_R = 54$.   (c) Population inversion at z = L for
molecular ensembles with different axial velocities v in units
of u_1, where $ku_1 = 1/T_2^*$. Note that molecules with $|v| < 2u_1$
can be coherently de-excited.

simulate spontaneous emission does not significantly affect the output pulses. Therefore, in the numerical fits, an input field envelope of constant amplitude $\mathcal{E}_{eff}$ is assumed to be incident upon the entry face of the medium (z = 0). $\mathcal{E}_{eff}$ is derived from Eq. (32) and depends on I and p. In terms of $\mu\mathcal{E}_{eff}/\hbar$, this parameter ranges from 10 to 50 sec^{-1}. As previously shown [Fig. 2(c)], after the level population is inverted over a period of $\sim 20 T_R$ duration, the output emission builds up in a time

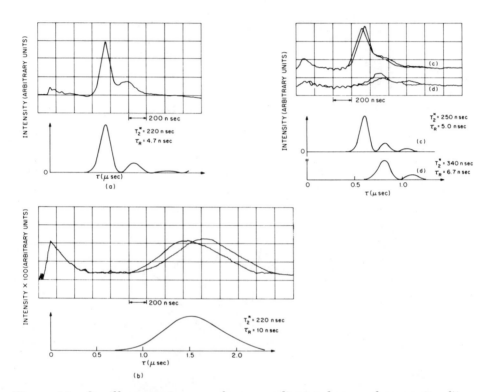

Figure 10. Oscilloscope traces of superradiant pulses and computer fits.
(a) J=3→2 transition at 84 μm pumped by $P_1(4)$ laser line.
I=2.2kW/cm^2, p=4.5 mTorr, κL=2.5(L=100 cm), giving T_2^*
= 220 nsec, T_R = 4.7 nsec. (b) Same as (a) except that
p=2.1 mTorr, giving T_2^* = 220 nsec, T_R = 10 nsec. Note increased delay and broadening of pulse. (c) Same transition
as (a), but pumped by $R_1(2)$ laser line. I=1.7kW/cm^2, p=1.2
mTorr, κL = 3.5(L=100 cm) giving T_2^* =250 nsec, T_R = 5.0
nsec. (d) Same as (c) except I=0.95 kW/cm^2, giving T_2^*
= 340 nsec, T_R =6.7 nsec. The same intensity scale is used
in fitting curves (a) and (b), and (c) and (d). Note the reproducibility of the oscilloscope traces in double exposure.

delay. The presence of random jumps in the phase of the input field to
$\sim 100 T_R$, when a pulse of width $\sim 20 T_R$ is emitted. Characteristic ringing
is also observed.

Figure 10 illustrates the effect of changing the sample cell pressure and the pump intensity. Only the pressure was changed (by about a
factor of 2) between Figs. 10(a) and 10(b). Note the greatly reduced intensity, the increased delay, and the broadening of the pulse at lower
pressure [Fig. 10(b)].   Figures 10(c) and 10(d) show the effect of changing
I by about a factor of 2. Note the smaller intensity, longer delay, and
wider pulse at lower intensity [Fig. 10(d)].   The output signals were fairly
reproducible, as can be seen from the double exposure in Fig. 10.

The following general features are evident from the numerical
results, and confirm the qualitative statements of Sec. III and IV:

1. The effects of relaxation are unimportant as long as the homogeneous relaxation time T_2 exceeds the pulse delay T_D. When T_2 becomes
comparable to T_D the pulses are reduced in size and the ringing is cut
down. At the mTorr pressures of the experiments T_2, determined by
collisions, is always much longer than T_D.

2. The effectiveness of the input field is limited to the first few
T_R 's. As mentioned above, the system behavior is insensitive to the
exact shape and phase of the input field.

3. The output pulses are insensitive to the specific time dependence of the population excitation. This is true even for excitation pulses
approaching the superradiant pulse delay in duration.

Three features of the results are of fundamental importance:

1. The area of the output pulse is determined by the gain αL and
the size of the input pulse θ_o, in accordance with the area theorem [Eq.
(26)]. Since in our case $\theta_o \sim 10^{-7}$, $\alpha L \gtrsim 20$ is needed for appreciable pulse
build-up. Such gains are available in HF at mTorr pressure.[17,18]

2. For a given value of $\alpha L \gg 1$ the time scale of the pulse evolution (i.e. delays, pulse shapes) is determined by T_R.

3. For a fixed αL and T_R, homogeneous and inhomogeneous
broadening, level degeneracy, and moderate variations in the value
of κL change the results in only minor ways. The weak influence
of these effects on the output pulse can be seen in Fig. 11, which

shows the changes in the pulse shape and delay caused by varying these parameters. The simple results obtainable by setting $1/T_2 = 1/T_2^* = \kappa L = 0$ and neglecting level degeneracy may be used for a qualitative understanding of the behavior. This simplified model gives the normalized emission curve of Fig. 7.

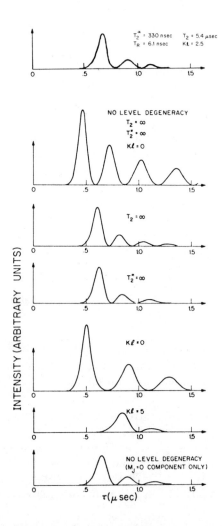

Figure 11. Computer results showing the influence of parameters on pulse evolution. The uppermost curve is the theoretical fit shown in Fig. 9 (b). All parameters have the same values as in this curve except when stated otherwise. The values of the modified parameters are indicated in the figure. The same intensity scale is used throughout.

VI. SIMPLIFIED THEORY AND CONNECTION WITH PREVIOUS WORK

The preceding computer analysis led to the conclusion that most of the input parameters of the system have very little effect on the output (Fig. 11). If we set $\gamma = ku = \kappa = \Lambda = \Lambda_P = 0$ in Eqs. (14) – (16) and ignore the effects of level degeneracy, a set of simplified equations is obtained:

$$\partial \mathcal{E}/\partial z = -2\pi k S, \tag{36}$$

$$\partial S/\partial T = -\mu_z^2\, n \mathcal{E}/\hbar, \tag{37}$$

$$\partial n/\partial T = \mathcal{E} S/\hbar. \tag{38}$$

Here, $S = -\mathcal{P}$ and n, $\mathcal{E}$, and $\mathcal{P}$ are all real. Equations (37) and (38) have the solution

$$n = n_o \cos \varphi, \tag{39}$$

$$S = -\mu_z n_o \sin \varphi, \tag{40}$$

$$\partial \varphi/\partial T = \mu_z \mathcal{E}/\hbar, \tag{41}$$

where $\varphi(z, T)$ is the "partial area" of the pulse,

$$\varphi = \int_{-\infty}^{T} (\mu_z/\hbar)\mathcal{E}(z, T')dT', \tag{42}$$

so that $\varphi(z, T = \infty) = \theta(z)$. In the geometrical representation,[36] φ is the "tipping angle" of a Bloch vector[27] of length n_o in the z-y $[n - (S/\mu_z)]$ plane. The initially inverted system, corresponding to the Bloch vector standing on end ($n = n_o$, $S = 0$), gradually evolves into a superradiant state ($n = 0$, $S/\mu_z = -n_o$), as described in Sec. III.

Combining Eqs. (36) and (41) gives an equation for $\varphi(z, T)$:

$$\frac{\partial^2 \varphi}{\partial z \partial T} = \left(\frac{1}{T_R L}\right) \sin \varphi. \tag{43}$$

This equation was studied for the case of an absorber by Burnham and Chiao,[37] who showed that for a delta function input field, the time dependence of $\varphi(z, T)$ and therefore [Eq. (41)] of $T_R \mathcal{E}(z, T)$ was a function of T/T_R only. Our case is that of an amplifier, and the time dependence of the delta function response is again a function of T/T_R only. Furthermore, since the input field is only important during the first $\sim T_R$, any input pulse of small area θ_o during the first $\sim T_R$ will give the same response as a delta function input pulse of the same area. This explains the scaling

properties of the normalized emission curve of Fig, 7. Essentially the
same scaling law is exhibited by the computer calculations, even when all
of the refinements of the theory are included (Fig. 11). The main requirement for the validity of the normalized emission curve is that T_R be much
shorter than all other times associated with the system ($T_R \ll T_2'$), so that
superradiant emission can occur before dephasing or relaxation set in.
It immediately follows from Eq. (25) that this condition is equivalent to
$\alpha L \gg 1$, so that any sufficiently high gain system can produce superradiant emission when suitably excited.

As a consequence of the long delay times which result from small
input pulses, φ remains small for almost all of the pulse evolution time.
Accordingly, the expression for $\mathcal{E}(z, T)$ in the small angle approximation
($\sin \varphi \approx \varphi$) can be used. (In the final stages of pulse evolution, φ approaches π and this approximation breaks down.) A number of useful results
can be derived from this approximation and the simplified equations
[Eqs. (36)–(38)].

Consider first the delay time T_D from the pump pulse to the peak
of the first lobe of the emitted radiation. In the small angle limit a small
step function input electric field with envelope $\mathcal{E}(z=0, T>0) = \mathcal{E}_o$ gives rise
to an output field at $z = L$ of the form

$$\mathcal{E}(L, T) \sim \mathcal{E}_o I_o (u), \tag{44}$$

where $I_n(x)$ is the modified Bessel Function of order n, and $u = 2\sqrt{T/T_R}$.
Then $\Phi(T) \equiv \varphi(z=L, T)$, the partial area at $z = L$, is given by

$$\Phi(T) = \int_{-\infty}^{T} (\mu_z/\hbar) \mathcal{E}(L, T') dT' \tag{45}$$

$$\sim \theta_o u I_1 (u), \tag{46}$$

where $\theta_o = \mu_z \mathcal{E}_o T_R/\hbar$. For a high gain system, $\Phi(T_D)$, the partial area
at the peak of the first lobe, is $\sim 2\pi$. Since Φ only deviates from Eq. (46)
during the last few T_R before $T=T_D$, only a small error is made by
approximating $\Phi(T_D)$ by Eq. (46). Setting $\Phi(T_D) = 2\pi$ and solving for T_D
in the limit $T_D \gg T_R$, one obtains

$$T_D \approx \frac{T_R}{4} (\ln \theta_o)^2. \tag{47}$$

In our experimental system, $\theta_o \sim 10^{-7}$, so $\ln \theta_o \sim -16$ and T_D is
insensitive to changes in θ_o. Therefore $T_D \sim 70 T_R$ is a convenient estimate
of the time required for the superradiant state to evolve. Since T_D is pro-

portional to T_R, T_D should be inversely proportional to the excitation density, and therefore the pressure, in the sample cell. This inverse proportionality and the estimate $T_D \sim 70 T_R$ are theoretical predictions which can be tested against the experimental results. The excellent agreement of these predictions with the experiment confirms the validity of our analysis. The linearity of the delay time as a function of inverse pressure is clearly evident in the curve of Fig. 12.

An estimate of the width, T_W, of the first lobe of the radiation pulse can be obtained from the $1/e$ width of $\Phi(T)$ at $T=T_D$, i.e. by solving the equation $\Phi(T_D - T_W/2) = e^{-1}\Phi(T_D)$ for T_W. If we assume that Eq. (46) is valid near $T = T_D$, this gives[30]

$$T_W \sim T_R \left| \ln \theta_o \right|, \tag{48}$$

since $T_W \ll T_D$. Equation (48) predicts a pulse width $\sim 16 T_R$ for our sys-

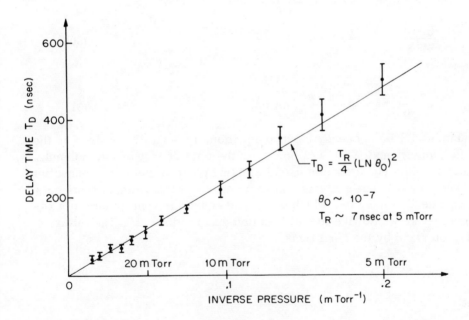

Figure 12. Time delay from pump pulse to first lobe of superradiant pulse at 84 μm ($J=3 \rightarrow 2$), pumped by the $P_1(4)$ laser line, as a function of the inverse HF pressure in the sample cell. Since $T_R \sim 7$ nsec at 5 mTorr, the equation of the straight line fit to the data is $T_D \sim 69 T_R$.

tem, which is somewhat smaller than the observed pulse widths, which are $\sim$20-40T_R (Figs. 9, 10). This is expected, since the growth of $\Phi(T)$ near T_D is actually slower than that given by Eq. (46).

This result has another interpretation. It was emphasized above that because the sample is optically thick (i.e. $\alpha L \gg 1$) the polarization envelope, $\mathcal{P}(z, T)$, varies over the medium and may undergo changes in sign (Fig. 6). The contributions to the electric field from regions of opposite sign tend to cancel one another, and, in effect, only a limited region of the sample of length L_{eff} (i.e. the characteristic length of the lobes of $\mathcal{P}(z, T)$) determines the output. Since L_{eff} is smaller than L, the pulses are wider than T_R. An estimate of L_{eff} can be obtained by writing $T_{sp}(8\pi/n\lambda^2 L_{eff}) = (T_R)_{eff} \sim T_w$. This gives

$$L_{eff} \sim L/|\ln \theta_o|. \tag{49}$$

We turn now to a comparison with previous theoretical work. The two equations (36) and (38) can be combined to give

$$\frac{\partial}{\partial z}(\frac{c}{8\pi}\mathcal{E}^2) = -\frac{\omega}{2}S\mathcal{E} = -\frac{\hbar\omega}{2}\frac{\partial n}{\partial T}, \tag{50}$$

or

$$I = \frac{c}{8\pi}\mathcal{E}^2 A = -\frac{\hbar\omega}{2}\frac{\partial}{\partial T}\int nA\ dz. \tag{51}$$

Equation (51) is an energy balance equation, $I = -dW/dT$, where I is the field intensity and $W = (\hbar\omega/2)\int nA\ dz$ is the total energy of the molecular system. This equation is a useful starting point to make the connection between the present analysis and some previous theoretical treatments[7,8] of an extended medium, which developed equations appropriate to a regime that we would like to call "limited superradiance." The latter is characterized by the conditions

$$T_R \ll T_{sp} \tag{52}$$

but

$$\alpha L \ll 1. \tag{53}$$

By neglecting spatial variations in the sample, the above treatments implicitly assume condition (53), since only in the low gain limit are spatial variations of φ, n, and S small over the entire length of the medium. Ignoring these spatial variations one can set

$$n = n_o \cos \varphi \tag{54}$$

and

$$W = \frac{1}{2} \hbar \omega \, n_o \, AL \cos \varphi, \tag{55}$$

so that, using Eqs. (40) and (41),

$$I = \frac{N \hbar \omega}{4 T_R} \sin^2 \varphi = \frac{N \hbar \omega}{4 T_R} \operatorname{sech}^2 [(t-t_0)/T_R]. \tag{56}$$

Hyperbolic secant-squared dependence of this type was predicted by Rehler and Eberly[7] and by Bonifacio, Schwendimann, and Haake.[8] Several features of this result should be noted. This output intensity does not ring. Instead, it consists of a single lobe of width $\sim T_R$. The lack of ringing in Eq. (56) is a consequence of the small gain assumption--ringing has been suppressed by not allowing n and S to vary over the length of the medium (cf. Fig. 7). This indicates that the results of Refs. 7 and 8 implicitly assume the thin sample limit and, accordingly, they describe the regime of limited superradiance. In fact, inspection of the assumptions of Ref. 7 shows that spatial variations in the medium are not taken into account in a consistent manner. Similar remarks apply to the results of Ref. 8. A set of equations almost identical to the superradiance rate equations of Ref. 8 can be derived from our approach in the thin sample limit.

Limited superradiance can be considered to be superradiance in the sense that $T_R \ll T_{sp}$, so that collective effects are important. However, since $\alpha L \ll 1$, it follows from Eq. (25) that $T_2' \ll T_R$, i.e. the radiative damping time exceeds the characteristic time after which incoherent processes become important. Therefore, the system can only superradiate for a time short compared to the width of the sech^2 pulse, after which the coherent polarization in the sample is destroyed by dephasing or collisional decay. Over this short time Eq. (55) correctly describes the collective emission in a thin sample, such as occurs in photon echo experiments, where dephasing is crucial. In contrast to this the regime of strong superradiance, which is studied in the present experiment, is characterized by the conditions

$$T_R \ll T_{sp} \tag{57}$$

and

$$\alpha L \gg 1. \tag{58}$$

The latter equation implies that $T_R \ll T_2'$, which insures that the sample can superradiate before decay processes set in. Accordingly, in this regime almost all of the energy stored in the sample can be emitted coherently.

In this connection the long delays preceding the observed output pulses deserve further discussion. For example, in Fig. 9(b) the delay time is greater than $100 T_R$, whereas T_2^* is only 50 T_R, so that $T_D > 2 T_2^*$. It is remarkable that fully developed pulses with ringing can evolve over such long times, despite the presence of dephasing processes. This behavior is unique to high gain amplifiers, where the high gain can overcome dephasing during the early stages of pulse evolution. The system eventually dephases, but the effective dephasing time is increased to $\alpha L T_2^*$, as can be established by considering the response of a high gain inhomogeneously-broadened amplifying medium to a short input pulse of small area θ_o. An analytical expression has been given by Crisp. [38] For a Lorentzian lineshape the output $\mathcal{E}$-field is of the form

$$\mathcal{E}(z = L, T) \approx \frac{\mathcal{E}_o}{2\pi} (T/T_R)^{-3/4} \exp[2\sqrt{T/T_R} - T/(T_R \alpha L)], \quad (59)$$

where $\mathcal{E}_o = \hbar \theta_o / \mu T_R$. Initially, the square-root term in the exponential dominates and $\mathcal{E}$ increases. Its maximum value is reached at $T \approx T_2^* \alpha L$, after which exponential decay due to dephasing sets in. Therefore, the effective dephasing time is

$$(T_2^*)_{eff} \approx T_2^* \alpha L. \quad (60)$$

When this expression is compared with Eq. (47) for the delay time, one finds that

$$\frac{T_D}{(T_2^*)_{eff}} \approx \left(\frac{\ln \theta_o}{2\alpha L}\right)^2 = (2\alpha L_{eff})^{-2}, \quad (61)$$

which is always smaller than unity for $\alpha L \gg |\ln \theta_o|$. Therefore, an inhomogeneously broadened system of sufficiently high gain will always superradiate before it can dephase. [39]

VII. CONCLUDING DISCUSSION

An experiment has been presented in which the energy stored in an extended optically thick sample is extracted by collective radiative processes. This effect can be described as "self-induced emission" or, in the terminology of Dicke, as "superradiance," and is based on the transient response of an amplifying medium. In the process a phase-coherent polarization is produced throughout the medium. The ensuing radiation process releases nearly all of the stored energy, leaving the system in a state of total de-excitation. This is in contrast to the usual "amplified spontaneous emission" and other stimulated processes in which the emission terminates when the level populations are equalized.

The theoretical analysis has shown the connection with Dicke's work and with other previous treatments. The nature of the superradiant emission process has been clarified, and the dependence of the effect on various experimental parameters has been established. It has been shown that the single most important parameter describing the decay is T_R, the collective radiation damping time; that this parameter is closely connected to the gain of the system; that in order to obtain rapid and efficient energy extraction, high gain is required; and that this requirement leads, by necessity, to spatial variations in the polarization produced in the medium.

One important consequence of the analysis is that inhomogeneous broadening of the transition does not necessarily inhibit the superradiant emission process. This awareness opens the possibility of extending these studies to Doppler broadened atomic and molecular systems in the visible and infrared regions, as well as to inhomogeneously broadened solids.

Although in our experiments care was taken to minimize feedback in order to obtain long delays, it is emphasized that the presence or absence of feedback is not in itself an essential consideration. (In related experiments we have observed similar output signals and long delays in the presence of intentional feedback, but at still lower pressures than those reported above.) The basic criterion for strong superradiance to occur is that $T_R \ll T_2'$. This condition insures that a macroscopic polarization can develop in the medium before dephasing destroys the phase coherence.   It follows that any system of sufficiently high gain will superradiate when properly prepared.

In order to obtain long delays it is necessary to avoid leaving a large macroscopic polarization in the medium. In our experiments this was achieved by exciting the system via a coupled transition. We believe that this and other indirect pumping schemes[40] can be used to advantage in producing superradiance in other systems.

As a final point, note that due to the transient nature of the emission process the width of the pulses is not bandwidth limited. An example confirming this statement is shown in Fig. 9 where the pulse is twice as narrow as the effective gain bandwidth $(\sim 1/T_2')$ would imply.

A more detailed version of this report is being submitted to Physical Review.

ACKNOWLEDGMENTS

We would like to thank Norman Kurnit for many useful discussions, Bill Ryan for his technical assistance, and Ali Javan for his support and encouragement throughout this work.

REFERENCES

1. R. H. Dicke, Phys. Rev. 93, 99 (1954).
2. A. Abragam, The Principles of Nuclear Magnetism, (Oxford U. P. , London, 1961); R. M. Hill, D. E. Kaplan, G. F. Herrmann and S. K. Ichiki, Phys. Rev. Letters 18, 105 (1967).
3. V. Ernst and P. Stehle, Phys. Rev. 176, 1456 (1968).
4. G. S. Aggarwal, Phys. Rev. A2, 2038 (1970).
5. R. H. Lemberg, Phys. Rev. A2, 883 and 889 (1970).
6. D. Dialetis, Phys. Rev. A2 , 599 (1970).
7. N. E. Rehler and J. H. Eberly, Phys. Rev. A3, 1735 (1971).
8. R. Bonifacio, P. Schwendimann, and F. Haake, Phys. Rev. A4, 302 (1971) and A4 , 854 (1971).
9. I. D. Abella, N. A. Kurnit and S. R. Hartmann, Phys. Rev. 141, 391 (1966); C. K. N. Patel and R. E. Slusher, Phys. Rev. Letters 20, 1087 (1968).
10. S. L. McCall and E. L. Hahn, Phys. Rev. 183, 457 (1969).
11. R. E. Slusher and H. M. Gibbs, Phys. Rev. A5, 1634 (1972).
12. G. B. Hocker and C. L. Tang, Phys. Rev. 184, 356 (1969).
13. E. B. Treacy and A. J. DeMaria, Phys. Letters 29A, 369 (1969).
14. H. P. Grieneisen, N. A. Kurnit and A. Szöke, Opt. Commun. 3, 259 (1971).
15. R. G. Brewer and R. L. Shoemaker, Phys. Rev. Letters 27, 631 (1971) and R. L. Shoemaker and R. G. Brewer, Phys. Rev. Letters 28, 1430 (1972).
16. N. Skribanowitz, I. P. Herman, J. C. MacGillivray and M. S. Feld, Phys. Rev. Letters 30, 309 (1973)
17. N. Skribanowitz, I. P. Herman, R. M. Osgood, Jr. , M. S. Feld and A. Javan, Appl. Phys. Letters 20, 428 (1972).
18. N. Skribanowitz, Ph. D. Thesis, M. I. T. , 1973 (unpublished).

19. See, for example, L. W. Casperson and A. Yariv, IEEE J. Quantum Electron. QE-8, 80 (1972), and J. H. Parks, in Fundamental and Applied Laser Physics: Proceedings of the Esfahan Symposium, edited by M. S. Feld, N. A. Kurnit and A. Javan (Wiley, New York, 1973), and references contained therin.

20. D. A. Leonard, Appl. Phys. Letters 7, 4 (1965).

21. R. H. Dicke, in Proceedings of the Third International Conference on Quantum Electronics, Paris 1963, edited by P. Grivet and N. Bloembergen (Columbia Univ. Press, New York, 1964), p. 35.

22. The connection between Dicke's formalism and the semiclassical approach has been discussed by F. T. Arecchi, E. Courtens, R. Gilmore, and H. Thomas [Fundamental and Applied Laser Physics: Proceedings of the Esfahan Symposium, edited by M. S. Feld, N. A. Kurnit and A. Javan (Wiley, New York, 1973)]. The semiclassical approach has also been applied to superradiance by Friedberg and Hartman [Phys. Lett. 38A, 227 (1972)]. A promising analysis whish is fully quantum mechanical has been given by Willis and Picard [Phys. Rev. A, to be published(Sep., 1973)].

23. N. Skribanowitz, I. P. Herman and M. S. Feld, Appl. Phys. Lett. 21, 466 (1972).

24. R. G. Wenzel and G. P. Arnold, IEEE J. Quantum Electron. QE-8, 26 (1972).

25. We are grateful to Bob Wenzel of Los Alamos Scientific Laboratory for providing plans for this laser.

26. J. D. Jackson, Classical Electrodynamics, (John Wiley and Sons, New York, 1962).

27. F. Bloch, Phys. Rev. 70, 460 (1946).

28. A. Icsevgi and W. E. Lamb, Jr., Phys. Rev. 185, 517 (1969).

29. See, for example, Ref. 28; F. A. Hopf and M. D. Scully, Phys. Rev. 179, 399 (1969); and Ref. 10. For a comprehensive list of references, see G. L. Lamb, Jr., Rev. Mod. Phys. 43, 99 (1971).

30. J. C. MacGillivray, I. P. Herman, N. Skribanowitz and M. S. Feld, Phys. Rev. A, to be published.

31. This formula is a consequence of the optical pumping process and is not in any way basic to the superradiant pulse evolution process.

32. R. Friedberg and S. R. Hartmann, Phys. Lett. 37A. 285 (1971).

33. C. K. Rhodes, A. Szöke and A. Javan, Phys. Rev. Letters 21, 1151 (1968).

34. J. A. Fleck, Jr. [Phys. Rev. B1, 84 (1970)] has considered a related problem in which spatial variations are neglected.

35. In a long sample where the transit time is greater than T_R, the sample can break up into independently radiating sections. This effect

is related to the cooperation length described by F. T. Arecchi and E. Courtens [Phys. Rev. A2, 1730 (1970)]. Further discussion of such effects will be given in Ref. 30.

36. R. P. Feynman, F. L. Vernon, Jr., and R. W. Hellwarth, J. Appl. Phys. 28, 49 (1957).

37. D. C. Burnham and R. Y. Chiao, Phys. Rev. 188, 667 (1969); see also G. L. Lamb, Jr., Phys. Letters 29A, 507 (1969), and F. T. Arecchi and E. Courtens (Ref. 35).

38. M. D. Crisp, Phys. Rev. A1, 1604 (1970).

39. Note that this conclusion applies only to an inhomogeneously broadened system, where de-excitation of the levels is unimportant. The population decay associated with homogeneous broadening cannot be counteracted by high gain.

40. N. Tan-no, K. Kan-no, K. Yokoto and H. Inaba, IEEE J. Quantum Electron. QE-9, 423 (1973).

Infrared-Microwave (Radiofrequency)

Two-Photon Spectroscopy

Takeshi Oka

Division of Physics
National Research Council of Canada
Ottawa, Ontario, Canada

This is a conference on laser spectroscopy, but so far I do not feel that we have heard much on systematic spectroscopic works. Being an associate of Dr.Herzberg, I would like to talk about our effort in this direction by using an infrared radiation from a 10 μm CO_2 or N_2O laser and a microwave or a radiofrequency radiation. Neither the infrared nor the rf radiation is directly resonant to molecular transitions but, by using the non-linearlity of molecular transition processes, we can observe two-photon transitions for which the sum or the difference of the frequencies of the two radiations correspond to molecular transitions. The idea is that, although the frequencies of the infrared laser lines are not tunable and they are not in resonance with molecular transitions in general, we can take care of the discrepancy by tuning the frequency of the "added" or "subtracted" rf frequency. An alternative way, of course, is to use some non-linear solid state element for a mixing of the two radiations[1],but since molecules themselves are good non-linear elements we can mix the two radiations right on the molecule.

The two-photon transitions are much weaker than the normal single photon transitions but, because of the high sensitivity of the laser spectroscopy, we can observe them with good signal to noise ratios. I shall discuss two cases; (case I) in which the energy levels involved in the experiment have unique parities and (case II) in which they have double parities. The NH_3

molecule is used as the former example and the CH_3F and
the PH_3 molecules are used as the latter example.

        Figure 1 indicates energy level systems of NH_3
used in the infrared-microwave two photon experiment.
Each rotational level of NH_3 with $K \neq 0$ splits into an
inversion doublet with opposite parity because of a
tunnelling motion in NH_3. The splitting in the ground
vibrational state is about 0.8 cm^{-1} and is in the convenient microwave region whereas that in the upper ν_2
vibrational state is about 35 cm^{-1}. The ν_2 vibrational
band appears in 10 μm region and can be studied by using a CO_2 or a N_2O laser. Fujio Shimizu has done systematic laser Stark spectroscopy using this band[2,3]; his
results were of great help for the present study.

        The two-photon processes are well understood pro-
cesses since the early years of quantum mechanics[4,5,6].
The transition moment M_{if} is expressed as

$$M_{if} = <1|\mu_p E_m|2><2|\mu_v E_1|3>/2h\Delta\nu \qquad (1)$$

where μ_p and μ_v are the permanent dipole moment and the

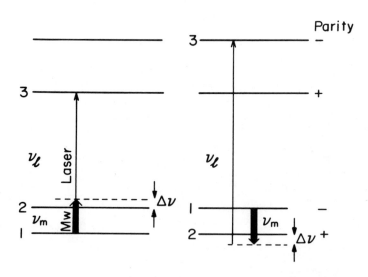

Fig. 1   Energy level scheme for infrared-microwave
         two-photon processes in NH_3

vibrational transition moment, E_m and E_l are the electric field of the microwave and the laser radiations, respectively, and $\Delta\nu$ is the difference between the laser frequency and the frequency of the molecular transition (see Fig. 1). It is seen from Eq.(1) that a two-photon transition is weaker than a normal vibrational transition by a factor of $|<1|\mu_p E_m|2>/2h\Delta\nu|^2$. If a microwave radiation of sufficient power is used, this factor is not very small even for relatively large $\Delta\nu$. Such two photon absorption has already been observed by using two microwave radiations[7,8].

A block diagram of the apparatus is shown in Fig.2. The NH_3 gas, with a pressure of 0.01 to 1 Torr in a K-band waveguide of 1 to 4 meters experiences the infrared and the microwave radiations simultaneously. A two photon transition can be detected by detecting either of the two radiations but obviously detecting higher energy photons is more efficient. This apparatus was the first one constructed[9] and since has been changed considerably. A frequency modulated OKI klystron with a power of about 1 Watt is used instead of the power modulated Elliott 12 TFK2 tube as the microwave source. The microwave field in the waveguide is about 70 V/cm and the value of $<1|\mu_p E_m|2>/2h\Delta\nu$ is typically 30 MHz. The Au-Ge detector in the figure is now replaced by a Pb-Sn-Te detector.

An example of two photon absorption is shown in Fig. 3.

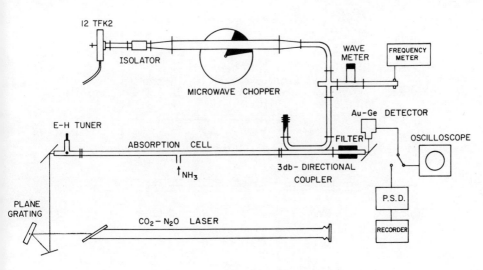

Fig. 2. Block diagram of the apparatus

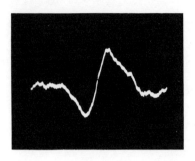

|←——50MHz——→|

Fig.3. An example of a
two-photon signal.
Transition,NH_3 $a^qQ(4,4)$
Laser line, N_2O P(11)
$\Delta\nu$ = 12,303 $\pm$ 5 MHz
Pressure, 2 Torr
Time constant, 30 msec.

In this example the frequencies of the laser line and that of the molecular absorption are separated by about 0.4 cm^{-1}, but the signal can be seen on an oscilloscope by using a relatively short time constant. The sharpness of the signal is not genuine but is caused by the narrow banded microwave circuit, because the intensity of a two photon signal is proportional to the microwave power. The matching of the microwave circuit over wide frequency range is rather difficult and therefore we cannot expect to be able to measure the frequency with an uncertainty less than 5 MH

For a more accurate frequency measurement, we used the Lamb dip technique by saturating two-photon transitions. A block diagram of the apparatus is shown in Fig. 4.

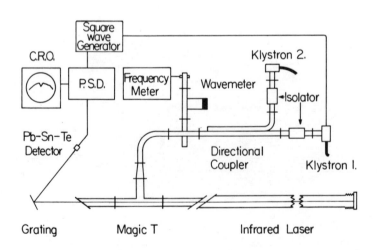

Fig. 4. Two-photon Lamb dip spectrometer

The spectrometer is similar to the normal two-photon
spectrometer except that the cell which is made of a
30 cm long oversized waveguide (RG-910) is now placed
in the cavity of the laser. Microwave power of about
1 Watt generated by an OKI 24V11 klystron (klystron 1
in Fig.4) was introduced into the cell through the E
arm of a magic T. The frequency of the microwave was
modulated and swept by applying a 20 kHz squarewave
and a low frequency sawtooth voltage to the reflector
of the klystron. Klystron 2 in   Fig.4 is another
OKI 24V11 klystron which was used for a double resonance
experiment described later (see Fig.7). The laser line
was stabilized at the maximum of the gain profile.
Using this construction we could observe about 30 two-
photon Lamb dips of NH_3.

An example of the two photon Lamb dip signal is
shown in Fig.5. For this example, the difference between
the laser frequency and the center of the molecular absorption is 1229.1 MHz. The line width of the two-photon
Lamb dip is about 0.8 MHz under the condition described
in the caption of   Fig.5, and therefore we can measure the infrared transition frequency with an accuracy
of microwave spectroscopy. An interesting result for
the $s^qR(5,K)$ transitions is shown in Fig.6. The centrifugal distortion and the J,K dependence of the inversion
splittings are such that these transitions with different

Mw-Ir Two Photon Lamb Dip
Spectrum of the sQ(3,3) Line

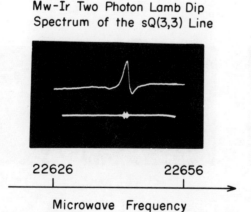

22626   22656

Microwave Frequency

Fig.5.   An example of microwave-infrared two-
photon Lamb dip signal.   Infrared transition; $\nu_2[s^qQ(3,$
$3)]$.   Laser line; N_2O R(36). $\Delta\nu$=1229.1 MHz. P=10mTorr.

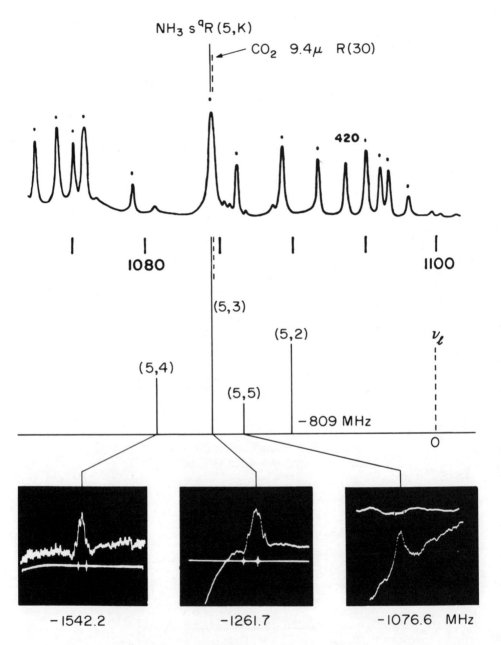

Fig.6. Two-photon Lamb dips for the NH_3 $s^qR(5,K)$ lines with K=5,4,and 3. These three lines appear within 500 MHz and are not resolved by a conventional high resolution spectrometer(top figure).

K values appear rather close and are not resolved by a
conventional high resolution spectrometer. By using
the two-photon Lamb dip technique, we can easily separate
them and measure the frequencies with   high accuracy.
The frequency scale for the spectral pattern in the
middle of Fig.6 is about 500 times magnified from that
of the top figure which shows the conventional spectrum,
and the frequency scale for the observed dip in the
bottom is magnified by another 100.

It is noticed from Fig.5 and more clearly from
Fig.6 that the detected lineshape by using a small frequency modulation is of the form of second derivative
indicating that the **original** form of the two-photon
Lamb dip signals is of the form of first derivative.
The lineshape was observed to change with the pressure
of the sample; at a lower pressure where the saturation
is more complete, the second derivative form changes
into a first derivative.

In order to assign the K number to the observed
dips, we carried out a double resonance experiment the
result of which is shown in Fig.7. In this experiment,
in addition to the two radiations ν_1 and ν_m for the
two-photon transition, we apply a second microwave
radiation ν_m'. When the frequency of the ν_m' is tuned
to the inversion splitting the Lamb dip splits into a
doublet as shown in Fig.7. By measuring the frequency
and looking up a microwave table, we can assign the line.

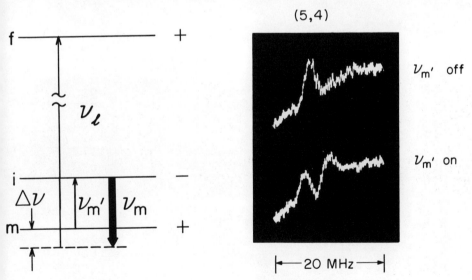

Fig.7. Coherency splitting of the Lamb dip

The two-photon Lamb dips so far observed in NH_3 are summarized in the table below. **Practically all near** coincidences to within 3.5 GHz have been used. We have not used particularly high microwave power nor a cavity so far; use of these will increase the range of difference easily to 10 GHz. Thus we have acquired a mean to do systematic infrared spectroscopy with a resolution of microwave spectroscopy. The accurate frequency measurement of the Lamb-dip to within 0.1 MHz is easy but the accuracy depends also on how well we can stabilize the laser. So far we have stabilized the laser only at the maximum of the gain profile. This might introduce errors of a few MHz in the measurement.

There are a few cases in the Table for which a check of internal consistency is possible. So far such tests have proved that the error is indeed within

Two-Photon Lamb Dips of $^{14}NH_3$

Laser Line		Ir Transition	Mw Frequency	$\nu_{NH_3} - \nu_{Laser}$
N_2O	P(15)	aQ(9,8)	25574.8	1917.3
N_2O	P(14)	aQ(11,9)	21699.6	628.9
N_2O	P(13)	aQ(8,7)	23239.6	7.4
N_2O	P(10)	aQ(6,5)	21238.0	−1494.5
CO_2	P(34)	aQ(8,6)	24330.8	3611.6
N_2O	P(9)	aQ(5,4)	21337.4	−1315.6
N_2O	P(8)	aQ(2,1)	24699.0	1600.2
"		aQ(3,2)	23173.0	338.9
N_2O	P(7)	aQ(5,3)	22826.6	1541.3
CO_2	P(32)	aQ(5,3)	22245.0	959.7
N_2O	P(6)	aQ(8,5)	19719.2	910.5
N_2O	R(27)	sQ(10,10)	27809.1	795.6
N_2O	R(30)	sQ(9,6)	19819.6	−1320.1
CO_2	R(2)	sQ(8,8)	23536.6	2982.3
N_2O	R(32)	sQ(7,7)	26827.8	−1112.7
CO_2	R(4)	sQ(7,5)	20172.4	632.4
CO_2	R(6)	sQ(5,5)	27511.5	−2978.6
"		sQ(5,4)	22085.0	568.0
N_2O	R(36)	sQ(3,3)	22641.0	1229.1
CO_2	R(8)	sQ(2,2)	22782.4	940.2
N_2O	R(37)	sQ(1,1)	23983.0	−288.5
CO_2	R(14)	aR(1,1)	22249.0	−1445.5
CO_2	R(30)	sR(5,4)	24195.2	−1542.2
		sR(5,3)	22547.0	−1261.7
		sR(5,5)	25609.5	−1076.6

the error of the laser frequency measurements for the
CO_2 laser[11] and the N_2O laser[12]. An interesting four
level system depicted in Fig. 8 could be used to determine the frequency   of a N_2O laser line.
The difference between the sQ(1,1) line and the N_2O
R(37) line and that between the aR(1,1) line and the
CO_2 R(14) line were measured accurately by using the
two-photon technique. We also measured the millimeter
wave transition in the excited state (circled in the[13]
figure) by a conventional millimeter spectrometer.
Combining these results with the ground state inversion
splitting which is well known, we determined the frequency of the N_2O R(37) line to be 29020121 ± 3 MHz.
This is 62 MHz off from the estimated frequency calculated by extrapolating Sokoloff and Javan's measurements.
This indicates that a straightforward extrapolation of
these measurements to high J lines does not give very
precise frequencies.

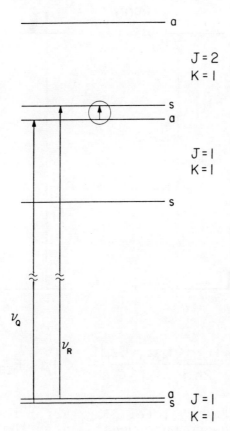

Fig. 8. A four level system in NH_3.

Now let us consider the case II in which the molecular levels involved in the two-photon processes are of double parity (see Fig.9). In this case a third level is not necessary and we can "add" or "subtract" the radiofrequency directly. Whether the radiofrequency is added (left in Fig.9) or subtracted (right in Fig.9) can be easily checked by manually increasing the frequency of the laser and seeing whether the radiofrequency satisfying the two photon condition decreases or increases. For each case there are two processes shown in Fig.9, **corresponding to whether** the radiofrequency photon is absorbed first or the infrared photon is absorbed first. Since these two processes are not distinguishable by the experiment, we have **an** interference of the two pro-

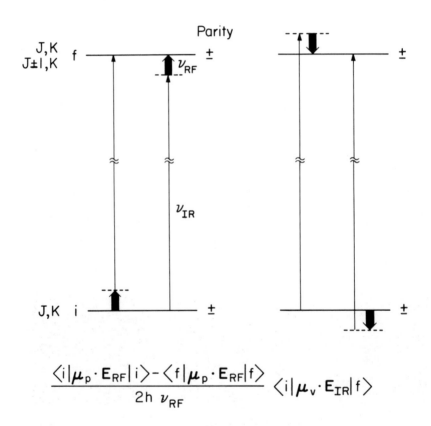

$$\frac{\langle i|\mu_p \cdot E_{RF}|i\rangle - \langle f|\mu_p \cdot E_{RF}|f\rangle}{2h\,\nu_{RF}}\,\langle i|\mu_v \cdot E_{IR}|f\rangle$$

Fig. 9   Energy level diagram for the infrared-radiofrequecy two-photon processes when the levels involved are of double parity.

cesses.   Unfortunately this is a destructive interference because the energies of the virtual states of the mole- cule-radiation system for the two cases have opposite signs.   The radiofrequency matrix elements appear in the transition moment (see Fig. 9) as a difference and almost cancel each other.   Still we could observe quite a few two-photon Lamb dips in CH_3F.   The apparatus used is similar to that for the infrared-microwave two-photon experiment (Fig. 4) except that the waveguide cell is now replaced by a coaxial radiofrequency cell.

An example of the observed results is shown in Fig. 10.   The $J=5 \leftarrow 4$ transitions of the ν_3 band (C-F stretching vibration) of $^{13}CH_3F$ appear very close to the P(32) line of the 9.4 μm band CO_2 laser[14,15].   These transitions are composed of 5 components corresponding to $K=4,3,2,$ 1, and 0 all within 500 MHz.   By using the two-photon

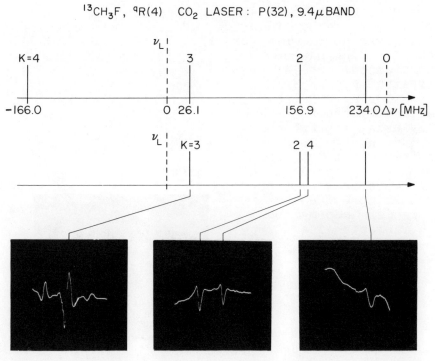

Fig.10   Infrared-radiofrequency two-photon Lamb-dips for the $\nu_3[qR(4)]$ transitions of $^{13}CH_3F$. $\Delta\nu = \nu_{laser} - \nu_{mol}$. The K=0 component is not observable because the levels have single parity.

technique, we could observe all the components except for the K=0 component for which the levels have a single parity. As seen from Fig. 10, the two photon Lamb dips are quite strong and are observable on the oscilloscope with a time constant of 10 msec or so. When the frequency of the laser was varied, the line corresponding to K=4 moved in anopposite way from the other lines indicating that the K=4 line is on the other side of the laser line. When the two-photon Lamb dip is within the Doppler profile of the absorption line (such as the K=3 line in Fig. 10), a multiplet structure is observed. This was not understood at the time of the Conference but was explained by F. Shimizu after the Conference[16]. The line width is less than 1 MHz and there is no problem in measuring the position of the two-photon Lamb dips to within 0.1 MHz. However because of aninstability of the locking of the laser used in the experiment, the frequency shown in Fig. 10 may be in error of 1 MHz. This instability of locking is partly due to the arrangement to have the rf cell inside the laser cavity and will be removed in the near future.

We have observed two-photon Lamb dips for other transitions of methyl fluoride and for other molecules. Fig. 11 gives an example for PH_3. It has a doublet structure, which has not been explained. The inversion splitting of PH_3 is much smaller even in the vibrationally excited state.

By using the two-photon technique, we can measure the infrared spectrum with the resolution and accuracy of microwave spectroscopy. The results provide us with precise information on the vibrational and rotational properties of molecules.

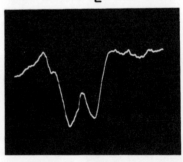

PH_3 $^qP(3,2)$

R(33) N_2O Laser

|← |←0.89 MHz

82.828 MHz

Fig. 11   Two-photon
Lamb dips in PH_3.
Pressure, 50 mTorr.
Time constant 10 msec.

The combination of infrared laser radiation and rf radiation can be used not only for increasing the resolution and the accuracy of spectroscopy but also the sensitivity. A remarkable example has been given recently in the observation of pure rotational transitions of CH_4[17,18]. CH_4 is a non-polar molecule and it is normally understood that a pure rotational transition is strictly forbidden. However when the molecule rotates, a small dipole moment appears as a result of the centrifugal distortion and the rotational transition is weakly allowed[19,20,21]. We have observed this weak transition by using the double resonance technique in the three level system shown in Fig. 12. A rf cavity containing about 20 mTorr of CH_4 is placed in the cavity of a He-Ne laser. The output power of the He Ne laser was observed while the rf frequency was varied. Fig. 13 shows an observed signal. By using the double resonance technique we are gaining sensitivity of at least a factor of 10^7.

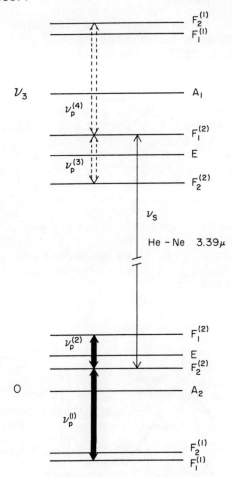

Fig.12.   The energy level scheme for the infrared-rf double resonance in CH_4

Fig. 13   The infrared-radiofrequency double resonance signal in CH_4. The cavity was tuned to 423.02 MHz. The sweep covers about 3 MHz.

So far I have discussed applications of the two-photon technique for high resolution spectroscopy. Now I would like to describe an application of the method for the study of molecular collisions. For the last several years many experiments of four-level microwave double resonance have been conducted and produced useful information on the selection rules governing collision-induced transitions and their relative transition probabilities (See Ref.22 for a summary). More recently, several experiments have been reported using the infrared-microwave double resonance method.[23] A natural extension of this method is to the infrared-infrared double resonance. Such experiments are almost impossible if we have to rely on rare coincidences between laser frequencies and molecular absorptions. However the two-photon technique described earlier makes this type of experiment possible. The observation of the two-photon Lamb dips indicates that we can pump a transition by the two radiations.

Fig. 14 is a block diagram of the apparatus. Infra-

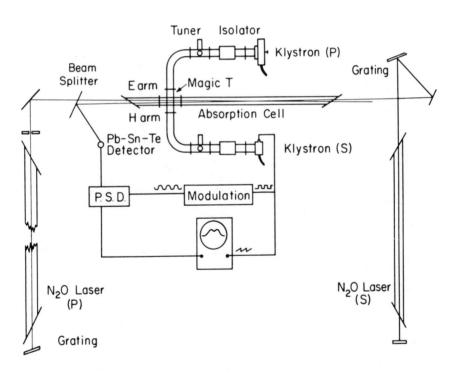

Fig. 14. Block diagram of the infrared double resonance apparatus. The N2O laser(P) and the klystron(P) are used for two-photon pumping and the laser(S) and the klystron (S) are used for monitoring the two-photon absorption.

red radiation of about 5 Watts generated by an N_2O laser (P) and microwave radiation of about 10 Watts generated by an Elliott 12 TKF2 klystron (P) were used for two-photon pumping, and infrared radiation of about 0.5 Watts with MW from an OKI 24V11 klystron(S) were used for the two-photon signal. All four radiations were linearly polarized in the same direction. More detail of the apparatus and the experimental procedure will be found in the original paper[24].

Fig.15 shows a result of an experiment in which both the two-photon pumping and the two-photon signal are tuned to the sQ(4,4) transition of $^{15}NH_3$ which is known to have a near coincidence (312 MHz) with the P(15) N_2O laser line. The frequency of the pumping microwave radiation ν_{PM} is fixed while that of the signal microwave radiation is swept for the oscilloscope display. By using this technique, we can monitor the "hole" burnt in the velocity profile of molecules in the lowest level in Fig.15a. The useful characteristics of infrared-infrared double resonence which does not exist in microwave double resonance is that we can also monitor the velocity of molecules.

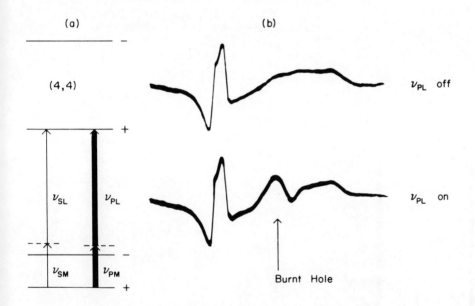

Fig. 15. Burning a "hole" in the velocity profile of molecules with two-photon pumping and monitoring the hole with two-photon absorption. The burnt hole (shown with an arrow) can be tuned both by tuning the microwave radiation and the laser radiation.

After monitoring the hole burnt by the two radiations for pumping, we shift the monitoring two photons
from the sQ(4,4) transition to the aQ(4,4) transition as
shown in Fig.16a. For this purpose, the signal N$_2$O laser is tuned to the P(28) N$_2$O line and the signal microwave is swept about 1250 MHz below the (4,4) inversion
frequency. The four-photon scheme in Fig.16a is rather
complicated but is essentially the same as the four-level double resonance scheme indicated in Fig.16b. The
two photon process pumps molecules from the parity +
level to + level thus introducing non-Boltzmann distributions to these two levels; these distributions are
then transferred to other levels by collisions as indicated with wavy arrows in Fig.16b. It is known from the
microwave experiments that when NH$_3$ molecules collide
with NH3 molecules, the collision-induced transitions
occur most frequently between inversion doublet levels.
    Fig.16c indicates a result of the experiment. The
lower recorder trace is of the two-photon signal without two-photon pumping. This profile represents a half
of the derivative shape of the two photon signal. When
the two-photon pumping is applied a sharp feature appeared on the profile. This indicates that the hole burnt
by the two-photon pumping in the Doppler profile of the
+ ↔ + transition is transferred to the - ↔ - transition

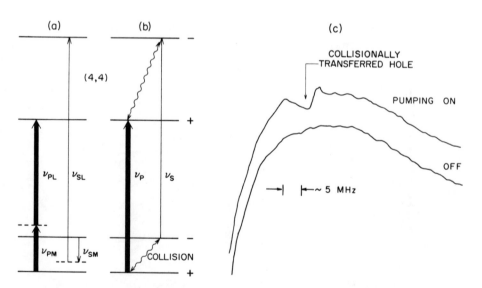

Fig.16. Transfer of the burnt hole to other levels due
to weak collisions. The hole burnt by the two-photon
pumping in the Doppler profile of the + ↔ + transition
is transferred to the - ↔ - transition by collision-
induced transitions.

by collision induced transitions.

Fig.17 indicates qualitatively the velocity profile of molecules in each level under conditions of
steady state pumping. The two photon pumping introduces
a hole and a spike in the velocity profiles of the level 1 and the level 3, respectively. The hole and spike
are then transferred to levels 2 and 4 by collisions.
The observation of collision-transferred hole in Fig.16
indicates that these transfers occur without much affecting the molecular velocity.
As seen in the result of this experiment the

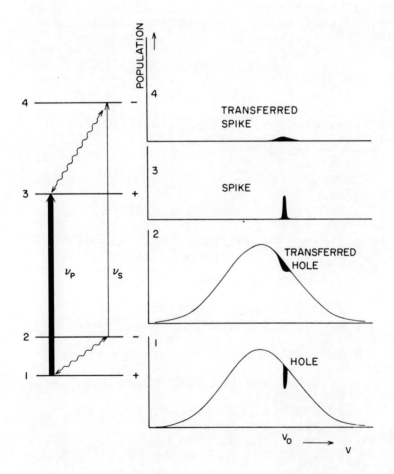

Fig. 17. Qualitative explanation of the collisional
transfer of the hole. The curves on the right indicate velocity profiles of molecules in individual
levels.

infrared-infrared double resonance is capable of giving
us information of velocity change simultaneously with
that of state change. This kind of information will be
very useful for the study of relatively strong collisions in which rotational energy is carried away as
translational energy.

Acknowledgement

The experiments described in this paper have been
conducted in collaboration with F.Y. Chu, R.F. Curl,
S.M. Freund, J.W.C. Johns, A.R.W. McKellar, and M.
Römheld. I am particularly indebted to S.M. Freund
whose ingenuity and perseverance in constructing ap-
paratuses and performing the experiment made these
works possible.

References

1. V.J. Corcoran, R.E. Cupp, J.J. Gallagher and W.T.
      Smith Appl. Phys. Lett. 16, 316 (1970)
2. F. Shimizu, J. Chem. Phys. 52, 3572 (1970)
3. F. Shimizu, J. Chem. Phys. 53, 1149 (1970)
4. H.A. Kramers and W. Heisenberg, Zs.f.Physik 31,
      681 (1925)
5. P.A.M. Dirac, Proc. Roy. Soc.(London) A114, 143,
      710 (1927)
6. M. Göppert-Mayer, Am. Physik 9, 273 (1931)
7. T. Oka, Can. J. Phys. 47, 2343 (1969)
8. T. Oka and T. Shimizu, Phys. Rev. A2, 587 (1970)
9. T. Oka and T. Shimizu, Appl. Phys. Lett., 19,
      88 (1971)
10. S.M. Freund and T. Oka, Appl. Phys. Lett. 21,
      60 (1972)
11. T.Y. Chang, Opt. Commun. 2, 77 (1970)
12. D.R Sokoloff and A. Javan, J. Chem. Phys. 56,
      4028 (1972)
13. F.Y. Chu and S.M. Freund, J. Mol. Spectrosc.,
      in press
14. T.Y. Chang, C.M Wang and P.K. Cheo, Appl. Phys.
      Lett. 15, 157 (1969)
15. R.G. Brewer and R.L. Schoemaker, Phys. Rev. Lett.
      27, 631 (1971)
16. F. Shimizu, to be published
17. R.F. Curl, T. Oka and D.S Smith, J Mol. Spectrosc.
      46, 518 (1973)
18. R.F. Curl, J. Mol. Spectrosc., to be published
19. J.K.G. Watson, J. Mol. Spectrosc. 4, 536 (1971)
20. K. Fox, Phys. Rev. Lett. 27, 233 (1971)

21. A. Rosenberg, I. Ozier and A.K. Kudian, J. Chem.
        Phys. 57, 568 (1972)
22. T. Oka, Advance in Atomic and Molecular Physics,
        Vol. IX, Academic Press, Oct. 1973
23. T. Shimizu and T. Oka, Phys. Rev. A2, 1177 (1970)
        J. Chem. Phys. 53, 2536 (1970)
        L. Frenkel, M. Marantz and T. Sullivan, Phys.
            Rev. A3, 1640 (1971)
        M. Takami and K. Shimoda, Jap. J. Appl. Phys.
            10, 658 (1971)
        J. Lemaire, J. Houriez, F. Herlemont and
            J. Thibault, Chem. Phys. Lett. 19, 373 (1973)
        W.A. Kreiner, M. Römheld and H.D. Rudolph,
            Z. Naturforsch. 28 (1973)
24. S.M. Freund, J.W.C. Johns, A.R.W. McKellar and
        T. Oka, J. Chem. Phys. 59, in press

OBSERVATION OF LAMB-DIP AND TRANSIENT NUTATION

AT MICROWAVE FREQUENCY

Tadao Shimizu

Department of Physics, University of Tokyo, Tokyo 113

JAPAN

1. INTRODUCTION

Several nonlinear effects at microwave frequency have been observed in interaction of molecules with a coherent radiation field. From the klystron, a microwave oscillator, a coherent radiation is available, which is as good as the laser radiation in spectral purity, photon flux density, directivity and so on. In the microwave frequency region may be found the counterparts of the various nonlinear phenomena which have been observed in the infrared and visible wavelength regions by laser spectroscopic methods. Those effects might be very usefully employed for an investigation of the fast process of rotational relaxation in molecules. In the first part of the present article an observation of saturation dip in the microwave absorption line is described. In the second part a transient change in the intensity of molecular absorption line caused by a fast switching of microwave field, will be discussed.

2. SATURATION-DIP IN MICROWAVE ABSORPTION LINE[1]

For an observation of saturation dip in the spectral line, the widths due to any other effects than the Doppler effect should be much smaller than the Doppler width, which is very small at microwave frequency and is of an order of several tens kilohertz. A large absorption cell containing very low pressure gas is used in the experiment, so that the pressure broadening, the wall collision broadening, and the uncertainty width due to finite interaction time of molecule with the microwave beam, may be permissible.

Since the saturation dip signal in the microwave transition
was first observed by Costain,[2] it turned out that the method was
promissing for a high resolution spectroscopy and for an achievement
of the frequency standard in microwave and millimeter wave regions.
A detailed description of the experimental setup was given in the
reference.[2]   In the present experiment, however, several parts in
the experimental system are improved.

A block diagram of experimental setup is shown in Fig. 1.
The absorption cell is composed of two sections of cylindrical
stainless steel tank, each of which is 1 m in diameter and 2 m in
length.   The tank is evacuated by a 4" oil diffusion pump and a
sorption pump.   The pressure of absorbing gas is measured by a diaphragm pressure gauge to the accuracy of 1×10^{-6} Torr.   Microwave power is fed into the tank by a horn on the side wall and is
collimated by an off-axis paraboloid antenna, 50 cm in diameter,
placed at an end of the tank.   An effective diameter of the collimated microwave radiation is 45 cm.   In the other end is placed
a microwave reflector with very low reflectivity (about 10%).
The reflector is made from lossy styrofoam containing carbon dust
and is covered by plexiglass.   The strong travelling microwave may
cause saturation in the transition of the molecule with a proper
velocity and the weak reflected wave probes the saturated absorption.   The sensitivity and the signal to noise ratio of the detecting system have been much improved by using the weak probing

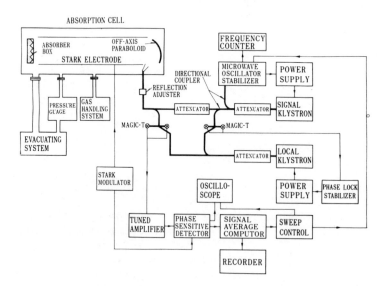

Fig. 1   Schematic diagram of the experimental system for observation
of the microwave saturation dip.

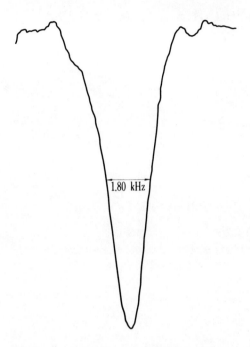

Fig. 2   A typical trace of the dip signal in the J = 3 ← 2 transition of OCS molecule.   Pressure: 0.028 mTorr, microwave power density: 7.5 nW/cm^2, time constant of the detecting system: 100ms,
scanning time: 2.5 sec, and integration: 8 scans.

radiation and a heterodyne detection.   A low frequency square wave
Stark modulation is employed to observe an undistorted line shape
on the recording.   An entire experimental system is vibration isolated.

The frequency of signal klystron is stabilized by a harmonics
of 15 MHz crystal oscillator and is scanned being phase-locked over
several tens kilohertz across the absorption line center.   The long
term stability is better than 1 × 10^{-8}.   The frequency of local
klystron is set 30 MHz apart from the signal frequency.   The saturation signal is amplified at 30 MHz, lock-in detected, and accu-
mulated over several tens scans by a mini-computor.

In Fig. 2 a typical trace of the saturation dip in the J = 3
← 2 transition of OCS molecule is shown.   The pressure and microwave power dependences of the dip width are carefully investigated
in the ranges of 1 to 30 × 10^{-5} Torr and of 30 to 200 µW.   As shown

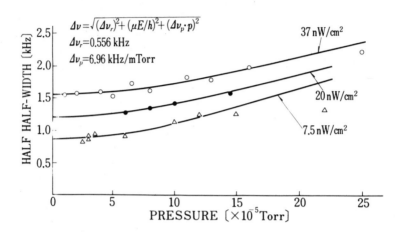

Fig. 3   Pressure and microwave power dependences of the saturation dip width in the J = 3 ← 2 transition of OCS.

in Fig. 3 the result will be well understood from an equation of

$$\Delta\nu = \sqrt{(\Delta\nu_r)^2 + (\bar{\mu}E/h)^2 + (\Delta\nu_p \cdot p)^2} \tag{1}$$

where $\Delta\nu_r$ is the residual width, $\bar{\mu}E/h$ the saturation broadening, and $\Delta\nu_p \cdot p$ the pressure broadening.   The residual width arises from

    i)   the spectral purity of microwave radiation : 300 Hz,
   ii)   the Stark modulation broadening : 200 Hz,
and
  iii)   the uncertainty width due to the finite interaction time
        : 300 Hz.
The pressure broadening parameter is determined to be 70 Hz/10^{-5} Torr in the very low pressure range, which agrees well with the value at relatively high pressure, 6 kHz/mTorr.[11]   The microwave power density in the cell estimated from the saturation broadening is of an order of 10 nW/cm^2 and is consistent with the measurement. The observed dip width (half width at half maximum) is as narrow as 0.78 kHz and is very close to the limitting value of the present experimental apparatus.

3. TRANSIENT NUTATION EFFECT IN MICROWAVE TRANSITIONS[3]

Recently Brewer and Shoemaker observed an optical nutation signal in absorption of the CO_2 laser radiation by CH_3F and NH_2D molecules.[4] In this case the absorption frequency of the molecules with the proper velocity was brought into the resonance with the laser frequency by applying pulsed Stark field. Steinfeld and his collaborators observed the transient nutation signal in a microwave transition of NH_3, which was subjected to resonant excitation of the infrared transition by a Q-switched N_2O laser.[5] Several time-resolved experiments in microwave spectroscopy have carried out with respect to studies of relaxation.[6~9] Since in microwave region a wide variety of molecular transitions appear and a tunable radiation source is available, a systematic study of the relaxation mechanism is possible. Another interest in the study of transient effects is to compare the results with the static measurements of the linewidths of microwave transitions.

In this paper we report the transient nutation effect in the $J = K$, $J = K + 1$, and $J = K + 2$ inversion transitions of NH_3 molecule observed by switching the microwave frequency out of the resonance. The decay time constant of the transient signal is analyzed and compared with the results of linewidth measurement.

A block diagram of the experimental setup is shown in Fig. 4.

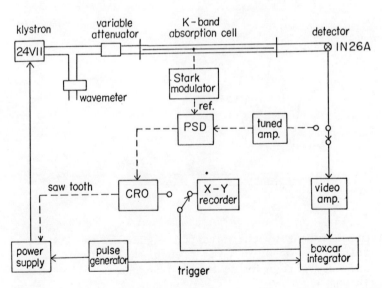

Fig. 4 Block diagram of the experimental system for observation of the microwave nutation.

The system is assembled on the basis of a conventional Stark modulation microwave spectrometer.   The microwave frequency is switched
by applying square wave pulse voltage on the repellor electrode of
a klystron.   The repetitive signal is integrated by the boxcar integrator, which output is displayed on a recorder chart.

The microwave field strength, E, in the absorption cell was
estimated to be 0.1 V/cm, so that the saturation parameter $\bar{\mu}E/h$ was
of an order of 50 kHz for most transitions investigated, where $\bar{\mu}$ was
the dipole matrix element of the transition.   The pulse voltage
applied on the repellor electrode was about 2 V, which corresponded
to detuning of the microwave frequency by about 1 MHz.   The repetition rate of the voltage pulse was 50 kHz.   The gate time of the
integrator was set at 50 ns.   The time scale on the recording was
calibrated by a standard oscillator, the frequency of which was
measured by a frequency counter to a accuracy of 10^{-8}.

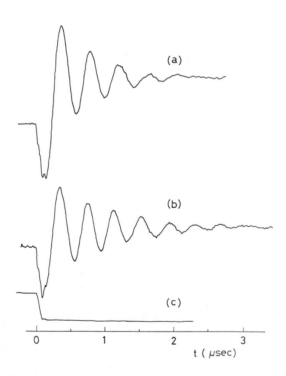

Fig. 5   Typical traces of the nutation signal in the J = 2, K = 1
inversion transition of NH_3.   (a) pressure: 23 mTorr, (b) pressure:
13 mTorr, and (c) voltage pulse applied on the repellor electrode
of the klystron.

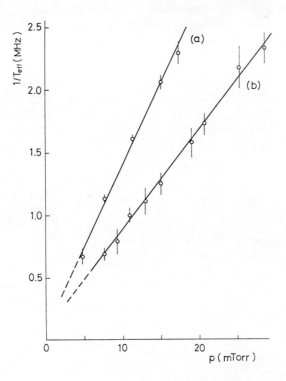

Fig. 6   The inverse of decay time constant as a function of pressure.   (a) J = 4, K = 3 transition, and (b) J = 2, K = 1 transition.

NH$_3$ gas with 99.5 % purity was used.   The pressure in the absorption cell was measured by a Pirani gauge.   Although the gauge
was carefully calibrated by a mercury manometer and an α-tron gauge,
the error of several percents in absolute pressure measurement is
still probable.

In Fig. 5 is shown the nutation signal in the J = 2, K = 1 inversion transition as a typical example.   The traces are signals
averaged over 10^5 scans.   Since $\bar{\mu}E/h$ is much smaller than the off
set frequency in the present experimental condition, the nutation
frequency is primarily determined by $|\omega_0 - \omega_m|$, where ω_0 is the cen-
ter frequency of the absorption line and ω_m is the microwave frequency.   An envelope of the damping oscillation is found to be well
fitted by a single exponential function.   The inverse of decay time
constant is shown in Fig. 6 as a function of pressure of absorbing
molecule.

A time dependent behavior of the two level molecule which is subjected to a coherent perturbation of

$$H^{(c)}(t) = - \mu E \cos \omega_m t \qquad (2)$$

and a random perturbation $H^{(r)}(t)$ can be described by

$$\frac{dr_1}{dt} = - \omega_0 r_2 - \Gamma r_1$$

$$\frac{dr_2}{dt} = \omega_0 r_1 - \left(- \frac{2\bar{\mu}}{\hbar} E \right) r_3 - \Gamma r_2 \qquad (3)$$

$$\frac{dr_3}{dt} = - \frac{2\bar{\mu}}{\hbar} E r_2 - 2\Gamma^n (r_3 - r_3^0) \ ,$$

where the pseudo vector $\underline{r}$ is defined by density matrix elements as

$$r_1 = \rho_{12} + \rho_{21}$$

$$r_2 = i(\rho_{21} - \rho_{12}) \qquad (4)$$

$$r_3 = \rho_{22} - \rho_{11} \qquad .$$

The relaxation is caused by the random perturbation and the parameters Γ'_s in Eq. (3) are explicitly expressed by

$$\Gamma = \Gamma^n + \Gamma^a$$

$$\Gamma^n = \frac{1}{2\hbar^2} \int_{-\infty}^{\infty} \left[< H_{12}^{(r)}(t) \ H_{21}^{(r)} \ (t - \tau) > e^{-i\omega_0 \tau} \right.$$

$$\left. + < H_{21}^{(r)}(t) \ H_{12}^{(r)}(t - \tau) > e^{i\omega_0 \tau} \right] d\tau \qquad (6)$$

$$\Gamma^a = \frac{1}{\hbar^2} \int_0^{\infty} < \left(H_{11}^{(r)}(t) - H_{22}^{(r)}(t) \right) \left(H_{11}^{(r)}(t - \tau) - H_{22}^{(r)}(t - \tau) \right) > d\tau$$

$$(7)$$

Γ^n may represent the decay constant of the molecular state for non-adiabatic relaxation, which causes the transition between two levels, while Γ^a the adiabatic relaxation constant due to dephasing collisions.   A transverse and a longitudinal relaxation time constants, which are conventionally used, are expressed by

$$\frac{1}{T_2} = \Gamma^n + \Gamma^a \tag{8}$$

and

$$\frac{1}{T_1} = 2\Gamma^n \ . \tag{9}$$

When the microwave field is not so large that

$$\frac{\bar{\mu}E}{\hbar} T_1 \ll 1 \ ,$$

$$\frac{\bar{\mu}E}{\hbar} T_2 \ll 1 \ ,$$

and

$$|\omega_0 - \omega_m| \gg \frac{\bar{\mu}E}{\hbar} \ ,$$

a solution of Eq. (3) can be simply expressed by

$$r_2 = r_2^{\ 0} \exp \left[-t/T_{eff}\right] \cos(\Omega t + \phi_0) \tag{10}$$

where

$$\Omega = [(\omega_0 - \omega_m)^2 + \left(\frac{\bar{\mu}E}{\hbar}\right)^2]^{1/2} \simeq |\omega_0 - \omega_m|$$

and

$$\frac{1}{T_{eff}} \simeq \frac{1}{T_2} + \frac{1}{2} \left(\frac{1}{T_1} - \frac{1}{T_2}\right) \frac{(\bar{\mu}E/\hbar)^2}{(\omega_0 - \omega_m)^2} \ . \tag{11}$$

A precise study of microwave power dependence of the decay time constant of the damped nutation signal may give the values of T_2 and T_1, independently.   Eq. (11) is compared with a expression of the linewidth of

Table I. Rotational quantum number dependences of the inverse of
decay time constant and the pressure broadening parameter.

(J.K)	$(2\pi T_{eff} \cdot p)^{-1}$ [MHz/Torr] (present observation)	$\Delta\nu_p$ [MHz/Torr] (reference 10)
(2.1)	10.5	
(3.1)	10.1	12.8
(3.2)	16.6	17.6
(3.3)	23.0	24.0
(4.3)	17.2	22.1
(4.4)	21.8	24.4
(5.4)	17.6	
(5.5)	21.8	24.6
(6.5)	18.2	22.0
(6.6)	21.8	24.0
(7.6)	19.8	
(7.7)	21.6	24.6
(8.7)	18.9	21.9
(10.9)	19.2	21.2

$$2\pi\Delta\nu = \left[\left(\frac{1}{T_2}\right)^2 + \left(\frac{\mu E}{h}\right)^2 \left(\frac{T_1}{T_2}\right)^2 \right]^{1/2} . \tag{12}$$

As shown in Eqs. (11) and (12), the observations of the relaxation parameters in the time and frequency domains may give essentially the same constant. The values of $1/2\pi T_{eff} \cdot p$ are determined for 14 inversion transitions of NH_3 by analyzing pressure dependence of the decay time constants. In Table I the results are compared with the previous measurement of pressure broadening parameter. The dependence of the parameter $1/T_{eff} \cdot p$ on the rotational quantum number J and K is found to be in reasonable agreement with that of the previously measured pressure broadening parameter, ν_p.[10] The absolute values of $1/T_{eff} \cdot p$, however, systematically smaller than ν_p being slightly over the experimental error. The pressure broadening parameters of several transions were remeasured by using the same absorption cell and the pressure measurement system employed in the observation of the transient phenomena. The measured pressure broadening parameters of 23 MHz/Torr for the J = 3, K = 3 transition and of 16 MHz/Torr for the J = 3, K = 2 transition closely agree with the corresponding values of $1/2\pi T_{eff} \cdot p$ in the table. The experiment has confirmed that the observation of transient nutation gives the same results with those of linewidth measurement with respect to the relaxation of molecular state.

ACKNOWLEDGEMENTS

The author is grateful to Dr. C. C. Costain (National Research Council of Canada) and Dr. T. Amano (University of Tokyo) for their collaborations and significant contributions to the saturation dip experiment and the microwave nutation experiment, respectively.

REFERENCES

1) T. Shimizu and C. C. Costain, "Observation of Saturation Dip in the Microwave Spectral Line", to be submitted.

2) C. C. Costain, Can. J. Phys. 47, 2431 (1969).

3) T. Amano and T. Shimizu, "Observation of Transient Nutation Effect in Microwave Transitions of Ammonia Molecules", to be published in J. Phys. Soc. Japan.

4) R. G. Brewer and R. L. Shoemaker, Phys. Rev. Letters 27, 631 (1971), Phys. Rev. A6 2001 (1972).

5) J. M. Levy, J. H. -S. Wang, S. G. Kukolich and J. I. Steinfeld, Phys. Rev. Letters 29, 395 (1972).

6) R. H. Dicke and R. H. Romer, Rev. Sci. Instr. 26, 915 (1955).

7) R. M. Hill, D. E. Kaplan, G. F. Herrmann and S. K. Ichiki, Phys. Rev. Letters 18, 105 (1967).

8) D. C. Lainé, D. Kakati, G. S. Uppal, G. D. S. Smart, and W. S. Bardo, Phys. Letters 29A, 376 (1969).

9) R. H. Schwendeman and H. M. Pickett, J. Chem. Phys. 57, 3511 (1972).

10) R. L. Legan, J. A. Roberts, E. A. Rinehart, and C. C. Lin, J. Chem. Phys. 43, 4337 (1965).

11) A. Battaglia, M. Cattani, and O. Tarrini, Nuovo Cimento 61B, 193 (1969).

QUANTUM BEAT SPECTROSCOPY USING TUNABLE PULSED DYE LASERS[*]

Serge Haroche[†] and Jeffrey A. Paisner

Department of Physics, Stanford University

Stanford, California 94305

ABSTRACT

The recently developed broadly tunable N_2 - laser pumped pulsed dye lasers are an ideal source of light excitation for Quantum Beat experiments. Using one of these lasers, we have observed optically induced hyperfine Quantum Beats in the $7^2P_{3/2}$ state of Cs . This experiment shows that the detection of dye laser induced Quantum Beats may be developed as a practical tool in high resolution spectroscopy without Doppler effect. Some speculations regarding future applications of the technique are considered.

Quantum beat modulations in spontaneous emission have been detected in various kinds of experiments over the past ten years.[1,2,3] These modulations are observed following a short pulse excitation whose spectral bandwidth Δ is large enough to prepare the atoms in a coherent superposition of excited states having different energies. By spontaneous emission the atoms subsequently return to the ground state (or to another level of lower energy) and the probability amplitudes of photon emission from the different excited states do interfere with each other in the fluorescence

---

[*] Work supported by the National Science Foundation under Grant GP-28415.

[†] Permanent address: Laboratoire de Spectroscopie Hertzienne de l'Ecole Normale Superieure, Paris, France.

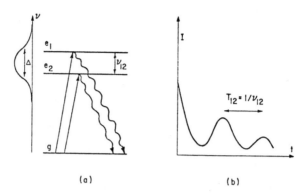

(a)   (b)

Fig. 1.   Quantum Beats in the case of a two-level excited state
system.   (a) Diagram of the relevant energy levels: the pulse
excitation, whose spectral bandwidth Δ is larger than the energy
splitting ν_{12} between the excited state e_1 and e_2 , prepares the
atoms in a coherent superposition of these states (full line arrows).
By spontaneous emission, the atoms then return to the ground state
g (wavy lines).   The amplitude corresponding to the two possible
scattering channels interfere with each other.   (b) Plot of the
intensity of polarized fluorescence I as a function of time: as a
result of the interference process, a beat at frequency ν_{12} appears
in the atomic fluorescence detected with a given polarization.

process (see Fig. 1a). As a result of this interference effect, a
beat pattern is superposed to the normal exponential decay curve
of the atomic fluorescence (see Fig. 1b). The beats, whose phase
and amplitude depend on the polarization of the detected light,
occur at the Bohr frequencies corresponding to the energy splittings
between the simultaneously decaying excited states.[4] Consequently,
a Fourier analysis of the frequencies yields, in a very simple way,
the energy structure of the emitting levels. Furthermore, the
modulations observed in a gaseous sample are not affected at all by
the Doppler effect, since the Doppler shifts of the different
transitions cancel each other in the interference process which
produces the beats. The definition of the modulation frequencies
is in consequence limited only by the natural linewidth of the
emitting levels. The Quantum Beat detection method appears thus as
a very interesting technique in high resolution spectroscopy of
gases, its resolution being basically the same as the one of double
resonance, level crossing and saturation spectroscopy. Furthermore,
it is a very simple and elegant method requiring no search or scan
of the line, since the excited atoms spontaneously display, without
being perturbed, their own spectrum.

The only requisite to develop such spectroscopic investigations is to be able to produce a strong resonant pulse whose length τ is shorter than the period of the expected beats. This condition obviously implies (through the relation $\Delta \tau \geqslant 1$) that the bandwidth Δ of the pulse is large enough to cover the whole structure of the energy levels under study. It is however a more restrictive condition, which has absolutely to be fulfilled in order to avoid partial or complete washing out of the beat pattern. In atomic beam experiments[3] this condition is satisfied by passing accelerated ions through a thin carbon foil in which the ions pickup electrons, emerging as excited atoms (beam foil spectroscopy). The excitation thus provided is short enough to prepare quasi-instantaneously the atoms in a coherent superposition of fine or hyperfine structure levels. The beat pattern is then detected by observing the spatial modulation of the fluorescence along the path of the atoms.

This interesting technique has been so far limited to a few atomic species and obviously does not apply to atomic vapors. In this latter case, optical pulses[1] produced by resonant spectral lamps shuttered by a Kerr cell or electron pulses[2] have been used to induce Quantum Beats. However, in all these experiments the intensity of the pulse excitation was very weak, resulting in a very low rate of photon emission. These intensity problems have limited the time resolution of the technique and only rather low frequency beat patterns (a few Mhz) arising from weak field Zeeman structure splittings have been studied in this way. Consequently, these experiments in which very long averaging techniques had to be used to extract the signal from the noise, were generally intended to be more a demonstration of the Quantum Beat phenomenon than an attempt at spectroscopic measurements of energy splittings.

The recent advent of widely tunable N_2 - laser pumped pulsed dye lasers[5] renews the interest of these earlier experiments and opens a new field of possible applications for the Quantum Beat technique in high resolution spectroscopy of gases. These lasers provide, indeed, an ideal source of light excitation for Quantum Beat spectroscopy. They deliver repetitive few nanosecond long pulses with a power in the kilowatt range; furthermore they can be tuned throughout the visible spectrum with a possible extension of their frequency range to the ultraviolet by frequency doubling.[6] It becomes thus possible to excite selectively and efficiently a very big number of energy levels in atoms and molecules and to study in these levels Quantum Beats arising from Zeeman,[7] fine or hyperfine[8] structures at frequencies up to a few hundred megacycles. It is the purpose of this paper to report the first application of this technique to the detection of hyperfine Quantum Beats[8] (observed in atomic Cs vapor) and to discuss briefly an experiment presently under progress at Stanford University, in which we intend to measure by Quantum Beat spectroscopy the still unknown hyperfine structures of high lying p states in alkali atoms.

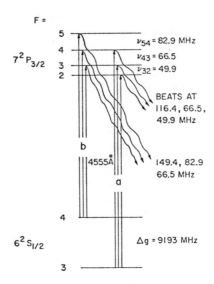

Fig. 2.   Hyperfine structure of the $6^2S_{1/2}$ and $7^2P_{3/2}$ levels
of cesium. Transitions starting respectively from levels F = 3   and
F = 4 in the ground state are labelled by letters (a) and (b). The
sets of Quantum Beat frequencies expected in each case (a) and (b)
are indicated.

We have studied the Quantum Beats emitted by the $7^2P_{3/2}$
level of Cs^{133} . The relevant energy levels for our experiment are
shown in Fig. 2. The laser light excites the $6^2S_{1/2}$ - $7^2P_{3/2}$
transition (second resonance line in the principle series at
4555Å). The ground state $6^2S_{1/2}$ is split in two hyperfine levels
(F = 3 and F = 4) separated by the hyperfine interval Δg = 9.193
Ghz which is resolvable by classical spectroscopic methods. On
the other hand, the excited state $7^2P_{3/2}$ is split in four hyperfine
levels whose separations ν_{54} = 82.9 Mhz , ν_{43} = 66.5 Mhz   and
ν_{32} = 49.9 Mhz are much smaller than the Doppler width (about 750
Mhz) and have been measured using double resonance or level crossing
techniques.[9]

Let us suppose that the duration τ and bandwidth Δ of the
laser pulse obey the following compatible conditions:

$$1/\tau > \nu_{52} = \nu_{54} + \nu_{43} + \nu_{32} \tag{1}$$

$$\Delta < \Delta g \quad . \tag{2}$$

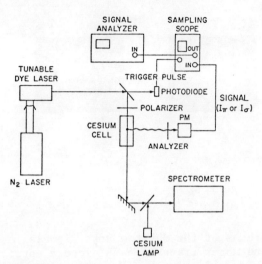

Fig. 3. Experimental setup for Quantum Beat observation and analysis.

Condition (1) implies that the pulses are short enough to induce
Quantum Beats in the $7^2P_{3/2}$ state. Condition (2) shows on the
other hand that the laser light is narrow-band enough to resolve the
two sets of transitions starting respectively from ground states
F = 3 (transitions a) and F = 4 (transitions b). In each case
a or b, the atoms are excited to a superposition of the three
hyperfine states which, according to the selection rule $\Delta F = 0, \pm 1$,
are coupled to the relevant hyperfine ground state. Thus, when the
laser is tuned to transitions a, one must observe beats at frequencies
ν_{43} , ν_{32} and $\nu_{42} = \nu_{43} + \nu_{32}$; on the other hand, when the laser
light is in resonance with transitions b, beats at frequencies
ν_{54} , ν_{43} and $\nu_{53} = \nu_{54} + \nu_{43}$ must be detected (see Fig. 2).

     The experimental set-up used for the observation of the beats
is shown in Fig. 3. The N_2 - laser-pumped-dye laser used for the
excitation of the Cs atoms has been thoroughly described
elsewhere.[5] The dye utilized in this experiment is a 7-diethylamino-
4-methyl coumarin solution in ethanol. The laser delivers, at a rate
of about 40 pps , pulses of length $\tau \approx 2$ nsec and spectral
bandwidth $\Delta \approx 1$ Ghz which fulfill conditions (1) and (2). The laser
is tuned to one of the two sets of resonances (a) or (b) by comparing
with a high resolution Jarrell-Ash spectrometer the pulse spectrum
to that of a Cs spectral lamp. The laser radiation is linearly
polarized and sent through a Cs cell, temperature regulated at
about 40°C . The fluorescent light emitted at right angles is
detected by a photomultiplier (RCA 1P28) which is sensitive only to
the transition back to the ground state (4555Å) and not to the

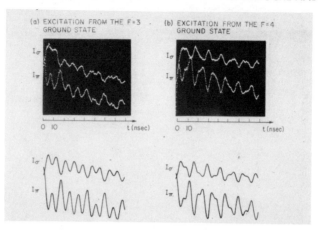

Fig. 4. Pictures of the Quantum Beat signals in cases (a) and (b). The lower and upper trace in each picture correspond respectively to I_π and I_σ. Each trace contains 500 sampling points and is averaged over 256 runs. The base line has been offset between the recordings of I_π and I_σ to avoid overlapping of the traces. The corresponding theoretical plots of the beats are shown under each picture, with the same time scale.

other possible decay channels of the $7^2P_{3/2}$ state (transitions back to the $7^2S_{1/2}$, $5^2D_{3/2}$ and $5^2D_{5/2}$ levels in the infrared). An analyzer allows the detection of the polarization component of the fluorescent light either parallel (I_π) or perpendicular (I_σ) to that of the incoming pulse. The repetitive output of the photomultiplier is sent to a sampling scope (Tektronix 661) triggered by the laser pulse (using a ITT photodiode). This detection method allows the reconstruction of the signal on a long term basis, approximately 10 sec instead of a few hundred nanoseconds real time, enabling us to store and process the data. The sampling technique introduces, however, a noise due to the pulse to pulse amplitude fluctuation of the dye laser. A Hewlett-Packard 5480A Signal Analyzer is used to enhance the signal to noise ratio by signal analyzing methods. Figure 4 shows the I_π and I_σ beat signals corresponding to resonances (a) and (b). One can compare them to the corresponding theoretical plots shown under each picture. The patterns of the beats in the different signals correspond very well to the theoretical predictions. It may in particular be shown quite generally that the modulated part of I_π and I_σ must be in phase opposition, the modulation in I_π being twice as big as the one in I_σ (this result follows from the fact that the Quantum Beat effect must vanish if one detects the sum of the light intensities emitted along three orthogonal polarizations,[10] this sum being obviously equal to $I_\pi + 2I_\sigma$). This relationship between I_π and I_σ is well verified in the recordings of Fig. 4.

The same beat patterns have also been recorded on a longer time scale, the plots being shown in Figs. 5a and 5b, corresponding respectively to cases (a) and (b). In each figure, one can see from

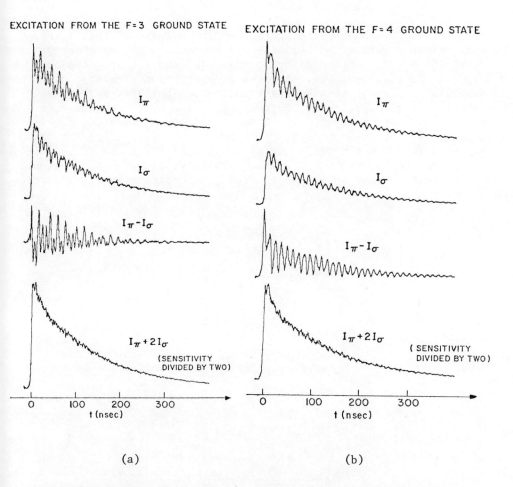

EXCITATION FROM THE F=3 GROUND STATE   EXCITATION FROM THE F=4 GROUND STATE

I_π

I_σ

$I_\pi - I_\sigma$

$I_\pi + 2I_\sigma$

(SENSITIVITY DIVIDED BY TWO)

(a)   (b)

Fig. 5. Quantum Beat recordings on a long time scale in cases (a) and (b). Signals I_π and I_σ have been recorded separately (500 sampling points; 256 runs). They have then been subtracted and added in order to give the $I_\pi - I_\sigma$ and $I_\pi + 2I_\sigma$ curves.

top to bottom, curves corresponding to the signals I_π , I_σ , $I_\pi - I_\sigma$
and $I_\pi + 2I_\sigma$. The difference and sum curves have been obtained by
using a Hewlett-Packard computer interfaced to the signal analyzer.
We have drawn the difference curves $I_\pi - I_\sigma$ in order to take
advantage of the phase opposition relationship between the I_π and I_σ
signals: by plotting the difference of the two signals, one reduces
the unmodulated part of the decay curves and enhances the modulation
depth of the beats. On the contrary, in the sum curves $I_\pi + 2I_\sigma$,
the modulations disappear as predicted by the theory and one gets
pure exponential decay curves giving a lifetime of 135 (5) nsec for
the $7^2P_{3/2}$ state which is in very good agreement with a previous
level crossing determination.[9]

At last, we have performed a Fourier analysis of the $I_\pi - I_\sigma$
beat patterns, using the Hewlett-Packard computer. The power spectra
$S_a(\nu)$ and $S_b(\nu)$ corresponding to cases (a) and (b) are displayed in
Fig. 6. In each spectrum, the zero frequency peaks correspond to the
unmodulated part of the fluorescent light. The three successive
peaks, indicated by arrows, correspond to the three eigenfrequencies
expected in each beat signal. From $S_a(\nu)$, one finds the hyperfine
intervals to be $\nu_{43} = 65.5 \pm 1.5$ Mhz , $\nu_{32} = 49.1 \pm 1.5$ Mhz and
$\nu_{42} = 114.6 \pm 2$ Mhz ; from $S_b(\nu)$, $\nu_{54} = 81.9 \pm 2$ Mhz , $\nu_{43} = 66.7$
± 1.5 Mhz and $\nu_{53} = 148.6 \pm 2$ Mhz . These values are in agreement
with those obtained in previous works.[9] (Compare with the expected

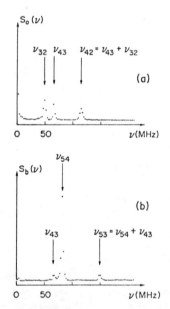

Fig. 6. Frequency analysis of the $I_\pi - I_\sigma$ signals in cases
(a) and (b). The frequency interval between points in the discrete
spectra $S_a(\nu)$ and $S_b(\nu)$ is 2.34(2) Mhz.

frequencies indicated in Fig. 2.) The quoted uncertainties come in part from the possible error in time calibration, partly from the relatively large intervals between points in the discrete Fourier spectra. The precision of the technique could no doubt be improved by sampling more points and recording the beats on a longer time scale. One could also, instead of Fourier analyzing a time resolved Quantum Beat signal, send directly the photomultiplier output to a frequency analyzer device and thus measure directly the Quantum Beat frequencies. This technique should lead to a considerable increase in the accuracy of the frequency measurements.[11]

The experiment described above shows clearly that the Quantum Beat technique using pulsed dye lasers may be developed as a practical spectroscopic tool. It is obvious that this method should be primarily used for the study of levels which cannot be conveniently investigated by other techniques such as level crossing and optical double resonance. In these latter techniques, one makes use generally of continuous light sources (spectral lamps and, since very recently, cw dye lasers) which emit strongly in the infrared and visible spectrum and (in case of the spectral lamps) on some strong ultraviolet transitions. It is in general very hard, with these sources, to reach directly high lying states of atoms, linked by weak uv transitions to the ground state. Thus, the spectroscopic study of these states by double resonance or level crossing is very difficult. For example, in the np series of each alkali atom, only a few low lying n states have been investigated (up to n = 10 in Cs).[12] The Quantum Beat technique described here makes the investigation of high lying levels much easier. As we already noted the high peak power available in the dye laser pulses allows frequency doubling to get a continuously tunable source of intense light pulses throughout the uv spectrum. These pulses should be strong enough to excite efficiently very weak transitions to high lying atomic p states. An experiment designed to investigate the $np_{3/2}$ Cs levels with n bigger than 10 is actually under progress at Stanford. We have already obtained a pulse source tunable between 3200 and 3400Å by doubling the output of a cresylviolet dye laser with an angle tuned KDP crystal. The 11, 12, 13 and 14p Cs levels lie in this wavelength range and will hopefully soon be studied using the technique described above for the 7p level.

One should say in conclusion that Quantum Beats are not the only interference effects which can be induced in an atomic vapor by short dye laser pulses. Instead of time dependent oscillations, one can also observe modulations as a function of an applied magnetic field in the fluorescent light emitted at a fixed delay time after the laser pulse.[13] The use of pulsed dye lasers to study all these related interference effects in atomic fluorescence certainly opens a promising new field in spectroscopy of gases. Useful information about lifetimes, g-factors, hyperfine structures should be obtained in such studies, complementing those already yielded by classical level-crossing experiments.

REFERENCES

[1]For optical pulse induced Quantum Beats see J. N. Dodd, R. D. Kaul, and D. M. Warrington, Proc. Phys. Soc. (London) 84, 176 (1964); E. B. Alexandrov, Opt. Spektrosk. 17, 957 (1969) [Opt. Spectrosk. 17, 522 (1964)]; J. N. Dodd, W. J. Sandle, and D. Zissermann, Proc. Phys. Soc. (London) 92, 497 (1967).

[2]For electron pulse induced Quantum Beats see: T. Hadeishi and W. A. Nierenberg, Phys. Rev. Letters 14, 891 (1965).

[3]For Zero-field Quantum Beats in Beam Foil Spectroscopy see: H. J. Andrä, Phys. Rev. Letters 25, 325 (1970); H. G. Berg and J. L. Subtil, Phys. Rev. Letters 27, 1103 (1971); K. Tillman, H. J. Andrä, and W. Wittman, Phys. Rev. Letters 30, 155 (1973).

[4]For a simple theory of Quantum Beat effect see: A. Corney and G. W. Series, Proc. Phys. Soc. 83, 207 (1964); see also P. Franken, Phys. Rev. 121, 508 (1961) (Appendix I).

[5]T. W. Hänsch, Appl. Opt. 11, 895 (1972).

[6]F. B. Dunning, E. D. Stokes, and R. F. Stebbings, Opt. Comm. 6, 63 (1972).

[7]W. Gornik, D. Kaiser, W. Lange, J. Luther, and H. H. Schulz, Opt. Comm. 6, 327 (1972).

[8]S. Haroche, J. A. Paisner, and A. L. Schawlow, Phys. Rev. Letters 30, 948 (1973).

[9]S. Svanberg and S. Rydberg, Z. Phys. 227, 216 (1969).

[10]This sum rule results from an orthogonality relationship between the Clebsch-Gordon coefficients which appear in the expression of the Quantum Beat signal.

[11]In order to get accurate values of the hyperfine splittings, one would also have to cancel the earth magnetic field which was applied to the atoms at the time the spectra of Fig. 6 were recorded. This field splits the hyperfine levels in several components and thus broadens the peaks in the Fourier spectra. This effect was overlooked in this experiment because its resolution was in any case limited by the existence of a large interval between points in the spectra. In the study of the high lying Cs p states described below, the magnetic field perturbation will become much more important, due to the smaller value of the hyperfine

coupling in these states. Consequently, the earth field will be
carefully cancelled out in these experiments.

[12]S. Rydberg and S. Svanberg, Physica Scripta $\underline{5}$, 209 (1972).

[13]P. Schenck, R. C. Hilborn, and H. Metcalf, Bull. Am. Phys. Soc.
$\underline{18}$, 611 (1973), and Phys. Rev. Letters, submitted for publication
(1973); see also H. Walther, paper in this Conference.

OBSERVATION OF SUPERHYPERFINE MODULATION AND QUANTUM BEATS

IN PHOTON-ECHO SPECTROSCOPY IN RUBY[*]

I. D. Abella[†]

Joint Institute for Laboratory Astrophysics

University of Colorado, Boulder, Colorado 80302

A. Compaan[††] and L. Q. Lambert

Department of Physics

University of Chicago, Chicago, Illinois 60637

Abstract

   Experiments are summarized of photon-echo spectroscopy in ruby
at low temperature in which the simple two-level picture is no
longer appropriate. The superhyperfine interactions of Cr-Aℓ in
ruby serve to form a series of nearly degenerate multilevel systems
from which the coherently excited macroscopic dipole moments
radiate at slightly different frequencies. The interference or
quantum beats between these radiation modes yield modulated photon
echoes from the solid, which are observed as a function of the time
between pulses. Since echo signals are proportional to N^2, the
sensitivity is enhanced over fluorescence studies of the same
material.

[*] Work supported by the National Science Foundation under grant
    GP-18622, and The Material Research Laboratory, University of
    Chicago.

[†] Visiting Fellow, JILA, 1972-73.
    Permanent Address: Ryerson Physical Laboratory, University of
    Chicago, Chicago, Illinois 60637.

[††] Present Address: Department of Physics, New York University.

I. D. ABELLA, A. COMPAAN, AND L. Q. LAMBERT

Photon echo spectroscopy has proved to be a powerful tool in studying the interactions between coherently excited atoms. Thus, we have many[1-9] examples of solids and gases which have been studied either by the method of direct application of two short laser pulses to a resonant medium or the recent technique[8] of applying cw laser excitation together with pulsed molecular resonance Stark shifts.

Although the simple two level picture[9] of photon-echo generation may be adequate for pedagogical purposes, it has been demonstrated to be more complicated in many of the cases already studied. We note, for example, that interacting atoms in a solid form a series of multilevel systems[1,10] which are nearly degenerate, and depending on the applied magnetic field can have a range of splitting up to the order of a hundred megahertz. In this case, optical pulses of suitable pulse duration and spectral content can coherently excite a group of states. These coherently excited multilevel systems have macroscopic electric dipole moments, which interfere in the observed radiation due to their differing frequencies. A series of papers on the theory of modulated photon echoes has appeared in

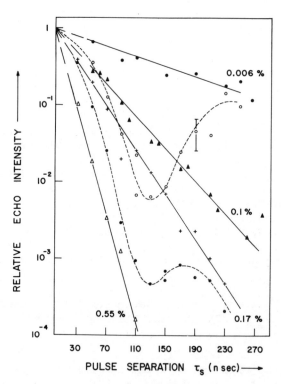

Fig. 1. Echo vs. pulse separation in 2.7 kG field. Solid lines for field ∥ c-axis; dashed lines for field angle 4° (Ref. 2).

the literature.[1,10,11] Quantum beats[12] have of course been studied
by methods other than laser excitation, such as beam foils,[13]
electron excitation,[14] and weak optical excitation.[15]

For the case of ruby which we have studied, Lambert et al.[2]
reported photon-echo modulation as a function of pulse separation,
when magnetic fields were applied to the sample at varying angles
to the optic axis. The appearance of modulation depends on the
details of the Cr-Aℓ magnetic interaction as a function of magnetic
field magnitudes and direction. On axis, these effects are negligible compared to the strong modulation at increasing field angle.

Figure 1 shows the effect of Aℓ modulation of the Cr electronic
energy levels. Classically this can be viewed as Aℓ nuclei preces-
sing in the local field and the resulting oscillatory magnetic field
modulating the Cr energy levels. A more detailed analysis has been
given[1,10] in terms of a density matrix treatment for nearly degenerate states. The decay as a function of Cr concentration has been
studied by Compaan[3] and will not be discussed here. The results of
Fig. 1 were obtained with plane polarized pulses, which are not
eigenpolarizations in ruby. Consequently the schematic experimental
design shown in Fig. 2 was used to produce circularly polarized
pulses and echoes appropriate to the transitions to be studied.
Temperature tuning of the laser source or magnetic tuning of the
sample were used to obtain the resonance condition.

Figure 3 shows the nomenclature for the three circularly
polarized transitions studied σ_1, σ_2, σ_3 for the $^4A \rightarrow {}^2E(E)$ transitions respectively $\sigma_1(3/2 \rightarrow 1/2)$, $\sigma_2(-1/2 \rightarrow -1/2)$ and $\sigma_3(1/2 \rightarrow 1/2)$.
With the aid of this experimental arrangement, it was then possible
to observe the circularly polarized echo signals as a function of
time and magnetic field angle for each of the separate eigenpolarizations. The echo modulation, or quantum beat signal, is typically
given for σ_2 and σ_3 in Fig. 4. What is shown here are the experimental points,[1] together with several machine calculated fits,
taking into account the various nearest-neighbor interaction constants in the ground and excited states, with the ratio of excited
to ground state constants a fitting[16] parameter. A discussion of
excited parameters is given by Lambert.[1] Included also in the calculation is the quadrupole splitting as well as the effect of newly
measured[17] constants.

The theoretical analysis proceeds along the lines developed by
Gordon et al.[5,16,18] and has been applied to the problem of nearly
degenerate levels and modulation.[1,10]

Equation (1) expresses the electric dipole moment of a single
atom evaluated in the reference frame "rotating" at the frequency
of the applied radiation field

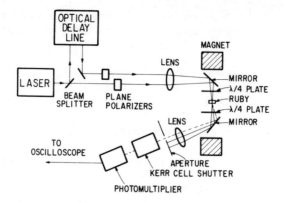

Fig. 2. Experimental schematic for production and detection of the circularly-polarized photon echoes, low temperature cryostat omitted.

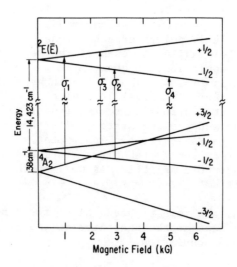

Fig. 3. Energy level diagram of dilute ruby with nomenclature for the various circularly polarized transitions for field ∥ optic axis.

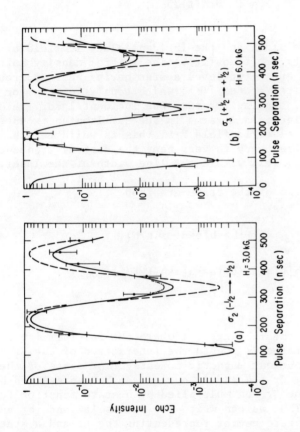

Fig. 4. Photon-echo intensity vs. pulse separation time, showing the superhyperfine modulation and quantum beats: (a) for the $\sigma_2(-\tfrac{1}{2} \rightarrow -\tfrac{1}{2})$ transition in 3.0 kG field at $\chi=8°$ and (b) for the $\sigma_3(\tfrac{1}{2} \rightarrow \tfrac{1}{2})$ transition in 6.0 kG field at angle $\chi=5°$. Solid curves are theoretical. Taken from reference 1.

$$\langle \vec{P} \rangle = \mathrm{Tr}[\vec{P}\rho(t)]$$

$$= \mathrm{Tr}[\vec{P}\exp(-i\mathcal{K}t/\hbar)\exp(i\vec{F}\cdot\vec{P})$$ (1)

$$\times \rho(0)\exp(-i\vec{F}\cdot\vec{P})\exp(i\mathcal{K}t/\hbar)] \quad ,$$

where $\mathcal{K}$ is the effective Hamiltonian in the "rotating frame" and

$$\vec{F} = \int_0^{\Delta\tau} dt\,\vec{\mathcal{E}}(t)/\hbar \quad ,$$

where $\Delta\tau$ is the pulse width and the incident electric field of the pulse is given by $\vec{E}(t) = \vec{\mathcal{E}}(t)\cos(\omega t)$. The initial density matrix $\rho(0)$ characterizes a quasi-two-level system having sets of ground states $\{|a\rangle\}$ and excited states $\{|b\rangle\}$ which consist of two (or more) nondegenerate levels. During the time the incident laser field is present, we have neglected all terms in the Hamiltonian except the interaction with the electric field $\vec{E}\cdot\vec{P}$; this is valid for those ions whose resonant frequency is very near the central frequency of the laser pulse, i.e., well within the line width of the laser.

The non-zero terms of the trace are then

$$\langle \vec{P} \rangle = \frac{1}{2}\sum_{ab} \langle a|\vec{P}|b\rangle\langle b|\exp(-i\mathcal{K}t/\hbar)|b\rangle$$

$$\times \langle b|\sin(2\vec{F}\cdot\vec{P})|a\rangle\langle a|\exp(i\mathcal{K}t/\hbar)|a\rangle$$ (2)

$$+ \text{ c.c.}$$

We now specialize to the ruby crystal where the multiplicity of ground and excited states is due to the interaction of the Cr spin with neighboring Aℓ nuclear magnetic moments. The state of the coupled system of the Cr ion and Aℓ nuclei can be represented by a state function for the Cr ion multiplied by a state function for the nuclear spins. Thus we can write the states $|a\rangle$ and $|b\rangle$ as the direct product of a state vector representing the Cr and a state vector representing the Aℓ neighbors,

$$|a\rangle = |\alpha\rangle|\{m_j\}\rangle$$

$$|b\rangle = |\beta\rangle|\{n_j\}\rangle \quad ,$$ (3)

where $|\alpha\rangle$ represents one of the four $^4A_2$ ground states, $|\beta\rangle$ one of the two ^2E excited states and $\{m_j\}$ and $\{n_j\}$ are sets of quantum numbers for the spins of the Aℓ neighbors, j. The electric dipole operator $\vec{P}$ operates only on the Cr ion part of the state function so when Eq. (3) is substituted into Eq. (2), we obtain

$$\langle\vec{P}\rangle = \frac{1}{2}\sum_{\alpha\beta} \langle\alpha|\vec{P}|\beta\rangle\langle\beta|\sin(2\vec{F}\cdot\vec{P})|\alpha\rangle$$

$$\times \left[\sum_{\{m\}\{n\}} \langle\{m_j\}|\{n_j\}\rangle\langle\{n_j\}|\exp(-i\mathcal{K}_\beta t/\hbar)|\{n_j\}\rangle \right. \tag{4}$$

$$\left. \times \langle\{n_j\}|\{m_j\}\rangle\langle\{m_j\}|\exp(i\mathcal{K}_\alpha t/\hbar)|\{m_j\}\rangle \right] .$$

Here $\mathcal{K}_\alpha = \langle\alpha|\mathcal{K}|\alpha\rangle$ and $\mathcal{K}_\beta = \langle\beta|\mathcal{K}|\beta\rangle$ are Hamiltonians corresponding to $^4A_2$ and ^2E states, respectively, and have been given explicitly. One defines a transformation matrix W whose elements are $\langle\{n_j\}|\{m_j\}\rangle$; this matrix transforms $A\ell$ eigenstates for the $^4A_2$ state to the $A\ell$ eigenstates for the ^2E state. Rewritten in terms of the matrix W, the electric dipole moment becomes

$$\langle\vec{P}\rangle = \frac{1}{2}\sum_{\alpha\beta} \langle\alpha|P|\beta\rangle\langle\beta|\sin(2\vec{F}\cdot\vec{P})|\alpha\rangle$$

$$\tag{5}$$

$$\times \left\{ \mathrm{Tr}[W^\dagger\exp(-iE_\beta t/\hbar)W\exp(iE_\alpha t/\hbar)] \right\} .$$

Here E_α and E_β are the diagonal operators representing the Hamiltonians $\mathcal{K}_\alpha$ and $\mathcal{K}_\beta$. The radiation intensity is obtained by transforming the electric dipole moment back into the laboratory reference frame, squaring and taking the average over times long compared to the optical frequencies. Assuming that only one of the possible $|\alpha\rangle$ to $|\beta\rangle$ transitions in resonant, the intensity is

$$I \sim \left| \langle\alpha|\vec{P}|\beta\rangle\langle\beta|\sin(2\vec{F}\cdot\vec{P})|\alpha\rangle \right|^2$$

$$\tag{6}$$

$$\times \left| \mathrm{Tr}[W^\dagger\exp(-iE_\beta t/\hbar)W\exp(iE_\alpha t/\hbar)] \right|^2 .$$

This equation which we evaluate later predicts modulations in the fluorescence from a ruby crystal which is coherently excited by a laser pulse, i.e. the so called quantum beat phenomenon.

This predicted fluorescence modulation is intimately related to the photon-echo modulation which we have observed in ruby, as we shall now show by calculating the photon-echo intensity vs. pulse separation. For a sequence of two pulses we retain only those terms of Eq. (1) which contribute to the echo at $t = 2\tau$. Following Ref. 10 we obtain for the electric dipole moment $\langle\vec{P}\rangle$ in the rotating frame:

$$\langle\vec{P}\rangle = \frac{1}{2} \sum_{aa'bb'} \langle a|\vec{P}|b\rangle\langle b|\exp(-i\mathcal{H}(t-\tau)/\hbar)|b\rangle$$

$$\times \langle b|\sin(\vec{F}_2\cdot\vec{P})|a'\rangle\langle a'|\exp(-i\mathcal{H}\tau/\hbar)|a'\rangle$$

$$\times \langle a'|\sin(2\vec{F}_1\cdot\vec{P})|b'\rangle\langle b'|\exp(i\mathcal{H}\tau/\hbar)|b'\rangle \tag{7}$$

$$\times \langle b'|\sin(\vec{F}_2\cdot\vec{P})|a\rangle\langle a|\exp(i\mathcal{H}(t-\tau)/\hbar)|a\rangle$$

$$+ \text{ c.c.}$$

If we again specialize to ruby by making the substitutions for
$|a\rangle$ and $|b\rangle$, and assume only one of the $|a\rangle$ to $|b\rangle$ transitions is
resonant, the result in terms of the matrix W is:

$$\langle\vec{P}\rangle = \frac{1}{2} \langle\alpha|\vec{P}|\beta\rangle\{\langle\beta|\sin(\vec{F}_2\cdot\vec{P})|\alpha\rangle\}^2\langle\alpha|\sin(2\vec{F}_1\cdot\vec{P})|\beta\rangle$$

$$\times \{\text{Tr}[W^\dagger\exp(-iE_\beta(t-\tau)/\hbar)W\exp(-iE_\alpha\tau/\hbar)$$

$$\times W^\dagger\exp(iE_\beta\tau/\hbar)W\exp(iE_\alpha(t-\tau)/\hbar)]\} \tag{8}$$

$$+ \text{ c.c.}$$

The result of Lambert's analysis,[1] is that at $t = 2\tau$, the trace of
Eq. (8) can be re-written, where $W = \prod_j W_j$

$$\langle\vec{P}(2\tau)\rangle \sim \prod_j \text{Tr}[W_j^\dagger \exp(-iE_{\beta,j}\tau/\hbar)W_j \exp(-iE_{\alpha,j}\tau/\hbar)$$

$$\times W_j^\dagger \exp(iE_{\beta,j}\tau/\hbar)W_j \exp(iE_{\alpha,j}\tau/\hbar)] \quad . \tag{9}$$

The resulting modulations are calculated in practice by a numerical
diagonalization to determine $E_{\alpha,j}$, $E_{\beta,j}$ and W_j from the ruby
Hamiltonians.

In the case of fluorescence following a single laser pulse,
substitution of the explicit Hamiltonians for ruby into Eq. (6)
gives an intensity proportional to

$$I \sim \left|\prod_j \text{Tr}[W_j^\dagger \exp(-iE_{\beta,j}t/\hbar)W_j \exp(iE_{\alpha,j}t/\hbar)]\right|^2 \quad . \tag{10}$$

A particularly simple application of the preceding formalism has yielded an expression[10] which is more easily recognized as a quantum beat phenomenon. We considered the case of the nearly-degenerate levels which occur near level crossings, such as in the ground state of ruby, the $^4A_2(+3/2, -1/2)$ at 2.06 kG. In off-axis magnetic fields, the $M_S = 3/2$ and $M_S = -1/2$ states are strongly mixed and a so-called anti-crossing occurs.

To a good approximation, the upper and lower energy states, respectively, may be expressed in the crossing region as

$$|a_1> = \alpha|3/2> + \beta|-1/2> \text{ and } |a_2> = \beta|3/2> - \alpha|-1/2> \quad (11)$$

where the coefficients α and β are real numbers less than unity, and are rapidly varying functions of the field $\vec{H}$ near the crossing. The α and β are determined by diagonalization of the ground state spin Hamiltonian.

Suppose one considers the $^4A_2(3/2) \rightarrow \overline{E}(^2E)(1/2)$ transition which arises only from the $|3/2>$ component of each of the mixed states of Eq. (11). Experimentally, this is assured by thermally tuning the laser source to the $\sigma_1(3/2 \rightarrow 1/2)$ resonance. Then,

$$<a_1|\vec{P}|b> = \alpha<3/2|\vec{P}|1/2> \text{ and } <a_2|\vec{P}|b> = \beta<3/2|\vec{P}|1/2> \quad (12)$$

and the matrix elements for the incident pulses can be evaluated, for example, the second pulse contributions are

$$<a_1|\sin(\vec{F}_2 \cdot \vec{P})|b> = \alpha<3/2|\sin(\vec{F}_2 \cdot \vec{P})|1/2> \quad ,$$
$$<a_2|\sin(\vec{F}_2 \cdot \vec{P})|b> = \beta<3/2|\sin(\vec{F}_2 \cdot \vec{P})|1/2> \quad . \tag{13}$$

The resulting photon-echo intensity is given by the expression,

$$I = I_0\left\{\left[\alpha^4 + \beta^4 + 2\alpha^2\beta^2\cos(\delta\omega\tau)\right]_{av}\right\}^2 \tag{14}$$

where the energy splitting is given by $\delta\omega = \omega_{a1}-\omega_{a2}$. The detailed behavior of echoes near the level crossings in ruby is complicated by spin-dependent decay and is discussed in reference 1 and elsewhere. The analysis of the level-crossing example was intended to derive the more familiar modulation equation shown in Eq. (14).

Evaluation of Eq. (10) typically gives a modulation of the fluorescent intensity when a magnetic field is applied to the sample. Finally, it is worth noting that although the modulation factors appear in both the echo and fluorescence expressions, the signal strength for the echo case is proportional to N^2. Thus, we have the possibility of enhanced experimental detection over the same experiment performed in fluorescence,[19] if dephasing times are sufficiently long for the echo to be observed within the order of the lifetime of the excited state.[20]

References

1. L. Q. Lambert, Phys. Rev. B 7, 1834 (1973).

2. L. Q. Lambert, A. Compaan, and I. D. Abella, Phys. Letters 30A, 153 (1969).

3. A. Compaan, Phys. Rev. B 5, 4450 (1972).

4. M. O. Scully, M. J. Stephen, and D. C. Burnham, Phys. Rev. 171, 213 (1968).

5. C.K.N. Patel and R. E. Slusher, Phys. Rev. Letters 20, 1087 (1968); B. Bolger and J. C. Diels, Phys. Letters 28A, 401 (1968).

6. J. P. Gordon, C. H. Wang, C.K.N. Patel, R. E. Slusher, and W. J. Tomlinson, Phys. Rev. 179, 294 (1969).

7. V. Kh. Kopvillem, V. R. Nagibarov, V. A. Pirozhkov, V. V. Samartsev, and R. G. Usmanov, Fiz. Tverd. Tela 14, 1794 (1972) [Sov. Phys.-Solid State 14, 1544 (1972)]; V. V. Samartsev and V. R. Nagibarov, Fiz. Tverd. Tela, 11, 3669 (1969) [Sov. Phys.- Solid State 11, 3085 (1970)]; A. I. Alekseev and I. V. Evseev, Zh. Eksp. Teor. Fiz. 57, 1735 (1969); 56, 2118 (1969); [Sov. Phys.-JETP 30, 938 (1970); 29, 1139 (1969)].

8. R. G. Brewer and R. L. Shoemaker, Phys. Rev. Letters 27, 631 (1971); 28, 1430 (1972); R. G. Brewer, Science 178, 247 (1972).

9. I. D. Abella, N. A. Kurnit, and S. R. Hartmann, Phys. Rev. 141, 391 (1966); N. A. Kurnit, I. D. Abella, and S. R. Hartmann, Phys. Rev. Letters 13, 567 (1964).

10. L. Q. Lambert, A. Compaan, and I. D. Abella, Phys. Rev. A 4, 2022 (1971).

11. G. Oliver and R. Levebre, J. Phys. $\underline{31}$, 761 (1970); A. Alexeyev, Phys. Letters $\underline{31A}$, 495 (1970); G. Oliver, Physica $\underline{51}$, 197 (1971).

12. G. W. Series, Physica $\underline{33}$, 138 (1967) and references therein.

13. J. Macek, Phys. Rev. Letters $\underline{23}$, 1 (1969); O. A. Keller and R. T. Robiscoe, Phys. Rev. $\underline{188}$, 82 (1969); H. J. Andra, Phys. Rev. Letters $\underline{25}$, 325 (1970); H. G. Berry and J. L. Subtil, Phys. Rev. Letters $\underline{27}$, 1103 (1971); D. J. Burns and W. H. Hancock, Phys. Rev. Letters $\underline{27}$, 370 (1971); K. Tillman, H. J. Andra and W. Wittman, Phys. Rev. Letters $\underline{30}$, 155 (1973).

14. T. Hadeishi and W. A. Nierenberg, Phys. Rev. Letters $\underline{14}$, 801 (1965).

15. J. N. Dodd, R. D. Kaul, and D. M. Warrington, Proc. Phys. Soc. (London) $\underline{84}$, 176 (1964); E. B. Alexandrov and V. P. Kozlov, Opt. Spektrosk. $\underline{16}$, 533 (1964); [Opt. Spectrosc. $\underline{16}$, 289 (1964)]. E. B. Alexandrov, Opt. Spektrosk. $\underline{17}$, 957 (1964), [Opt. Spectrosc. $\underline{17}$, 522 (1964)]; J. N. Dodd, W. J. Sandle and D. Zisserman, Proc. Phys. Soc. (London) $\underline{92}$, 497 (1967).

16. D. Grischkowsky and S. R. Hartmann, Phys. Rev. B $\underline{2}$, 60 (1970).

17. P. F. Liao and S. R. Hartmann, Phys. Letters $\underline{38A}$, 295 (1972).

18. C. V. Heer and R. H. Kohl, Phys. Rev. A $\underline{1}$, 693 (1970); $\underline{2}$, 549 (1970); C. K. Rhodes, A. Szöke and A. Javan, Phys. Rev. Letters $\underline{21}$, 1151 (1968).

19. S. Haroche, J. A. Paisner, and A. L. Schawlow, Phys. Rev. Letters, $\underline{30}$, 948 (1973), and paper this conference.

20. A recent beam foil study of hyperfine modulation was made in Li in which a fast fluorescent decay obtains: H. G. Berry, J. L. Subtil, E. H. Pinnington, H. J. Andra, W. Wittmann, and A. Gaupp, Phys. Rev. A $\underline{7}$, 1609 (1973).

TUNABLE LASERS III

SPIN-FLIP RAMAN LASER AND APPLICATIONS

C. K. N. PATEL

Bell Telephone Laboratories, Incorporated

Holmdel, New Jersey 07733

ABSTRACT

This paper reports on the most recent advances in the spin-flip Raman lasers and their applications to spectroscopy, low level pollution detection and nonlinear optics.

I. INTRODUCTION

In this paper we will report on some of the latest advances in the spin-flip Raman (SFR) lasers and their applications to spectroscopy, pollution detection and nonlinear optics. Since the first operation of the SFR laser in 1970,[1] a very large number of significant advances have taken place in the field. However, we will not review all of these because a number of review papers already exist which have summarized the status of the field from time-to-time.[2-4] Thus here we will report only on the most recent advances which include the operation of a SFR laser with a closed cycle cooler[5] (for keeping the InSb SFR laser sample at $T \sim 12K$) which removes the necessity of having cryogenic liquids, measurements of the spontaneous SFR scattering linewidth by measuring the SFR gain,[5] (which is an important parameter governing both the SFR laser linewidth and the SFR laser fine tuning), improvements in nitric oxide pollution measurement capability and nonlinear optics to generate tunable coherent radiation at frequencies other than those directly available from the spin-flip Raman laser.

The spin-flip Raman laser consists of a pump laser at a fixed frequency, ω_o, and a sample of appropriately doped InSb sample with plane parallel surfaces kept at low temperature in a magnetic

471

field. At high enough intensity of the pump radiation, the process
of spontaneous spin-flip Raman scattering[6-8] from the electrons in
InSb in the magnetic field B can be made to go over into the
region of stimulated SFR scattering i.e., SFR laser. The SFR
laser frequency, ω_s, is given by

$$\omega_s - \omega_o - g\mu_B B \tag{1}$$

where g is the g-value of the electrons in InSb, and μ_B is the
Bohr magneton. The SFR laser in InSb has been operated with a
pulsed CO_2 laser at $\sim 10.6\mu$ (Q-switched[1,9] or pulsed TEA variety[10,11])
or a cw[12,13] or pulsed[14] CO laser at $\sim 5.3\mu$. The tunability of the
SFR laser radiation in the first case is seen to cover from $\sim 9\mu$
to $\sim 14.6\mu$ and the latter case from $\sim 5\mu$ to $\sim 6.5\mu$ using magnetic
fields of up to ~ 100 kG.[4] One of the recent major advances in the
SFR laser field occurred with the operation of the SFR laser at
magnetic fields as low as ~ 400G.[13] The lowest field achieved
until that time was ~ 15 kG. This low field operation was achieved
with the use of a rather low carrier concentration InSb sample
($\sim 1\times 10^{15}$ cm^{-3} as compared with the $\sim 1\times 10^{16}$ cm^{-3} samples used until
that time.) The low field operation immediately removed the
necessity of a superconducting solenoid for providing the magnetic
field. A conventional electromagnet with its ease of operation
and field control now replaced the superconducting magnet. A
conventional electromagnet, even with its considerably lower
magnetic field available as compared to that available from a
superconducting magnet, still provides a sizeable tuning range for
the SFR laser because of the large number of closely spaced pump
frequencies available from a CO_2 laser[15] (spacing ~ 2 cm^{-1}) and
from a CO laser[15] (spacing ~ 4 cm^{-1}). Thus whenever one reaches the
maximum tuning available with a given magnetic field, the pump
frequency is "jump changed" (by adjusting the grating which
controls the output wavelength from the CO_2 or the CO laser). This
is best seen by an examination of Eq. (1).

The low field operation, further permitted the measurement of
the linewidth of the SFR laser by heterodyne techniques.[16] The
measurements indicated a SFR laser linewidth of considerably less
than 1 kHz in agreement with the calculated values. It is clear
that the very narrow linewidth of the SFR laser together with its
sizeable tunability and high cw as well as pulsed power output
makes the SFR laser a very desirable tool of coherent narrow band
radiation for such diverse applications as spectroscopy, local
oscillators, pollution detection, sources for nonlinear optics, etc.

In Section II we will describe the operation of a SFR laser
with a closed cycle cooler. It will be seen that this experiment
has gone a long way towards making the SFR laser a practical tool
for potential applications. In Section III we will describe the

recent experimental studies of measurements of the spontaneous SFR scattering linewidth, $\Delta\nu_{spont}$, at low magnetic fields in lightly doped InSb samples. We find that the spontaneous SFR scattering linewidth in $n_e = 10^{15}$ cm^{-3} InSb sample is $\lesssim$ 300 MHz. We will give some preliminary experimental results of the variation of $\Delta\nu_{spont}$ with sample temperature. We will also show the variation of g-value with magnetic field in this sample. The spontaneous SFR scattering linewidth is a very important parameter in description of the gain of the SFR scattering process, the SFR laser linewidth and the fine tuning characteristics of the SFR laser. These will also be discussed. In Section IV we will discuss the recent results of the advances in pollution detection of NO. Earlier techniques have been improved, and detection of No concentrations of as low as 10^9 molecules/cm^3 has been demonstrated. In Section V we will very briefly touch upon the new area of nonlinear optics using the SFR laser as a source of intense coherent tunable radiation. The most significant of the experiments to date involves generation of tunable far infrared radiation in the 100μ range.[17] Finally, in Section V we will summarize the paper and give some comments about the future of spin-flip Raman lasers.

II.   ADVANCES IN SFR LASER OPERATION

With the operation[13] of the SFR **laser** with a InSb sample having a carrier concentration of $\sim 10^{15}$cm^{-3}, the field of tunable Raman lasers took a sizeable step towards making the SFR laser a practical tool for numerous applications. The advantage gained by using the low carrier concentration sample was that the lowest magnetic field at which the SFR laser would operate has been reduced to approximately 200 G compared to a minumum magnetic field of approximately 15 kG required for the previously used 1×10^{16} cm^{-3} concentration InSb sample. The reasons for the lowering of the magnetic field limit is reasonably well understood and has its origin in the quantum limit considerations as described in Ref. 1. This low field operation clearly gets around the requirement of using a superconducting magnet for tuning the SFR laser which was necessary in the earlier work. Now, with only a small loss of tuning one can operate the SFR laser with a conventional electromagnet capable of delivering magnetic field of less than 5-10 kG. The complete loss of tunability as compared to that available from a 100 kG magnet is avoided by the virture of the fact that the CO laser as well as the CO$_2$ laser which are used as a pump source for the SFR laser possess closely spaced lines which can be step-tuned to extend the tuning range of the SFR laser using a smaller magnetic field. In all of this, however, the InSb sample has to be kept at a temperature of $\gtrsim$ 10 K. The reason for the requirement of low temperature is two-fold. First, it is necessary to keep all the electrons in the conduction band in the lowest spin-state and avoid thermal excitation to the upper

spin-state because that would cause a lower gain. Second, lower
temperature represents a lower loss for the SFR scattered radiation
which is resonant inside the InSb sample. Because of the low
temperature requirement one has to use cryogenic liquids such as
liquid helium to assure temperatures in the vacinity of 10 K.
Recently we have used a closed cycle helium refrigerator to
provide the required cooling capacity. The regrigerator is a
two-stage closed cycle helium heat exchanger which is connected to
a compressor with flexible metal hoses. The only input to the
system is electrical power and no maintenance in terms of cryogenic
liquid service is required. The InSb sample SFR laser sample is
mounted inside a radiation shield which is attached to the first
stage of the refrigerator. The InSb sample is mounted on the
second stage which allows the temperatures in the region of 9 K
under no-load condition and approximately 11-13 K with the pump
laser power impinging upon the sample. The sample can be maintained
at these temperatures indefinitely with no operator control or
attention. The remaining part of the experimental experiment
includes a cw CO laser with an intracavity grating to obtain cw
pump power of approximately 3 W at a number transitions near 5.3μ.
We used the 1888.32 cm^{-1} line, the $P_{9-8}(12)$ transition of CO for
demonstrating the usefulness of the closed cycle cooler for
operating the SFR laser and for describing its operating
characteristics. The InSb sample was $2 \times 2 \times 8$ mm parallel piped
with 2 mm × 2 mm faces parallel and optically polished. The
electron concentration of the sample was 1×10^{15} cm^{-3}. The
sample was situated between the pole faces of a 4" Varian electromagnet. Pump radiation was focused into the 2 mm × 2mm face
forming the SFR laser cavity so that the SFR laser radiation was
resonated collinearly with the pump radiation and was propagating
perpendicular to the magnetic field. The pump radiation for this
experiment was polarized parallel to the magnetic field which
indicated that the SFR laser radiation should be polarized[6]
perpendicular to the magnetic field. The emerging SFR laser
radiation was analyzed with a 3/4 meter grating spectrometer and
detected with a Au:Ge (77K) photoconductor.

A spectrometer analysis of the output from the SFR laser at
three different magnetic fields is shown in Fig. 1. In Fig. 1 (a)
the trace for B=0 shows only the pump radiation coming out from
the exit face of the InSb sample. The emerging pump power is
approximately 2W. Figures 1 (b) and 1 (c) analyze the output at
magnetic fields of 2,280 and 2,876 G showing cw operation of the
anti-Stokes, I-Stokes, II-Stokes, III-Stokes, IV-Stokes SFR laser
as well as a small amount of unused pumped radiation coming from
the sample. There is a considerable depletion of the pump as the
magnetic field is turned on. It should be pointed out that in
going from Fig. 1 (a) to 1 (b) and 1 (c) no polarizers or filters
were introduced between spectrometer and the exit face of the InSb

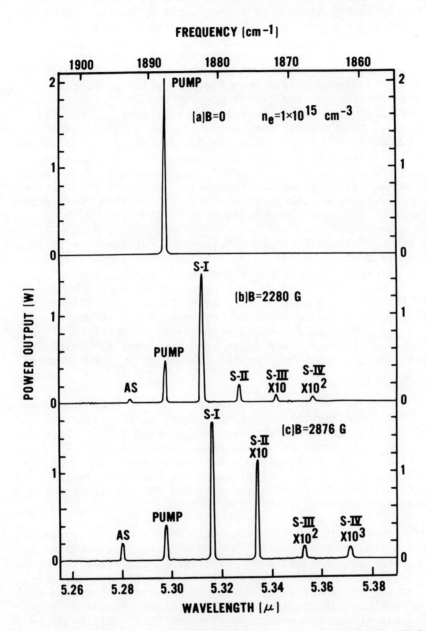

Figure 1   Spectrum of the low field SFR laser at magnetic fields of (a) 0 G, (b) 2280 G, and (c) 2786 G. The pump laser is the $P_{9-8}(12)$ transition of CO ($\sim$3 W).

sample, assuring us that the scale on the left hand side of Fig. 1
is correctly represented in cw power output in watts. About 80%
of the pump laser power is converted into the I-Stokes SFR laser
output. It was also seen that with very tight focusing of the
pump power into the sample, the conversion of pump laser power
into tunable SFR laser power was reduced. This reduction in the
conversion efficiency arises from the saturation process discussed
in earlier references[2] and has to do with the inability of the
electrons excited to the upper spin state to relax to the lowest
spin state so that another pump photon can be scattered into a
I-Stokes photon. This saturation is concentration dependent and
is seen to become more critical as one goes to lower electron
concentration InSb samples. However, as can be seen from Fig. 1
proper adjustment of focusing parameters for the pump radiation
into the InSb sample assures us that we still have a conversion
efficiency which exceeds about 80%.

In Fig. 1 (b) and 1 (c) we should point out that the IV-Stokes
SFR laser is being reported to operate under cw conditions for
the first time. This arises because of the high conversion
efficiency to the I-Stokes and subsequently to the II and III
Stokes SFR lasers. At magnetic field of 2,876 G the III-Stokes
SFR laser power emerging from the InSb sample allows us to
estimate the threshold for the IV-Stokes SFR laser to be ~15-20 mW.
This threshold for SFR laser is in good agreement with earlier
reports[18] ~5-10 mW obtained for the operation for the I-Stokes
SFR laser.

The tuning curves for the SFR laser reported here are shown
in Fig. 2. The disappearance of the IV-Stokes SFR laser at
1.75 kG and that of the III-Stokes SFR laser at 3.75 kG is
attributed to the decrease in the resonance enhancement factor
as the frequencies of the various SFR lasers move away from the
sharp bandgap edge in InSb with increasing magnetic field. A
conversion efficiency of 0.5% was obtained for the IV-Stokes
SFR laser at a magnetic field of ~570 G.

The operation of the SFR laser without any cryogenic liquids
and with a electromagnet makes this tunable laser important in
numerous practical applications. Because of the reliability
of the closed cycle coolers they need no operator control or
attention and after a preliminary cool down time the system will
operate continuously for long periods of time. At the present
time the aspects of the experimental setup which require great
deal of attention are the pump lasers. Sealed off CO_2 and CO
lasers have been devised and it is expected that with further
improvements in the pump lasers, the low field operation and
the closed cycle cooling will make the SFR laser reliable for
routine operation.

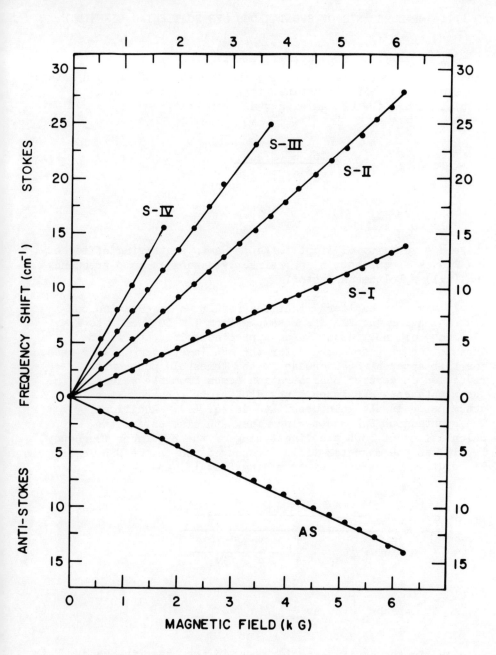

Figure 2 Tuning characteristics of the low field SFR laser for
I, II, III and IV Stokes as well as anti-Stokes components.

III.   MEASUREMENTS OF SPONTANEOUS SFR SCATTERING LINEWIDTH

The spontaneous SFR scattering linewidth, $\Delta\nu_{spont}$, is an important parameter in describing the operation of the SFR laser. The spontaneous SFR scattering linewidth determines in a crucial manner the SFR gain, g_s, as seen below[1]

$$g_s = \frac{16\pi^2 c^2 (S/\ell d\Omega)}{\hbar\omega_s^3 n_p^2 (\bar{n}+1)\Delta\nu_{spont}} \tag{2}$$

where c = velocity of light, $S/\ell d\Omega$ = Raman scattering efficiency, $\hbar$ = Planck's constant, n_p = refractive index at pump frequency and $(\bar{n}+1)$ = Boltzmann factor.

It was seen earlier[13] that the reduction in electron concentration for the SFR laser sample from 1×10^{16} cm^{-3} to 1×10^{15} cm^{-3} did not result in an appreciable increase in the threshold pump power needed for the SFR laser. One of the reasons for this apparent nonincrease in the threshold power with reduction in carrier concentration comes from the reduction in the spontaneous SFR laser linewidth which appears in the denominator in the gain described in Eq. (2). Additionally the spontaneous SFR scattering linewidth also enters the calculations of both the fine tuning of the SFR laser frequency, $d\nu_L/dB$, as the magnetic field is changed and in the SFR laser linewidth $\Delta\nu_L$, as seen by examining Eqs. (3) and (4).

$$\frac{d\nu_L}{dB} = \frac{\Delta\nu_c \dfrac{\partial\nu_s}{\partial B} - \dfrac{\nu_s \Delta\nu_{spont}}{n} \times \dfrac{\partial n(\nu_L,B)}{\partial B}}{\Delta\nu_c + \Delta\nu_{spont}\left[1 + \dfrac{\nu_L}{n}\dfrac{\partial n(\nu_L,B)}{\partial\nu_s}\right]} \tag{3}$$

$$\Delta\nu_L = 8\pi h\nu_L \left[(\Delta\nu_c \Delta\nu_{spont}/(\Delta\nu_c + \Delta\nu_{spont})\right]^2 P^{-1} \tag{4}$$

Here $\Delta\nu_c$ is the cavity linewidth, ν_L is the laser frequency, P is the power output, n is the refractive index and ν_s is the linecenter of the SFR gain curve. We see that with a decrease in the spontaneous Raman scattering linewidth the expected laser linewidth will be narrower. Thus for comparison between experiment and

theory it is important that a good measure of the spontaneous SFR
scattering linewidth be available. In the fine tuning characteristics
of the SFR laser as mentioned earlier[19] the change in frequency with
magnetic field is not continuous because of the existence of the SFR
laser cavity modes. Examination of Eq. (3) shows that the rate of
change of the frequency of the SFR laser with a change in magnetic
field depends upon a number of parameters including the cavity
linewidth and the spontaneous SFR linewidth. Equation (3) can be
simplified considerably if one assumes that $\Delta\nu_{spont}$ is very much
narrower than the cavity linewidth in which case the rate of change
of the SFR laser frequency very nearly equals the rate of change
of the frequency of the SFR gain line. The typical cavity linewidth
for SFR lasers without any high reflectivity coatings on the surfaces
is ~ 0.1-0.2 cm^{-1}. In the InSb SFR samples used in earlier experiments the carrier concentration was approximately 1×10^{16} cm^{-3}
where the spontaneous SFR linewidth was of ~ 0.2 cm^{-1}. Here it can
be seen from Eq. (3) that the tuning rate for the SFR laser frequency
is ~ 30-40% of the tuning rate for the spontaneous SFR gain line.
Thus there will be about 60% holes in the tuning characteristics or
in other words the SFR laser frequency will tune continuously only
over about 40% of the range between successive cavity resonances.
If, for example, the SFR spontaneous linewidth is 0.01 cm^{-1}, then
we can have a tuning rate which is approximately 95% of the SFR
gain tuning rate indicating that there will be at most 5%
inaccessible regions in tuning frequency for a SFR laser of a given
sample dimension.

Studies of spin resonance[20] in InSb having different electron
concentrations have shown that for electron concentration of
approximately 1×10^{16} cm^{-3} the spin resonance linewidth is
approximately 0.5 cm^{-1}. This linewidth reduces to ~ 0.15 cm^{-1}
when $n_e \sim 2 \times 10^{15}$ cm^{-3} and to about 0.05 cm^{-1} with
$n_e \approx 4 \times 10^{14}$ cm^{-3}. In the situation where all the electrons are
located in the lowest spin state, the spontaneous SFR scattering
linewidth will be very nearly equal to the electron spin resonance
linewidth. The spontaneous SFR scattering linewidth heretofore
has been measured by analyzing the spontaneous SFR scattered light
with a spectrometer. This, however, has its limitation in the
available resolution. It was mentioned earlier in the description
of low field SFR lasers that the reduction in carrier concentration
is expected to give us a reduction in the spontaneous SFR scattering
linewidth and as seen from the ESR data this should be true. In
the absence of other broadening mechanisms such as nonparabolicity
effects of the conduction band of InSb, the spontaneous SFR
scattering linewidth will be limited by the spin relaxation time
which is expected to be of the order of 10^{-9} sec. This implies a
linewidth of the order of 160 MHz which is clearly beyond the
resolution obtainable with conventional grating spectrometers.

Thus a new technique is required by which these narrow linewidths can be measured. In other words we have to somehow carry out high resolution spectroscopy of the InSb SFR laser sample itself. One simple technique of investigating this narrow linewidth is to use a second SFR laser which tunes its radiation through the gain curve of the first SFR laser sample. Thus, instead of measuring the spontaneous SFR scattered light and analyzing its subsequently we measure the SFR gain. From Eq. (2) it is seen that the linewidth associated with the SFR gain curve will be the linewidth associated with the spontaneous SFR scattering. Thus measurement of gain can unequivocally yield the spontaneous SFR scattering linewidth. The technique mentioned above for measuring SFR gain is, of course, unduly complicated since it requires two separate magnets and two separate cryogenically cooled InSb samples - one sample whose SFR gain curve is being measured and second sample which supplies the tunable laser radiation. There is an alternate technique for measuring the SFR laser linewidth that takes advantage of the fact that the CO laser is capable of generating a number of laser transitions separated by $\sim$4 cm^{-1}. Thus one alternative is to use two separate CO lasers - one for pumping the InSb sample to produce gain and the second one as a signal source whose amplification through the InSb sample is monitored. Instead of tuning the signal radiation past the gain curve we can tune the gain curve past the fixed frequency of the second CO laser. Thus we can see that the intensity variation of the second CO laser will represent the gain curve arising from the SFR process due to the first CO laser pump. Yet an easier technique utilizes the same CO laser for producing both the pump laser line as well as the signal laser line. This can be done in a manner similar to that used by us for measuring the SFR laser linewidth.[16] The CO laser is made to oscillate simultaneously on the $P_{9-8}(12)$ transition at 1888.32 cm^{-1} and on the $P_{10-9}(6)$ transition at 1885.74 cm^{-1}. The grating which controls the CO laser operation is adjusted so that the $P_{9-8}(12)$ transition is approximately 5 to 10 times stronger than the P_{10-9} (6) transition. The $P_{9-8}(12)$ transition acts as the pump while the $P_{10-9}(6)$ transition will be the signal whose amplification we will measure. The pump power emerging from the CO laser is polarized at 45° to the magnetic field thus both the pump radiation as well as the signal radiation contain components which are polarized parallel to the magnetic field and perpendicular to the magnetic field. (The polarization selection rules for SFR scattering and SFR gain are satisfied[6] for this geometry.) The intensity of the pump radiation at 1888.32 cm^{-1} is adjusted so that no SFR laser oscillation takes place. A spectrometer is set to look at $P_{10-9}(6)$ transition. When the magnetic field is tuned to $\sim$1,116G the SFR gain occuring from the $P_{9-8}(12)$ transition sweeps past the $P_{10-9}(6)$ transition which is being monitored. Figure 3 shows amplification trace of $P_{10-9}(6)$ function of magnetic field. The pump laser radiation ($P_{9-8}(12)$) was so adjusted that

a peak amplification of a factor of ~2 occurred. The half height
width of the gain curve is seen to be ~5 G which yields as
spontaneous SFR scattering linewidth of the order of 350 MHz. The
lineshape seen does not appear to be a Lorentzian indicating in
that the linewidth seen in Figure 3 is not limited by spin
relaxation time.

This technique gives us a very powerful tool for doing
spectroscopy of an InSb SFR scattering sample itself. We have made
preliminary measurements of the variation of the SFR spontaneous
scattering linewidth as a function of temperature. Figure 4 shows
a series of traces of the gain measured at five different temperatures.
The linewidth is seen from Fig. 5 to change as the temperature is
increased and it goes through minimum. It is known that at higher
temperature, the spontaneous SFR scattering linewidth can be
considerably wider. However, it is puzzling that the linewidth
does not monotonically go to some small value as the temperature
is reduced but goes to a pronounced minimum. Another aspect of
the gain curves shown in Fig. 4 is that there is a shift in the
position of the SFR gain curve as the temperature is changed. It
is seen that the effective g-value plotted in Fig. 5 is becoming
smaller as we go to higher temperatures. These results are free
from the cavity mode effects which would be present if we were

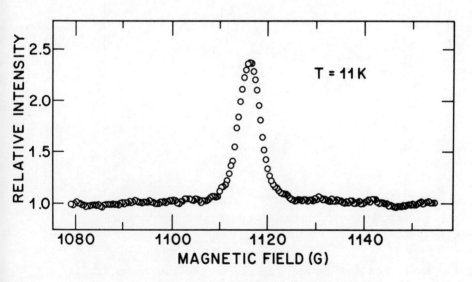

Figure 3 Amplification for $P_{10-9}(6)$ at 1885.74 cm^{-1} due to Stokes
SFR gain from pump at 1888.32 cm^{-1} [$P_{9-8}(12)$] as a function of
magnetic field which tunes the SFR gain curve through the $P_{10-9}(6)$
transition.

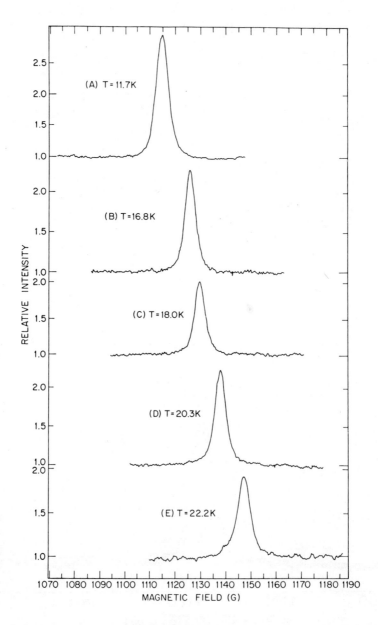

Figure 4 Amplification curves similar to the one shown in Figure 3 at (A) T = 11.7 K, (B) T = 16.8 K, (C) 18 K, (D) 20.3 K, and (E) 22.2K.

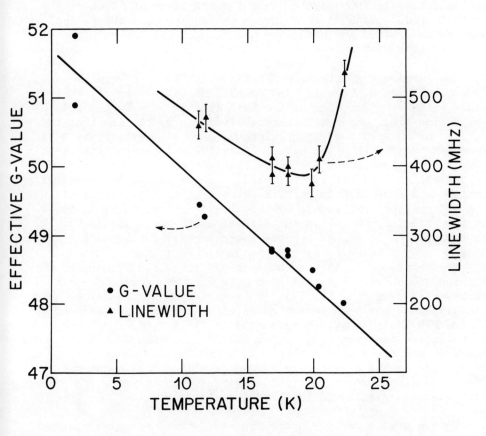

Figure 5   Effective g-value and spontaneous SFR linewidth as a
function of temperature for 1×10^{15} cm^{-3} InSb sample.

measuring the frequency of the SFR laser. Thus we are measuring
the accurate variation of the g-value of the electron as the
temperature is changed. We know that the g-value of electrons in
InSb at room temperature is of the order of -70. Instead of the
g-value increasing monotonically from about -51 to -70 from 4.2°K
to room temperature, it appears to become smaller at first and
then increase subsequently. (This apparent reduction in g-value
as the temperature is increased was noticed[21] earlier in higher
concentration in InSb samples when the SFR laser at 10.6μ was
investigated.) Both of this phenomona require more theoretical
work to understand the shape and position of the SFR gain curve
as a function of temperature.

The narrowest spontaneous SFR linewidth that we have measured
from the SFR gain curve is ∿3 G or ∿210 MHz (0.007 cm^{-1}). This is
still somewhat wider than a linewidth limited only by the spin
relaxation time. From this narrow linewidth we can see that the
earlier measurements of spin resonance gave a linewidth which was
somewhat broader than the actual linewidth. A consequence of this
very narrow spontaneous SFR linewidth is that in the lower carrier
concentration InSb samples the gain is not going to be necessarily
smaller. A reduction in carrier concentration by a factor of 10
has reduced the spontaneous SFR scattering linewidth by a factor
of approximately 20. Thus the effective SFR gain seen in Eq. (2)
is expected to be higher for the lower carrier concentration sample.

The reduction in the spontaneous SFR linewidth with reduced
carrier concentration should also make the SFR laser linewidth
narrower as seen from Eq. (4). This variation in SFR laser
linewidth has not been investigated since the only measurements
of the SFR laser linewidth were made in the 1 × 10^{15} cm^{-3} sample
which is the same sample in which the SFR spontaneous SFR linewidth
measurements are reported here. It might be worthwhile perhaps
to go back to the higher carrier concentration sample and measure
the linewidth there to see if the increased spontaneous SFR linewidth becomes visible in the SFR laser linewidth.

We can also clearly see the impact of these extremely narrow
spontaneous SFR linewidths on the fine tuning characteristics of
the SFR laser. For the 1 × 10^{16} cm^{-3} sample Brueck and Mooradian[19]
had reported a fine tuning curve which showed large gaps due to
mode jumping as the magnetic field was tuned. In the present case
with the 1.1 × 10^{15} cm^{-3} sample which has a spontaneous SFR
linewidth ∿0.007 cm^{-1} Eq. (3) shows that the rate of tuning of the
SFR laser frequency should be approximately 98% of the value
g$μ_B$B or in other words 98% of the rate at which the SFR gain curve
changes. There will be only about 2% holes for a given dimension
of the SFR laser cavity. Thus the lower carrier concentration

sample used for the SFR lasers now has the advantage of lower SFR laser threshold, smaller SFR laser linewidth and considerably extended fine tuning characteristics.

Another consequence of the very narrow spontaneous SFR linewidth is to be seen by examining the expected linewidth for a superradiant (i.e., nonresonant) SFR laser,[2] given below:

a) for Lorentzian spontaneous SFR gain curve

$$\Delta\nu_{sup} = \Delta\nu_{spont}/[(\alpha\ell/\ln 2)-1]^{1/2} \tag{5}$$

b) for Gaussian spontaneous SFR gain curve

$$\Delta\nu_{sup} = \frac{\Delta\nu_{spont}}{(\ln 2)^{1/2}}\left[\ln\left(\frac{\alpha\ell}{\alpha\ell-\ln 2}\right)\right]^{1/2} \tag{6}$$

where $\Delta\nu_{sup}$ is the superradiant SFR laser linewidth and $\alpha\ell$ is the total SFR gain. For the presently measured $\Delta\nu_{spont} \approx 210$ MHz, we can see that for an SFR laser (nonresonant) output of ∿1 W, (i.e., taking $\alpha\ell \approx 30\text{-}50$ nepers as described in Ref. 2), the linewidth, $\Delta\nu_{sup}$, of the output radiation will be ∿25 MHz (∿8×10^{-4} cm^{-1}) for the case of Lorentzian SFR gain curve and ∿29 MHz (∿10^{-3} cm^{-1}) for Gaussian SFR curve. While these are extremely narrow output linewidths, they clearly do not compare well with the resonant SFR laser linewidth[16] of <<1 kHz. However, with no feedback and only the superradiant SFR laser emission, the fine tuning of the superradiant SFR laser frequency is given directly by the tuning of the SFR gain curve [see Eq. (1)]. There are no cavity mode effects as described in Eq. (3) which will give mode hopping. This continuous tunability with <10^{-3} cm^{-1} linewidth and power output in excess of 1 W should be of immense technical importance in various applications of SFR lasers where the ∿1 Hz linewidth available for resonant SFR lasers[16] is not needed. (It should be pointed out that the nonresonant SFR laser now appears to be a practical tool because of the narrow spontaneous SFR linewidths available from the low concentration InSb laser [13] samples.)

IV. ADVANCES IN LOW LEVEL NITRIC OXIDE DETECTION

Earlier we had used the SFR laser in conjunction with an acousto-optic detection scheme for monitoring low level pollutant concentration of NO in air.[22] The earlier experiments involved the use of a CO laser pumped SFR laser operated with a super-conducting magnet. The opto-acoustic detection technique is a

technique which is well suited to measuring low concentrations
of absorbing species of molecules because it is essentially a
calorimetric technique. Using conventional techniques for
measuring absorption which require monitoring a change in the
transmitted intensity through an absorption cell the experimental
techniques put a lower limit on the measured absorption to $\sim 1:10^5$.
By using modulation techniques this limit can be decreased to
$\sim 1:10^8$. We have used modulation spectroscopy techniques[4] which
involve the use of a small air cored magnetic coil in addition
to the primarily magnetic field which tunes the SFR laser. This
additional coil is excited with an alternating source of current
which gives rise to a time dependent frequency shift of the SFR
laser. The time dependent magnetic field provided by the modulation
coil is usually kept at less than 1 G (the tuning rate of the
SFR laser is approximately 70 MHz/G). We will not go into the
details of the modulation spectroscopy and our results obtained
on measuring small absorption. It was found that the modulation
technique provides us with an extremely useful application of the
SFR laser for low level measurement of absorptions.[4]

     The opto-acoustic technique, since it is a calorimetric
technique, is limited by three factors in its capability for
detection of small absorptions. The first is absorption in the
transmitting window through which the laser radiation enters the
opto-acoustic absorption cell. Second, the amount of light
which is intercepted by the walls of the absorption cell and
thirdly, by the amount of tunable laser power available. The
capacitance microphone used in our experimental measurements
has shown that we can detect $\sim 10^{-8}$-10^{-9}W of absorbed power in
the gas in the opto-acoustic cell. Thus, to be able to measure
absorption of $\sim 1:10^8$ we need a SFR laser power of 1 W. Our
earlier attempts to measure small quantities of NO were limited
by the amount of SFR laser power that we could generate. With
a SFR laser power of ~ 10-100 mW, we had shown a capability of
detecting 10 ppB of NO in air at atmospheric pressure. As seen
from Section II, we have now improved our SFR laser to produce
SFR power output in excess of 1W. This would imply that we
should see a considerable improvement in our detection capability
of NO. We have carried out experiments with lower concentrations
of NO in nitrogen to see the limits of the NO detection capability
using the low field SFR laser and the opto-acoustic setup.
Figure 6 shows the results for acousto-optic signals obtained
from 100 ppB samples of NO in nitrogen at a total pressure of
approximately 24 Torr. We see a signal-to-noise ratio in excess
of 100. The two major lines seen at magnetic fields of ~ 2050 and
~ 2110 G correspond to the abosrption lines[4] of NO for $\Omega = 1/2$ and
$\Omega = 3/2$ for m = 4.5. (The line at 2120 G is approximately twice
as strong as the line at 2050 G because the former line consists
of a doublet due to Λ splitting which has been resolved earlier

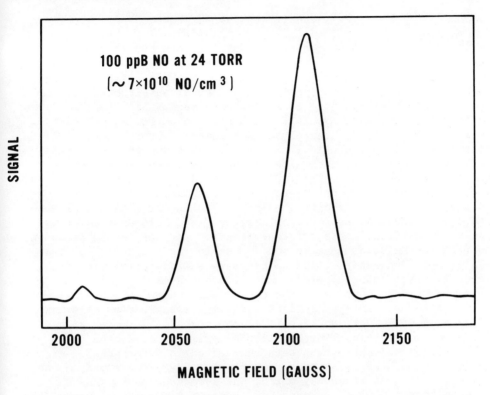

Figure 6 Absorption spectrum of 100 ppB NO in 24 Torr of N_2 (NO concentration is 7×10^{10} molecules/cm^3).

using the SFR laser and reported in Reference 4.) Thus it is seen that we can easily detect a NO concentration of 7×10^8 molecules cm^{-3}. It should be pointed out that the results in Fig. 6 are obtained with a measurement time constant of approximately 1 sec. Increasing the time constant will result in an improved detection capability of NO. Earlier we had applied this technique to NO pollution measurements.[22] The improvement in detection capability since that time has considerably strengthened the conviction that SFR lasers could be a very useful tools in detection of atmospheric pollutants. It should be mentioned also that the low level NO detection capability shown in the present paper points to a very tantalizing possibility of measuring NO concentration in stratosphere (at altitude between 18 kM and 30 kM) where the expected NO concentration lies in the region of 1-10 ppB at the ambient pressure which varies from ∿20 Torr to ∿80 Torr. This corresponds to a NO concentration in the region of 10^9 to 10^{11} molecules cm^{-3}.

V.   NONLINEAR OPTICS USING SFR LASERS

It can be seen from earlier discussion that the SFR laser has one very serious drawback and that is the limited tuning range. The present tuning range is ∿5μ to 6.5μ and ∿9μ to 14.6μ. While it is true that over this particular range of wavelengths SFR laser is an excellent source of coherent, narrow band tunable radiation, large gaps are left which need to be filled in order to increase the usefulness of the SFR laser. The SFR laser in both the modes of operation, namely, CO laser pumping as well as CO_2 laser pumping produces sizeable amounts of cw as well as pulsed power outputs respectively. This indicates that one can use further nonlinear optics to generate tunable radiation at wavelengths not produced by the SFR laser itself. This involves the use of one SFR laser and/or another fixed frequency laser to generate sum or difference frequencies (or perhaps second harmonic) to fill in the voids that we see in the tuning capability of the SFR laser. Nonlinear optics to generate additional wavelengths severely bites into the efficiency of conversion of radiation from the primary sources to the needed wavelength region, however, it is felt that for applications involving spectroscopy this low conversion efficiency would not be detrimental. A number of different possible nonlinear optics experiments have been contemplated and these have been summarized in Reference 3. Table 1 shows a set of typical experiments involving sum frequency mixing, different frequency mixing and second harmonic generation using either a CO laser pumped SFR laser or a CO_2 laser pumped SFR laser together with a CO or a CO_2 laser. We can see that with nonlinear optics we can now cover a wavelength range from about 2.5μ to ∿1 mm.

Some of the experiments listed in Table 1 have already been carried out and these include the generation of sum frequencies[23] using a pulsed SFR laser at 10.6μ and a pulsed CO_2 laser at 10.6μ, different frequency generation[3] for obtaining radiation near 15μ using a cw operating SFR laser pumped with a CO laser together with a CO_2 laser operating at 9.6μ, and tunable different frequency generation in the 100μ range[17] by using a CO_2 laser pumped SFR laser together woth a CO_2 laser radiation itself. The last experiment appears to be extremely exciting because it affords us a tunable source of radiation in the hitherto difficult region of spectroscopy in the submillimeter range region. The experimental setup used for the 100μ tunable radiation generation includes a SFR laser sample consisting of InSb pumped with a CO_2 laser at 10.6μ. The radiation coming out through the sample includes the ⁺ ⁻ble I-Stokes SFR laser and unused pump radiation at 10.6μ. ⁻diation then enters into another InSb sample in which ⁻e frequency mixing takes place between the SFR laser ⁺ $\omega_s = \omega_o - g\mu B$ and the unused pump radiation at frequency

Table I   A partial list of Nonlinear Optics Experiments to extend
the wavelength range of tunable radiation available from the SFR
laser.

FUTURE (TOGETHER WITH NONLINEAR OPTICS)

PRIMARY SOURCES		RESULTANT TUNABILITY
TUNABLE SOURCE	**FIXED FREQUENCY SOURCE**	
SFR CO_2 9.0μ - 14.6μ		SHG 4.5μ - 7.3μ
SFR CO 5.3μ - 6.5μ		SHG 2.65μ - 3.25μ
SFR CO_2 9.0μ - 14.6μ	CO_2 9.2μ - 10.8μ	DIFF: 30μ - 1000μ
SFR CO_2 9.0μ - 14.6μ	CO 5.0μ - 6.7μ	SUM: 3.2μ - 4.6μ
SFR CO 5.3μ - 6.5μ	CO_2 9.2μ - 10.8μ	DIFF: 3.56μ - 22.8μ
SFR CO_2 9.0μ - 14.6μ	H_2O 27.974μ	DIFF: 13.3μ - 30.5μ SUM: 6.76μ - 9.6μ

ω_o due to the second order nonlinearity of the electron spins[24]
themselves. Since this nonlinearity is resonant at the same
magnetic field at which the SFR laser operates for a given difference
frequency generation at $\omega_{dif} = \omega_o - \omega_s = g\mu_B B$, both the samples of
InSb, the SFR laser and the difference frequency generation sample
are kept in the same magnetic field. The radiation emerging
from the second InSb sample contains the different frequency
radiation at $\omega_{dif} = g\mu_B B$. By changing the magnetic field ω_{diff}
is tuned. A preliminary experiment gave tunability from ∼90 microns
to ∼100 microns. (A detailed report of different frequency
generation experiments is being presented at this conference by
T. J. Bridges and V. T. Nguyen in the following paper.) Spectroscopy
of CO has also been carried out in the 100μ region using ω_{diff}
generated above. Without going through details it is fair to
say that the techniques of nonlinear optics are going to be
extremely powerful in extending the tunability region of the SFR
laser and will allow spectroscopy and other experiments in
wavelength regions not directly accessible with the SFR laser.

VI. CONCLUSIONS

In this paper we have very briefly summarized some of the recent experimental results delineating the advances in SFR laser operation and their applications. We have seen that the field of SFR lasers is maturing and more and more new applications are becoming apparent. Still, however, a significant factor remains and that has to do with the fact that while the SFR laser in InSb has been pumped at a number of different wavelengths covering the 5μ region and the 10μ region, as yet no other materials have been found to produce SFR laser action. A number of obvious choices exist and these should allow us to extend the ranges of tunable radiation available from the SFR laser. These materials include $Pb_xSn_{1-x}Te$ and $Hg_xCd_{1-x}Te$ whose bandgaps can be tailored by changing the relative concentrations of Pb and Sn in PbSnTe and of Hg and Cd in HgCdTe. This tailoring of bandgaps should result in cw SFR lasers at newer pump frequencies.

REFERENCES

1. C. K. N. Patel and E. D. Shaw, Phys. Rev. Letters $\underline{24}$, 451 (1970).

2. C. K. N. Patel and E. D. Shaw, Phys. Rev. $\underline{B3}$, 1279 (1971).

3. C. K. N. Patel in Proceedings of the Symposium on Laser Physics and Applications, Esfahan, Sept. 1971 (Wiley, New York, to be published).

4. C. K. N. Patel in "Coherence and Quantum Optics" Ed. L. Mandel and E. Wolf (Plenum Press, 1973) pp. 567-593 and references cited therein.

5. C. S. DeSilets and C. K. N. Patel, Appl. Phys. Letters, 543 (1973).

6. Y. Yafet, Phys. Rev. $\underline{152}$, 858 (1966).

7. P. A. Wolff, Phys. Rev. Letters $\underline{16}$, 225 (1966).

8. R. E. Slusher, C. K. N. Patel and P. A. Fleury, Phys. Rev. Letters $\underline{18}$, 77 (1967).

9. E. D. Shaw and C. K. N. Patel, Appl. Phys. Letters $\underline{18}$, 215 (1971).

10. R. L. Allwood, S. D. Devine, R. G. Mellish, S. D. Smith and R. A. Wood, J. Phys. $\underline{C3}$, L186, 1970.

11. R. B. Aggarwal, B. Lax, C. E. Chase, C. R. Pidgeon, D. Limpert and F. Brown, Appl. Phys. Letters 18, 383 (1971).

12. A. Mooradian, S. R. J. Brueck and F. A. Blum, Appl. Phys. Letters 17, 481 (1970).

13. C. K. N. Patel, Appl. Phys. Letters 19, 400 (1971).

14. C. K. N. Patel, Appl. Phys. Letters 18, 274 (1971).

15. C. K. N. Patel in "Advances in Lasers" Ed. A. K. Levine (M. Dekker, New York, 1969) pp. 1-183.

16. C. K. N. Patel, Phys. Rev. Letters 28, 649 (1972).

17. V. T. Nguyen and T. J. Bridges, Appl. Phys. Letters (to be published) see also proceedings of this conference.

18. A. Mooradian in Proceedings of the Symposium on Laser Physics and Applications, Ed. A. Javan and M. Feld (Wiley, New York , in Press).

19. S. R. J. Brueck and A. Mooradian, Appl. Phys. Letters 18, 229 (1971).

20. B. D. McCombe and R. J. Wagner, Phys. Rev. B4, 1285 (1971); B. D. McCombe (Private Communication).

21. E. D. Shaw and C. K. N. Patel (unpublished results).

22. L. B. Kreuzer and C. K. N. Patel, Science 173, 45 (1971).

23. C. R. Pidgeon, B. Lax, R. L. Aggarwal and C. E. Chase, Appl. Phys. Letters 19, 333 (1971).

24. V. T. Nguyen and T. J. Bridges, Phys. Rev. Letters 29, 359 (1972); T. A. Brown and P. A. Wolff, ibid, 362 (1972).

HIGH RESOLUTION INFRARED SPECTROSCOPY USING TUNABLE SEMICONDUCTOR LASERS*

F. A. Blum† and K. W. Nill

Lincoln Laboratory, Massachusetts Institute of Technology

Lexington, Massachusetts 02173

The application of cw, continuously tunable lead-salt diode lasers to ultra-high resolution infrared spectroscopy has been significantly extended to shorter wavelengths (3.5 μm to 7 μm). Rotational fine structure including Λ-type doubling, nuclear hyperfine splitting and Zeeman splitting have been fully resolved for the first time in infrared molecular spectra. Spectra of some atmospheric water vapor lines have shown anomalous, narrow linewidths, and the gain lineshape of a CO gas laser has been measured.

The potential of the application of cw, tunable semiconductor diode lasers to infrared molecular spectroscopy was first demonstrated by Hinkley[1] in 1970. Subsequent[2-5] work concentrated on the use of PbSnTe lasers in the 8 to 11 μm spectral region to study molecules such as NH_3, SO_2 and SF_6. Over the past two years we have attempted to expand the application of tunable lead-salt lasers to cover shorter wavelengths and some of the simple diatomic molecules.[6-14] Table I lists the gases studied, the effects observed, and the lasers used in our work which has covered the 3.5 μm to 7 μm region.

---

*This work was sponsored by the U.S. Department of the Air Force.

†Present Address: Central Research Laboratories, Texas Instruments Incorporated, Dallas, Texas 75222

Table I. High Resolution Molecular Spectroscopy (3.5 μm to 7 μm)

Molecule	Laser	Wavelength (μm)	Effects Observed (Reference)
CO	$PbS_{0.82}Se_{0.18}$	4.7	Linewidths and Strengths (6), Isotope Shifts (A)
	$PbS_{0.6}Se_{0.4}$	5.3	Laser Gain Lineshape (8)
NO	$PbS_{0.6}Se_{0.4}$	5.3	Zeeman, Λ-type Doubling, Hyperfine Structure, Isotope Shifts, Magnetic Rotation (7, 9, 12, 13)
H_2O	$PbS_{0.6}Se_{0.4}$	5.3	Line Strengths, Anomalous Linewidths (11)
CO_2	PbS	4.2	Isotopic Lines, Atmospheric Lines (A)
O_3	$PbS_{0.8}Se_{0.2}$	4.7	(A)
CH_4	PbTe	6.5	(10, A)
NO_2	PbTe	6.2-6.5	(10, A)
CS_2	$PbS_{0.8}Se_{0.2}$	4.6	(A)
H_2CO	$Pb_{0.98}Cd_{0.02}S$	3.6	(14)
HCl	$Pb_{0.98}Cd_{0.02}S$	3.6	(14, A)

A Unpublished results, most of which are discussed in Solid State Research Reports, Lincoln Laboratory, MIT, 1971-72.

Additional work in this wavelength range has been done by Eng, et al.,[15] who studied H_2O and Antcliffe, et al.,[16] who studied NO. In our work several fundamental effects have been fully resolved for the first time in infrared molecular spectra. These include Λ-type doubling,[7] Zeeman splitting,[7] nuclear hyperfine splitting,[9] and magnetic rotation spectra.[12] We have also studied[8] the gain line-shape of a CO molecular laser and observed[11] surprisingly narrow linewidths for some lines in the ν_2 band of water vapor in ambient air. Most of our measurements were made with $PbS_{1-x}Se_x$ diode lasers which can be compositionally tuned between[14] 4 μm and 8 μm. However, PbTe lasers[10] and $Pb_{1-x}Cd_xS$ lasers[14] have also been used. The use of cw $Pb_{0.98}Cd_{0.02}S$ lasers operating near 3.6 μm is particularly significant since this wavelength approaches that of the C-H stretch vibration of the hydrocarbons.

    The lead-salt diode lasers were fabricated from both vapor grown and Bridgman grown single crystals whose nominal composition was chosen to yield emission at the wavelength of interest. Diffused p-n junctions were formed by type conversion via deviation from stoichiometry. Fabry-Perot laser cavities were formed by cleavage along $\langle 100 \rangle$ planes yielding devices nominally 0.2 x 0.2 x 0.5 mm. Low resistance contacts were usually formed by evaporation of gold. The lasers were mounted on the cold finger of liquid helium dewars at temperatures of 10 to 15 K (at low bias currents). Threshold currents for cw operation were generally in the range of 100 to 300 mA ($J_{th} \sim 100$ A/cm^2) and cw operation could often be achieved at currents ten times that at threshold. The laser output typically consisted of 3 to 5 longitudinal modes with a power output per mode ranging from 10 μw to 1 mw. With these power levels the estimated linewidths[17] were between 10 kHz and 1 MHz. However, our effective resolution was limited to $\sim$ 1 MHz due to power supply ripple. For the spectroscopic measurements the output of the lasers was usually collected and collimated with an infrared lens (BaF_2), passed through a 5 to 10 cm long gas cell and a grating spectrometer for mode isolation and absolute calibration, and detected with an InSb or Ge:Cu cryogenically cooled photodetector.

    Tuning of the lasers was accomplished by compositional selection in the ternary alloys, temperature adjustment (bias-current induced joule heating and external heaters) and variation of applied magnetic fields.[10] Alloy composition selection gave the nominal laser emission wavelength. Applied magnetic fields were used to grossly tune the laser to wavelengths not ordinarily covered by a given alloy composition. Finally, bias-current (temperature) tuning allowed continuous fine tuning of individual modes over a range of 1 to 2 cm^{-1} (30 to 60 GHz). It is this continuous fine tuning and the lasers' narrow linewidths which permit one to obtain infrared molecular spectra with unprecendented resolution.

Gross, noncontinuous tuning bandwidths of 50 to 100 cm^{-1} have been obtained using applied magnetic fields and/or bias current tuning. Figure 1 shows the temperature and magnetic field tuning curves of a $PbS_{0.82}Se_{0.18}$ diode laser. The points shown are the approximate centers of the observed mode patterns.

The magnetic field tuning curve was taken with laser operating in a cold finger liquid-helium dewar at a temperature of about 10 K in the bore of a 100 kG superconducting magnet. The wavenumber range shown covers about a dozen lines of the CO fundamental band centered at 4.7 μm. We have used gross magnetic field adjustment and fine current tuning to obtain spectra of many of these lines. Due to the multimode nature of the laser output it is generally possible to achieve coincidence, mode hopping notwithstanding, via the dual adjustment of magnetic field and current. The temperature

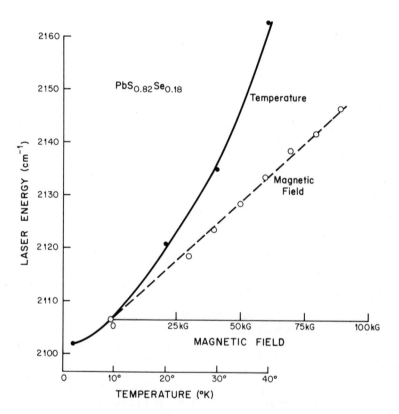

Figure 1. Temperature and magnetic field tuning of a $PbS_{0.82}Se_{0.18}$ diode laser. The points locate the approximate center of the mode patterns for each set of conditions.

tuning curve of Figure 1 was actually taken with the laser operating on a pulsed basis in an immersion variable temperature dewar.
However, on selected lasers we have observed gross current tuning
bandwidths as high as 90 cm^{-1} ($\sim$ 5%) in cw operation. For such
lasers we estimate the effective junction temperature to be $\gtrsim$ 55 K.
Such results suggest that operation at liquid nitrogen temperatures
may not be so far off.

The calibration of tunable infrared lasers remains a persistent
problem. Grating spectrometers were used for absolute wavelength
calibration with an accuracy of about 0.02%. This accuracy is not
particularly good when one is considering molecular fine structure;
however, it is often sufficient to facilitate line identification.
The inability to calibrate the wavelength with high accuracy leads
one to frequency calibration techniques which have much higher
precision, but poor absolute accuracy. We have used air-spaced,
confocal Fabry-Perot cavities and polished solid germanium etalons
for this purpose. Using a confocal Fabry-Perot with a finesse of
about 50 and a free spectral range of 3 GHz, we have calibrated the
relative frequency of individual modes with an accuracy of $\pm$ 2.5%.
However, commercial Fabry-Perots are difficult to keep aligned and
it is awkward to couple laser radiation through cavities with such
a high finesse. We have found that the Ge etalons with a much lower
finesse and a free spectral range of 2 GHz are much simpler to use
and yield accuracies approaching $\pm$ 5%. We have also used the
known width of isolated, Doppler broadened molecular lines for
frequency calibration. This technique is particularly simple and
convenient since coupling to a resonant cavity is not involved.
Bias current tuning rates measured using these techniques ranged
from 20 to 1000 MHz/mA with 300 MHz/mA being a median value. This
wide variation in tuning rates is not surprising since the rate
is very sensitive to diode mounting and thermal environment.
Detailed studies of the variation in the tuning rate itself were
not made. However, they were usually found to vary by an amount
less than the error in the rate over the range of interest in the
measurements (typically < 1 GHz). Magnetic field fine tuning rates
were found to be in the range 0.2 to 1 MHz/G, but this technique
for fine tuning was not often used.[10]

The fundamental properties of an isolated spectral line
are its location (center wavelength), intensity, and lineshape.
Present tunable laser technology has not affected the status of
absolute calibration techniques. As discussed above, we still use
conventional techniques for determination of line locations. On
the other hand, a line's intensity and lineshape (and width) can be
directly and accurately measured with a tunable laser. Even the
best grating spectrometers cannot directly measure these quantities
at low gas pressures. Although we made no careful intensity mea-

surements (using ratio techniques, for example, our measured strengths have usually agreed within ± 20% with calculated intensities or those determined by indirect techniques. At low pressures, collisions are unimportant and the lineshape is Gaussian[18] with a Doppler width of

$$\Delta w \approx \frac{214}{\lambda} \sqrt{\frac{T}{MW}} \quad MHz, \tag{1}$$

where λ is the wavelength in μm, T is the gas temperature in Kelvin, and MW is the molecular weight of the molecule in atomic units. Middle infrared Doppler widths range between 50 and 200 MHz. We have generally found close agreement between our measured and cal-culated Doppler widths, although no detailed studies of the Gaussian lineshape have been made. At higher gas pressures, collisions become dominant and the lineshape is Lorentzian. We have measured[11] the linewidths of several lines in the 6.3 μm $\vee$ band of H_2O in the ambient atmosphere. Some of these lines showed anomolously narrow widths which were much smaller (one by a factor of four) than those calculated using accepted line broadening theory. Further studies of these and other water vapor lines have been made by Eng, et al.[15] We have also measured the gain lineshape of a CO molecular laser.[8] The measured line center gains were in reasonable agreement with those measured using a matched gas laser, amplifier pair. By measuring the gain linewidth we estimated the gas temperature in the bore center to be 170 K as compared to the wall temperature of 77 K.

Two types of rotational fine structure which we have observed in our studies[7,9,12,13] of NO are shown in Figure 2(a). This is an absorption spectrum[9] of the first line in the Q branch of NO at 1876.1 cm^{-1}: $Q(1/2)_{1/2}$. These data were taken using a PbS$_{0.6}$Se$_{0.4}$ diode laser. The gross splitting into symmetric pairs of lines is due to Λ-type doubling. This Λ-doubling results from rotational uncoupling of the electronic orbital angular momentum in the $2\pi_{1/2}$ states from the internuclear axis. The smaller splitting in the symmetric pairs results from magnetic dipole coupling to the spin $(I = 1)$ of the N^{14} nucleus of $N^{14}O^{16}$. The transitions involved are illustrated schematically on the right of Figure 2 for one of the Λ-doublet pairs. Here, $F = J + I$. A calculated[9] absorption spectrum is shown in Figure 2(b). There are a total of eight hyperfine lines whose relative strengths and positions are given by the vertical lines. The overall agreement between the measured and calculated spectra is excellent.

We have also studied[7] Zeeman splitting in NO. The measured Zeeman line positions (open circles) of the $R(15/2)_{3/2}$ line of NO are shown in Figure 3. The data were taken with a $PbS_{0.6}Se_{0.4}$ laser. There are 14 major Zeeman components due to the high J's involved and the strong dependence of g_J on J for $^2\pi_{3/2}$ states.

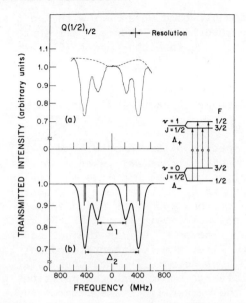

Figure 2.   Absorption spectrum of the $Q(1/2)_{1/2}$ line of NO
illustrating Λ-type doubling and nuclear hyperfine splitting: (a)
experimental and (b) theoretical.   The inset shows the energy
levels and associated transitions for one Λ-doublet pair.   (See
Ref. 9).

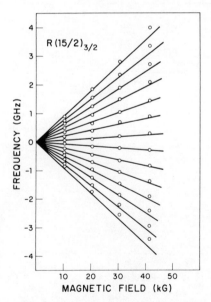

Figure 3.   For the $R(13/2)_{3/2}$ line of NO: Zeeman line frequencies
vs. magnetic field given by experiment (open circles) and linear
Zeeman theory (solid lines).

The data are in reasonably good agreement with the calculated solid lines (linear Zeeman theory), but show a small systematic deviation at high fields, perhaps due to higher order (B^2, etc.) Zeeman terms. In a related experiment we obtained[13] the magnetic rotation spectrum of R(13/2)$_{1/2}$ which is determined by both the circular birefringence and dichroism of NO in a magnetic field. We were able to observe the true complex lineshape characteristic of this phenomena for the first time.

Figure 4 shows spectra of CO$_2$ near 4.3 μm taken with a current tuned PbS laser. The upper trace was taken with the laser radiation propagated over an atmospheric path of ~ 4 m. The lower trace is the same path interrupted by a 10 cm long cell containing 1 Torr of CO$_2$. The low pressure lines appear as sharp spikes on top of the broad atmospheric lines. The main line near the center of the spectrum is the P(35) line of $C^{12}O^{16}$ in the hot band $[(01^11) \leftarrow (01^10)]$.[19] The other lines are fundamental and hot band lines of both $C^{12}O^{16}$ and $C^{13}O_2^{16}$ as noted. The measured relative line positions are in reasonable agreement with the positions given by Oberly, et al.[19] Note that the weaker lines are completely washed out by atmospheric broadening. The laser tuning rate for these data was measured using a Ge etalon and is estimated to be accurate and constant to ± 5% over the frequency range of Fig. 4.

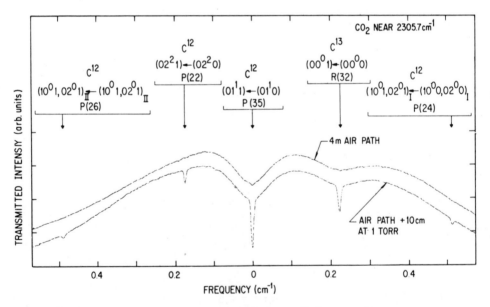

Figure 4.   Absorption spectra of CO$_2$ near 4.3 μm.   Upper trace: atmospheric path of ~ 4 m.   Lower trace: same path interrupted by a 10 cm cell containing 1 Torr CO$_2$.

In summary, ultra-high resolution spectra of a number of molecules have been obtained for the first time in the 3 to 7 μm wavelength range using cw, tunable lead-salt lasers. Both Doppler-
limited and atmospheric broadened lines have been observed.
Rotational fine structure not previously observed in infrared molecular spectra has been fully resolved. Tuning bandwidths of 5% at
5 μm have been achieved using temperature and/or magnetic field
tuning. Useful extensions of this work include cw operation at
liquid nitrogen temperature to achieve wider cw tuning bandwidths,
and cw operation at shorter wavelengths so that spectra of the
hydrocarbon C-H stretch bands at 3.0-3.5 μm can be obtained.

We are indebted to W. E. DeFeo and R. E. Reeder for the helpful technical assistance. We are also grateful to A. J. Strauss
and T. C. Harman for growth of the crystals from which the lasers
were made, and to A. R. Calawa for fabrication of the $PbS_{1-x}Se_x$
lasers used in the early portions of this work.

References

1. E. D. Hinkley, Appl. Phys. Letters 16, 351 (1970).
2. E. D. Hinkley and P. L. Kelley, Science 171, 635 (1971); E. D.
    Hinkley, J. Opto-electronics 4, 69 (1972).
3. E. D. Hinkley, A. R. Calawa, P. L. Kelley and S. A. Clough,
    J. Appl. Phys. 43, 3222 (1972).
4. P. Norton, P. Chia, T. Braggins and H. Levinstein, Appl. Phys.
    Letters 18, 158 (1970).
5. G. A. Antcliffe and J. S. Wrobel, Appl. Optics 11, 1548 (1972).
6. K. W. Nill, F. A. Blum, A. R. Calawa and T. C. Harman, Appl.
    Phys. Letters 19, 79 (1971).
7. K. W. Nill, F. A. Blum, A. R. Calawa and T. C. Harman, Chem.
    Phys. Letters 14, 234 (1972).
8. F. A. Blum, K. W. Nill, A. R. Calawa and T. C. Harman, Appl.
    Phys. Letters 20, 377 (1972).
9. F. A. Blum, K. W. Nill, A. R. Calawa and T. C. Harman, Chem.
    Phys. Letters 15, 144 (1972).
10. K. W. Nill, F. A. Blum, A. R. Calawa and T. C. Harman, Appl.
    Phys. Letters 21, 132 (1972).
11. F. A. Blum, K. W. Nill, P. L. Kelley, A. R. Calawa and T. C.
    Harman, Science 177, 694 (1972).
12. F. A. Blum, K. W. Nill and A. J. Strauss, J. Chem. Phys.
    58, 4968 (1973).
13. K. W. Nill, F. A. Blum, A. R. Calawa, T. C. Harman and A. J.
    Strauss, J. Mol. Spectry. (to be published).
14. K. W. Nill, A. J. Strauss and F. A. Blum, Appl. Phys. Letters
    22, 677 (1973).

15. R. S. Eng, A. R. Calawa, T. C. Harman, P. L. Kelley and
     A. Javan, Appl. Phys. Letters 21, 303 (1972); R. S. Eng,
     P. L. Kelley, A. Mooradian, A. R. Calawa and T. C. Harman,
     Chem. Phys. Letters (to be published).
16. G. A. Antcliffe, S. G. Parker and R. T. Bate, Appl. Phys.
     Letters 21, 505 (1972).
17. E. D. Hinkley and C. Freed, Phys. Rev. Letters 23, 277 (1969).
18. S. S. Penner, Quantitative Molecular Spectroscopy and Gas
     Emissivities (Addison-Wesley, Reading, Mass., 1959), Chap. 1.
19. R. Oberly, K. N. Rao, Y. H. Hahn, and T. K. McCubbin, J.
     Mol. Spectry. 25, 138 (1968).

FAR INFRARED GENERATION*

Benjamin Lax[+] and R. L. Aggarwal

Francis Bitter National Magnet Laboratory[‡]
Massachusetts Institute of Technology
Cambridge, Massachusetts 02139

INTRODUCTION

One of the exciting challenges in recent years was to develop a coherent radiation source which would bridge the energy region between the infrared and the microwaves. We have achieved this objective with the use of two high power CO_2 TEA lasers and a non-collinear phase matching technique in a crystal of GaAs.[1] Our results represent one of several methods which may be used to do this.[2] At present, our technique gives a broadband spectrum which spans over two decades from 70μ m to 1 cm, with resolution of the order of 10^{-2} cm^{-1} and capable of peak power levels which vary from $\sim$ 1 watt to $\sim 10^{-4}$ watt from the short to the long wavelength regions, respectively.

Our original plan for bridging the I.R.-microwave gap was based upon our work on the high power spin-flip Raman laser which we developed earlier.[3] With a homemade CO_2 TEA laser as a pump, we generated an average peak power of $\sim$ 1 kW in the Stokes radiation at a saturation level of 50 kW in the pump. The actual pulse power may have exceeded this by a factor of 10 due to the self-mode locking of our TEA lasers. Our scheme was to mix the tunable output of the spin-flip Raman laser with that of a fixed frequency of a second TEA laser to obtain the sum and difference frequency. The former was achieved by the use of a tellurium crystal which is non-linear and birefringent. By rotating the crystal as we varied the magnetic field, and hence the sum frequency, we obtained a phase matched output from 5.4 to 5.8 microns.[4] We attempted a similar experiment for

---
*Work supported in part by the Office of Naval Research.
+Also Department of Physics, MIT.
‡Supported by the National Science Foundation.

the difference frequency with tellurium, but discovered that it absorbed too strongly in the far I.R. region. Not having a suitable non-linear birefringent crystal which would operate in the submillimeter region, we decided to employ a non-collinear technique for mixing in either GaAs or CdTe for obtaining tunable difference frequencies.

NON-COLLINEAR PHASE MATCHING

The usual scheme for mixing two frequencies is to use a non-linear crystal which is birefringent so as to obtain phase matching of the three collinear electromagnetic waves. The output power at the phase matched $k_1 - k_2 = k_3$ condition is given by the expression

$$\frac{P_3}{area} = 2 \sqrt{\frac{\mu_0}{\epsilon}} \; \omega_3^2 \; d_{eff}^2 \; E_1^2 \; E_2^2 \; L^2 \qquad (1)$$

However, one can also use a polar crystal with cubic symmetry such as the III-V and II-VI semiconductors which lack inversion symmetry along the [111] direction. It is then a simple matter to match non-collinearly, i.e.,

$$\vec{k_1} - \vec{k_2} = \vec{k_3} \qquad (2)$$

To achieve this the crystal must possess certain properties. It must be transparent in the 9 to 13 micron region as well as in the far I.R. to microwave regions. To satisfy Eq. (2) the crystal must have anomalous dispersion between the short and long wavelength regions at which it transmits. For GaAs these requirements are satisfied as shown in Fig. 1 which has a restrahlen absorption and anomalous dispersion at ~36 μm.

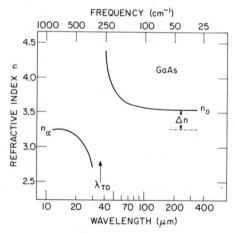

Fig. 1. Refractive index of high resistivity GaAs at 8° K (After Johnson, Sherman, and Weil, Appl. Optics 8, 1667 (1969)).

The indices of refraction at the low and high frequency limits are $n_0 = 3.59$ $n_\infty = 3.28$. By simple algebraic manipulation we can show that the angles θ and φ have the approximate values given by

$$\theta \approx \frac{\omega_1 - \omega_2}{\sqrt{\omega_1 \omega_2}} \sqrt{\frac{2 \Delta n}{n}}$$

$$\varphi \approx \sqrt{\frac{2 \Delta n}{n}} \approx \text{constant} \qquad (3)$$

$$\Delta n = n_0 - n_\infty$$

This result shows that the angle φ is nearly fixed and, hence, allows for a broadband mixing arrangement.

The original experiments were carried out by using two fixed frequency CO_2 TEA lasers; one operating at $9.6\mu m$, the other at $10.6\mu m$. The two lasers were directed at the same point on one face of the GaAs crystal with an external $\theta_E \approx n_\infty \theta \approx 8^\circ$ where $\theta \approx 2.1^\circ$. The arrangement in Fig. 2 also shows grating adjustments for tuning both lasers. In order to adjust the external angle for phase matching condition, as the frequency varied, mirror M_2 was mounted to provide two degrees of freedom; one a lateral motion and the other a rotational one.

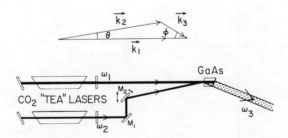

Fig. 2. Schematic diagram of experimental arrangement for non-collinear phase-matched FIR difference frequency generation in GaAs; the wavevector triangle shows the direction of propagation of the incident beams and that of the difference frequency radiation inside the nonlinear crystal.

For the two fixed frequencies above the adjustment about $\sim 8°$ was such that the total half-power width was $\sim 0.2°$. The maximum could be adjusted within less than 1% of the external angle. The output face of the crystal was cut at angle φ to prevent total internal reflection and to have the phase front of the far I.R. radiation parallel to it. The arrangement of the two overlapping beams are shown in Fig. 3.

THEORY

The theory of the collinear phase matching arrangement is a one-dimensional problem which is relatively simple. The non-collinear case becomes a much more complicated two-dimensional problem. For the odd-shaped geometry shown in Fig. 3 the calculations become algebraically complex. First we rotate the coordinate system about the [111] axis which is perpendicular to the wave vectors so that one of the axes is parallel to the output wave vector k_3. We then solve the electromagnetic problem for the nonlinear source in the three sections indicated. The wave front for the phase matched condition is always parallel to y' or perpendicular to k_3 all throughout the crystal. As the wave emerges, its amplitude is nearly trapezoidal as shown. The analysis represents the solution of Maxwell's equations with the nonlinear driving term

$$\nabla^2 E_3 - \frac{n^2}{c^2} \frac{\partial^2}{\partial t^2} E_3 = \mu_0 \frac{\partial^2}{\partial t^2} P_{NL} (\vec{r}, t)$$

where

$$P_{NL} (\vec{r}, t) = 2 d_{eff} E_1 E_2^* e^{i[(\vec{k}_1 - \vec{k}_2) \cdot \vec{r} - (\omega_1 - \omega_2)t]} \quad (4)$$

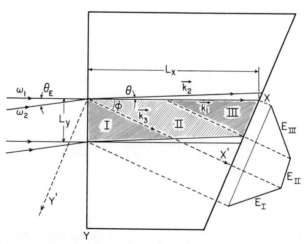

Fig. 3. Schematic diagram showing the interaction region for noncollinear mixing and the variation of the electric field E_{30} for the difference frequency radiation across the output face of the GaAs crystal.

where d_{eff} is the nonlinear coefficient of the crystal. If we take the solution to be of the form

$$\vec{E}_3 = \vec{E}_{30}(x',y') e^{i(\vec{k}_3 \cdot \vec{r} - \omega_3 t)} \tag{5}$$

and neglect the second derivative which varies very slowly, we then obtain

$$i\,\vec{k}_3 \cdot \nabla \vec{E}_{30} = \mu_o \omega_3^2 d_{eff}\, E_{10} E_{20}^* e^{i(\vec{k}_1 - \vec{k}_2 - \vec{k}_3) \cdot \vec{r}}$$

or

$$i\,k_3\, \frac{\partial E_{30}}{\partial x'} = \mu_o \omega_3^2\, d_{eff}\, E_{10} E_{20}^* e^{i\Delta k_{x'} x'}\, e^{i\Delta k_{y'} y'} \tag{6}$$

where

$$\Delta k_{x'} = k_1 \cos\varphi - k_2 \cos(\theta + \varphi) - k_3$$

$$\Delta k_{y'} = -k_1 \sin\varphi + k_2 \sin(\theta + \varphi) \tag{7}$$

We can integrate Eq. (6) along a filament as specified by given y' coordinate. The result for the amplitude of the electric field E_{30} for phase matched condition $\Delta k_{x'} = \Delta k_{y'} = 0$ at the output face for the three sections is:

I) $\quad E_3(y') = -2i\mu_o \omega_3^3 d_{eff}\, \dfrac{E_{10} E_{20}^*}{2k_3} \left\{ \begin{array}{l} -y'[\cot(\theta + \varphi) + \tan\varphi] \\ +L_y[\cos\varphi\,\cot(\theta + \varphi) + \sin\varphi] \end{array} \right\}$

II) $\quad E_3(y') = -2i\mu_o \omega_3^3 d_{eff}\, \dfrac{E_{10} E_{20}^*}{2k_3} \left\{ \begin{array}{l} -y'[\cot(\theta + \varphi) - \cot\varphi] \\ +L_y[\cos\varphi\,\cot(\theta + \varphi) + \sin\varphi] \end{array} \right\}$

III) $\quad E_3(y') = -2i\mu_o \omega_3^3 d_{eff}\, \dfrac{E_{10} E_{20}^*}{2k_3} \quad (y'\cot\varphi + L_x \cos\varphi) \tag{8}$

This is so far exact. We can solve for the total power in the far-field or the near-field approximations. From conservation of energy the two are equivalent as indicated in Born and Wolf.[5] Mathematically it is simpler to calculate the power for the near-field. Then

$$P_3 = \frac{1}{2}\sqrt{\frac{\epsilon}{\mu_o}}\, L_z \int E_3(y')\, E_3^*(y')\, dy' \tag{9}$$

if the summation is carried out for all the filaments parallel to k_3 in the three regions. By making use of the fact that $\theta \ll \varphi$ and neglecting θ we obtain the approximate result for a trapezoidal amplitude

$$P_3 = 2\sqrt{\frac{\mu_o}{\epsilon}}\, \omega_3^2\, d_{eff}^2\, E_1^2\, E_2^2 \left(\frac{L_y}{\sin\varphi}\right)^2 \left[\frac{L_x}{L_y} \sin\varphi - \frac{1 + \sin^2\varphi}{3\cos\varphi}\right] L_y L_z \tag{10}$$

From this expression we can obtain an effective length equivalent to that of the collinear geometry as in Eq. (1).

$$L_{eff} = \frac{L_y}{\sin \varphi} \sqrt{\frac{L_x}{L_y} \sin \varphi - \frac{1 + \sin^2 \varphi}{3 \cos \varphi}} \tag{11}$$

For the parameters in our experiment $\varphi = 21°$, $L_x = 1.0$ cm, $L_y = 0.3$ cm, $L_{eff} = 0.75$ cm.

Using the above results we can calculate the power output, knowing the input of the two lasers and the measured value of the nonlinear coefficient. This we have done for power levels $P_1 \approx P_2 = 200$ kW, $L_y L_z \approx 0.1$ cm^2, and obtained ~ 15 W at 100 μm wavelength. Our measurements were of the order of ~ 30 milliwatts. The discrepancy is attributed to self-modelocking of the two lasers.

BROADBAND SPECTRA

The first application of the above non-collinear technique was to mix two TEA lasers whose frequencies can be varied with suitable gratings. These lasers can produce about 80 different wavelengths corresponding to the rotational vibrational transitions of the P and R branches of the 10.6μm and 9.6μm vibrational bands of the CO_2 molecules. If these are appropriately combined about 3000 lines can be produced between 70μm and 1 cm. We can span essentially the whole spectrum from the far infrared into the microwave region. This capability has been experimentally demonstrated recently[6] where we have obtained approximately 60 lines from 70μm to 2 mm, using a fixed frequency laser operating at 10.6μm and another laser whose frequency was varied with a grating to provide the above spectral range. More recently[7] the results have been extended by employing gratings on both lasers. This arrangement now permits us to cover the entire range.

These results are shown in Fig. 4 where we have plotted the phase matching angle, θ_E, versus the difference frequency, ω_3. The plot shows approximately 100 lines which span the entire spectral range. The curvature at the higher frequencies is due to the dispersion of the dielectric constant as the difference frequency begins to approach the restrahlen region of GaAs. We are in the process of repeating this experiment using CdTe as the nonlinear element. This has the advantage that the nonlinear conversion efficiency in the far infrared should be an order of magnitude larger than that of GaAs.

FUTURE PLANS

We have demonstrated that it is possible to obtain coherent radiation in the far infrared and submillimeter regions beyond 70μm. Although the

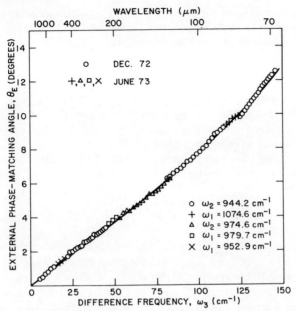

Fig. 4.   External phase matching angle θ_E vs. the difference frequency of the FIR radiation. The symbols o, +, etc. represent the experimental data and the solid line represents the theoretical curve.

present system has the possibility of 3000 lines, this can be substantially increased by using one or two isotopic species of CO_2 to increase the total by one order of magnitude. The present arrangement can give a mean spectral separation of 10^{-2} cm^{-1}; the isotopic laser would have 10^{-3} cm^{-1}. Such sources can be used for high resolution spectroscopy by a judicious programming of the grating drives coupled with the motion of the movable mirror. At present the peak power output of the system is of the order of tens of milliwatts at 100μm and hundreds of microwatts in the millimeter region. However, these can easily be increased by at least three orders of magnitude by eliminating the self-modelocking of the two lasers, reducing the time jitter between them, and increasing the size of the mixing crystal. Such improvements should result in average power output of tens of microwatts in the 100μm region and a fraction of microwatt in the milli- meter region. These would exceed present tunable broadband sources by many orders of magnitude accompanied by the high resolution.

By using high pressure CO_2 lasers it would be possible to obtain continuous tuning over the two decades of frequencies. The disadvantage would be the loss of resolution which would be determined by the gratings of the CO_2 lasers. However, for many applications this would not be a serious drawback. Another way of obviating the loss of resolution and still obtain continuous tuning would be to use a high power spin-flip Raman laser as one of the sources. This would have a limited range since they operate at highest powers between 40 and 60 kilogauss. However, by combining the step tunable capability of the CO_2 lasers and the spin-flip Raman laser judiciously, it would be possible to obtain continuous tuning in the submillimeter range. The CO_2 laser pumping the spin-flip Raman laser would be restricted to operating in the $10.6\mu m$ branch.

Another approach to the nonlinear mixing problem is that suggested by Yacoby et al.[8] By combining the lamellar scheme of Bloembergen[9] and with a non-collinear arrangement as shown in Fig. 5, we can easily build a CW technique of mixing in either GaAs or CdTe. Each layer is alternately of $[111]$ and $[\bar{1}\bar{1}\bar{1}]$ orientation. This is the equivalent of inserting a periodic variation into the nonlinear coefficient. It is then easily shown that

$$\vec{k_1} - \vec{k_2} \mp \vec{k_o} = \vec{k_3} \qquad (12)$$

where $k_o = \pi/d$ and d is the thickness of the slab. The advantage of this technique is that the angle between $\vec{k_1}$ and $\vec{k_2}$ can be substantially reduced so that they are nearly collinear. By making k_3 collinear with k_o and varying the angle of k_1 and k_2 we can produce a broadband nonlinear element with this technique. As a practical matter, the internal angle of k_1 and k_2 would be limited to $\sim 19^o$ for CdTe and $\sim 22^o$ for GaAs, determined by internal reflection. With suitable choice of slab dimensions two or three sets of lamella can cover the spectral range from the far infrared through the millimeter range. Since the two beams internally are nearly collinear, it is possible to use CW laser for mixing. They could be collimated to much smaller beam sizes than for the non-collinear case. Hence, CW operation would be facilitated by the higher electric fields that would result.

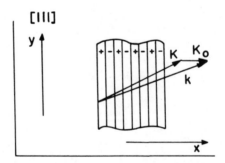

Fig. 5. Schematic diagram of lamellar device for phase matching by the periodic variation of the nonlinear coefficients.

APPLICATIONS

The purpose of these mixing schemes is to provide sources for resonance spectroscopy to match the high magnetic fields that are now available. Thus, not only cyclotron resonance would benefit but magnetic resonance of all types would also be possible. We also plan to do resonance studies of ionized plasmas and plasma diagnostics with these tunable sources. Molecular spectroscopy and lattice vibration studies are obvious candidates for far infrared tunable sources. A simple double heterodyne system using the step-tunable mixing arrangement would be very useful to radio astronomy in the submillimeter region.

ACKNOWLEDGMENTS

We are extremely grateful to Dr. C. E. Chase whose skillful development of our TEA lasers formed the backbone of this program. We also wish to express our appreciation to Dr. N. Lee and G. Favrot for help with the experiments.

REFERENCES

[1] R. L. Aggarwal, B. Lax, and G. Favrot, Appl. Phys. Letters 22, 329 (1973).

[2] G. D. Boyd, T. J. Bridges, C. K. N. Patel, and E. Buehler, Appl. Phys. Letters 21, 553 (1972); K. H. Yang, J. Morris, P. L. Richards, and Y. R. Shen, Bull. Am. Phys. Soc. 18, 350 (1973); Van Tran Nguyen and T. J. Bridges, Bull. Am. Phys. Soc. 18, 400 (1973).

[3] R. L. Aggarwal, B. Lax, C. E. Chase, C. R. Pidgeon, and D. Limbert, Appl. Phys. Letters 18, 383 (1971).

[4] C. R. Pidgeon, B. Lax, R. L. Aggarwal, C. E. Chase, and F. Brown, Appl. Phys. Letters 19, 333 (1971).

[5] See, for example, M. Born and E. Wolf, Principles of Optics, Third Revised Edition (Pergamon Press, New York, 1965), p. 385.

[6] B. Lax, R. L. Aggarwal, and G. Favrot, in Conference Digest of the 1973 IEEE Conference on Laser Engineering and Applications, Washington, D.C., 1973.

[7] B. Lax, R. L. Aggarwal, and G. Favrot, submitted to Appl. Phys. Letters.

[8] Y. Yacoby, R. L. Aggarwal, and B. Lax, to appear in the July 1973 issue of the J. Appl. Phys.

[9] J. A. Armstrong, N. Bloembergen, J. Ducuing, and P. S. Pershan, Phys. Rev. 127, 1918 (1962).

TUNABLE FAR-INFRARED GENERATION BY DIFFERENCE-FREQUENCY MIXING IN INDIUM ANTIMONIDE

Van Tran Nguyen and T. J. Bridges

Bell Telephone Laboratories, Incorporated

Holmdel, New Jersey   07733

ABSTRACT

A strong resonant optical nonlinearity due to the conduction electron spins has been observed when the frequency (~ 100 cm^{-1}) of far-infrared (FIR), generated by difference frequency mixing of two orthogonally polarized CO_2 laser beams is tuned to the spin-resonance frequency in InSb. This effect allows us to take full advantage of the availability of the tunable radiation from the spin-flip Raman laser (SFR) in the generation of tunable FIR from 90 cm^{-1} to 110 cm^{-1}. A successful attempt to solve the problem of "jump tuning" of FIR due to the presence of longitudinal cavity modes of SFR laser by replacing it with a nonresonant superradiant spin-flip Raman laser (NRSSFR) is also discussed in this paper.

I.   INTRODUCTION

In this paper, we review experimental and theoretical results of the generation of tunable far-infrared (FIR) radiation by difference-frequency mixing using the electron-spin nonlinearity in InSb. In Section II, we describe briefly the second-order resonant optical nonlinearity of the conduction electron-spins of InSb. By using a spin flip Raman laser as a source, tunable FIR can be generated as discussed in Section III. We will also include in this paper in Section IV recent results of the generation of continuously tunable FIR by difference mixing between the radiation from a nonresonant superradiant spin-flip Raman laser (NRSSFR) and the pump radiation of 10.6 μ.

II.   CONDUCTION ELECTRON SPIN NONLINEARITY

The problem of finding suitable nonlinear material for difference-mixing to generate tunable FIR is tantalizing because the recent development of the spin-flip Raman laser[1] (SFR) has made available an excellent tunable source in the near infrared. This search has led us to the discovery of the strong resonant optical nonlinearity due to the conduction electron spins in InSb.[2] This second order nonlinear effect directly related to the spin-flip Raman scattering process has been observed in InSb at 15°K by utilizing the difference frequency mixing of two orthogonally polarized CO_2 laser beams at frequencies ω_1 and ω_2 to produce a FIR at frequency $\omega_3 = \omega_1 - \omega_2 \sim 90$ cm^{-1}.   The experimental setup is described in Ref. 2.   It is important to point out right now that the geometry of the experimental setup is identical to one which has been used for the observation of the spin-flip Raman scattering process[3] if we consider the radiation at frequency ω_1 as the pump and the radiation at frequency ω_2 as the STOKES frequency[3].   This remark has important implications in the generation of tunable FIR which will be discussed in Section III.   In Fig. 1, we show the FIR power output P_{ω_3} as a function of the magnetic field H.   A sharp peak of P_{ω_3} is observed at H = 44.63 kG where $\omega_3 = \omega_1 - \omega_2 = \omega_s$.   The frequency ω_s is the spin-resonance frequency. Away from the resonance, the signal is due to the conventional nonlinearity which is present in InSb.[2]   The FIR at frequency ω_3 is created by the radiation from a forced magnetization wave of the spin system of the conduction electrons which is driven by an effective magnetic field at frequency ω_3 resulting from the vector product of the magnetic vector potential[2] of the incident radiation

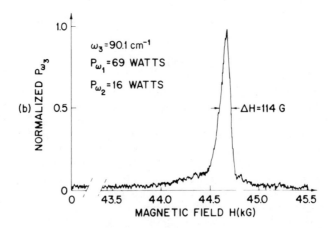

Fig. 1.   Far infrared power output $P(\omega_3)$ as a function of magnetic field, for fixed input frequencies.

at frequencies ω_1 and ω_2. Because of the narrow linewidth of the spin-resonance in InSb,[4] a large enhancement of FIR is obtained at the spin-resonance frequency ω_s. The theory of this conduction electron spin nonlinearity is described in Ref. 5. The effective nonlinear coefficient d^s_{eff} of the spin process is

$$d^s_{eff} = \frac{1}{2} n_3 C(\omega_3) \qquad (1)$$

where n_3 is the refractive index at FIR frequency and $C(\omega_3)$ is related to the spin magnetization $M(\omega_3)$ at frequency ω_3 and to the electric fields $E_{\omega_{1,2}}$ of incident radiations by:

$$M(\omega_3) = C(\omega_3) E_{\omega_1} E_{\omega_2} \qquad (2)$$

and defined as:

$$C(\omega_3) = n_0 \mu^* \frac{e^2}{m_s c^2} \frac{\hbar\omega_1 E_G}{E_G^2 - (\hbar\omega_1)^2} \frac{c^2}{\omega_1 \omega_2} [2\hbar(\omega_3 - \omega_s + i\gamma/2)]^{-1}. \qquad (3)$$

where n_0 is the electron density, μ^* is the magnetic moment of electrons in InSb, E_G is the InSb energy gap, m_s is the spin mass and γ is the linewidth (full width at half maximum) of the spin resonance. As it was described in Ref. 2, the orientation of crystallographic axis of the InSb sample was chosen in such a way that a direct measurement of the ratio of the effective d^s_{eff} of the spin process to the effective conventional nonlinear coefficient d^e_{eff} $(d^c_{eff}=d_{14}/\sqrt{3})^2$ was possible. We have found for a linewidth of 114 Gauss ($\sim.2$ cm^{-1}) the ratio $d^s_{eff}/d^c_{eff} = 9.4$ or $d^s_{eff} = 5.4 \, d_{14}$. This is in good agreement with the theoretical value calculated from Eq. (3) $d^s_{eff}/d^c_{eff} = 14$.

III. GENERATION OF TUNABLE FIR

It is clear that using the scheme of the mixing experiment discussed in Section II, the generation of tunable FIR could be obtained by replacing the two frequencies ω_1 and ω_2 by the pump and tunable Stokes radiation from a SFR laser[1] which automatically differ by the spin resonant frequency ω_s. It was also pointed out in Section II that because of the close relation between the SFR laser mechanism and the mixing phenomenon the experimental

arrangement is identical for both.[2] In principle, the generation
of tunable Stokes radiation and the difference mixing between the
pump and the Stokes frequency could be combined in the same crystal,
and so a SFR laser could itself be a generator of tunable FIR.
However, optimal tunable Stokes radiation from an SFR laser requires
a higher electron concentration than will allow the phase matching
of the mixing process. We have consequently used separate and
adjacent crystals of InSb of different doping concentrations about
10^{16} cm^{-3} for the SFR laser and 3.9×10^{15} cm^{-3} for the mixer
crystal. The details of the experiment for the generation of
tunable FIR is described in Ref. 6. The solid curve of Fig. 2(a)
shows FIR output versus wave number producing by slowly scanning

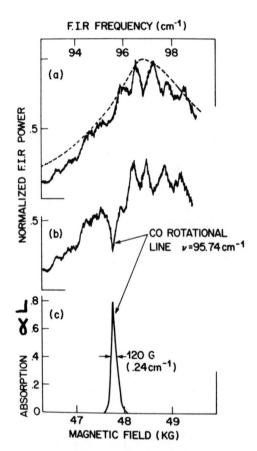

Fig. 2. Tunable far infrared output from combination of a spin
slip Raman laser and a mixer crystal. a) As a function of
frequency. b) Showing absorption line due to 400 torr of
CO. c) Showing apparent linewidth of CO absorption line.

the magnetic field. Efficient mixing is obtained only when the
interacting waves are phase matched. Since InSb is not birefringent, conventional methods of phase matching are impossible.,
However, the electron plasma itself may be used to adjust the
refractive index at the FIR frequency ω_3 and obtain phase matching.
Previous experiments on phase matching with the electron plasma
used the <u>magnetoplasma</u> effect in the Voigt or Faraday configuration.[7] By contrast, in the present geometry where $\vec{E}_{\omega_3} \perp H$, the
contribution of the electron plasma is independent of magnetic
field strength, and can be adjusted at a particular frequency only
by adjusting the carrier concentration.[8] It is fortunate that for
the FIR frequencies of interest in this experiment, phase matching
may be obtained at carrier concentrations ($n_e \sim 3.9 \times 10^{15}$ cm^{-3}),
compatible with strong electron spin nonlinearity.[2]

The FIR power generated can be estimated from Eq. (1) of
Ref. 2. Since many of the parameters are approximately constant
over the present range of operation, P_{ω_3} becomes proportional to
$\omega_3^2 I$, where I is the coherence factor of Ref. 7 which takes into
account the effects of phase mismatch and absorption losses. We
have computed $\omega_3^2 I$ as a function of ω_3, and this is plotted in
Fig. 2(a) (dotted line), normalized at the peak power. There is
approximate agreement between the theoretical and experimental
curves. Periodic fluctuations in the experimental curve have a
free spectral range corresponding to Fabry-Perot resonances in the
mixing crystal at the FIR wavelength. In estimating an absolute
output power from Eq. (1) of Ref. 2, we used values for P_{ω_1} = 2 kW,
P_{ω_2} = 10 W, ω_3 = 98 cm^{-1}, obtaining a value of 6 µW peak power for
P_{ω_3}. This is in reasonable agreement with our experimental value
of $\sim$2µW.

In previous work[1] the output of the pulsed SFR laser was
shown to have a linewidth of $\sim$0.03 cm^{-1}, and we estimate the linewidth of the pump laser at < .001 cm^{-1}. The FIR linewidth should
be no worse than $\sim$0.03 cm^{-1}, the combination of the beating linewidths. We found, however, by a simple spectroscopic experiment,
that our apparent FIR linewidth was as great as $\sim$0.24 cm^{-1}. In
this experiment we observed the pure rotational absorption lines
of CO gas. These lines are simple and widely spaced and so provide an uncomplicated test of the source linewidth. In Fig. 2(b)
the v = 0, J = 24 absorption[9] at 95.74 cm^{-1} is shown and in
Fig. 2(c) the measurement is converted to absorption coefficient.
The measured linewidth of 0.24 cm^{-1} was found to be independent of
pressure up to $\sim$400 Torr. Since the true pressure broadened

linewidth is proportional to pressure, with a value[10,11] of
0.05 cm^{-1} at 400 Torr, we conclude that the measured linewidth is
due to the source, and is nearly an order of magnitude larger than
expected.

The reason for this behavior was discovered by analyzing the
Stokes output of the SFR laser at $\sim$844 cm^{-1} with a scanning
Fabry-Perot interferometer. It was found that as the gain line
was tuned by scanning the magnetic field, the SFR laser oscillation
did not tune continuously but occurred on 2 or 3 longitudinal
modes at one time. The spacing (0.18 cm^{-1}) and number of the modes
was in agreement with that calculated from the oscillator cavity
length and consistent with a Raman gain linewidth of about 0.5 cm^{-1}.
It is clear that the failure to measure the correct linewidth in
Fig. 2(b) is due to the presence of the multiple longitudinal
modes of SFR laser.

IV.   GENERATION OF CONTINUOUSLY TUNABLE FIR

One way to resolve the problem of the presence of multiple
longitudinal modes of SFR laser is to destroy the resonant cavity
effect in SFR laser or in other words to obtain nonresonant
superradiant spin-flip Raman laser (NRSSFR). Superradiance has
been previously discussed as an explanation of linewidth measurements at 10.6[1] and 5.3μ[12]. We have successfully obtained a NSSFR
laser by using a folded path superradiant amplification of spontaneous scattered Stokes radiation. The shape and dimensions of
the crystal (electron concentration $\sim$10^{16} cm^{-3}) for NRSSFR laser is
shown in the Fig. 3. The input and output polished surfaces were
made nonparallel (5 degrees of angle) in order to destroy the
resonant cavity effect. To generate FIR, we have used the same
mixer crystal as one already used in the generation FIR experiments
discussed in Section III. The arrangement of NRSSFR crystal and
the mixer crystal is shown in Fig. 3 together with the polarization
of electric fields of the pump, Stokes radiation and the FIR.
Using the scanning Fabry-Perot interferometer technique, we have
studied the linewidth and the tuning characteristic of the NRSSFR
and the FIR radiations. Details of the interferometer used for
each wavelength will be published elsewhere. The results of
linewidth measurement are shown in Fig. 4(a) for NRSSFR together
with the pump line of CO_2 laser and in Fig. 4(b) for the FIR. The
linewidth of .09 cm^{-1} for both measurements is in agreement with
the theoretical estimation of linewidth of NRSSFR given by Patel
et al.[1] considering the spontaneous linewidth of InSb samples and
pump intensity used in the present experiment. In order to be certain that the resonant cavity effect has been suppressed and the
NRSSFR radiation is continuously tunable with the magnetic field,
we have used the Fabry-Perot interferometer with fixed spacing and

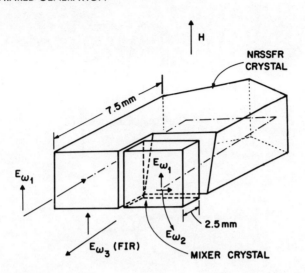

Fig. 3.   Schematic of nonresonant superradiant spin flip Raman
          laser and mixer crystal.

scan the magnetic field from 38.9 kG to 56 kG. The output from the
Fabry-Perot interferometer is shown in Fig. 4(c). This result
implies that the continuous tuning of NRSSFR radiation and consequently of the FIR which results from the mixing of the NRSSFR and
the pump, is beyond any doubt. The tuning rate between 38.9 kG
and 56.1 kG deduced from this result is about 1.8 cm^{-1}/kG.
Presently, we are trying to reduce the linewidth of FIR.

V.   CONCLUSIONS

We have demonstrated a new type of difference frequency
mixing using the spin resonance of conduction electrons in n-type
InSb. This effect combines naturally with the tunable spin flip
Raman (SFR) laser in InSb to produce a tunable far infrared (FIR)
source. The narrow linewidth of the SFR laser results in a FIR
linewidth potentially much less than 0.1 cm^{-1}.

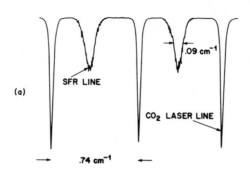

Fig. 4.   a) Fabry-Perot scanning interferometer analysis of
          tunable superradiant spin flip Raman laser output set
          to ~ 850 cm^{-1}.   b) Fabry-Perot scanning interferometer
          analysis of FIR output at ~ 97 cm^{-1}.   c) Stationary
          Fabry-Perot interferometer analysis of scanning fre-
          quency of superradiant spin flip Raman laser; tuning
          from ~ 867 to 837 cm^{-1}.

          At present, experiments have been confined to pulsed operation
of the SFR laser with a 10.6μ wavelength pump.   Continuous wave
operation of the SFR laser has been previously demonstrated[13]
using a pump wavelength near to 5.3μ.   This possibility is due to
the large resonant enhancement of the Raman scattering cross-
section when the pump energy approaches the band gap of InSb.   The

same resonant enhancement is found in the nonlinear coefficient $C(\omega_3)$ as will be seen in Eq. (3). For these reasons we believe that cw operation of the tunable FIR difference mixing source will be possible. Such a narrow line cw continuously tunable source will be invaluable in the further investigation of the far infrared region.

VI. ACKNOWLEDGMENT

We are grateful for the collaboration of P. A. Wolff and T. L. Brown. Technical assistance was provided by W. N. Wittnebert and J. S. Hasiak. We would like to thank C. K. N. Patel, W. F. Brinkman and A. Ashkin for continued interest in this work.

REFERENCES

1. C. K. N. Patel and E. D. Shaw, Phys. Rev. B 3, 1279 (1971).
2. V. T. Nguyen and T. J. Bridges, Phys. Rev. Letters 29, 359 (1972).
3. R. E. Slusher, C. K. N. Patel and P. A. Fleury, Phys. Rev. Letters 18, 77 (1967).
4. B. D. McCombe and R. J. Wagner, Phys. Rev. B 4, 1285 (1971).
5. T. L. Brown and P. A. Wolff, Phys. Rev. Letters 29, 362 (1972).
6. T. J. Bridges and V. T. Nguyen, Appl. Phys. Letters, July 15, 1973.
7. V. T. Nguyen and C. K. N. Patel, Phys. Rev. Letters 22, 463 (1969). C. K. N. Patel and V. T. Nguyen, Appl. Phys. Letters 15, 189 (1969).
8. E. D. Palik and G. B. Wright in Semiconductors and Semimetals, edited by R. K. Willardson and A. C. Beer, (Academic, New York 1967), Vol. 3, pp. 421-458.
9. K. D. Moller and W. G. Rothschild, "Far-Infrared Spectroscopy" (Wiley-Interscience, New York, 1971), p. 51.
10. Full width at half height.
11. D. A. Draegert and D. Williams, J. Opt. Soc. Am. 58, 1399 (1968).
12. R. A. Wood, R. B. Dennis and J. W. Smith, Opt. Com. 4, 383 (1972).
13. A. Mooradian, S. R. J. Brueck, and F. A. Blum, Appl. Phys. Letters 17, 481 (1970).

KILOWATT SPIN-FLIP OUTPUTS IN THE 5 μm REGION: APPLICATION TO CHEMICAL SPECTROSCOPY AND RESONANT NON-LINEAR MIXING

S.D. Smith, C.R. Pidgeon, R.A. Wood, A. McNeish and
    N. Brignall
Department of Physics

Heriot-Watt University, Edinburgh, Scotland

I. HIGH POWER SFRL OPERATION NEAR 5 μm

Continuous wave operation of the spin-flip Raman laser (SFRL), using a CO pump, has demonstrated the importance of the resonance enhancement obtained by pumping near the band edge of InSb[1-3]. Conversion efficiencies in excess of 50% and threshold powers of less than 50mW have been achieved[1,3]. However, up till the present time, output powers have been limited to about 1W.

The combination of kilowatt powers and microsecond pulse lengths is of interest in time resolved molecular experiments such as fluorescence and flash photolysis. A further application of the SFRL, which is of considerable current interest, is to far-infrared generation by difference mixing in InSb. In particular, a resonant optical non-linearity due to conduction electron spins in InSb has been observed[4] and used to generate far-infrared difference frequency radiation[5] between 93 and 100 cm^{-1}. It has been noted that this would be doubly enhanced[4] if the primary pump frequencies were near the band gap at 5 μm. This doubly-resonant process can theoretically generate milliwatt levels of tunable difference frequency between the pump and Stokes radiation - 100 times more intense than from 10μm mixing. This advantage could also lead to generation directly from the SFRL crystal itself.

We report a very large (two orders of magnitude) increase in the SFRL output power near 5μm achieved by the relatively obvious application of a CO_2 TEA (transversely excited

atmospheric pressure) laser, frequency doubled with Te oriented
at the phase match angle, as the pump source.   Thus, we have
combined the advantages of high intensity bulk cavity excitation
(previously used in the SFRL near 10μm[6]) with the resonant
enhancement derived from pumping near the band gap.   Stokes
output power levels close to 1kW have been achieved near 5μm,
corresponding to a conversion efficiency of $\sim$ 45%.

This mode of operation of the SFRL - which is well above
threshold for a large cavity volume - has enabled us to
demonstrate much more directly than hitherto the quantum
mechanical nature of the process, by the direct observation of
the oscillatory output from higher Landau level transitions
together with the associated pump depletion.   Continuous tuning
has been obtained over a range of 52 cm^{-1}, using 13 different
(doubled) CO_2 pump lines and a simple electromagnet giving only
7 kG.   Very satisfactory spectroscopy has been achieved, using
a double beam pulse ratioing system, with linewidth of $\sim$ 0.02 cm^{-1}.
This linewidth is significantly less than the 0.2 cm^{-1} most
recently reported by Bridges and Nguyen near 10μm [5].   Our
carrier density was 2.5 x 10^{15}cm^{-3} implying, from Brueck and
Mooradian[7], a spontaneous linewidth $\sim$ 0.05cm^{-1} at 7kG.   The
realisation of continuous tuning with a linewidth $\sim$ 1/10 of the
axial mode spacing of the cavity is both striking and useful for
spectroscopy.

Fig. 1 shows schematically the complete SFRL double beam

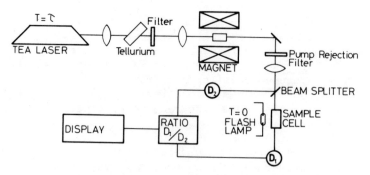

Fig. 1.   Schematic diagram of SFRL double beam spectrometer.

spectrometer.   The TEA CO_2 laser operated at 10 pps and produced
more than 100 KW on 10 lines between P(10) and P(34) selected with
an intra-cavity diffraction grating.   Partial transverse mode
control was obtained with an iris stop;   the output was weakly
focussed onto an area $0.25cm^2$ at the Te surface, giving a power
density of $\sim$ 5 MW/cm^2.   Recent advances in the quality of Te
crystals and surface preparation have raised the damage threshold
to 50MW/cm^2 [8] – in the present work we generated $\sim$ 3KW of second
harmonic radiation ($\sim$ 5% efficiency) well below damage threshold.
The harmonically doubled radiation was imaged onto an InSb cavity
(normally with N = 2.5 x $10^{15}cm^{-3}$) prepared by Edinburgh
Instruments Ltd.   A 120 kG superconducting magnet was used for
high field work but the practical cut-off for this sample was 50kG.
Spectroscopy was more conveniently conducted with the 7kG
electromagnet.

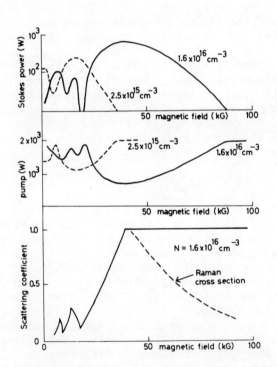

Fig. 2.   a) Magnetic field dependence of SFRL Stokes output, and
b) Magnetic field dependence of 5.3μm pump transmission
(i.e.) pump depletion) for N = 2.5 x $10^{15}cm^{-3}$ and 1.6 x
$10^{16}cm^{-3}$;   c) associated scattering coefficient and
relative Raman cross-section, calculated by method of
ref. 12, for N = 1.6 x 10^{16} cm^{-3}.

SFRL output was observed up to 40kG for the cavity mentioned and up to 80kG for a second cavity with $N = 1.6 \times 10^{16} cm^{-3}$. Smooth tuning curves were observed with no sign of mode hopping[6,9] to a frequency precision less than 0.02 cm^{-1} as demonstrated by molecular spectroscopy (see below). The form of the output is shown in Fig. 2a, the associated strong pump depletion in Fig. 2b and the calculated cross section[11,12] for the higher concentration in Fig. 2c. The prominent fall in SFRL output below the quantum limit and subsequent oscillations at lower field were predicted by Wherrett and Harper[11] and further studied by Dennis et. al.[12], from which calculations Fig. 2c was constructed. The results clearly demonstrate the Fermi-level blocking effect; the lower field peaks occur when the initial and final states of spin-flip excitation, corresponding to the ↑ and ↓ states with $n = 1$ and $n = 2$, emerge from the Fermi sea. For the lower concentration sample strong emission occurs as low as 1kG which has the useful practical consequence that spectroscopy can be readily carried out using a simple electromagnet[13].

The magnetic field variation of the output makes it desirable to use a double beam system for spectroscopy. An even stronger reason concerns the pulse to pulse instabilities of the pump laser, typically of the order of a few percent. We make use of a pulse ratio system[9], using two Cu:Ge, detectors, in which the measuring and reference pulses, after passing through pre-amplifiers, pulse stretching and sample and hold circuits, are individually divided. Critical trigging is avoided by use of pulse stretchers which trigger off the signal input. The system operates with a dynamic range of 50:1 and is linear to 0.1%.

II.   SPECTROSCOPY

Stimulated spin-flip Stokes scattering was observed with the electro-magnet using TEA CO_2 pumping lines P(10) to P(34) inclusive, corresponding to harmonically doubled pump beams ranging from 1906 to 1862cm^{-1}, with adjacent line separations of $\sim$ 3.7cm^{-1}. The SFRL tuning variation of $\sim$ 1.8$cm^{-1}kG^{-1}$ was sufficient to achieve overlap between the tuning ranges corresponding to every third pump line, and a total continuous tuning range of 52cm^{-1} was observed with the maximum field of 7kG available (Fig. 3[13]). Longer wavelength TEA CO_2 lines than P(34) are also available, but are considerably weaker in power. The short wavelength limit to the SFRL tuning range is set by interband absorption in the InSb SFRL crystal. With the carrier concentration employed in the present measurements, only very weak pump transmission was observed using the harmonically doubled P(12) and P(10) TEA CO_2 lines, and stimulated recombination radiation[9] was emitted close to 1899cm^{-1}. Stokes emission was only observed for magnetic fields corresponding to Stokes wavelengths longer than 1901cm^{-1} (Fig. 3).

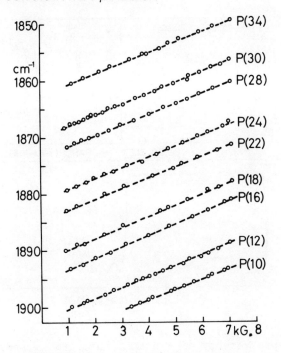

Fig. 3.   Stokes output tuning curves for a variety of doubled CO_2
          TEA laser pump lines (ref. 13).

        The linewidth of the Stokes radiation was determined by
tuning the SFRL output through R branch absorption lines of nitric
oxide, and deuterium bromide (Fig. 4) contained in a low-pressure
gas cell.   A pulse to pulse amplitude stability of better than 1%
was achieved, probably limited by mode instabilities in the pump
laser.   A radiation linewidth (FWHM) of 0.02cm^{-1} was determined
by adjusting gas cell pressures.   The spectra show no sign of
irregularity associated with mode-hopping although this linewidth
is $\sim$ 10 times narrower than the axial mode separation of the SFRL
cavity.   This is in contrast to recent results[5] at 10μm, and
suggests that the device is operating in a single 'mode' not
strongly influenced by the cavity.   The time-bandwidth limit is
$\sim 0.001 \text{ cm}^{-1}$ so that it may be possible to improve our linewidth by
using purer cavity crystals with smaller spontaneous linewidths[7].

        The spectra in Fig. 4 show the isotope splitting of the $J = 4$
$\rightarrow J = 5$, R branch line of deuterium bromide due to the bromine
isotopes $Br^{78.91}$ (50.5% abundant) and $Br^{80.91}$ (49.6% abundant) at
1880.0 and 1879.5cm^{-1} respectively.   The accuracy of the double
beam system is just sufficient to indicate the differing

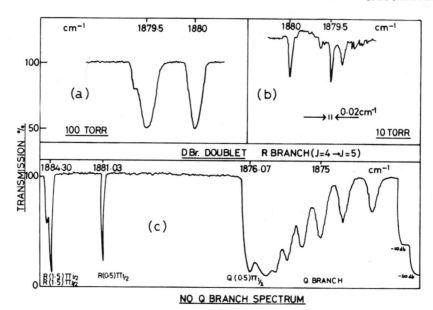

Fig. 4.   SFRL spectroscopy of R branch absorption lines of DBR and
          NO. a) Isotope splitting of J = 4 → J = 5, R branch DBr
          line (double beam operation at 100 torr), b) same DBr line
          with single beam operation at 10 torr, c) part of No
          Q-branch spectrum and spin-orbit-split J = 0 → J = 1 R
          branch line (double beam operation).

concentration of the two isotopes (Fig. 4a).   Single beam spectra
are presented for comparison Fig. 4b - in this case the cell
pressure was reduced to 15 torr to demonstrate the 0.02cm^{-1} linewidth.   Fig. 4c shows the Q branch of NO - clearly demonstrating
the anharmonic effects responsible for the structure as well
the first J = 0 → J = 1, R branch line split by spin-orbit effects.
These spectra demonstrate the stability of the double beam
measurement.

     The above results clearly demonstrate satisfactory
spectroscopy averaging over only a few 200 nanosecond pulses,
and indicate that the system is sufficently developed for time
resolved experiments.   The simplest experiment where time
resolution is important is to study the fluorescence of a gas at
right angles to the SFRL beam.   We are at present investigating
fluorescence associated with the decay of rotational (∿ 1μsec)
and vibrational (∿ 100μsec) modes of Q and R-branch lines of NO
and DBr, by tuning the SFRL to exact coincidence with appropriate

lines of the molecular spectrum in the 5-6μm region.   Further
work is planned on various kinds of flash photolysis and double
resonance experiments, particularly in carbonyl radicals.

III. MEASURING ACCURACY

The relatively very high power available can be used either
in frequency resolution or time resolution, or in measuring
accuracy which includes the case of small changes in highly
absorbing systems - e.g. photolysis effects in liquids.   In the
present state of the art, detector noise is not approached and
the limitations are set by amplitude and mode instabilities of
the SFRL output.   In general modulation techniques are used to
overcome this problem and yield the same S/N ratio for the
measured quantity as that from the SFRL output.   For experiments
where this is not possible, ratioing techniques have to be used.
The present system is capable of resolving 1 part in 3000,
averaging over 100 pulses.   In a transient experiment (e.g. flash
photolysis) one would sample the transmission of the SFRL through
the medium, at given fixed delay times after the flash, for ∿ 100
pulses.   An interesting comparison can be made between the pulsed
SFRL, the cw SFRL and a Nernst source using a Cu-Ge detector
$(D^* ∿ 10^{10})$ for a measuring time of 1 μsec, Table 1.   Detector,
photon random arrival time, laser jitter and overall S/N are
shown.   Both laser systems are very much superior to the Nernst.
If the beam attenuation is >100:1 the present pulsed SFRL is
superior for short measuring times, and this performance is likely
to be improved by at least an order of magnitude (i.e. making the
pulsed SFRL better than the cw SFRL for attenuations >10:1).

TABLE 1.   Signal/Noise for 1 μsec sample time.

	Detector	Shot	Jitter	Overall Noise
Nernst $(1cm^{-1})$	1	10^3	-	1
cw SFRL	10^7	5×10^6	3×10^3	10^4
pulse SFRL	2×10^9	10^8	3×10^2	10^3

IV.   DOUBLY RESONANT DIFFERENCE MIXING IN InSb.

Tunable far infrared radiation produced by mixing has been reported[5] using a Q-switched CO_2 laser at $\omega_1 = 944.2 cm^{-1}$ with a peak power of 5kW as a pump.   This pump was mixed with approximately 10 watts of Stokes radiation at a frequency ω_2 varying between 840 and $854 cm^{-1}$.   Output powers produced between $93-100 cm^{-1}$ are of the order of microwatts.   Using the high output powers of our 5μm SFRL we have the possibility of mixing 3kW of pump power with 1kW of tunable Stokes power, in addition to a strong resonant enhancement at the band gap.   Equation (1)[4] shows the frequency dependence of the far infrared intensity, $I(\omega_3)$, due to difference mixing

$$ I_{(\omega_3)} = M \left| \sum_t \frac{\langle t | \underline{\varepsilon}_1^* \cdot \underline{p} | \uparrow \rangle \langle \downarrow | \underline{\varepsilon}_2 \cdot \underline{p} | t \rangle}{\hbar\omega_1 - (E_\downarrow - E_t)} - \frac{\langle t | \underline{\varepsilon}_2 \cdot \underline{p} | \uparrow \rangle \langle \downarrow | \underline{\varepsilon}_1^* \cdot \underline{p} | t \rangle}{\hbar\omega_2 + (E_\downarrow - E_t)} \right|^2 $$

$$ \times \left| g^* \beta H - \hbar\omega_1 + \hbar\omega_2 + i\Gamma/2 \right|^{-2} I_{(\omega_1)} I_{(\omega_2)} \left\{ (1-2f)N \right\}^2 , $$

$$ (1) $$

where M is a constant, and conventional notation is used[12]. The second factor shows the resonant enhancement at the spin-flip frequency exploited by Bridges and Nguyen[5].   the first factor has the same form as the Raman coefficient and resonates at the band gap.   In the final term of equation (1), f is the fractional population of the upper spin level;   this term indicates the effect of spin saturation on the emitted intensity.   Mixing in the same crystal in which the Stokes radiation is generated is possible, but is diminished by the existence of this saturation effect.   Whilst the saturation, and the necessity to approach phase-match for the mixing situation, required use of 2 samples to produce far infrared radiation from initial 10 micron pumps, it appears that the added enhancement at 5 microns and the fact that a low concentration sample can produce high power Stokes oscillation should enable mixing to be achieved in a single sample for 5 micron excitation.   Comparing the two situations of mixing near 10μm with two crystals[5]   and near 5μm with one crystal ($N \sim 10^{15} cm^{-3}$), the latter is favoured in respect of:

(i)   power - 3kW and 1kW at ω_1 and ω_2 leads to $\sim$ 100-fold advantage.

(ii)   band gap resonant enhancement - $\sim$ 100, estimated from pulsed SFRL measurements at 10.6 and 5.3μm[6,13]

(iii) spin flip resonant enhancement - $\sim$ 10-20 enhancement over the two-sample configuration where the g-factor mis-match resulting from two different carrier concentrations reduces the non-linearity;

but   (iv)   saturation, on the other hand, would tend to reduce the advantage by a factor $\sim$ 100.

(v)   phase matching may require non-collinear geometry.

Overall, the one-sample 5μm process should be favoured by a factor $\sim$ 1000, if reasonable phase matching can be achieved.

V.   ACKNOWLEDGEMENTS

We are grateful to Dr B. Wherrett for helpful discussions concerning the doubly-resonant difference mixing process in InSb, and to the Science Research Council, U.K., for financial support.

REFERENCES

1. A. Mooradian, S.R.J. Brueck and F.A. Blum, Appl. Phys. Lett., 17, 481 (1970); S.R.J. Brueck and A. Mooradian, Appl. Phys. Lett., 18, 229 (1971).

2. S.R.J. Brueck and A. Mooradian, Phys. Rev. Lett., 28, 161, (1972).

3. C.K.N. Patel, Appl. Phys. Lett., 19, 400 (1971).

4. Van Tran Nguyen and T.J. Bridges, Phys. Rev. Lett., 29; 359 (1972);
    T.L. Brown and P.A. Wolff, Phys. Rev. Lett., 29, 363 (1972).

5. T.J. Bridges and Van Tran Nguyen, Appl. Phys. Lett., July (1973).

6. R.L. Aggarwal, B. Lax, C.E. Chase, C.R. Pidgeon, D. Limbert
    and F. Brown, Appl. Phys. Lett., 18, 383 (1971); C.R. Pidgeon,
    B. Lax, R.L. Aggarwal and C.E. Chase, Appl. Phys. Lett., 19,
    333, (1971).

7. S.R.J. Brueck and A. Mooradian, Phys. Rev. Lett., (to be
    published).

8. D.C. Hanna, B. Luther-Davies, H.N. Rutt, R.C. Smith and
    C.R. Stanley, I.E.E.E. Journ. of Quant. Elect. QE-8, 317
    (1972).

We are extremely grateful to R.C. Smith and C. Hammond of
Southampton University for provision of a crystal of Te
oriented at the phase-match angle.

9. R.A. Wood, R.B. Dennis and J.W. Smith, Opt. Commun., 4,
    383 (1972); R.G. Mellish, R.B. Dennis and R.L. Allwood,
    ibid, 4, 249 (1971); R.A. Wood, A. McNeish, C.R. Pidgeon
    and S.D. Smith, J. Phys. C. 6, L144 (1973).

10. H. Kogelnik and T.Li, Proc. I.E.E.E., 54, 1312 (1966);
    A. Nurmikko, T.A. Detemple and S.E. Schwarg, Appl. Phys.
    Lett., 18, 130 (1971).

11. B.S. Wherrett and P.G. Harper, Phys. Rev. 183, 692 (1969).

12. R.B. Dennis, C.R. Pidgeon, S.D. Smith, B.S. Wherrett and
    R.A. Wood, Proc. Roy. Soc. Lond. A 331, 203 (1972).

13. R.A. Wood, A. McNeish, N.L. Brignell and C.R. Pidgeon,
    Appl. Optics (to be published).

SPECTROSCOPY IV

LASER MAGNETIC RESONANCE SPECTROSCOPY

K. M. Evenson
Quantum Electronics Division
Institute for Basic Standards
National Bureau of Standards

Carleton J. Howard
Aeronomy Laboratory, NOAA
Environmental Research Laboratories
Boulder, Colorado 80302

Abstract

Laser magnetic resonance spectroscopy provides one of the most sensitive techniques ever devised for the detection of free radicals such as OH, CH, HO_2, and HCO. Paramagnetic molecules inside a laser cavity are Zeeman tuned into coincidence with the laser frequency and a decrease in laser power is detected. With this technique, the reaction rates of free radicals can be measured in a reaction flow system.

Rotational levels of paramagnetic molecules can be Zeeman shifted into coincidence with far infrared laser lines to produce resonance absorption of laser radiation. This technique, called laser magnetic resonance, was first demonstrated with O_2[1,2,3] and has subsequently been successfully applied to other stable molecules such as NO[4] and NO_2[5]. Recently, the Zeeman rotational spectra of the transient free radicals OH[6,7], CH[8], HO_2[9], and HCO[10] have also been observed.

The LMR spectrometer used in most of these studies is shown schematically in Fig. 1. The laser oscillates in the cavity between mirrors C and D in a single longitudinal mode. The cavity is divided into two parts by a dielectric beam splitter; the sample or absorption region is on the left and the active laser medium, a dc discharge, is on the right. The beam splitter serves three purposes: (1) it provides a vacuum seal between the absorption cell and the laser discharge, (2) it acts as a reflector to couple a small fraction of laser power out of the cavity to detectors A and B (for this purpose its angle with respect to the axis of the laser is adjustable near the Brewster angle), and (3) it restricts the laser radiation to linear polarization which may be oriented either parallel or perpendicular with respect to the magnetic field in the absorption cell.

Detector A is a Golay cell which is normally used for tuning the laser to the desired wavelength and power level. The laser can be operated at 78, 79, or 119 μm with H_2O vapor in the discharge tube and at 72, 84, or 108 μm with D_2O. The power level is normally between 1 and 10 μW at the detectors. Detector B is a liquid helium cooled bolometer which is used to detect the modulated absorption signal during spectroscopic measurements.

A 15 inch magnet with 5.5 inch ring shimmed pole tips and a 2 inch air gap was used to produce fields up to 23.5 kG in the absorption cell. An adjustable field modulation of up to 100 G at 270 Hz was used to generate the modulated absorption signal which was measured using phase sensitive detection. The active absorption region is a small volume (about 2 cm^3) defined by the intersection of the mutually perpendicular laser beam, homogeneous magnetic field, and a flow tube which delivers the paramagnetic molecules and is not shown in Fig. 1.

The sensitivity of the system is greatly increased compared with conventional absorption spectrometers because the absorption cell is inside the laser cavity. In fact, OH radical densities as low as 10^8 radicals per cm^3 have been detected using a 1 second integration time.

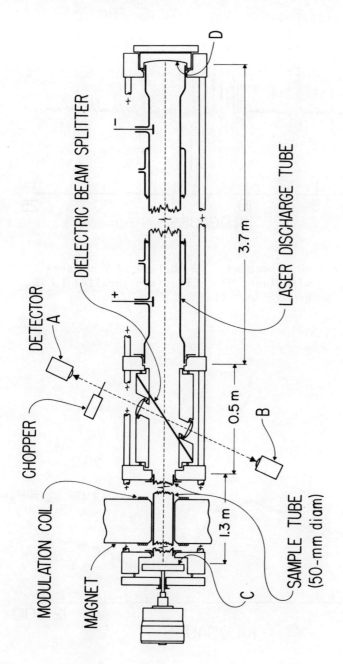

Figure 1. Schematic diagram of the LMR apparatus.

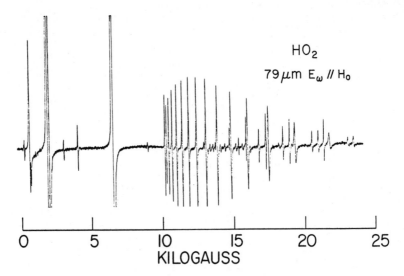

Figure 2. HO_2 spectrum - 79 μm line of H_2O laser
with laser electric field parallel to the
dc magnetic field.

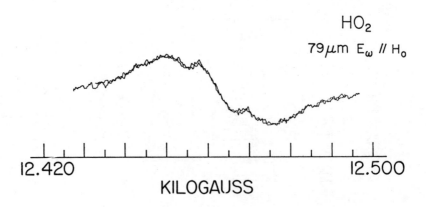

Figure 3.   Saturated absorption spectra of 12.46 kG HO_2
line.

Special procedures must be followed to form the transient free radicals in _situ_ by means of chemical reactions. In most cases the desired radicals can be formed by the reaction of stable molecules with other radicals or atoms formed in an electrodeless microwave discharge. For example,

1) OH is formed by the reaction of H atoms from a discharge with NO_2,

$$H + NO_2 \rightarrow OH + NO.$$

2) HO_2 is formed by the reaction of H atoms with O_2,

$$H + O_2 + M \rightarrow HO_2 + M.$$

3) HCO is formed by the reaction of fluorine atoms with formaldehyde,

$$F + H_2CO \rightarrow HCO + HF.$$

Figure 2 shows an HO_2 spectrum taken with the 79 μm water vapor laser line. In this case the electric field of the laser radiation was polarized parallel to the dc magnetic field.

At very low pressures and with a low modulation amplitude it is possible to increase the resolution by the saturated absorption process (i.e. to observe the Lamb-Dip). The apparent hyperfine components of the HO_2 lines were observed at pressures of about 0.1 Torr and are shown in Fig. 3.

The high sensitivity and versatility of the laser magnetic resonance spectrometer has made it desirable to combine it with a flow reaction system for the study of free-radical reaction kinetics. Measurements are now being made on several important reactions of OH, HO_2, and CHO. This is the first time it has been possible to directly study the kinetics and mechanisms of numerous radical reactions which are necessary for a complete understanding of the chemistry of discharges, flames and other combustion processes, and the atmosphere. In particular, the application of this work to problems of normal and polluted atmospheres can be very important in assessing the impact of technology on the environment.

References

1. K. M. Evenson, H. P. Broida, J. S. Wells, R. J. Mahler, and M. Mizushima, Phys. Rev. Lett. _21_. 1038 (1968).

2. M. Mizushima, J. S. Wells, K. M. Evenson, and William M. Welch, Phys. Rev. Lett. _29_, 831 (1972).

3.   K. M. Evenson and M. Mizushima, Phys. Rev. A6, 2197 (1972).

4.   M. Mizushima, K. M. Evenson, and J. S. Wells, Phys. Rev. A5, 2276 (1972).

5.   R. F. Curl, Jr., K. M. Evenson, and J. S. Wells, J. Chem. Phys. 56, 5143 (1972).

6.   K. M. Evenson, J. S. Wells, and H. E. Radford, Phys. Rev. Lett. 25, 199 (1970).

7.   P. Kasuya and K. Shimoda, Japan J. Appl. Phys. 11, 1571 (1972).

8.   K. M. Evenson, H. E. Radford, and M. M. Moran, Jr., Appl. Phys. Lett. 18, 426 (1971).

9.   H. E. Radford, K. M. Evenson, and C. J. Howard [Paper in preparation].

10   R. F. Curl, Jr., C. J. Howard, and K. M. Evenson [Paper in preparation].

INFRARED-MICROWAVE DOUBLE RESONANCE MEASUREMENTS OF T_1 IN METHYL FLUORIDE AND METHYL CHLORIDE

J. C. McGurk, C. L. Norris, T. G. Schmalz, E. F. Pearson and W. H. Flygare

University of Illinois
Noyes Laboratory, Urbana, Illinois 61801

In the experiments described here we pulse-excite a vibration-rotation transition with a Q-switched CO_2 laser where the laser pulse is short relative to any relaxation processes in the system. We then observe the effect of the population change caused by the laser pulse by observing the time dependent change in the absorption coefficient of a rotational transition which has a rotational J level in common with the two level system which was excited by the laser. We treat this three level system by a two stage solution of the Bloch equations. First we solve the Bloch equations to describe the effects of the infrared laser pulse. We then use the infrared pulse excited solution as the initial condition for the solution of the Bloch equations to describe the microwave absorption.[2]

In general we start with the Hamiltonian given by

$$\mathcal{H} = \mathcal{H}_0 - 2\hat{\underset{\sim}{\mu}} \cdot \hat{e} \, \mathcal{E} \cos(\omega t - kz) \tag{1}$$

where $\mathcal{H}_0$ is the time independent Hamiltonian of the two level system with eigenvalues E_a and E_b corresponding to the lower and upper states, respectively. $\underset{\sim}{\mu}$ is the dipole moment operator and

$$\underset{\sim}{E}(t) = 2\hat{e} \, \mathcal{E}(z,t) \cos(\omega t - kz) \tag{2}$$

is the electric field, polarized in the $\hat{e}$ direction. Using the density matrix formalism, the time dependent Schrödinger equation may be solved to yield the electric dipole analog of the NMR Bloch equations when $\Delta M \neq 0$ rotational transitions are unimportant.[2]

$$\frac{\partial P_r}{\partial t} + \Delta\omega P_i + P_r/T_2 = 0$$

$$\frac{\partial P_i}{\partial t} - \Delta\omega P_r + \varkappa^2 \mathcal{E} \left(\frac{\hbar\Delta N}{4}\right) + P_i/T_2 = 0$$

$$\frac{\partial}{\partial t}\left(\frac{\hbar\Delta N}{4}\right) - \mathcal{E} P_i + \frac{\hbar}{4}\frac{(\Delta N - \Delta N_0)}{T_1} = 0$$

$$\Delta\omega = \omega_0 - \omega$$

$$\omega_0 = (E_b - E_a)/\hbar \tag{3}$$

ω is the radiation frequency, P_r and P_i are the real and imaginary components of the macroscopic polarization of the gas, $\Delta N = N_a - N_b$ is the population difference in the two state system with initial value $\Delta N_0 = (N_a)_0 - (N_b)_0$.

We interpret T_1 in Eq. (3) as the relaxation of the population difference after a sudden disturbance. These relaxations require a change in J, the rotational state. T_2 is the relaxation of the induced polarization. Contributions to T_2 arise from three mechanisms: 1) Those processes which change J, 2) those processes which change only M, and 3) those processes which change only the phase of the polarization. For a weakly absorbing gas it is sufficient to substitute for $\mathcal{E}(z,t)$ in Eqs. (3) the incident field $\mathcal{E}(t)$. We can then replace $\frac{\partial}{\partial t}$ by $\frac{d}{dt}$. Under this condition it can be shown from Maxwell's equations that the absorption coefficient is given by[3]

$$\gamma = -\frac{4\pi\omega}{c}\frac{P_i}{\mathcal{E}} \tag{4}$$

This can be seen simply by using the well known relation for the absorption coefficient in terms of the imaginary part of the molar electric polarizability, α_i,

$$\gamma = -\frac{4\pi\omega}{c}\alpha_i$$

and noting that

$$\alpha_i = \frac{P_i}{\mathcal{E}} \quad .$$

When considering the translational motion of the molecules in a weakly absorbing system it is sufficient to solve Eqs. (3) for p_i using for $\Delta\omega$ the Doppler shifted expression

$$\Delta\omega = \omega_0 - \omega + \omega\frac{v}{c} \tag{5}$$

for molecules moving with velocity v in the z direction. The resulting expression for $\gamma(\omega_0 - \omega + \omega \frac{v}{c}, t)$ is then integrated over the velocity distribution function. We can write this convolution in the usual way as

$$\gamma(t) = \frac{q}{\pi^{1/2}} \int_{-\infty}^{\infty} e^{-q^2 \omega'^2} \gamma(\omega_0 - \omega + \omega', t) d\omega' \qquad (6)$$

where

$$\omega' = \frac{\omega v}{c}$$

and

$$q = \frac{\sqrt{\ln 2}}{\Delta \omega_D} \qquad (7)$$

with the Doppler half width at half height, $\Delta \omega_D$, given by

$$\Delta \omega_D = \frac{\omega}{c} \sqrt{\frac{2kT\ln 2}{m}} \quad . \qquad (8)$$

We are now prepared to apply the above equations to describe the results of the double resonance experiment described previously. First we must solve Eqs. (3) to determine the effect of the laser pulse on the lower level being perturbed by the laser. These results will give the initial conditions to describe the microwave absorption after the infrared pulse excitation has ceased. We shall assume for simplicity a square pulse of amplitude $2\mathcal{E}_p$ and pulse duration t_p. Since $t_p \ll T_1, T_2$ we can neglect the collisional damping terms in Eqs. (3) and solve instead the equations

$$\frac{dP_r}{dt} + \Delta \omega P_i = 0$$

$$\frac{dP_i}{dt} - \Delta \omega P_r + \varkappa^2 \mathcal{E} \left(\frac{\hbar \Delta N}{4}\right) = 0$$

$$\frac{d}{dt}\left(\frac{\hbar \Delta N}{4}\right) - \mathcal{E} P_i = 0 \quad . \qquad (9)$$

The solution to these equations gives[2]

$$\frac{\Delta N_p}{\Delta N_{op}} = \frac{(\Delta \omega_p)^2}{\varkappa_p^2 \mathcal{E}_p^2 + (\Delta \omega_p)^2} + \frac{\varkappa_p^2 \mathcal{E}_p^2 \cos \Omega t_p}{\varkappa_p^2 \mathcal{E}_p^2 + (\Delta \omega_p)^2} \qquad (10)$$

where

$$\Omega = [\varkappa_p^2 \mathcal{E}_p^2 + (\Delta \omega_p)^2]^{1/2} \quad . \qquad (11)$$

We use the subscript p to denote a variable involved in the pulsed laser transition and sub-zeros indicate equilibrium values.

We must now solve Eqs. (3) for the microwave transition. The weak microwave power will have a negligible effect on the population difference between the energy levels compared to the effect of the laser pulse. It should then be a good approximation to neglect the $-\mathcal{E}p_i$ term in the last of Eqs. (3) and solve instead the equations

$$\frac{dP_r}{dt} + \Delta\omega P_i + \frac{P_r}{T_2} = 0$$

$$\frac{dP_i}{dt} - \Delta\omega P_r + \varkappa^2\mathcal{E}\left(\frac{\hbar\Delta N}{4}\right) + \frac{P_i}{T_2} = 0$$

$$\frac{d}{dt}\left(\frac{\hbar\Delta N}{4}\right) + \frac{\hbar}{4}\frac{(\Delta N - \Delta N_0)}{T_1} = 0 \quad . \tag{12}$$

The exact solution for P_i is given by[2]

$$\frac{P_i}{\Delta N_0} = -\frac{\hbar\varkappa^2\mathcal{E}}{4}(m_0-1)\left\{\frac{(\frac{1}{T_2}-\frac{1}{T_1})}{(\frac{1}{T_2}-\frac{1}{T_1})^2+(\Delta\omega)^2}\left(e^{-t/T_2}\cos\Delta\omega t\right)\right.$$

$$\left.+\left[1-\frac{(\frac{1}{T_2}-\frac{1}{T_1})^2}{(\frac{1}{T_2}-\frac{1}{T_1})^2+(\Delta\omega)^2}\right]e^{-t/T_2}\frac{\sin\Delta\omega t}{\Delta\omega}\right\}-\frac{\hbar\varkappa^2\mathcal{E}}{4}\frac{\frac{1}{T_2}}{\frac{1}{T_2^2}+(\Delta\omega)^2} \tag{13}$$

where m_0 is the value of $\Delta N/\Delta N_0$ at time t=0 (immediately after the laser pulse). We now need an expression relating (m_0-1) to N_p and N_{op}.

Consider the three level system shown below.

infrared, vibration-rotation

microwave, rotation

$(N_b)_0 - (N_b)_{t=0} = (N_c)_{t=0} - (N_c)_0$

$(N_a)_0 = (N_a)_{t=0}$

where the sub 0 indicates equilibrium and the sub t=0 indicates the value immediately after the pulse. According to the diagram

$$\Delta N_{t=0} = (N_a)_{t=0} - (N_b)_{t=0}$$

$$\Delta N_o = (N_a)_o - (N_b)_o$$

$$\Delta N_p = (N_b)_{t=0} - (N_c)_{t=0} \qquad (14)$$

$$\Delta N_{op} = (N_b)_o - (N_c) \quad .$$

Eqs. (14) and the equations in the diagram can then be used to show that

$$\left(\frac{\Delta N}{\Delta N_o} - 1\right) = (m_0-1) = \frac{\Delta N_{op}}{2\Delta N_o}\left(1 - \frac{\Delta N_p}{\Delta N_{op}}\right) \quad . \qquad (15)$$

Eq. (15) is derived for the case of the upper microwave level being part of the vibration-rotation transition being excited. When the lower microwave level is coincident with the vibration-rotation level being excited with the laser, we rewrite Eq. (15) to give

$$(m_0-1) = \frac{\Delta N_{op}}{2\Delta N_o}\left(\frac{\Delta N_p}{\Delta N_{op}} - 1\right) \quad . \qquad (16)$$

Returning now to the case shown in the diagram, Eq. (15) can be used to rewrite Eq. (10) in the form

$$(m_0-1) = \left(\frac{\Delta N_{op}}{2\Delta N_o}\right)\frac{\varkappa_p^2 \mathcal{E}_p^2(1-\cos\Omega t_p)}{\varkappa_p^2 \mathcal{E}_p^2 + (\Delta w_p)^2} \quad . \qquad (17)$$

Since in our experiments the directions of the laser and microwave beams were to a good approximation perpendicular (see Fig. 1) we must integrate the absorption coefficient over the two dimensional velocity distribution function with Δw_p given by

$$\Delta w_p = w_{op} - w_p + w_p \frac{u}{c} \qquad (18)$$

and Δw given by

$$\Delta w = w_0 - w + w \frac{v}{c} \qquad (19)$$

where u and v are the velocity components in the directions of the laser and microwave beams, respectively. Since u and v are independent variables we can perform the integrations separately. That is, we use in Eq. (13) the value of (m_0-1) obtained by integrating Eq. (17) over the u velocity distribution function. By analogy with Eq. (6) we have

$$(m_0-1)_D = \frac{q_p}{\pi^{1/2}} \left(\frac{\Delta N_{op}}{2\Delta N_0}\right) \int_{-\infty}^{\infty} \frac{e^{-q_p^2\omega'^2}\; \varkappa_p^2 \mathcal{E}_p^2(1-\cos\Omega t_p)}{\varkappa_p^2 \mathcal{E}_p^2 + (\omega_{op}-\omega_p+\omega')^2}\, d\omega'$$

$$(20)$$

where from Eq. (11)

$$\Omega = [\varkappa_p^2 \mathcal{E}_p^2 + (\omega_{op}-\omega_p+\omega')^2]^{1/2}$$

$$\omega' = \frac{\omega u}{c} \qquad\qquad\qquad\qquad\qquad\qquad (21)$$

For molecules moving with velocity v the absorption coefficient is, from Eq. (4),

$$\Upsilon(v) = -\frac{4\pi\omega}{c}\frac{P_i(v)}{\mathcal{E}}\; . \qquad\qquad\qquad (22)$$

Before integrating over the velocity distribution function we shall make some simplifying assumptions. We shall assume that

$$\left(\frac{1}{T_2}-\frac{1}{T_1}\right) \ll \Delta\omega_D\; . \qquad\qquad\qquad (23)$$

For the experiments described in this paper, Eq. (23) imposes no restrictions on the relative sizes of T_1 and T_2 since our measurements were made in the Doppler broadened limit $\left(\frac{1}{T_2} \ll \Delta\omega_D\right)$.

Now it is a good approximation to put

$$\frac{\left(\frac{1}{T_2}-\frac{1}{T_1}\right)}{\left(\frac{1}{T_2}-\frac{1}{T_1}\right)^2 + (\omega_0-\omega+\omega')^2} \cong \pi\delta(\omega_0-\omega+\omega') \qquad (24)$$

where δ is the Dirac δ-function and

$$\frac{\left(\frac{1}{T_2}-\frac{1}{T_1}\right)^2}{\left(\frac{1}{T_2}-\frac{1}{T_1}\right) + (\omega_0-\omega+\omega')^2} \cong 0\; . \qquad\qquad (25)$$

In the limit where $\left(\frac{1}{T_2}-\frac{1}{T_1}\right) = 0$, Eqs. (24) and (25) are exact.

We shall neglect the steady state term in Υ and compute only the transient term Υ_t. Combining Eqs. (13), (22), (24) and (25) we have, after convolution with the velocity distribution function

$$Y_t = \frac{\pi^{1/2} \hbar \varkappa^2 \Delta N_0 (m_0-1) D^q}{c} [e^{-t/T_1} I_1 - e^{-t/T_2} I_2 + e^{-t/T_2} I_3] \quad (26)$$

where

$$I_1 = \pi \int_{-\infty}^{\infty} e^{-q^2 \omega'^2} \delta(\omega_0 - \omega + \omega') d\omega' \quad (27)$$

$$I_2 = \pi \int_{-\infty}^{\infty} e^{-q^2 \omega'^2} \delta(\omega_0 - \omega + \omega') \cos(\omega_0 - \omega + \omega') t \, d\omega' \quad (28)$$

$$I_3 = \pi \int_{-\infty}^{\infty} e^{-q^2 \omega'^2} \frac{\sin(\omega_0 - \omega + \omega') t}{(\omega_0 - \omega + \omega')} d\omega' \quad . \quad (29)$$

The integrals I_1 and I_2 may be evaluated immediately with the result

$$I_1 = I_2 = \pi e^{-q^2(\omega_0-\omega)^2} \quad . \quad (30)$$

The integral I_3 may be approximated in two important cases:[1]

Case (a) $t \ll 2q$

$$I_3^{(a)} = \frac{\pi^{1/2}}{q(\omega-\omega_0)} \sin(\omega-\omega_0) t \quad (31)$$

Case (b) $t \gg 2q$

$$I_3^{(b)} = \pi e^{-(\omega-\omega_0)^2 q^2} \quad . \quad (32)$$

Substituting Eqs. (30), (31) and (32) into Eq. (26) we have for the absorption coefficient in the two cases:

Case (a) $t \ll 2q$

$$Y_t^{(a)} = \frac{\pi^{3/2} \hbar \varkappa^2 \Delta N_0 (m_0-1) D^q}{c} \{ e^{-q^2(\omega-\omega_0)^2} [e^{-t/T_1} - e^{-t/T_2}]$$

$$+ \frac{\pi^{-1/2}}{q(\omega-\omega_0)} e^{-t/T_2} \sin(\omega-\omega_0) t \} \quad (33)$$

Case (b) $t \gg 2q$

$$Y_t^{(b)} = \frac{\pi^{3/2} \hbar \varkappa^2 \Delta N_0 (m_0-1) D^q}{c} e^{-q^2(\omega-\omega_0)^2} e^{-t/T_1} \quad . \quad (34)$$

In Eq. (33) we note that as $|\omega-\omega_0|$ increases the first term decreases very much faster than the second. A few Doppler widths off resonance then, for $t \ll 2q$, the absorption will be a damped sine function of frequency $\omega-\omega_0$ with a characteristic relaxation time T_2. For $t \gg 2q$ Eq. (34) shows that the absorption will be a maximum when $\omega = \omega_0$ and will decay exponentially with the relaxation time T_1.

We have observed the microwave absorption as described in Eq. (33) in $^{13}CH_3F$ and $^{12}CH_3^{35}Cl$ both pulse excited by a CO_2 laser. The experimental apparatus is shown in Fig. 1. The CO_2 laser, consisting of a water cooled discharge tube and external mirrors, was of the conventional flow design and was not stabilized. The laser cavity was formed by a 4 meter radius gold-coated spherical mirror and a plane diffraction grating. A 0.8 mm hole was drilled through the center of the mirror allowing a small amount of the laser radiation to be coupled out and passed into a 0.5 meter Perkin-Elmer monochromator. This monochromator was calibrated with a He-Ne laser and was used to determine which CO_2 vibration-rotation transition was oscillating. The diffraction grating (1800 lines/inch) was oriented so that the first order diffracted beam reflected back onto the spherical mirror forming the laser cavity. Laser oscillation could be obtained on any desired CO_2 vibration-rotation transition by rotating the grating. An iris placed in the cavity near the spherical mirror helped to limit the laser oscillation to a single transition and also confined the oscillation to the lowest transverse mode. About 5-10% of the incident power on the grating was reflected in zero order out of the cavity. This zero order reflected beam was diverged by a spherical mirror with a one meter focal length and directed into the microwave absorption cell through a NaCl window. The laser was mechanically Q-switched by rotating the mirror which forms one end of the cavity. In this mode of operation the average power was of the order of 0.1 watts with a peak pulse power of 2000 watts for a 0.5 microsecond pulse. The Q-switched laser pulses were directly observed using a liquid nitrogen cooled gold-doped germanium detector.

The microwave absorption cell consisted of a 14 in. length of 4 in. diameter Corning brand conical pipe. Quartz windows one eighth inch thick were mounted on the cell to transmit the microwave radiation. In order to increase the overlap of the laser radiation with the microwave radiation inside the cell, a pair of gold plated plane mirrors ~6 cm apart were used to achieve multiple reflections of the laser beam. Sample pressure in the cell was measured by a McLeod gauge which was separated from the cell by a $-70^{\circ}C$ trap for $^{13}CH_3F$ and by a differential capacitance monometer in the CH_3Cl experiment. The cell was capable of maintaining a pressure of less than 10^{-3} torr when sealed off for 30 minutes.

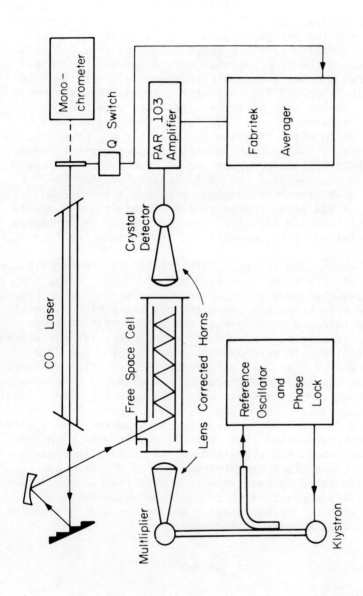

Figure 1. Block diagram for the time resolved infrared-microwave double resonance experiments.

Microwave radiation for the probe signal was derived from
standard klystron harmonic generators. The output of the multiplier
was coupled into the cell by a lens corrected horn with a 3 cm
aperture. A similar arrangement was used to couple the signal to
the detector. A point contact silicon detector followed by a PAR
model 103 amplifier was used to observe the time dependent double
resonance signals. The detector operated as a voltage generator
with no external bias and care was taken to insure that the detector was not limiting the time response of the system. For pressures greater than 2 mtorr the double resonance signals could be
observed directly on an oscilloscope with a signal to noise ratio
of 3-4. Actual measurements of relaxation times were made by signal averaging on a Fabritek 1074. Typically each measurement consisted of from two to eight thousand scans. In the $^{13}CH_3F$ experiment
the range of pressures over which relaxation times were measured
was limited to 0.25-5.0 mtorr by the one microsecond per channel
dwell time of the averager and the cell dimensions. For CH_3Cl a
Biomation model 610B transient recorder with a 0.1 microsecond per
channel dwell time was added to the system between the PAR amplifier and the Fabritek (see Fig. 1). The measurement of relaxation
times was then limited by the duration of the laser pulse itself.

The $^Q R_3(4)$ transition in the ν_3 vibrational band of $^{13}CH_3F$ was
pumped by the P(32) CO_2 laser line at 9.657 microns.[4] Using the
laser in the Q-switched mode we observed appreciable time dependent
double resonance effects on the two microwave transitions directly
connected to the level being pumped by the laser. Fig. 2 shows
the energy level structure, the relevant transitions for our experiment, and a typical double resonance signal for the J=3→4; K=3
transition in $^{13}CH_3F$. The maximum increase in absorption, which
was observed when the microwave frequency was directly on resonance,
was about 4 percent and all relaxation curves were fit with a single
exponential decay. Double resonance signals for the J=4→5; K=3
transition showed decreases in absorption, as expected, and yielded
pressure dependent relaxation times agreeing quite well with those
obtained from the J=3→4; K=3 transition. The relaxation times obtained from both transitions are plotted in Fig. 3. The major
source of experimental uncertainty in this data lies in the determination of the sample pressure which could be measured to no
better than 10% with our McLeod gauge. A least squares fit of the
data in Fig. 3 yields a pressure dependence of

$$\frac{1}{2\pi T} = 3.5 \pm 0.6 \text{ kHz} + (15.2 \pm 0.8 \text{ kHz/mTorr}) P \quad .$$

The pressure independent contribution of 3.5 kHz (45 μsec)
corresponds to the molecular transit time across the microwave
beam. The corresponding pressure dependent relaxation time is
T = (10.5 $\pm$ 0.6) μsec·mtorr.

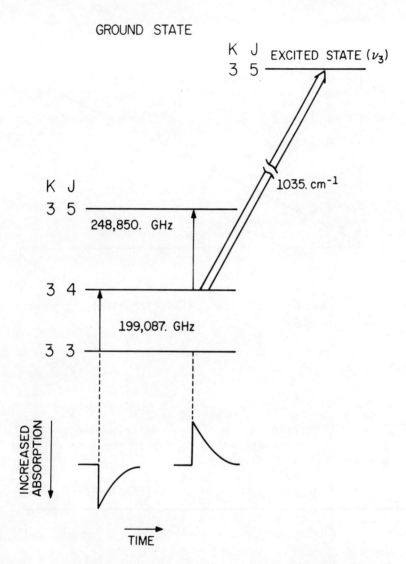

Figure 2. Energy level diagram and expected double resonance signals for the infrared-microwave double resonance experiments described here in $^{13}CH_3F$. The 1035 cm^{-1} infrared transition is excited by the ν_3 $^Q R(4)$ vibration-rotation transition in the CO_2
laser.

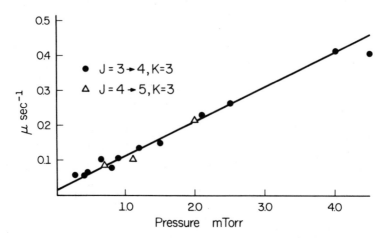

Figure 3. A plot of reciprocal relaxation times $(\frac{1}{T})$ versus pressure for $^{13}CH_3F$.

Table I. The relaxation times measured in the J=5→6, K=3, F = 13/2 → 15/2 transition in $CH_3{}^{35}Cl$ as described in the text. The uncertainties are estimated as 10% in the case of pure CH_3Cl gas and 15% in all other cases.

Foreign gas	$(1/2\pi T)$ kHz/mTorr
CH_3Cl	10.9
OCS	5.4
He	2.3
CH_3F	7.3
Ar	2.1

We also searched for double resonance effects in transitions of $^{13}CH_3F$ not directly connected to the level being pumped by the laser. For the J=3→4, K=0 transition a detectable decrease in absorption was observed. This effect was at least two orders of magnitude below that observed for the levels directly connected to the pumped level. No affect was observed for the K=1 and K=2 lines. Other transitions with K=3 could not be observed because they were beyond the frequency range of the spectrometer.

We also observed transient effects in $^{12}CH_3{}^{35}Cl$ where the $^RQ_3(6)$ vibrational transition was pumped by the P(26) CO_2 laser line.[5] We observed transient effects in the J=5→6, K=3 and J=6→7, K=3 rotational transitions which appeared very similar to the results diagramed in Fig. 2 on methylfluoride. The J=5→6, K=3 methylchloride (^{35}Cl) transition is actually a quartet due to the ^{35}Cl nuclear quadrupole coupling. The double resonance effect and subsequent relaxation measurement was observed in each member of the quartet at the following frequencies:

J=5→6, K=3

F = 7/2 → 9/2	159,479.46 MHz
F = 13/2 → 15/2	159,478.42 MHz
F = 9/2 → 11/2	159,474.70 MHz
F = 11/2 → 13/2	159,473.66 MHz

The relaxation times in each $\Delta F = +1$ transition were identical with the results shown in Table I.

We also list the relaxation times for foreign gas broadening with OCS, He, CH_3F, and Ar in Table I. According to our previous discussion in the case of methylfluoride, the on-resonance experiments described in this paper (Eq. (33)) give T_1.

For the J=5→6, K=3 transition in CH_3Cl the third harmonic of the klystron was used to provide the microwave probe signal. This resulted in a factor of between 5 and 10 increase in power over that used in the $^{13}CH_3F$ measurements (4th and 5th klystron harmonics). It was found that the decay rates increased with increasing microwave power. At the highest available power the relaxation curves were markedly non-exponential. When the power was decreased by a factor of two the relaxation curves could be fitted by an exponential though there was still some curvature in the semi-log plots. A plot of reciprocal relaxation time against pressure was linear and yielded the pressure dependent rate of

$$\frac{1}{2\pi T} = 34.2 \text{ kHz} + 10.8 \text{ kHz mtorr}^{-1}$$

We note the intercept is unrealistically large for a molecular transit time across the microwave beam (cf. the $^{13}CH_3F$ intercept of 3.5 kHz). When the power was decreased by a further factor of two the relaxation curves became good exponentials although there was no appreciable difference in the derived rates. A similar effect was noted for the weakest quadrupole component $F = 7/2 \rightarrow 9/2$ which has a line strength about a factor of two less than the $F = 13/2 \rightarrow 15/2$ transitions. This is to be expected as the power dependent terms will enter the theory in the form of power x line strength $(\mu^2 \mathcal{E}^2)$. When the power was attenuated by about a factor of ten from the maximum, the pressure dependent rate of

$$\frac{1}{2\pi T} = 11.9 \text{ kHz} + 10.9 \text{ kHz mtorr}^{-1}$$

was observed. Here the intercept is much lower than the high power data but the slope is unchanged. A similar behavior was noted in the foreign gas measurements. The intercept but not the slope was power dependent. We are in the process of examining a number of theoretical reasons for these microwave power dependent rates and they will be fully reported at a later date.

Acknowledgement: The support of the National Science Foundation is gratefully acknowledged.

REFERENCES

1. H. Jetter, E. F. Pearson, C. L. Norris, J. C. McGurk, and W. H. Flygare, J. Chem. Phys. 00, 0000 (1973).

2. J. C. McGurk, T. G. Schmalz, and W. H. Flygare, "A Density Matrix, Bloch Equation Description of Infrared and Microwave Transient Phenomena," Advances in Chemical Physics, Ed. I. Prigogine and S. A. Rice (John Wiley and Sons, New York, 1973).

3. S. L. McCall and E. L. Hahn, Phys. Rev. 183, 457 (1969).

4. R. G. Brewer and R. L. Shoemaker, Phys. Rev. Lett. 27, 631 (1971).

5. L. Frenkel, H. Marantz, and T. Sullivan, Phys. Rev. A3, 1640 (1971).

ACCURATE ROTATIONAL CONSTANTS, FREQUENCIES, AND WAVELENGTHS FROM $^{12}C^{16}O_2$ LASERS STABILIZED BY SATURATED ABSORPTION

F. R. Petersen, D. G. McDonald
J. D. Cupp, and B. L. Danielson

Quantum Electronics and Cryogenics Divisions
National Bureau of Standards
Boulder, Colorado 80302

Abstract

New experimental measurements of the frequency separations of 30 pairs of $^{12}C^{16}O_2$ laser lines in the 10.4 um band and 26 pairs in the 9.4 um band have been made with lasers stabilized to Lamb-dip-type resonances observed in the 4.3 um fluorescent radiation. The use of a Josephson junction as the frequency mixing element simplified the measurements. Uncertainties in existing rotational constants for the laser vibrational levels were reduced 20 to 30 times and an additional rotational constant, H_v, was determined for the first time. Frequency and wavelength tables with estimated uncertainties are calculated for both bands with the new constants.

The determination of rotational constants for excited vibrational states in molecules has undergone dramatic improvement as the experimental methods have changed from wavelength metrology to frequency metrology. Wavelength methods were limited primarily by instrumental linewidths of absorption spectrometers to about 0.1 cm^{-1} or 3 GHz. For CO_2, the first big improvement was by Bridges and Chang[1] who used laser sources and measured the difference frequencies of pairs of laser transitions. The uncertainties in the rotational constants were reduced by 25 to 200 times with the accuracy being primarily limited by how well the laser could be set to the center of the Doppler and pressure broadened gain profile of the laser ($\sim$ 100 MHz wide).

With the discovery of standing-wave saturation resonances observed in the 4.3 μm fluorescent radiation,[2] the lasers could easily be servo-controlled to a spectral feature 100 times narrower than previously on every lasing line in the CO_2 spectrum. Thus, it appeared experimentally profitable to re-determine the rotational constants as well as the absolute frequencies. Further motivation is provided by the fact that the CO_2 laser is an excellent secondary frequency standard[3] in the wavelength band from 9 to 11 μm since the frequency of a line in each of the two laser bands has been accurately measured relative to the primary frequency standard.[4] In addition, it can be used as a highly accurate length standard since the accuracy of the speed of light is limited mostly by the uncertainty in the primary length standard.[5] A third motivation and the principal reason for using the Josephson junction was to study the Josephson effect at CO_2 frequencies. Results of this investigation will be reported separately.

Beams from two CO_2 lasers oscillating on adjacent rotational lines were focused onto a Josephson junction and mixed with the nth harmonic ($3 \leq n \leq 6$) of an applied X-band frequency. Thus, the junction served both as the microwave harmonic generator and as the infrared mixer. This technique allowed us to measure laser difference frequencies ranging from 32 to 63 GHz with a single microwave oscillator. The nominally 60 MHz beat frequency from the junction was amplified with a narrow band amplifier and observed on a spectrum analyzer. Due to residual spurious frequency modulation of the lasers, the beat signal spectrum was digitally averaged for 100 seconds to obtain a more accurate beat frequency. The averaged 60 MHz beat frequency, F_B, was measured relative to simultaneously recorded frequency markers by linear interpolation as illustrated in Fig. 1. The X-band klystron was stabilized by a standard phase-lock technique, and its frequency was counted directly by a counter referenced to the NBS atomic time scale via a quartz crystal oscillator. The laser difference frequency is then given by $F_D = n \times F(\text{X-band}) + F_B$.

The CO_2 lasers have been described previously.[4] Pressure in the internal absorption cells was the same for both lasers in every measurement (either 0.030 or 0.040 Torr) and no difference in the beat frequency was measurable over this pressure range. The laser frequency was locked to the zero-slope point on the dip in the 4.3 μm fluorescent radiation. Typical saturated absorption lines were approximately 1 MHz full width at half maximum. Mirror dither produced an FM envelope linewidth of approximately 400 kHz for a single laser, but proper phase and amplitude adjustment of the dither voltage on the two lasers reduced the beat signal linewidth to the range of 20 to 30 kHz (full width at half maximum). On certain weak laser lines, the width increased and was probably related to feedback coming from the vicinity of the junction. For most of the measurement, however, feedback was not an appreciable problem.[6]

Various types of effects, such as pressure shifts, power shifts, base line slope of saturated absorption line, and servo-electronic offsets, can shift the stabilized laser frequency from the unperturbed molecular frequency. These effects, although not at present investigated in great detail, are small and in this experiment were assumed to affect adjacent lines by identical amounts. Hence, no correction was applied. The two lasers when stabilized to the same saturated line normally differed in frequency by less than 2.5 kHz. Consequently, a 2.5 kHz uncertainty is included in the estimate of experimental error for this effect. Other inputs to the estimated experimental error include uncertainty in the X-band frequency, nonlinearity in the spectrum analyzer and data recording system, and uncertainty in determining the center frequency of the averaged beat note including small line asymmetry effects.

Table I gives the measured linewidths, beat frequencies, and estimated experimental uncertainties for each of the 30 lines observed in the 10.4 μm band and 26 lines observed in the

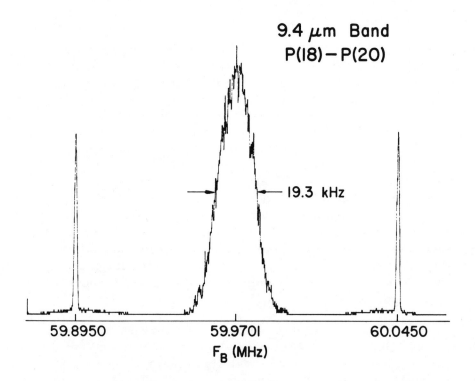

Figure 1.   Signal averaged beat note between transitions P(18) and
             P(20) in the 9.4 μm band with frequency markers. The spec-
             trum analyzer was set for linear response with a pre-detec-
             tion bandwidth of 1 kHz. The post-detection bandwidth,
             as determined by the averaging time of the digital average
             was 0.1 s/channel with 1024 channels for the above spectrum

9. 4 μm band. Our observations agreed well with the frequency
differences predicted with the aid of Bridges' and Chang's
rotational constants except at the higher rotational numbers.
The large shifts from the predicted beat frequencies for certain
lines[1] were not observed.

The term values for the vibrational-rotational energy
levels in CO_2 are given by[7]

$$T(v, J) = G(v) + B_v J(J+1) - D_v J^2(J+1)^2$$

$$+ H_v J^3(J+1)^3 - L_v J^4(J+1)^4 + \ldots \quad , \tag{1}$$

where the quantities involved are defined in their usual sense.
Expressions for the beat frequencies in terms of the rotational
constants and quantum numbers are derived by taking appropriate differences with the aid of Equation (1). Data reduction can
then be done with equations which in matrix notation have the
form

$$F = X\beta + \epsilon \quad . \tag{2}$$

The elements f_i of F are the measured beat frequencies and
the elements ϵ_i of ϵ are the unknown measurement errors,
which are assumed to be uncorrelated and to have a zero mean
and a common variance. The elements x_{ij} of X are the exactly
known functions of J, and the β_j are the rotational constants.[8]
The method of least squares was[9] used to estimate β and its
uncertainty and to propagate the uncertainties in forming linear
combinations of β_j.

Data analysis was performed on a digital computer which
maintained 25 significant figures throughout the calculations.
The input data were not weighted since the experimental uncertainties were of comparable magnitude. It was estimated
that at the higher J-values terms involving rotational constants
through L_v would be significant compared to the 3 kHz estimated
experimental uncertainty in the measured beat frequencies.
Therefore, various numbers of constants were tried to test the
goodness of the fit. Figure 2 shows a plot of the residuals for

Table I.   Observed and calculated laser difference frequencies. ΔF_B is the full width at half intensity of the 60 MHz beat frequency, f_i is the experimental measurement of the laser difference frequency, δf_i is the estimated experimental uncertainty in f_i, and r_i is the residual, i.e., the observed minus the predicted frequency. All frequencies are in MHz.

	00°1 – [10°0, 02°0]$_I$ Band (10.4 μm)				00°1 – [10°0, 02°0]$_{II}$ Band (9.4 μm)			
	ΔF_B	f_i	δf_i	r_i	ΔF_B	f_i	δf_i	r_i
P(42)-P(44)	0.1610	62511.8642	0.0029	0.0006				
P(40)-P(42)	0.0983	61747.9649	0.0039	-0.0015				
P(38)-P(40)	0.0881	60987.9862	0.0050	0.0027				
P(36)-P(38)	0.0627	60231.6550	0.0035	-0.0024				
P(34)-P(36)	0.0499	59478.7329	0.0039	-0.0021	0.0332	60275.3251	0.0027	0.0010
P(32)-P(34)	0.0365	58728.9700	0.0027	0.0020	0.0227	59527.3102	0.0026	-0.0021
P(30)-P(32)	0.0268	57982.1158	0.0026	0.0036	0.0281	58774.1189	0.0029	0.0009
P(28)-P(30)	0.0375	57237.9238	0.0030	-0.0037	0.0210	58015.9809	0.0028	0.0014
P(26)-P(28)	0.0316	56496.1807	0.0032	0.0031	0.0197	57253.1375	0.0029	-0.0022
P(24)-P(26)	0.0308	55756.6255	0.0027	-0.0043	0.0217	56485.8477	0.0027	0.0020
P(22)-P(24)	0.0293	55019.0548	0.0026	0.0002	0.0217	55714.3462	0.0027	-0.0028
P(20)-P(22)	0.0286	54283.2255	0.0028	-0.0001	0.0256	54938.9063	0.0033	0.0016
P(18)-P(20)	0.0287	53548.9212	0.0030	0.0018	0.0193	54159.7720	0.0027	0.0003
P(16)-P(18)	0.0220	52815.9168	0.0027	0.0019	0.0288	53377.2118	0.0029	0.0000
P(14)-P(16)	0.0220	52083.9924	0.0028	-0.0012	0.0230	52591.4887	0.0028	-0.0011
P(12)-P(14)	0.0247	51352.9398	0.0027	0.0008	0.0249	51802.8745	0.0027	0.0012
P(10)-P(12)	0.0239	50622.5349	0.0029	-0.0016	0.0286	51011.6321	0.0027	0.0001
R(12)-R(10)	0.0280	42201.6916	0.0026	-0.0014	0.0197	41812.5978	0.0029	-0.0005
R(14)-R(12)	0.0333	41461.7219	0.0027	0.0022	0.0253	41011.7838	0.0030	-0.0024
R(16)-R(14)	0.0259	40719.5614	0.0026	-0.0019	0.0223	40212.0685	0.0029	0.0006
R(18)-R(16)	0.0248	39975.0115	0.0031	0.0021	0.0197	39413.7140	0.0032	0.0006
R(20)-R(18)	0.0328	39227.8396	0.0027	-0.0022	0.0207	38616.9922	0.0027	0.0019
R(22)-R(20)	0.0246	38477.8438	0.0027	0.0017	0.0213	37822.1630	0.0030	-0.0007
R(24)-R(22)	0.0227	37724.7889	0.0026	-0.0007	0.0243	37029.4931	0.0031	-0.0027
R(26)-R(24)	0.0354	36968.4624	0.0027	0.0014	0.0236	36239.2524	0.0027	0.0070
R(28)-R(26)	0.0395	36208.6288	0.0031	-0.0015	0.0191	35451.6679	0.0029	-0.0003
R(30)-R(28)	0.0294	35445.0688	0.0034	0.0004	0.0247	34667.0134	0.0026	-0.0023
R(32)-R(30)	0.0248	34677.5481	0.0026	0.0053	0.0208	33885.5331	0.0026	-0.0025
R(34)-R(32)	0.0381	33905.8081	0.0029	-0.0095	0.0254	33107.4698	0.0026	-0.0013
R(36)-R(34)	0.0416	33129.6573	0.0034	0.0043	0.0346	32333.0632	0.0026	0.0024

Std. Dev. of Residuals = 0.0033 MHz.   Std. Dev. of Residuals = 0.0024 MHz.
Degrees of Freedom = 24   Degrees of Freedom = 20

both bands with two constants (B_v, D_v) and with three constants (B_v, D_v, H_v) used in the analysis. The systematic trend of the residuals with two constants is greatly reduced by the addition of H_v. The addition of L_v increased the uncertainties in the other coefficients without significantly reducing the residuals, and the estimated uncertainty in L_v was comparable to L_v itself. An F-type confidence test[9] indicated that H_v was highly significant while L_v was not statistically different from zero.

Results for the rotational constants when determined from the individual bands are given in the first two columns of Table II, and the residuals along with their standard deviations are given in Table I. The 3 kHz standard deviation of the residuals is comparable to the estimated experimental error and is about 100 times smaller than the previous best results — an

improvement probably best attributed to the factor of 100 reduction in the linewidth. Since our two independent measurements of the rotational constants for the $00^{\circ}1$ level agreed well
within the estimated error, we next fit constants to both bands
taken simultaneously. These results are given in the third
column of Table II. The standard deviation of the residuals in
this case was 2. 8 kHz, and the estimated error was reduced
by 20 to 40%. The simultaneous fit, therefore, is taken as the
best determination of the constants from our experimental data.
A comparison between the new values and previous best values[1]
for the rotational constants shows a 20 to 30 times reduction in
estimated standard errors for B_V and D_V and a statistically
significant measurement of H_V for the first time.

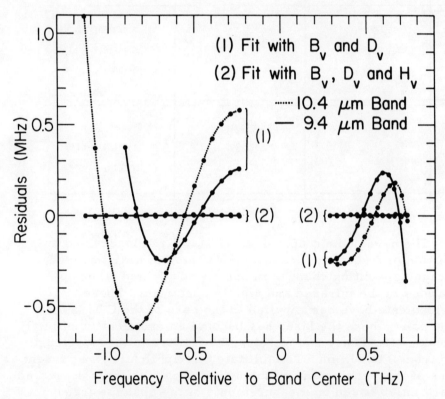

Figure 2.   Residuals plotted relative to the band center with (1) B_V
             and D_V and (2) B_V, D_V, and H_V used in the analysis. Stan-
             dard deviations of the residuals in the type (1) analysis
             are 416 and 214 kHz for the 10.4 and 9.4 μm bands respec-
             tively versus 3.3 and 2.4 kHz respectively for the type
             (2) analysis.

Table II.   Rotational constants for $^{12}C^{16}O_2$ derived from a least squares fit with 1-standard-deviation estimates of uncertainties. To obtain 95% confidence interval estimates, the 1-standard-deviation uncertainties for the 10.4 μm, 9.4 μm, and simultaneous fits should be multiplied by 2.064, 2.086, and 2.014, respectively, where these constants have been obtained from Student's t distribution for 24, 20, and 47 degrees of freedom. Constants uncertain in the last three significant figures are given to minimize rounding off errors when frequency and wavelength tables are calculated.

	00°1 − [10°0, 02°0] 10.4 μm Band I (MHz)	00°1 − [10°0, 02°0] 9.4 μm Band II (MHz)	Simultaneous Fit Both Bands (MHz)
$B_{00°1}$	11606.207194(530)	11606.207271(419)	11606.207253(335)
$D_{00°1}$	$3.988269(298) \times 10^{-3}$	$3.988226(248) \times 10^{-3}$	$3.988264(193) \times 10^{-3}$
$H_{00°1}$	$5.260(829) \times 10^{-10}$	$4.926(716) \times 10^{-10}$	$5.150(546) \times 10^{-10}$
B_I	11697.569699(521)		11697.569760(329)
D_I	$3.446178(280) \times 10^{-3}$		$3.446179(181) \times 10^{-3}$
H_I	$5.7380(741) \times 10^{-9}$		$5.7297(488) \times 10^{-9}$
B_{II}		11706.364891(419)	11706.364873(335)
D_{II}		$4.711530(249) \times 10^{-3}$	$4.711568(197) \times 10^{-3}$
H_{II}		$6.9615(732) \times 10^{-9}$	$6.9839(575) \times 10^{-9}$

The development of molecular stabilized lasers and the extension of frequency measurements into the near infrared are causing exciting changes in spectroscopic and other applications in the infrared and visible spectrum. These two developments have been applied in the case of the CO_2 laser to the point where the laser has become an excellent (unofficial) secondary frequency and wavelength standard in the wavelength band from 9 to 11 μm. To facilitate this application we present tables of frequencies and wavelengths along with their estimated uncertainties based on measurements of the absolute frequency of a line in each band and on the new rotational constants. In both of these basic measurements, $^{12}C^{16}O_2$ lasers stabilized to the standing-wave saturation resonance observed in the 4.3 μm fluorescent radiation from a low-pressure internal CO_2 absorption cell were used.

The input data contains the absolute frequencies of the reference lines which were recently measured to be 29. 442 483 315 (25) THz for the 10. 18 μm R(30) transition and 32. 134 266 891 (24) THz for the 9. 33 μm R(10) transition.[4] Frequencies, vacuum wavenumbers, and vacuum wavelengths for lines P(60) through R(58) are calculated relative to these absolute frequencies with the aid of the rotational constants in Tabe II. The vacuum wavenumbers and wavelengths also utilize the value 299 792 458 m/s ($\delta c/c = \pm 4 \times 10^{-9}$) for the speed of light[10] which is 100 times less uncertain than the previously accepted value, with the accuracy now being mostly limited by the uncertainty in the ^{86}Kr 6058 Å primary length standard.[5]

The frequencies are calculated from formulas which in matrix notation have the form

$$F = F_R + X\beta \ , \tag{3}$$

where the elements f_i of F are the predicted CO_2 laser frequencies, F_R is the reference frequency, the elements x_{ij} of X are the appropriate functions of the rotational quantum number, and the elements β_j of β are the rotational constants. Since the uncertainties in β are correlated, the estimated standard deviation of a predicted frequency relative to the reference line is given by[9]

$$\delta(X\beta)_h = s \left[\sum_i \sum_j x_{ih} x_{jh} c_{ij} \right]^{\frac{1}{2}} , \tag{4}$$

where s is the standard deviation of the residuals and c_{ij} is an element of the inverse matrix, $(X'X)^{-1}$, of the normal equations. The uncertainty in the absolute frequency of the reference line, δf_r, is assumed to be independent of $\delta(X\beta)_h$ and combined in the normal way to produce the total uncertainty in f_h. The total uncertainties in the other quantities in the table are calculated from

$$\frac{\delta(\lambda^{-1})}{\lambda^{-1}} = \frac{\delta\lambda}{\lambda} = \left[\left(\frac{\delta f}{f}\right)^2 + \left(\frac{\delta c}{c}\right)^2 \right]^{\frac{1}{2}} . \tag{5}$$

Tables III and IV present the results of the frequency and wavelength calculations. The format follows that of our predecessors.[11-13] The tables are terminated at J = 60 since the reliability of the model was demonstrated only for J ≤ 44 and one can anticipate the need for additional constants at high J-values. The relative uncertainty for the frequency is given in addition to the total uncertainty since it is anticipated that new measurements of the absolute frequencies of R(30) and R(10) will soon reduce the uncertainties of the reference lines. The additional frequency uncertainty incurred in propagating from the reference line is less than or equal to the uncertainty in the reference frequency for the band with J ≤ 42. In this same band, uncertainties in the vacuum wavenumbers and wavelengths are due almost exculsively to the uncertainty in the speed of light (4 parts in 10^9).

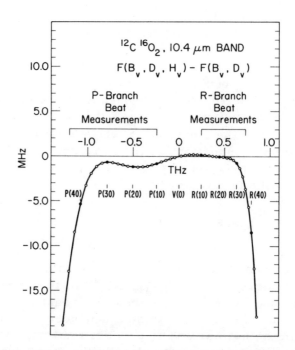

Figure 3. Differences between CO_2 frequencies estimated with constants given in this paper and the previous best constants. For the 10.4 μm band, the difference increases rapidly above J = 40 and reaches -148.9 MHz for P(60) and -130.5 MHz for R(58). The plot for the 9.4 μm band has a similar appearance with a difference of -200.2 MHz for P(60) and -172.1 MHz for R(58).

Table III. Frequencies and wavelengths for the 00°1 - [10°0, 02°0]$_I$ (10.4 μm) band.

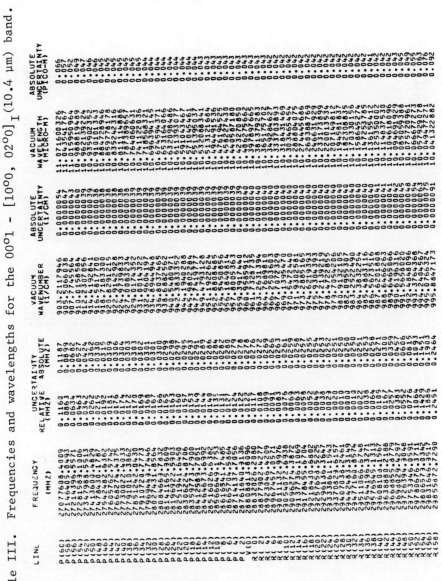

Table IV.　Frequencies and wavelengths for the $00^{0}1 - [10^{0}0, 02^{0}0]_{II}$ (9.4 μm) band.

LINE	FREQUENCY (MHZ)	UNCERTAINTY RELATIVE	UNCERTAINTY ABSOLUTE	VACUUM WAVENUMBER	ABSOLUTE UNCERTAINTY	VACUUM WAVELENGTH (MICROMETER)	ABSOLUTE UNCERTAINTY

Finally, figure 3 points out some possible limitations concerning reliability of the tables. Here differences between frequencies predicted with our rotational constants (B_v, D_v, and H_v) and those predicted with Bridge's and Chang's constants (B_v and D_v only) are plotted on a frequency scale relative to the band center. Predictions from the 2-constant and 3-constant molecular models agree approximately over the range of experimental measurement but differ greatly beyond this range. Moreover, the divergence is very rapid. One is led to conclude that over the region in which the data for our rotational constants was taken (roughly, for $J < 44$), the uncertainties should be approximately those given by the table. For $J > 44$, however, higher order terms in the model used to represent the vibrational-rotational energy (such as the $L_v J^4 (J+1)^4$ term, for example) may become important. Since our model uses terms only through $H_v J^3 (J+1)^3$, the extrapolation will probably be worse than the model predicts, and therefore, caution should be exercised. Nevertheless, these new tables should be useful for most of the commonly used $^{12}C^{16}O_2$ laser lines with the uncertainties being not appreciably different from their estimated values.

The authors acknowledge with appreciation many fruitful discussions with Eric Johnson, Daniel Albritton, and Arthur Schmeltekopf concerning statistical analysis of the data and the molecular physics in the experiment. We also thank Matthew Lojko for his advice, patience, and perseverance in producing the computer programs for the data analysis.

References

1. T. J. Bridges and T. Y. Chang, Phys. Rev. Lett. 22, 811 (1969).

2. Charles Freed and Ali Javan, Appl. Phys. Lett. 17, 53 (1970).

3. For example, Freed et al. have measured the frequencies of many laser lines in some of the less abundant CO_2 isotopes relative to $^{12}C^{16}O_2$ [Charles Freed, A. H. M.

Ross, Robert G. O'Donnell, To be published. Also discussed in a following paper.].

4.   K. M. Evenson, J. S. Wells, F. R. Petersen, B. L. Danielson, and G. W. Day, Appl. Phys. Lett. 22, 192 (1973).

5.   K. M. Evenson, J. S. Wells, F. R. Petersen, B. L. Danielson, G. W. Day, R. L. Barger, and J. L. Hall, Phys. Rev. Lett. 29, 1346 (1972).

6.   In a separate investigation, radiation was deliberately reflected back into one of the lasers by a vibrating mirror. With low driving frequencies (5 to 50 Hz), the amplitude of the driving voltage was adjusted such that the full width at half intensity of the beat frequency increased by a factor of 4 to approximately 100 kHz. There was no measurable shift in the center frequency.

7.   G. Herzberg, Molecular Spectra and Molecular Structure: I. Spectra of Diatomic Molecules (D. Van Nostrand Company, Inc., Princeton, N.J., 1950), 2nd ed., and Molecular Spectra and Molecular Structure: II. Infrared and Raman Spectra (D. Van Nostrand Company, Inc., New York, 1945).

8.   The fit was done initially using sums and differences of rotational constants as in Bridges and Chang[1] and later using the constants directly. There was no difference in the results.

9.   Experimental Statistics, National Bureau of Standards Handbook 91 (August 1, 1963), Superintendent of Documents, U. S. Government Printing Office, Washington, D. C. 20402, (pp. 6-1 to 6-17).

10.   Recommendation of the Comite Consultatif pour la Definition du Metre, 5th Session, June 1973, at BIPM, Serves, France - discussed in the preceding paper: K. M. Evenson, F. R. Petersen, and J. S. Wells.

11.   T. Y. Chang, Optics Communications 2, 88 (1970).

12.   K. M. Baird, H. D. Riccius, and K. J. Siemsen,
       Optics Communications 6, 91 (1972).

13.   Gerhard Schiffner, Opto-electronics 4, 215 (1972).

PICOSECOND STUDIES OF ULTRAFAST MOLECULAR PROCESSES IN LIQUIDS

T. J. Chuang and K. B. Eisenthal

IBM Research Laboratory

San Jose, California 95193

ABSTRACT: Picosecond light pulses generated from a mode-locked Nd-glass or ruby laser have been used to study a number of ultrafast physical and chemical processes. Experiments on two such phenomena, namely, orientational relaxation and electron transfer in liquids are described here. The effects of hydrogen bonding and the structure of the liquids on the molecular rotational motion as well as the kinetics of the excited charge-transfer complex formation are discussed.

I.   INTRODUCTION

Since the discovery of mode-locked lasers[1] capable
of generating light pulses of picosecond duration,
there has been increasing interest in applying these
lasers to study very fast molecular processes. Phenomena such as self-focusing,[2] multi-photon processes,[3,4]
electronic energy transfer,[5] vibrational relaxation,[6]
and radiative and nonradiative electronic relaxation[7,8]
have been studied. In our laboratory, we have been
studying some physical as well as chemical phenomena,
such as orientational relaxation,[9] electron transfer
reactions,[10] electronic energy transfer[4] and cage effect
in photodissociation,[11] etc. In this discussion, we
will focus mainly on the first two subjects. In these
experiments, either a mode-locked Nd^{+3}-glass laser at
1.06μ or ruby laser at 0.6493μ is used. Typically,
although not exclusively, light pulses at these or their
harmonic frequencies are used to excite a molecular
system. The dynamics of the system following the
excitation pulse is monitored by a weak probe pulse at
the same frequencies or other frequencies generated
from the laser pulses. The probe pulse is delayed in
time with respect to the excitation pulse. By varying
the delay time, one measures a formation or decay
curve of transmission of the perturbed systems, from
which kinetic or relaxation parameters are determined.

II.   ORIENTATIONAL RELAXATION

To study the orientational motion of molecules in
liquids, we excite the molecular system by a linearly
polarized picosecond light pulse. In this way, a
nonuniform orientational distribution with respect to
the transition dipole moments of the ground and excited
state molecules are produced. This anisotropy can decay
in time due to the rotational motion of the molecules
and consequently leads to a depolarization of the
emitted radiation[12] and a decay of the induced dichroism.
We have utilized the dichroic method to study the
orientational relaxation of rhodamine 6G in various
liquids with a mode-locked Nd-glass laser (see Fig. 1).
Rhodamine 6G was selected because of its strong and
highly polarized transition at the laser frequency and
more importantly because of its characteristic hydrogen
bonding interactions with a number of solvent molecules.[13] Specifically, we want to examine the effects
of hydrogen bonding on the rotational motion of the

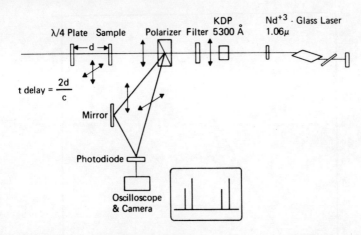

Fig. 1 Experimental arrangement for orientational relaxation studies. Linearly polarized light pulses at 5300A are used to excite the sample solution. A few percent (in intensity) of these pulses are reflected by a $\lambda/4$ wave plate and used as probe pulses. The probe pulses are delayed in time by 2d/C, where d is the distance between the wave plate and sample, and C is the light velocity in air. There are two polarization components in the probe pulse, namely, parallel and perpendicular components with respect to the polarization of the excitation pulse. Orientational relaxation times are obtained from the decay functions of the relative intensities of these two components.

solute molecule. Therefore, we measured the orientational relaxation time of rhodamine 6G in a series of alcohols, chloroform, formamide and ethylene glycol. These solvents represent a wide range of different interactions with the solute molecule because of the differences in molecular shape, size, strength of hydrogen bonding, basic liquid structures and hydrodynamic properties, etc. The experimental results are shown in Fig. 2.

We observed a linear relation between the orientational relaxation time and the solvent viscosity as predicted by the Debye-Stokes-Einstein hydrodynamic model[14] for liquids like chloroform, formamide, and the linear alcohols from methanol to octanol. This agreement with the hydrodynamic model seems rather surprising since according to this model, orientational relaxation time scales linearly with shear viscosity and with the

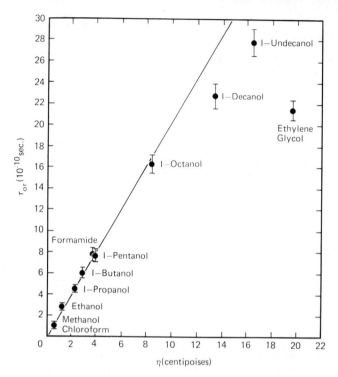

Fig. 2   Orientational relaxation time vs. solution viscosity for rhodamine 6G in various solvents.

hydrodynamic volume of the particle. Although the volumes of the solute-solvent complex vary considerably in the series of liquids from chloroform to octanol, the relaxation times scale linearly with viscosity thus indicating that the hydrodynamic volumes are the same in these liquids. Furthermore the liquids chloroform and methanol, which have the same viscosities, have different strengths of hydrogen bonding with rhodamine 6G and yet yield the same relaxation times. Similar arguments can be applied to the rotational motion of rhodamine 6G in the liquids formamide and 1-pentanol. To account for this apparent insensitivity of the relaxation times to the different volumes and strength of hydrogen bonding interactions between the solvent and solute we recognize that the hydrogen bonded complex cannot be described as a rigid particle. The

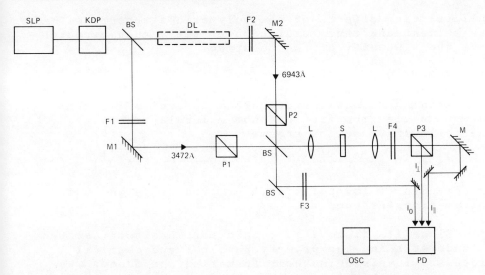

Fig. 3 Schematic of the setup for electron-transfer experiments.
SLP-single ruby pulse at 6943A from the laser oscillator, single
pulse selector and laser amplifier systems; KDP-second harmonic
crystal; BS - beam splitter; F - filters; DL - optical delay;
M - mirror; P - polarizer; L - lens; S - sample; PD - photodiode;
OSC - oscilloscope and camera.

hydrogen bond has some rotational freedom[15] so that a
small angular jump of the solute could not cause a
marked strain on the solute-solvent hydrogen bonding.
Furthermore, the complex is dynamic in that the hydrogen bonds are forming and breaking with the same or
other solvent molecules sufficiently fast and thus this
process could also reduce the strain of the bonding.
Therefore, the rotational motion of rhodamine 6G is
insensitive to its formation of hydrogen bond with the
solvent molecules.

The deviation of the relaxation times of rhodamine
6G in 1-decanol and 1-undecanol from the linear viscosity dependence as seen in Fig. 1 is not surprising.
The solvent molecules are much longer than the solute
molecule and the hydrodynamic model cannot be expected
to be adequate. We interpreted the deviation of
rhodamine 6G in ethylene glycol as to be due to extensive solvent-solvent aggregation by hydrogen bonding
interaction. The solute molecule does not suffer the

full friction effects of the polymeric structure of
the solvent and thus rotates more freely than expected
from the viscosity of the solution.

III. ELECTRON-TRANSFER REACTIONS

When an aromatic hydrocarbon is excited it can
accept an electron from a donor molecule according to
the following reaction scheme.

$$^1A + {}^1D \xrightarrow{\;h\nu\;} {}^1A^* + {}^1D \xrightarrow{\;K_{CT}\;} {}^1(A^-D^+)$$

In solvents of low polarity, the product is an excited
charge-transfer complex.[16]

Since its discovery, much effort has been devoted
to understanding the primary physical and chemical
processes involved in their formation and decay. We
have applied the picosecond technique to study the
kinetics of this reaction. Anthracene in its first
excited singlet state served as the electron acceptor
and N,N'-diethylaniline as the electron donor. N-hexane
was used as solvent. The experimental arrangement is
shown in Fig. 3. A single picosecond pulse at 6943Å
was extracted from the output train of a mode-locked
ruby laser and used as a probe pulse. The second
harmonic pulse at 3472Å was used as an excitation pulse
which excited anthracene into its ^1L_a state. The probe
pulse delayed with respect to the excitation pulse monitored the concentration of the charge-transfer complex
according to the transition $(A^-D^+) \longrightarrow (A^{-*}D^+)$.[17]
As discussed in the previous section, an anisotropy
is generated by the excitation. To measure the rate of
electron transfer, it is necessary to separate the
change in absorption due to orientational relaxation
from that due to the formation of the complexes. This
is readily done by taking the sum of the absorbance
for the probe light polarized parallel and polarized in
the two orthagonal directions. This sum is independent
of the rotational motion of the complex and depends only
on the population of the complexes at any given time.

The concentration of anthracene was fixed at
2.5×10^{-3} M and the concentration of diethlylaniline
(DEA) was varied from 0.1M to 6M (pure liquid). Typical
formation curves of the charge-transfer complexes are
shown in Fig. 4. We observed that at the lowest concentration of DEA, the long time behavior yielded a value of
K_{CT} close to the steady-state diffusion rate constant.[18]

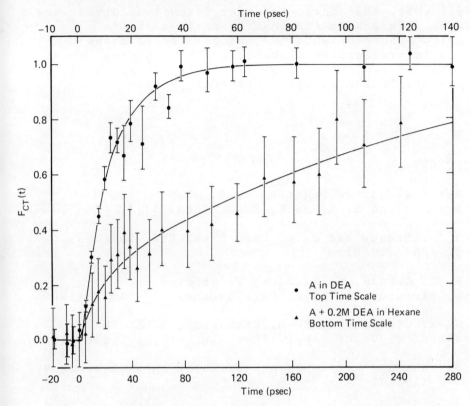

Fig. 4   Formation curves of charge-transfer complexes.

However, in the earliest and intermediate time regions,
the data shows transient behavior found in the
Smoluchowski diffusion model or the Noyes molecular
pair model.[19]   As the concentration of DEA is increased,
the transient effect becomes even more important.   Computer analysis indicates that formation curves calculated from the solution of diffusion model, including
all transient terms, seem to fit reasonably well for
all experimental data except those in the highest concentration regions ($\geq$3M).   The analyses in terms of
diffusion model yielded the same value of 8Å for the
encounter distance and the same rate constant of $10^{11}\ell$
mole^{-1} sec^{-1} for the reaction of DA* pairs at this
distance for all of the systems studied in the range of
0.1M to 1M DEA.   At the higher DEA concentrations of
3M and 6M, where the diffusion and Noyes models were

inapplicable, the charge transfer formation curves were the same. This suggests that there is a static quenching sphere of about 8Å, calculated from the average intermolecular distance at 3M. Apparently, at these higher concentrations, translational motion is not significant in the electron transfer process.

REFERENCES

1. See e.g., A. J. DeMaria, W. H. Glenn, Jr., M. J. Brienza and M. E. Mack, Proc. IEEE, 57, 2 (1969).

2. R. G. Brewer and C. H. Lee, Phys. Rev. Letters, 21, 267 (1968).

3. R. L. Carman, M. E. Mack, F. Shimizu and N. Bloembergen, Phys. Rev. Letters, 23, 1327 (1969).

4. J. A. Giordmaine, P. M. Rentzepis, S. L. Shapiro and K. W. Wecht, Appl. Phys. Letters, 11, 216 (1967).

5. D. Rehm and K. B. Eisenthal, Chem. Phys. Letters 9, 387 (1971).

6. D. Ricard, W. H. Lowdermilk and J. Ducuing, Chem. Phys. Letters, 16, 617 (1972); D. von der Linde, A. Lanbereau and W. Kaiser, Phys. Rev. Letters 26, 954 (1971); R. R. Alfano and S. L. Shapiro, Phys. Rev. Letters 26, 1247 (1971).

7. J. W. Shelton, J. A. Armstrong, IEEE J. Quantum Electr. QE-3, 696 (1967); M. A. Duguay and J. W. Hansen, Opt. Comm. 1, 254 (1969).

8. P. M. Rentzepis and C. J. Mitschele, Anal. Chem. 42, 20 (1970) and references therein.

9. T. J. Chuang and K. B. Eisenthal, Chem. Phys. Letters, 11, 368 (1971); K. B. Eisenthal and K. H. Drexhage, J. Chem. Phys. 51, 5720 (1969).

10. T. J. Chuang and K. B. Eisenthal, J. Chem. Phys. 59, 2140 (1973).

11. T. J. Chuang, G. W. Hoffman and K. B. Eisenthal (In preparation).

12. T. J. Chuang and K. B. Eisenthal, J. Chem. Phys. $\underline{57}$ 5094 (1972).

13. L. V. Levshin and D. M. Akbarova, J. Appl. Spectr. (USSR) $\underline{3}$, 326 (1965) (Eng.).

14. P. Debye, Polar Molecules, Dover Publication, London, 1929, p. 84.

15. G. E. Pimentel and A. L. McClellan, The Hydrogen Bond, Freeman, San Francisco, 1960, p. 267.

16. H. Leonhart and A. Weller, Ber. Bunsenges, Phys. Chem. $\underline{67}$, 791 (1963); A. Weller, Pure Appl. Chem. $\underline{16}$, 115 (1968).

17. R. Potashnik, C. R. Goldschmidt, M. Ottolenghi and A. Weller, J. Chem. Phys. $\underline{55}$, 5344 (1971).

18. D. Rehm and A. Weller, Ber. Bunsenges, Phys. Chem. $\underline{73}$, 834 (1969).

19. R. M. Noyes, Progr. Reaction Kinetics, $\underline{1}$, 129 (1961).

LASER NEGATIVE ION SPECTROSCOPY

W. C. Lineberger[*]

Joint Institute for Laboratory Astrophysics and Dept.

of Chemistry, University of Colorado, Boulder, Colorado

The electron affinities of atoms, free radicals, and small molecules are parameters of fundamental interest to the practicing chemical kineticist. Moreover, since the electron affinity is typically of the order of the correlation energy, a precise determination of the electron affinity of an atom or molecule provides a very sensitive test of the accuracy of quantum calculations.

For these and other reasons there have been many attempts to determine electron affinities over the past forty years. Among the many techniques employed[1] are ab initio calculations, isoelectronic extrapolations, studies of thermodynamic cycles, surface ionization, endothermic charge transfer, pair production in photoionization, studies of continuous absorption in plasmas, and photodetachment. In spite of such extensive studies, by 1970 there were reliable data or calculations for only about ten atoms or molecules. Direct calculations were reliable for only a few simple systems, most notably atomic hydrogen.

The experimental techniques (with the exception of photoabsorption and photodetachment) generally suffered either because approximations had to be made which could not be readily verified, or because of problems resulting from the fact that the electron affinity was not the quantity being directly measured. The photoabsorption studies have a distinct advantage in that the electron affinity is essentially directly determined by the wavelength of light at which absorption begins. However, in order to build up a sufficient negative ion density to do a traditional absorption

---

[*] Alfred P. Sloan Foundation Fellow, 1972-1974.

measurement one must generate negative ions in a rather hot plasma
with attendant problems resulting from plasma radiation, impurities,
etc. This technique, however, has been successfully applied to the
halogens in a very elegant set of measurements by Berry[2-4] and co-
workers. In order to circumvent these problems, one can perform a
crossed beam experiment in which a mass selected negative ion beam
is crossed by an energy resolved photon beam. One measures a cross
section for production of electrons as a function of photon energy.
Several very elegant such experiments were performed by Branscomb[5-7]
and co-workers in the 1960's. The inherently low signal-to-noise
ratio obtainable with conventional light sources, however, restricted
study to those ions for which copious beams could be produced. The
low intensity of these light sources also required that, in order
to have sufficient photon flux to see a signal, the photon energy
resolution be no better than approximately 50-500 Å in the visible
spectrum.

The advent of lasers has removed the serious light source restrictions previously encountered and has made a qualitative change
in the accuracy with which electron affinities can be measured and,
by removing the requirement for intense negative ion beams, has
made it possible to study ions which were not previously accessible.
Two such experimental techniques will be discussed: 1) a study of
photodetachment cross sections of negative ions using a tunable
dye laser as a light source, and 2) studies of the photoelectron
spectra of negative ions excited by a fixed wavelength laser. In
many ways, these two techniques are complementary; the former provides more accurate electron affinities while the latter is more
likely to provide the means for successful interpretation of the
structure of molecular ions.

EXPERIMENTAL TECHNIQUES

The photoelectron spectroscopy technique developed by Brehm,
Gusinow and Hall[8] has been previously described in detail.[9] In
brief, a 680 eV beam of negative ions is crossed by the linearly
polarized intra-cavity beam of an argon ion laser (4880 Å). The
interaction region is otherwise field free. Those electrons which
are photodetached into the 0.006 sr acceptance angle of a hemispheri-
cal electrostatic condenser are energy analyzed and detected. The
ion beam, laser beam, and electron collection directions are mutually orthogonal. The polarization of the laser can be rotated by
means of a half wave plate, allowing measurement of the photodetached electron angular distributions. Since contact potentials
are not well known in this apparatus, electron energies are determined relative to the energy of electrons detached from negative
ions whose electron affinities have been determined absolutely. By
a least-squares fitting procedure to the 40 meV fwhm transmission

profile of the analyzer, electron energy differences can be determined to approximately 1 meV in favorable cases.

A schematic diagram of the tunable laser photodetachment
machine[10] is shown in Fig. 1. A several kilovolt mass analyzed
negative ion beam is crossed with the focused output of a pulsed
tunable dye laser and we detect the fast neutrals resulting from
laser photodetachment. The neutral atoms are detected by a 15-stage
multiplier located 20 cm beyond the laser beam intersection region.
Just beyond the intersection region, the remaining negative ions
are electrostatically deflected into a Faraday cup, such that only
neutral atoms, produced by charge stripping on the background gas
and by laser photodetachment, reach the neutral detector. The
quantity measured is thus a relative cross section for production
of neutral atoms as a function of the laser wavelength, the latter
determined to ≤1 Å by a 0.35 m monochromator.

The light source[11] is a flashlamp-pumped, tunable organic dye
laser consisting of a linear xenon flash lamp and a dye cell located
at the foci of an elliptical cavity. The laser output is a 300 nsec
pulse, with an energy content of 1-10 mJ per pulse at a line width
typically 1-2 Å; typical repetition rates are 5 Hz. The wavelength
region employed in the experiments described here is 4500-7000 Å.

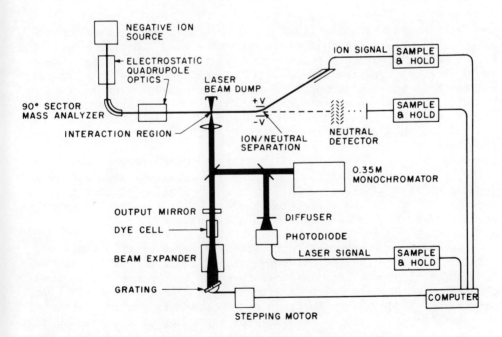

Fig. 1.   Schematic diagram of a tunable laser photodetachment
          apparatus.

For each shot of the dye laser, analog signals corresponding to
the neutral signal, laser intensity, and ion beam current are digi-
tized and a small computer calculates an apparent photodetachment
cross section. At each selected wavelength, the laser is fired
approximately 600 times and the computer produces an average cross
section. A more detailed description of the apparatus and data-
taking procedure will appear elsewhere.[12]

ATOMIC NEGATIVE IONS

The laser techniques described above have been used to study
photodetachment of He$^-$,[8] C$^-$,[13] O$^-$,[13] Ag$^-$,[14] Cu$^-$,[14] S$^-$,[10] Au$^-$,[15]
Pt$^-$,[15] and Se$^-$.[12] As an example of what one expects to see in a
threshold photodetachment cross section measurement, consider the
Se$^-$ case, in which one is detaching a bound p-electron into an
s-wave or d-wave continuum state. The threshold energy dependence
for such a process was derived in the classic paper[16] by Wigner,
concerned with processes in which pairs of particles are formed in
the final state. For the cross section σ, he derives the general
result

$$\sigma \propto k^{2L+1} \propto (E - E_{thr})^{\frac{2L+1}{2}}$$

for the case in which the final state interactions (other than the
centrifugal potential) fall off faster than r^{-2}. In this case,
k is the free electron momentum and L is its orbital angular momentum with respect to the atom. Thus, in the case of Se$^-$, we expect to
see six $E^{1/2}$ thresholds corresponding to detachment from Se$^{-2}P_{1/2,3/2}$
to Se$^3P_{2,1,0}$. Since the neutral atom spin-orbit splittings are
well known, identification of the thresholds is straightforward and
the wavelength corresponding to the $^2P_{3/2}$-$^3P_2$ channel opening gives
the electron affinity directly. Such threshold behavior is illustrated in Fig. 2. From a detailed analysis of these various thresholds, we find that EA(Se) is (2.0206 $\pm$ 0.0004) eV, and that the
doublet splitting in Se$^-$ is 2279 $\pm$ 2 cm^{-1}. Similar studies[15] on Au$^-$
ions have verified the Wigner threshold law for the case of photodetachment of a bound s-electron into a p-wave continuum.

In addition to these direct threshold measurements, there
exists the possibility of observing interference effects in photodetachment at photon energies corresponding to the opening of a new
channel which competes with an already open channel. This fact was
pointed out in detail by Norcross and Moores[17] in the case of Na$^-$
photodetachment at photon energies corresponding to the opening of
the lowest ^2P exit channel in the neutral. From a measurement of
the photon energy corresponding to the cusp at the channel opening
we can determine the electron affinities of the alkalis.

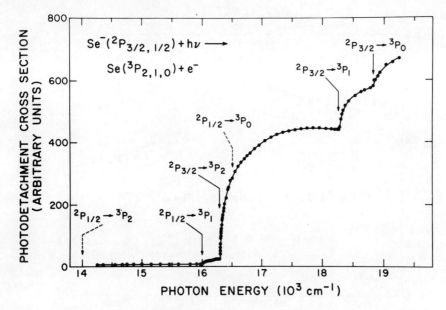

Fig. 2.   Se⁻ photodetachment cross section in the energy range
          14,000 – 19,000 cm⁻¹. The individual fine structure tran-
          sition thresholds are labeled.

     Figure 3 shows Na⁻ photodetachment data[18] in the region of the
Na($3^2P_{1/2,3/2}$) thresholds. In this case, the wavelength resolution
of the tunable laser was not adequate to resolve the individual
fine structure thresholds. The cross section shape near threshold
corresponds to a Wigner cusp, and is essentially in agreement with
calculations performed by Norcross and Moores.[17] These data, for
various experimental reasons, were quite noisy, and, as a result,
there is no reason to believe that the oscillations at long wavelengths are real. From an analysis of these data we find EA(Na)
to be (0.543 ± .010) eV.

     The K⁻ ion is a much more favorable case for experimental study
because the expected threshold falls in a spectral region corresponding to a bright laser and the spin–orbit splitting in the excited
state is much larger than for Na. Figure 4 shows K⁻ photodetachment
data[19] in the vicinity of the $^2P_{1/2,3/2}$ thresholds. In this case,
the individual thresholds are well resolved and we find EA(K) =
(0.5012 ± 0.0005) eV. That the observed structure is not a cusp is
not bothersome since the general treatment of threshold behavior
predicts only a discontinuity in the derivative at the channel
opening, but not necessarily a derivative sign change at the threshold.

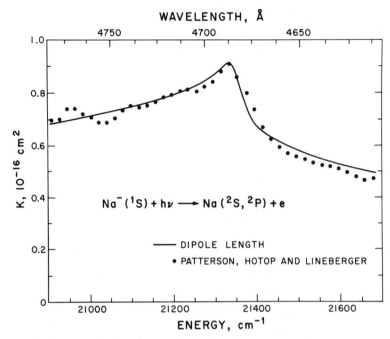

Fig. 3. Na⁻ photodetachment cross section in the region of the
Na $3^3P_{1/2,3/2}$ thresholds. The cusp occurs at the opening
of these ^2P exit channels. The solid line is a close
coupling calculation by Norcross and Moores, convolved
with our experimental energy resolution.

In some very recent measurements[20] of Cs⁻ photodetachment, the
threshold behavior is even more unexpected, as seen in Fig. 5.
The sharp drop in the total Cs⁻ cross section near 6682 Å goes, as
far as we can tell, to zero. Our present bound on the minimum is
3×10^{-3} of the fully developed cross section, and the width of the
minimum is governed solely by the 1 Å line width of the laser. Note
that there also appears to be a discontinuity in the derivative of
the total cross section near 6673 Å. Similar behavior is seen near
the $^2P_{3/2}$ threshold, except that the width of the hole is approximately a factor of 2 greater than in the $^2P_{1/2}$ case and the minimum
is not nearly so small. We have no a priori way of deciding whether
the $P_{1/2}$ channel opening occurs at 6682 Å or 6673 Å (corresponding
to electron affinities of 0.469 eV and 0.472 eV respectively).
Kasdan, however, has used the photoelectron spectroscopy apparatus
to measure the electron affinity of Cs and finds EA(Cs) = (0.471 ±
0.003) eV.[21] His result is suggestive that the higher energy feature
is the correct one but not conclusive proof. Norcross[22] has performed preliminary three-state close coupling calculations on Cs⁻
photodetachment in the vicinity of the ^2P threshold; these preliminary calculations[22] suggest that there is a doubly excited negative

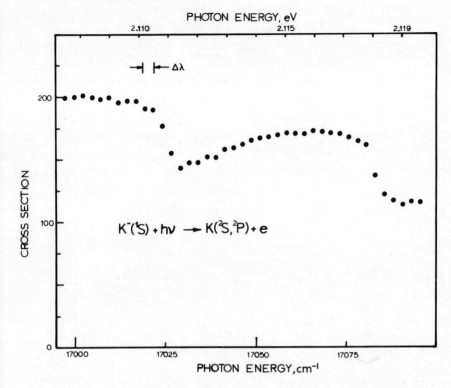

Fig. 4. K⁻ photodetachment data near the ²P thresholds.

ion configuration which lies very near the ²P threshold. Such a
state would give rise to a Fano line shape[23] with a true zero in
the total cross section. The opening of the ²P inelastic channel
would interrupt the development of the Fano line shape and give
rise to the higher energy cusp observed. If this interpretation is
correct, then the higher energy cusp in Cs⁻ is understood, as is
the fact that the $P_{3/2}$ minimum is less dramatic, as well as the
qualitative trend of the threshold behaviors through the alkali ion
series. As one progresses from sodium to cesium, a temporary negative ion state appears and moves further below the ²P threshold;
the development of the resonance line shape corresponding to this
state is then interrupted by the opening of the inelastic channel.
If this explanation is correct, then we would expect to see a Rb⁻
behavior which is intermediate between K⁻ and Cs⁻. Such an experiment is now in progress. These temporary negative ion states are
frequently seen in electron scattering from neutrals just below
excited state thresholds. It may well turn out that cesium represents a typical case, rather than a puzzling exception.

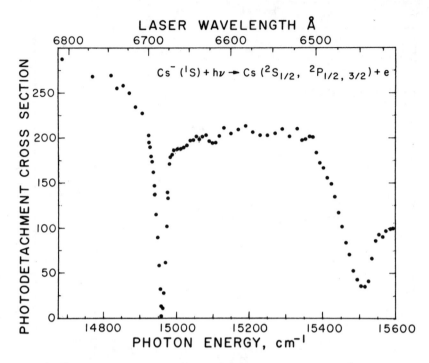

Fig. 5. Cs⁻ photodetachment cross section near the ^{2}P thresholds.
The minimum near 6682 Å is broadened by our wavelength
resolution. A possible interpretation of this result is
discussed in the text.

MOLECULAR NEGATIVE IONS

The detailed interpretation of the tunable laser atomic negative ion photodetachment data has been a relatively straightforward
task; molecular negative ions are a very different case. As a result of the manifold of rotational transitions involved and the
fact that the exact form of the threshold law is both unknown and
rotational-energy dependent, one can confidently predict that the
detailed interpretation of molecular ion photodetachment cross
section data will be extremely difficult.

The simplest case one can consider is photodetachment of OH⁻,
for which it has been established[24-26] that the potential curves
of the neutral molecule and negative ion are virtually identical.
Moreover, one expects[27] that the individual rotational thresholds
will be at least as sharp as $E^{1/2}$. The OH⁻ photodetachment cross
section[28,29] in the range (7000-6450) Å is depicted in Fig. 6.
The sharp onset, covering a range of 50 cm⁻¹, is due to the opening

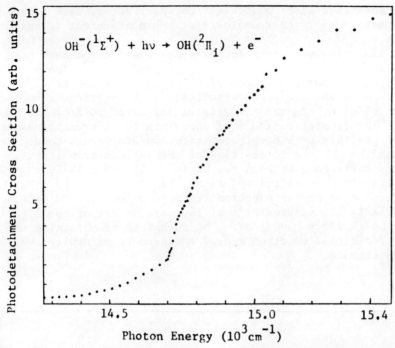

$$OH^-(^1\Sigma^+) + h\nu \rightarrow OH(^2\Pi_i) + e^-$$

Fig. 6.   OH$^-$ photodetachment cross section near threshold. The
          sharp onset is due to the opening of the nearly degenerate
          Q-rotational branches for transitions from OH$^-$ to OH.

of what we shall call the Q-rotational branches for transitions
from OH$^-$($^1\Sigma$) to OH($^2\Pi_{3/2}$). We were not able to identify clearly
individual rotational thresholds. Thus, we have been forced to
analyze these data by constructing synthetic photodetachment cross
sections based on the known spectroscopic constants of the neutral
molecule and the measured equilibrium internuclear separation of the
negative ion, assuming some form for the threshold law. Based on
such fits, we find[29] EA(OH) = (1.825 ± 0.002) eV and EA(OD) =
(1.823 ± 0.002) eV. The observed isotope effect [(2.5 ± 1) meV] on
the electron affinities can almost entirely be ascribed[29] to differences in the position of the ground rotational state in OH($^2\Pi_{3/2}$)
and OD($^2\Pi_{3/2}$). It has not proven necessary to invoke breakdown of
the Born-Oppenheimer approximation in order to explain our data.

     The substantial problems encountered in analysis of the "simple"
OH$^-$ data indicate that it will, in general, be very difficult to
analyze in detail such cross sections for diatomic and triatomic
negative ions. Some success has resulted, however, in tunable
parametric oscillator studies[30] of NH$_2$$^-$ photodetachment. The basic
difficulty, of course, is that the cross sections obtained are
effectively integral spectra of the transitions involved. In this

respect, the fixed laser photoelectron spectroscopy technique has a large advantage over the tunable laser photodetachment technique.

The laser photoelectron spectroscopy machine has been used to study NO^-,[9] O_2^-,[31] OH^-,[26] NH_2^-,[26] SO_2^-,[26] and SO^-.[32] The photodetached electron energy spectrum of SO^- (Fig. 7) consists of a series of peaks which are characteristic of transitions from the $v'' = 0$ level of SO^- to the vibrational levels of $SO(X^3\Sigma)$ and $SO(a^1\Delta)$. The identification[32] of the vibrational transitions involved is possible by isotopic substitution and Franck-Condon factor analysis of the intensities of the various transitions. The energy difference between the $SO^-(v'' = 0)$ and $SO(X^3\Sigma, v'' = 0)$ states, the electron affinity[32] of SO, is (1.126 ± 0.013) eV, measured relative to the electron affinity of O. In addition, one obtains the first measurement[32] of the term energy of the $SO(a^1\Delta)$ state to be (0.730 ± 0.003) eV. The advantage of obtaining the direct differential electron energy spectrum is painfully obvious from this example.

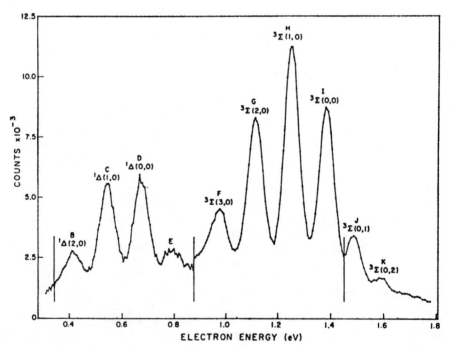

Fig. 7.   Photodetached electron energy spectrum of SO^-, resulting from absorption of a 2.54 eV photon. The individual vibrational peaks are identified by a process discussed in the text. Note that both the SO $^3\Sigma$ and $^1\Delta$ final states are clearly observable.

MULTIPHOTON PROCESSES

One possible way in which we could preserve the inherently high
energy resolution of the tunable laser technique, and at the same
time, directly obtain differential spectra would be to make use of
multiphoton processes.[33,34] If a molecular ion possesses a bound
electronically excited state, then one can study two-photon photodetachment of the ion, using the bound electronically excited state
as a real intermediate state. For such states which lie at energies
greater than 1/2 of the vertical detachment energy but below the
single-photon photodetachment threshold, the apparent photodetachment
signal is the product of the bound-bound excitation function and the
relatively smooth single photon photodetachment cross section from
the intermediate state. At our typical laser fluxes there would be
a detectable photodetachment signal only in the case of near resonant pumping of real intermediate states. The C_2^- ion is an ideal
candidate for the first such study, since C_2 has an electron affinity of approximately 3.5 eV,[35] and there is evidence of a C_2^{-*}
state[36-38] lying approximately 2.3 eV above the ground state.

Figure 8 shows the C_2^- apparent single-photon photodetachment
cross section near the origin[36] of the Herzberg-Lagerquist band
system. At the laser fluxes employed in obtaining these data, the
photodetachment signal depended linearly upon the laser flux, because
the bound-bound excitation was saturated. Only at lower laser fluxes
were we able to demonstrate the I^2 intensity dependence expected for
an unsaturated two photon process. The data shown in Fig. 8 look
like a low resolution spectrum of a $\Sigma-\Sigma$ absorption, and in fact it
is possible to analyze these spectra to obtain molecular parameters
for both of the states involved. One could easily improve the wavelength resolution of the tunable laser to the point that individual
rotational transitions would be resolved.

FUTURE APPLICATIONS

Among the more obvious extensions of this technique are those
involving extended wavelength regions and increased resolution.
The photoelectron spectroscopy technique is limited by the 2.54 eV
photons presently available. Recent increases in argon ion laser
output power make it possible to do the same experiments using a
frequency doubled argon ion laser; thus the "terminal" negative
ions frequently encountered in many chemical applications can be
studied.

At present, one can easily build tunable lasers with line
widths of several hundred megahertz,[39,40] sacrificing very little
power in the reduction from 1 Å line width. Thus, it is possible
to measure electron affinities to accuracies of the order of micro-

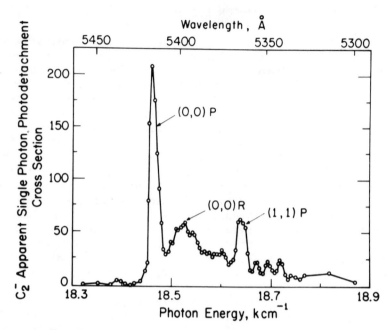

Fig. 8. C_2^- apparent single photon photodetachment cross section near the origin of the Herzberg-Lagerquist band system. At the photon fluxes employed for these data, the bound-bound excitation was saturated.

electron volts. Moreover, with such a laser, it is possible to saturate bound-bound transitions with oscillator strengths of order 10^{-6}. The two photon technique could then be used to do spectroscopy on partially forbidden transitions in positive and negative ions. Similar techniques can also be used to study the photodissociation of both positive[41] and negative ions.

ACKNOWLEDGMENTS

It is a pleasure to acknowledge the vital contributions of my colleagues and co-workers, Dr. H. Hotop, Mr. T. A. Patterson, Dr. R. A. Bennett, Dr. A. Kasdan, and Dr. J. L. Hall. Dr. D. Norcross has been of invaluable assistance both in his calculations on the alkalis and in explaining theoretical points to me. This work was supported by the National Science Foundation and by the Advanced Research Projects Agency of the Department of Defense and was monitored by U. S. Army Research Office-Durham, under Contract No. DAHCO4 72 C 0047.

REFERENCES

1. For review of work prior to 1970 see the following: R. S.
    Berry, Chem. Rev. <u>69</u>, 533 (1969); B. Steiner in <u>Case Studies</u>
    <u>in Atomic Physics II</u>, edited by E. W. McDaniel and M. R. C.
    McDowell (North Holland, Amsterdam, 1972), pp. 483-545; S. J.
    Smith, in <u>Methods in Experimental Physics</u>, Vol. 7A, edited by
    B. Bederson and W. L. Fite (Academic, N.Y., 1968), pp. 179-
    208; D. M. Smirnov, High Temperature <u>3</u>, 716 (1965); L. M.
    Branscomb, in <u>Atomic and Molecular Processes</u>, edited by D. R.
    Bates (Academic, N.Y., 1962), pp. 100-140.

2. R. S. Berry, C. W. David and J. C. Mackie, J. Chem. Phys. <u>42</u>,
    1541 (1965).

3. R. S. Berry, J. C. Mackie, R. L. Taylor and R. Lynch, J. Chem.
    Phys. <u>43</u>, 3067 (1965).

4. R. S. Berry and C. W. Reimann, J. Chem. Phys. <u>38</u>, 1540 (1963).

5. L. M. Branscomb, D. S. Burch, S. J. Smith and S. Geltman, Phys.
    Rev. <u>111</u>, 504 (1958).

6. L. M. Branscomb, S. J. Smith and G. Tisone, J. Chem. Phys. <u>43</u>,
    2906 (1965).

7. B. Steiner, M. L. Seman and L. M. Branscomb, J. Chem. Phys.
    <u>37</u>, 1200 (1962).

8. B. Brehm, M. A. Gusinow and J. L. Hall, Phys. Rev. Letters
    <u>19</u>, 737 (1967).

9. M. W. Siegel, R. J. Celotta, J. L. Hall, J. Levine and R. A.
    Bennett, Phys. Rev. A <u>6</u>, 607 (1972).

10. W. C. Lineberger and B. W. Woodward, Phys. Rev. Letters <u>25</u>,
    424 (1970).

11. D. A. Jennings and D. L. Baldwin, National Bureau of Standards
    Technical Note 603 (1971), available from U.S. Government
    Printing Office, Washington, D.C. 20402.

12. H. Hotop, T. A. Patterson and W. C. Lineberger, Phys. Rev. A
    (August 1973).

13. J. L. Hall and M. W. Siegel, J. Chem. Phys. <u>48</u>, 943 (1968).

14. H. Hotop, R. A. Bennett and W. C. Lineberger, J. Chem. Phys.
    <u>58</u>, 2373 (1973).

15. H. Hotop and W. C. Lineberger, J. Chem. Phys. 58, 2379 (1973).

16. E. P. Wigner, Phys. Rev. 73, 1002 (1948).

17. D. W. Norcross and D. L. Moores, in Atomic Physics 3, Proceedings of the Third International Conference on Atomic Physics, August, 1972, Boulder, Colorado, edited by S. J. Smith and G. K. Walters (Plenum, N. Y., 1973), pp. 261-67.

18. H. Hotop, T. A. Patterson and W. C. Lineberger, Bull. Am. Phys. Soc. 17, 1128 (1972).

19. H. Hotop, T. A. Patterson and W. C. Lineberger, to be published.

20. T. A. Patterson and W. C. Lineberger, to be published.

21. A. Kasdan, private communcation; to be published.

22. D. Norcross, private communication; to be published.

23. U. Fano and J. W. Cooper, Phys. Rev. 137, A1364 (1965); P. G. Burke, in Advances in Atomic and Molecular Physics, edited by D. R. Bates and I. Estermann (Academic, N. Y., 1968), Vol. 4, p. 173; G. J. Schulz, Rev. Mod. Phys. (July, 1973).

24. L. M. Branscomb, Phys. Rev. 148, 11 (1966).

25. P. E. Cade, J. Chem. Phys. 47, 2390 (1967).

26. R. J. Celotta, R. A. Bennett and J. L. Hall, to be published.

27. T. F. O'Malley, Phys. Rev. 137, A1668 (1965).

28. W. C. Lineberger, in Theoretical Chemistry, Proceedings of the Summer Research Conference on Theoretical Chemistry, Boulder, June, 1972 (Wiley, N. Y., in press).

29. H. Hotop, T. A. Patterson and W. C. Lineberger, to be published.

30. K. C. Smyth, R. T. McIver, Jr., J. I. Brauman and R. W. Wallace, J. Chem. Phys. 54, 2758 (1971); K. C. Smyth and J. I. Brauman, J. Chem. Phys. 56, 4620 (1972).

31. R. J. Celotta, R. A. Bennett, J. L. Hall, M. W. Siegel and J. Levine, Phys. Rev. A 6, 631 (1972).

32. R. A. Bennett, Thesis, University of Colorado, 1972; to be published.

33. W. C. Lineberger and T. A. Patterson, Chem. Phys. Letters 13, 40 (1972).

34. R. V. Ambartzumian and V. S. Letokhov, Appl. Opt. 11, 354 (1972).

35. D. Feldman, Z. Naturforsch. 25a, 621 (1970).

36. G. Herzberg and A. Lagerquist, Can. J. Phys. 46, 2363 (1968).

37. D. E. Milligan and M. E. Jacox, J. Chem. Phys. 51, 1952 (1969).

38. R. P. Frosch, J. Chem. Phys. 54, 2660 (1971).

39. H. Walther and J. L. Hall, Appl. Phys. Letters 17, 239 (1970).

40. T. W. Hänsch, Appl. Opt. 11, 895 (1972).

41. J.-B. Ozenne, D. Pham and J. Durup, Chem. Phys. Letters 17, 422 (1972).

TIME EVOLUTION OF UNIMOLECULAR PROCESSES IN
ISOLATED MOLECULES BY CROSSED LASER AND
MOLECULAR BEAMS

Stephen J. Riley,[*] Robert K. Sander and Kent R. Wilson

Department of Chemistry, University of California-
San Diego, La Jolla, California 92037

ABSTRACT

A new technique, double absorption photofragment spectroscopy (DAPS), has been developed in order to follow the time evolution of unimolecular processes in isolated molecules. Molecules removed from collisions in a molecular beam are prepared in well defined initial states by an initiation laser pulse As this state distribution evolves, it is monitored as a function of time by a second, delayed, probe laser pulse which, by exciting the molecules to a higher dissociative state, forces them to emit information-carrying particles. These particles, molecular fragments from photodissociation, are detected by photofragment spectroscopy, which can be used to derive information about the time evolution of the unimolecular process by following changes in such parameters as population, angular distribution, fragmentation pattern, and energy partitioning. As an initial demonstration of this technique, the predissociation of the iodine B state has been studied. The lifetime for the B state was found to vary between 770 and 1600 nsec depending on the $v'J'$ population distribution prepared by the initiation laser pulse. In addition, the characterization of a state, invisible in the ordinary absorption spectrum because transition to it is parity forbidden, is illustrated through the sequential absorption of two photons.

---

[*]Present address: Department of Chemistry, Yale University, New Haven, Connecticut 06520.

I. INTRODUCTION

Unimolecular processes, the changes in internal properties which occur within single molecules, are fundamental to chemistry. Many examples exist, including configuration change such as dissociation and rearrangement (variation in conformation or isomer), vibrational change such as transfer of excitation among modes, and electronic change such as internal conversion (radiationless transition between electronic states of the same spin multiplicity), and intersystem crossing (radiationless transition between electronic states of different multiplicity). There is considerable theoretical interest in unimolecular processes,[1] but many basic assumptions involved in the various models have remained untested by decisive experiments.

The ideal experiment would prepare the molecule in a well defined initial state ψ_i at time $t = 0$ and probe the isolated molecule as it evolves toward a final state ψ_f. The probe time t_d should be variable over a wide time scale to match that of the various possible internal processes. The two requirements of isolation and timing have made such experiments difficult. To prevent intermolecular collisions which would perturb the evolving molecule at longer times, the density of molecules must be reduced to the point where absorption spectroscopy becomes difficult as a probe. More sensitive techniques such as detection of spontaneous emission of fluorescence photons or other particles can be used, but then the molecule can only be probed if it emits and then only when it emits. A more powerful technique is needed which allows the study of molecular states at times chosen by the experimenter rather than by the molecule.

What measures can be taken to avoid the difficulties of perturbing collisions and provide a wide time scale for probing? First, the molecules can be isolated from each other and from walls by forming them into a molecular beam. Second, the desired molecular state can be prepared at $t = 0$ by photons from an initiation laser pulse whose duration can be as short as a few picoseconds. Third, the molecules can be studied at time t_d with a second, probe laser pulse which forces the molecules to emit particles (photons, electrons, ionic or neutral molecular fragments).[2] The experimental time scale during which the molecule evolves can be varied from approximately 10^{-12} seconds, the present lower limit of laser pulse durations, to approximately 10^{-3} seconds when the molecular beam finally hits the wall of the vacuum chamber. Thus the experiment involves crossing a molecular beam with two pulsed laser beams as shown in Fig. 1. The initiation laser pulse prepares the molecules in the state of interest. They then unimolecularly evolve until the second probe pulse forces them to emit information-carrying particles which are detected. The specific emitted particles we detect are neutral molecular fragments from photodissociation.

Photofragment spectroscopy[3] is the technique we use to extract detailed information from the fragments about the molecular states from which they have come. The information which

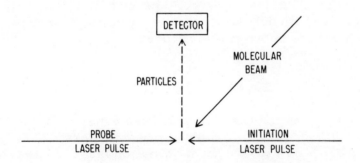

Fig. 1. Schematic of generalized laser-molecular beam experiment for the study of the time evolution of isolated molecules. The initiation pulse prepares the molecules in a state which evolves in time until the probe pulse excites the molecules to a higher unstable state from which particles are emitted. These information-carrying particles (photons, electrons, ionic or neutral photofragments) are detected as they recoil from the intersection volume.

photofragment spectroscopy provides includes:

i)   symmetry and structure of the molecular states from
       the fragment angular distribution, i.e., the number of
       fragments detected as a function of the angle θ be-
       tween the recoil direction and the polarization direction
       (electric vector) of the light pulse,

ii)   number of molecules in each state from the number of
       detected fragments,

iii)   identity of the fragments from their mass spectrum,

iv)   distribution of total translational energy in the recoiling
       fragments from their flight time distribution, and

v)   distribution of total fragment internal energy from
       energy balance.

In a time evolution experiment this data can be measured as a
function of time between the two laser pulses. Thus, if the prepared molecular state evolves into another state, which upon
photodissociation by the probe gives a different angular distribution, product distribution, or recoil energy distribution, the
evolution of the unimolecular process may be monitored as a
function of the time between laser pulses.

II. EXPERIMENTAL

The apparatus for carrying out these experiments is shown
schematically in Fig. 2. A molecular beam enters a high vacuum
chamber where it is crossed by an initiation laser pulse whose
polarization direction can be varied with respect to the fragment
recoil direction. From the opposite side of the chamber the
probe laser pulse, also with variable polarization direction, is
fired into the same interaction volume after a time delay t_d. A
quadrupole mass spectrometer detects the recoiling molecular
fragments resulting from photodissociation. The number of fragments detected is counted during each of 400 one microsecond intervals following the first laser pulse, and stored in an on-line time-
shared computer. The computer also controls the polarization
direction for both the probe and initiation pulses as well as controlling and monitoring the time t_d between laser pulses.

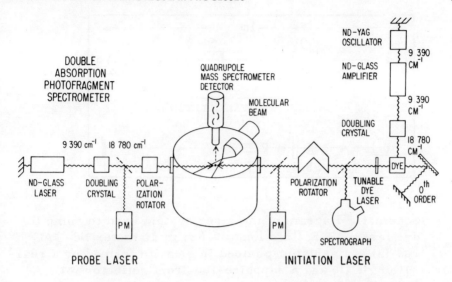

Fig. 2. Schematic of double absorption photofragment spectrom-
         eter. Molecules are prepared in a molecular beam to
         isolate them from collisions. The initiation pulse comes
         from a tunable dye laser, shown on the right, which is
         pumped by the second harmonic of a Nd-glass amplifier
         fed by a Nd-YAG oscillator. The dye laser's wavelength
         is controlled by a diffraction grating in the cavity and is
         monitored by a Bass-Kessler spectrograph. Its polariza-
         tion direction is set by two rotatable Fresnel rhombs.
         The probe pulse comes from the second harmonic of a
         Q-switched neodymium glass laser, shown on the left,
         and its polarization is set by a rotatable half-wave plate.
         Two fast photomultiplier tubes are used with associated
         circuitry to measure the delay time between the two laser
         pulses. Photodissociation fragments resulting from
         absorption of initiation and probe pulses are detected by
         a quadrupole mass spectrometer mounted above the
         laser-molecular beam intersection region.

    The initiation pulse is produced by a tunable dye laser,
using kiton red S, which in turn is pumped by the second harmonic
of a large Nd oscillator-amplifier laser system[4] shown in Fig. 3.
Dye laser pulses of 0.05 J in 20 nsec F.W.H.M. with a bandwidth
of 8 cm^{-1} are produced. The probe pulse is the 0.1 J second
harmonic output of a Q-switched Nd-glass laser.

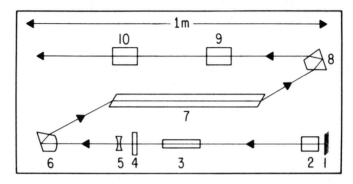

Fig. 3.   Schematic diagram of the laser system used to pump the
dye laser.   A 5.1 cm long, 0.63 cm diameter Nd-YAG
rod (3) is flashlamp-pumped in a cavity formed by a rear
reflector (1) and a sapphire-flat front reflector (4).   A
Pockels cell and Glan prism comprise the Q-switching
unit (2); the resulting oscillator produces 150 mJ at
9390 cm^{-1} (10 648 Å). The emergent beam, after being
expanded to a diameter of 1.5 cm by the combination of
a negative lens (5) and telescopic folding prism (6),
passes into a flashlamp-pumped Nd-glass amplifier rod
(7) through a Brewster angle face.   Amplification yields
a 10 J infrared pulse which preserves the narrow linewidth and low divergence of the YAG oscillator.   The
folding prism (8) diverts the beam into a temperature
controlled KDP frequency doubling crystal (9), yielding
3 J at 18 780 cm^{-1} (5324 Å).   This second harmonic output is used to pump the dye laser of Fig. 2.   For ultraviolet operation an ADP crystal (10) frequency doubles
the green radiation to a 1 J pulse at 37 550 cm^{-1} (2662 Å).

III.   RESULTS

We have demonstrated the feasibility of this new double
pulse technique,[5] which we call Double Absorption Photofragment
Spectroscopy (DAPS), by studying a relatively simple and well
known unimolecular process, the predissociation of the I_2 B
$0^+u(^3\Pi)$ state by the $1u(^1\Pi)$ state.[6-8]

This predissociation has been independently studied by a
variety of other techniques, including electron spin resonance,[9]
high resolution absorption spectroscopy,[10,11] fluorescence life-

time,[12,13] and photofragment spectroscopy.[3b] The iodine absorption spectrum and relevant molecular states are shown in
Fig. 4. In our experiments, different sets of vibration-rotation
levels of the B state are sampled by tuning the dye laser. As the
excited molecules evolve in time, the $v'J'$ dependent coupling
between the B state and the dissociative $1u(^1\Pi)$ state (this coupling may be viewed as leading to "intersystem crossing") causes
predissociation of the molecules into two ground state $I(^2P_{3/2})$
atoms. Simultaneously, B state molecules also radiatively decay

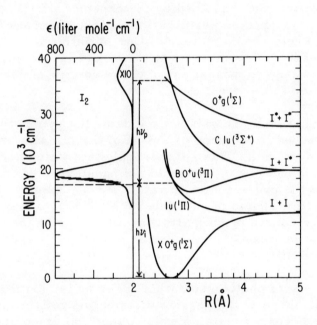

Fig. 4. I_2 absorption spectrum (from Ref. 8) and potential curves
(from Ref. 6). The main visible absorption is due to
transitions to the B $0^+u(^3\Pi)$ and $1u(^1\Pi)$ states. In the
present experiment, the initiation photon, $h\nu_i$, prepares
molecules in the B state, which then undergo two unimolecular processes, either predissociating through the
$1u(^1\Pi)$ state, giving two ground state atoms, or radiating
back to the $0^+g(^1\Sigma)$ ground state. The probe photon,
$h\nu_p$, excites the remaining B state molecules to a higher
$0^+g(^1\Sigma)$ state which dissociates to two excited atoms.

back to the $0^+g(^1\Sigma)$ ground state. The second laser pulse samples the remaining B state population as a function of time by exciting it into a higher $0^+g(^1\Sigma)$ state which dissociates, emitting two excited $I(^2P_{1/2})$ atoms, as we have previously shown.[14]

We use the fragment translational energy distribution to identify the atoms which result from the double photon absorption. The number of B state molecules remaining versus delay time t_d between initiation and probe pulses is plotted in Fig. 5 for three different initiation pulse wavelengths. The B state population decays at different rates depending on the $v'J'$ levels excited, which in turn depends on the wavelength of the initiation pulse. The 1/e decay times are 770 nsec at 16 929 cm^{-1}, 1.4 μsec at 16 943 cm^{-1}, and 1.6 μsec at 16 972 cm^{-1}. This large variation may be explained by the expected[7] $J'(J'+1)$ dependence of the squared perturbation matrix element connecting $\Omega = 0$ and $\Omega = 1$ electronic states.[15] Our initiation laser pulse overlaps several v' levels of the B state, and increasing the laser photon energy decreases the average rotational quantum numbers excited in these same v' levels.

By a "rotational analysis" of the data, the symmetry of the final state reached in the time evolution of the initially prepared state can be identified. Thus, if the final state which perturbs the B state also had $\Omega = 0$, we would not have expected much J' dependence of the predissociation rate. By changing v' levels that are excited it should be possible to use DAPS to map out the potential energy curve of the perturbing state, using the variation with v' of the matrix element between vibrational wavefunctions of the prepared and perturbing states.[7, 16, 13]

The theory of photodissociation angular distributions[3c,17,18] shows that the shape of the angular distribution reflects information about the prepared state as well as the final dissociative state produced by the probe pulse. This information includes:

i) change in the angle between the breaking bond and the transition dipole moment.

ii) rotation of the molecule due to a non-negligible lifetime on the molecular rotation time scale, and

iii) the symmetry of the electronic states involved.

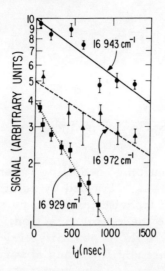

Fig. 5.   The I_2 B state decay curves obtained in DAPS experi-
          ments at the three different initiation pulse photon ener-
          gies shown.   The initiation pulse prepares the molecules
          in $v'J'$ levels of the B state at time $t = 0$.   The number
          of detected fragments, the vertical coordinate, is pro-
          portional to the B state population remaining at the time
          t_d of the probe laser pulse.   For the measurements
          shown, both lasers were polarized parallel to the recoil
          direction of the detected fragments.   The data for each
          of the three initiation pulse wavelengths have been arbi-
          trarily scaled to place all three curves in the same fig-
          ure.   The error bars are ± one standard deviation.

There are three useful angular distribution experiments that can
be performed:

   i)   rotating only the initiation pulse polarization,

  ii)   rotating only the probe pulse polarization, and

 iii)   rotating both polarizations together.

All three experiments should be done in order to completely analyze the photodissociation dynamics, but partial information can

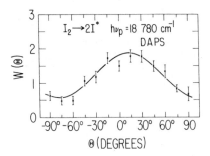

Fig. 6. Angular distribution of recoiling I atoms from the double
absorption process, obtained by rotating only the probe
pulse polarization, with initiation pulse polarization
fixed parallel to the recoil direction. The vertical scale
is proportional to the probability of detecting a photodissociation fragment following absorption of the probe
pulse at 18 780 cm^{-1}. The horizontal axis is the angle
Θ between the electric vector of the probe pulse and the
recoil direction of the detected I atoms. The initiation
pulse is at 16 943 cm^{-1}. Peaking near 0° implies that
the second transition is predominantly $\Delta\Omega = 0$ and thus
a 0^+g state is being reached by the final transition.
Error bars are ± one standard deviation.

be obtained from just one experiment. For example, Fig. 6 shows
an angular distribution obtained by rotating the probe pulse polarization with the initiation pulse polarization fixed parallel to the
recoil direction. The observed peaking at 0°, along the recoil
direction, implies that the transition caused by the probe pulse is
predominantly $\Delta\Omega = 0$, and since the intermediate B state has
$\Omega = 0$, the final state reached by the probe pulse transition also has
$\Omega = 0$. The electric dipole one photon selection rules u $\rightarrow$ g and
+ $\rightarrow$ + imply that the final dissociative state is 0^+g. Mulliken[6]
has predicted a 0^+g($^1\Sigma$) dissociative state correlating with two
$^2P_{1/2}$ excited I atoms in this energy region, just as we observe.[14]
Thus, using DAPS, we are able to characterize a state which,
since transitions to it are parity forbidden from the ground state,
is invisible in the ordinary absorption spectrum.

Interestingly enough, the one photon absorption spectrum
does show a continuum in this same energy region (see Fig. 4)
which, since the ground state is g, must be due to a u state.
We have demonstrated by photofragment spectroscopy that this so

called C state is a 1u state correlating with one ground state and one excited atom.[19] Thus one photon and two photon photofragment spectroscopy are complementary techniques for examining a manifold of excited states, particularly in molecules with a center of symmetry.

IV. CONCLUSIONS

In summary, we have developed a new technique, double absorption photofragment spectroscopy, in which we cross a molecular beam with an initiation laser pulse followed, after a delay time, by a probe laser pulse and observe the recoiling photodissociation fragments with a mass spectrometer. This technique should allow the observation of the time development of unimolecular processes in isolated molecules over a time span from picoseconds to milliseconds.

A first demonstration has been performed, the observation of the time evolution of predissociation in I_2 B state molecules as a function of their vibrational and rotational state distribution. More penetrating experiments appear possible. For example, in larger molecules it should be possible to observe one or all of the following:

i) variation of the fragmentation pattern as molecular structure and bonding evolve in time,

ii) change in angular distribution of the photofragments as the electronic state changes or bond angles and distances alter, and

iii) variation of the energy partitioning between translational and internal excitation of the recoiling fragments as the prepared molecular state evolves.

These types of measurements should provide new and deeper insights into the details of unimolecular processes, hopefully allowing the testing of basic theoretical assumptions.

Support by the National Science Foundation, by the Office of Naval Research (contract number N00014-69A-0200-6020) and computer usage supported by the National Institutes of Health and the Office of Computing Activities of the National Science Foundation are gratefully acknowledged.

REFERENCES

1. J. Jortner, S. A. Rice, and R. M. Hochstrasser, Advan. Photochem. 7, 149 (1969); E. W. Schlag, S. Schneider, and S. F. Fischer, Ann. Rev. Phys. Chem. 22, 465 (1971); V. Lawetz, G. Orlandi, and W. Siebrand, J. Chem. Phys. 56, 4058 (1972).

2. Other experiments which have used two laser pulses to excite an emitting state include: W. C. Lineberger and T. A. Patterson, Chem. Phys. Lett. 13, 40 (1972); V. S. Letokhov, Science 180, 451 (1973); C. B. Moore, Accounts Cheml. Resh. 6, 323 (1973).

3. A description of the apparatus is found in G. E. Busch, J. F. Cornelius, R. T. Mahoney, R. I. Morse, D. W. Schlosser, and K. R. Wilson, Rev. Sci. Instr. 41, 1066 (1970). Specific molecules which we have studied include a) Cl_2 and Br_2: K. R. Wilson, in Symposium on Excited State Chemistry, edited by J. N. Pitts, Jr. (Gordon and Breach, New York, 1970); b) I_2: R. J. Oldman, R. K. Sander, and K. R. Wilson, J. Chem. Phys. 54, 4127 (1971); c) NO_2: G. E. Busch and K. R. Wilson, J. Chem. Phys. 56, 3626 and 3638 (1972); d) NOCl: G. E. Busch and K. R. Wilson, J. Chem. Phys. 56, 3655 (1972); e) alkyl iodides: S. J. Riley and K. R. Wilson, J. Chem. Soc. Faraday Disc. 53, 132 (1972); f) ICN: G. Hancock and K. R. Wilson, in Fundamental and Applied Laser Physics: Proceedings of the Esfahan Symposium August 29-September 5, 1971, edited by M. S. Feld, A. Javan, and N. Kurnit (John Wiley and Sons, New York, 1973); g) acetone: G. Hancock and K. R. Wilson, Proceedings of the Fourth International Symposium on Molecular Beams, Cannes, France, July, 1973.

4. The Nd oscillator-amplifier system was designed and built under contract by Wilhelm Hagen at American Optical Co.

5. K. R. Wilson, Bull. Am. Phys. Soc. 18, 30 (1973).

6. R. S. Mulliken, J. Chem. Phys. 55, 288 (1971).

7. J. Tellinghuisen, J. Chem. Phys. 57, 2397 (1972).

8. J. Tellinghuisen, J. Chem. Phys. 58, 2821 (1973).

9. E. Wasserman, W. E. Falconer, and W. A. Yager, Berh. Bunsenges. physik. Chem. 72, 248 (1968).

10. A. Chutjian and T. C. James, J. Chem. Phys. 51, 1242 (1969).

11. D. G. Youmans, L. A. Hackel, and S. Ezekiel, J. Appl. Phys. 44, 2319 (1973).

12. K. C. Shotton and G. D. Chapman, J. Chem. Phys. 56, 1012 (1972).

13. G. A. Capelle and H. P. Broida, J. Chem. Phys. 58, 4212 (1973).

14. G. E. Busch, R. T. Mahoney, R. I. Morse, and K. R. Wilson, J. Chem. Phys. 51, 837 (1969).

15. G. Herzberg, Spectra of Diatomic Molecules (D. Van Nostrand Company, Inc., Princeton, New Jersey, 1950).

16. G. D. Chapman and P. R. Bunker, J. Chem. Phys. 57, 2951 (1972); M. S. Child, J. Mol. Spectry. 45, 293 (1973).

17. T. R. Bader and A. Gold, Phys. Rev. 171, 997 (1968); C. Jonah, J. Chem. Phys. 55, 1915 (1971); and R. N. Zare, Mol. Photochem. 4, 1 (1972).

18. R. K. Sander and K. R. Wilson, to be published.

19. R. D. Clear and K. R. Wilson, J. Mol. Spectry. 47, 39 (1973).

SELECTIVE TWO-STEP DISSOCIATION OF AMMONIA MOLECULES

R. V. Ambartzumian, V. S. Letokhov, G. N. Makarov, and
A. A. Puretzky

Institute of Spectroscopy, Academy of Sciences
Moscow, USSR

INTRODUCTION

A coherent frequency-tuned light beam may become a precision
instrument to study the structure and the composition of any substance
in various states. Its potential ability to affect material structure and
composition, using differences in absorption spectra, is less evident.
It is just this region where all the potentialities of laser light can show
up, among them are included monochromatisity, frequency tuning, directivity, short duration, etc. This region, linking up quantum electronics, spectroscopy and chemistry, is expected to give new methods
of selective action on matter. As an example, we can mention the following important problems within this region: separation of atoms and
molecules which are equal or close in their chemical properties (isotopes,
nuclear isomers, ortho- and para-molecules, etc.), selective destruc-
tion of particular molecules in a condensed medium (biological macro-
molecules, specifically), stimulation of particular chemical reactions
under non-equilibrium conditions, etc.

It should be said, that the first attempts to carry selective photo-
catalysis of chemical reactions (for a more complicated case of molecules
differing in isotope composition) were undertaken half a century ago. [1]
Though selective prelaser photochemistry is covered by dozens of papers
all of them are associated, basically, with the same idea, that is an
excitation of atomic or molecular electron states that increases the
chemical reaction rate. Surely, this idea can be realized more easily
with the help of laser light. Nevertheless, till recently an attempt to
conduct such an experiment has been made only in one work. [2] The
creation of lasers with a wide spectral range, from ultraviolet to infrared,
and with various power levels has made it possible to overstep the limits

of this idea and to develop fundamentally new methods of selective action of laser emission on matter. This report presents results of recent studies, carried out at the Institute of Spectroscopy, Academy of Sciences USSR, on the method of selective two-step photodissociation of molecules by laser radiation and also on the application of this method to successfully separate isotopes.

The above method was put forward and realized in 1970 by the authors of reference 3. The ammonia molecule, being one of the few spectroscopially well studied molecules, was chosen to study the method in detail. Experiments with this molecule[4] permitted in 1972 the separation of the isotopes ^{14}N and ^{15}N by the method of selective two-step photodissociation.[5] The two-step dissociation is an example of highly selective molecular chemical reaction stimulated by a laser emission. The common photodissociation is not selective toward molecules with identical structures (isotopic molecules) due to overlapping of continuous photodissociation absorption bands. A preliminary excitation of a molecule in the discrete narrow absorption line of the electronic or vibrational spectrum makes it possible to displace the red boundary of photodissociation and to create a large difference between spectra of absorption bands for excited and unexcited molecules. This difference is used by the second laser beam to photodissociate only the excited molecules. Like any monomolecular chemical reaction the two-step photodissociation requires no collisions, and so its rate may exceed considerably the rate of selectivity loss due to collisions. This is important to obtain high selectivity when separating isotopes.

There are three potential types of monomolecular chemical reactions suitable for a high-selectivity action of laser emission on matter:

1. Two-step photodissociation
2. Photopredissociation
3. Photoisomerization

The first process seems to be more complicated, but it is a universal way to perform selective chemical reactions with molecules in any (gas, liquid, solid) phase of substance. As opposed to it, the photopredissociation process needs only one laser beam but it is suitable for a limited class of molecules. Application of this method was suggested in reference 6. In reference 7 the method was used independently for a separation of hydrogen isotopes through photopredissociation of formaldehyde molecule. The same is true for the photoisomerization process which can be also highly selective under particular excitation conditions. The problem of using lasers to control selective chemical reactions was more generally studied by one of the authors in reference 8, while the problem of isotope separation was treated by C. B. Moore.[9]

TABLE I

Molecules	Absorption line of NH_3	ν of NH_3 (cm^{-1})	Emission line of CO_2 laser	ν of laser (cm^{-1})	$\nu NH_3 - \nu$ Las. (cm^{-1})
$^{14}NH_3$	aQ(5,3)	933,02 (10)	P(32) $(\nu_3 - \nu_1)$	932,961	0,03 [a]
	aR(1,1)	971,92 (10)	R(14) $(\nu_3 - \nu_1)$	971,930	-0,01
$^{15}NH_3$	aR(00)	970,80 (11)	P(16) $(\nu_3 - \nu_1)$	947,742	0.06
	SR(3,2)	1041,06 (11)	P(26) $(\nu_3 - 2\nu_2)$	1041,279	-0,229 [b]

a) F. Shimizu J. Chem. Phys. 52, 3572 (1970)
b) F. Shimizu J. Chem. Phys. 53, 1149 (1970)

INFRARED AND ULTRAVIOLET SPECTRA AND
AMMONIA PHOTOCHEMISTRY

Investigation of the selective two-step photodissociation the ammonia molecule was chosen here. First, because the vibrational levels of the ^{14}N and ^{15}N molecules can be selectively excited by CO_2 laser emission; second, by the possibility of exciting the unstable (predissociation) electronic state of NH_3 by the emission in an accessible ultraviolet region; third, by a considerable quantity of data available on the spectrum and photochemistry of the NH_3 molecule.

In a region of 10μ the $^{14}NH_3$ and $^{15}NH_3$ molecules have an absorption band caused by the transition $V'' = 1 \leftarrow V'' = 0$ at the ν_2 band to a vibrational level of lower energy. The absorption bands of $^{14}NH_3$ and $^{15}NH_3$ overlap each other, yet the spectral structure consists of hundreds of rotation-vibration lines. It has been studied in reference[10] for $^{14}NH_3$ in reference 11 for $^{15}NH_3$. In the spectra of the ν_2 band there are some non-coincident rotation-vibration lines of $^{14}NH_3$ and $^{15}NH_3$, their frequencies closely coinciding with those of rotation-vibration lines of the CO_2 laser emission. Table I lists such line pairs.

In the ultraviolet region the absorption spectrum of $^{14}NH_3$ is an electronic vibrational progression where a fully symmetric vibration of ν_2 appears, whereas it is active in the infrared absorption. The progression starts in the region of 2168 Å and extends to the short-wave region.[12] The electronic state $\tilde{A}$ is unstable due to the predissociation[13]. The absorption lines caused by transitions from the ground electronic state to the excited electronic state are shown schematically in Fig. 1 (a,b). For the vibrationally-excited NH_3 molecules to the level $V'' = 1$ of the ν_2 band, the electronic-vibrational spectrum differs somewhat in structure. The most substantial difference is a new absorption line arising due to the electronic-vibrational transition ($\tilde{A}$, $V' = 0$) $\leftarrow$ ($\tilde{X}$, $V'' = 1$). This line has a "red" shift the value of which equals the vibrational quantum energy. When the vibrational transition $V' = 1 \leftarrow V'' = 0$ is fully saturated by the infrared laser emission the molecules are distributed equally between the levels $V'' = 0$ and $V'' = 1$. Figure 1B illustrates a theoretical distribution of intensity in the electronic-vibrational spectrum of ammonia absorption for the case, when all the molecules are in the ground state $\tilde{X}$, $V'' = 0$, and for the case, when the molecules are equally distributed between the levels $\tilde{X}$, $V' = 0$ and $\tilde{X}$, $V'' = 1$. The calculations were carried out with the use of the Franck-Condon factors from reference[14]. The new electronic-vibrational line lying at the frequency 45250 cm^{-1} (in Figure 1B it is marked by an index), may be used for selective photodissociation of ammonia molecules. Thus, it follows from the spectroscopic data that there is a possibility of finding wavelengths of two frequencies for realizing the initial selective photochemical process.

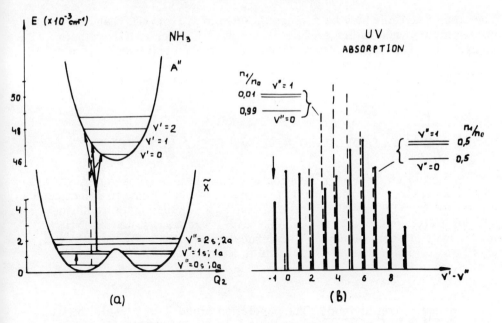

Fig. 1. a) Diagram of electron-vibrational transitions from the
ground (V" = 0) and excited (V" = 1) vibrational states
of NH_3 molecule.

b) Theoretical intensity distribution in the electronic-
vibrational spectrum of ammonia absorption for the
case, when all molecules are in the ground state
$\widetilde{X}$, V" = 0 (dotted lines), and for the case, when all
molecules are equally distributed between the levels
$\widetilde{X}$, V" = 0 and $\widetilde{X}$, V" = 1.

Selectivity of the secondary photochemical processes is also required for isotope separation. The photochemistry of ammonia has been
studied in many papers, the data of these studies is presented in a review[13].
The process of photochemical decomposition of ammonia into nitrogen and
hydrogen comes about using the following scheme[13]:

$$NH_3 + h\nu \rightarrow NH_2 + H$$

$$NH_2 + NH_2 \rightarrow N_2H_4$$

$$N_2H_4 + H \rightarrow N_2H_3 + H_2$$

$$2N_2H_3 \rightarrow 2NH_3 + N_2 \qquad (1)$$

The other reactions may be neglected. It follows from (1), that reactions occur only between radicals while unexcited molecules of NH_3 remain uninvolved. This means, that secondary photochemical processes are also selective in their nature, and it is hoped that a selective photodissociation may be carried out for, say, $^{15}NH_3$ molecules in the $^{14}NH_3 + ^{15}NH_3$ mixture.

EXCITATION OF AMMONIA VIBRATIONAL LEVELS

The data on the V-T relaxation in ammonia in various papers differs by 2-3 orders[15]. The rate of this process is of importance to select conditions for the vibrational excitation of NH_3 by a pulsed laser radiation without heating. There is no data on the V-V exchange in the ν_2 NH_3 mode, which is very essential when choosing conditions for the selective excitation of molecules with a particular isotopic composition. Because of this, the rates of the V-T and V-V relaxation processes had to be measured.

Measurement method. The population of the V" = 1 level, the rate of the V-T and V-V relaxation processes were measured by observing the amplitude and the time dependence of the ultra violet absorption of a new line for the transition V' = 1 ← V" = 0. The occurrence of this line is associated with the presence of molecules at the vibration level V" = 1 [16].

The experimental unit diagram is given in Figure 2. The emmission of the CO_2 laser (0.4 J 250 nsec FWHM) with a selectable frequency was directed into a cell with ammonia. The cell diameter (20 mm) was equal to the beam diameter. At the same time the cell was illuminated with ultraviolet radiation in a perpendicular direction. The monochromator selected radiation at the frequency of the transition V' = 0 ← V" = 1 (ν = 45250 cm^{-1}) and, was recorded by a photomultiplier.

Table I shows that some lines of the CO_2 laser emission and of the infrared absorption of $^{14}NH_3$ and $^{15}NH_3$ coincide closely, yet the detuning value exceeds the line widths. In order for the lines to coincide they were broadened by increasing the pressure of the ammonia and buffer gas. The pressure broadening values for various lines of the ν_2 NH_3 band have been measured previously[17,18]. They agree with the data obtained by radiospectroscopic methods[19]. The natural broadening of the NH_3 lines ranges from 20 to 25 MHz/Torr, and the broadening induced by buffer gases ranges from 1 to 3 MHz/Torr.

Population of the vibrational level V" = 1, $\tilde{X}$. The characteristic shape of the CO_2 laser pulse and the shape of the ultraviolet absorption

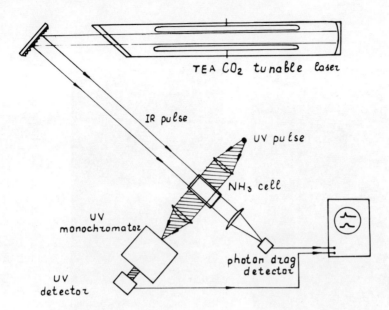

Fig. 2. Experimental set-up for measurements of population
values of the level V" = 1 and rates of the V-T and V-V
relaxations.

pulse at the reference transition are shown in Figure 3. Measuring the
amplitude of the ultraviolet absorption pulse at the probing transition
45250 cm^{-1} and knowing the relation between the amplitude and the number of molecules at the level V" = 1, we can calculate the number of
molecules at the level V" = 1 when exciting by laser radiation. The
calibrating curve for the relation between the number of molecules at
the level V" = 1 and the ultraviolet absorption value was obtained by
measuring the intensity of hot bands in the ultraviolet spectrum. The
data on hot vibrational bands in the ultraviolet spectrum of NH_3 is listed
in reference[12].

Figure 4 shows experimental relations between the number of
$^{14}NH_3$ molecules, excited to the level V" = 1, and the pressure of
different buffer gases; the pulse energy at the line P(32) 10.6 μ of the
CO_2 laser being constant. The regions where the broadening of the absorption line is approximately equal to the detuning of the emission and
absorption line are marked by dotted circles on the curves. Similar
relations are illustrated in Figure 5 for one buffer gas (Ar) and for
various lines of the $^{14}NH_3$ and $^{15}NH_3$ absorption. A characteristic increase in the number of excited particles and their saturation, as the

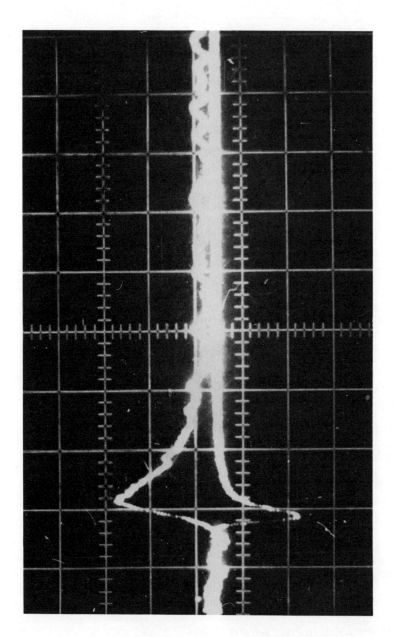

Fig. 3.   Characteristic shape of radiation pulse of the CO_2 laser and the shape of absorption pulse at the frequency 45250 cm^{-1} (transition V' = 0 ← V'' = 1). 6 Torr NH_3 + 680 Torr Xe. The time-scale is 0.5 μsec/div.

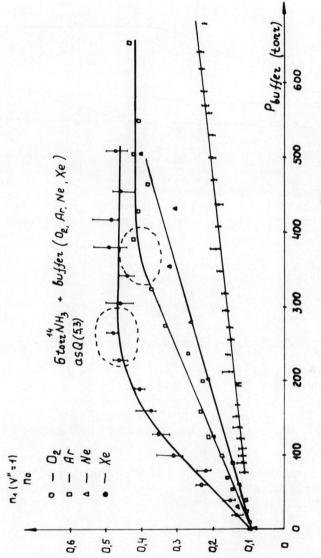

Fig. 4. Experimental results on the number of $^{14}NH_3$ molecules at the level $V'' = 1$ and the pressure of various buffer gases. The pumping was through the absorption line of ammonia aQ (5,3).

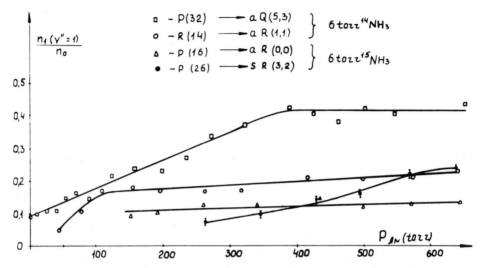

Fig. 5. Experimental results on dependence of the number of
NH₃ molecules at the level V" = 1 and the pressure of
buffer gas (Ar), with various absorption lines of ammonia
pumped.

buffer gas pressure increases, are contributed to by the following effects:

1. An increase in the overlapping of the CO_2 laser emission line
with NH_3 absorption lines. The role of this effect becomes evident when
correlating the region of pressures, under which the absorption line
broadening compensates for the detuning of pumping and absorption line,
with the region of pressures, under which the curves in Fig. 4 and 5
approximate a saturation.

2. The bottle-neck effect when the vibration-rotation band is excited by a monochromatic radiation pulse. This effect resides in the
fact that one half the molecules can be excited under the condition[20]:

$$T_p \geqslant \frac{T_{rot}}{q_t} \qquad\qquad (1)$$

where T_p is the pulse duration, T_{rot} is the time of molecular rotational
relaxation, q_t is the relative population of the lower rotational sublevel

of absorption transition which is determined by the expression:

$$q_t = \frac{g_t}{Z_{rot}} \exp{(-E_t^{rot}/kT)} \qquad (2)$$

Here, g_t is the degeneracy of level and those absorption lines coinciding with the pumping line; Z_{rot} is the statistical sum of rotational states; E_t^{rot} is the level rotational energy. If condition (1) is not complied with, then even during a strong saturation the maximum number of excited molecules doesn't exceed the value

$$N_{max} = 0.5N_0 \, (T_p/T_{rot}) \, q_t \qquad ,$$

where N_0 is the total number of molecules per unit volume.

Two parameters in (1) and in (2) depend on the buffer gas pressure, and, therefore, "the bottle-neck" effect gives the N_{max} pressure relation by virtue of two mechanisms:

a) An increase in the number of NH_3 absorption lines coinciding with one line of CO_2 laser emission, that is an increase in q_t with increasing pressure. This effect shows up vividly for the line P(32) 10.6 μm of CO_2 laser, that attains the maximum population of the $V'' = 1$ level.

b) An increase in the number of rotational levels which are capable of contributing to the interaction with one emission line in the T_p time, that is a decrease of T_{rot} with increasing pressure. According to reference[21] the time $T_{rot} = 0.92$ nsec/atm in pure NH_3, and the effect shows up at pressures of dozens of torr. It is of importance in part for the absorption line aR (0,0) $^{15}NH_3$ for which the q_t factor is too small.

Rate of the V-T relaxation. Measurements of the time dependence of an ultraviolet absorption pulse and the probing transition (45250 cm^{-1}) after a laser pulse may give information on the time of the V-T relaxation. During such measurements the rear edge of a CO_2 laser pulse was cut when the pulse was transmitted through a telescopic system. In the focus of this system a breakdown in air was performed. Such a breakdown absorbs completely the CO_2 laser emission, and so the pulse passing through the telescopic system has a very sharp rear edge (shorter than 10 nsec). The pulse of such a shape (Fig. 6) is rather convenient for relaxation measurements. The pulse shape of an ultraviolet absorption in ammonia at 45250 cm^{-1}, when being excited by a cut pulse of the CO_2 laser, is depicted in Figure 6a. Measurements of decay time for an absorption pulse give immediately the time value of the V-T relaxation for vibrations of ν_2 in ammonia: pT = 3.2 $\pm$ 0.5 nsec atm for $^{14}NH_3$.

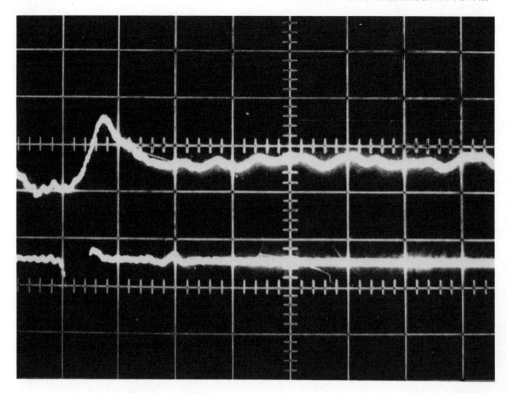

Fig. 6.   Characteristic shape of the CO_2 laser pulse, used in experiments for measuring the V-T relaxation rate, and the
form of absorption pulse at the transition $V' = 0 \leftarrow V'' = 1$.
The rate of exponent damping is similar to the rate of the
V-T relaxation. The pressure is 23 Torr. The scale is
250 nsec/div.

This value agrees with recent measurements obtained in reference[21],
where pT = 2.5 nsec atm.

Rate of the V-V relaxation.   Experiments on measuring the relation between the population of the level $V'' = 1$, $\tilde{X}$ and pressure have
shown, that in the region of asymptotic behavior of the curves (Fig. 4, 5)
the population of the level $V'' = 1$ ceases to depend on buffer gas pressure. This can be realized with the pressure of ammonia ranging within 3-15 Torr and with an excess of buffer gas for some absorption lines.
This phenomenon was used in measuring the rate of the V-V exchange
between $^{14}NH_3$ and $^{15}NH_3$ in the ν_2 mode. A mixture of $^{14}NH_3$ and
$^{15}NH_3$ in the ratio of 1:1 was irradiated using the unit depicted in Figure 2.
Only $^{314}NH_3$ was exposed to infrared excitation, and the total population

of the vibrational level V" = 1 for $^{14}NH_3$ and $^{15}NH_3$ was measured through ultraviolet absorption (Figure 7a). An excessive population of the level

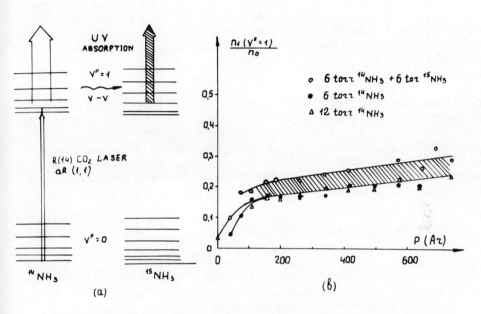

(a)   (6)

Fig. 7. Diagram of transitions and the experimental dependence of population of the level V " = 1 when $^{14}NH_3$ is under the pumping: the cell contains only $^{14}NH_3$ with mixed $^{15}NH_3$. The shaded area is the value of excessive population for the two-component gas compared to the single-component gas.

V " = 1, when compared to the case of pumping the single-component mixture of $^{14}NH_3$ under a pressure similar to the partial pressure of $^{14}NH_3$ in the mixture, suggests that throughout a laser pulse the vibrational excitation is transmitted from $^{14}NH_3$ to $^{15}NH_3$. The portion of transmitted energy is quantitatively given by the excess population. Figure 7b shows a typical experimental relationship. Comparing experimental curves with those computed we obtain the value for the time of the V-V exchange:

$$T^{V-V}_{^{14}NH_3 \rightarrow {}^{15}NH_3} = \left[3 \pm 1\right] T_{v-t} \quad ; \quad (3)$$

where T_{v-t} is the time of vibrational relaxation of the mixture $^{14}NH_3$ + $^{15}NH_3$ + Ar, which appears to be:

$$\frac{1}{T_{v-t}} = \frac{1}{T_{v-t}(NH_3, NH_3)} + \frac{1}{T_{v-t}(NH_3, Ar)} \quad ; \quad (4)$$

In making calculations the T_{v-t} values for $^{14}NH_3$ and $^{15}NH_3$ were assumed to be equal. The fact, that the rate of the V-V exchange turned out to be smaller than that of the V-T relaxation, was a surprise for us. Yet the number of collisions, necessary for the V-V exchange in NH_3 (25-40 collisions), coincides in order of magnitude with the results for other molecules given in reference[22].

Under optimum experimental conditions, with Ar as buffer gas, the portion of vibrational energy in $^{15}NH_3$ accounted for about 1/6 of vibrational energy in $^{14}NH_3$. This value limits the selectivity of two-step photodissociation of $^{14}NH_3$ or $^{15}NH_3$ in the mixture. It is obvious, that the photodissociation selectivity may be raised provided the pulse duration of ultraviolet radiation, that results in photodissociation of vibration-excited molecules, is shorter than the CO_2 laser pulse.

TWO-STEP SELECTIVE PHOTODISSOCIATION
OF NH_3 MOLECULE

Several studies on the selective laser excitation of vibrational levels of ammonia with various nitrogen isotopes have made it possible to observe directly the two-step photodissociation selectivity through the separation effect of nitrogen isotopes. The experimental set-up is shown in Figure 8. The $^{15}NH_3$ molecule was excited by a pulse of the CO_2 laser at the line P(16) 10.6 μ. This isotopic molecule was chosen for photodissociation because it was convenient to analyze the dissociation products $^{14}NH_2$, $^{14}N^{15}N$ and $^{15}N_2$. The least error can be realized during measurements at the mass spectrometer lines with m/e = 30 and 29, corresponding to the $^{15}N_2$ and $^{14}N^{15}N$ molecules. It is difficult to take measurements at the line with m/e = 28 because of a high value of background current of the mass spectrometer. Analysis of mass spectral lines with m/e = 29 and 30 allows us to determine completely the isotopic composition of photodissociation products.

A spark pulse source with continuous spectrum was used to produce ultraviolet radiation at the f requency of the new absorption line V' = 0, $\tilde{A}$ ← V" = 1, $\tilde{X}$ (45250 cm^{-1}). The full pulse duration of the ultraviolet source was 400 nsec. Synchronization of the spark was attained by firing the spark gap with a portion of the CO_2 laser pulse. Particular attention was paid upon the elimination of any delay of the UV pulse about the IR pulse. To remove completely from the spark continuous spectrum radiation that can be absorbed by the unexcited molecules

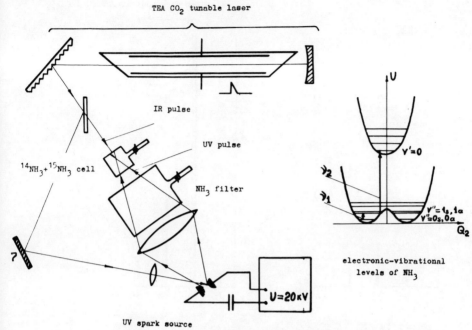

Fig. 8. Scheme of experimental unit for separating nitrogen
isotopes by the method of two-step dissociation of ammonia molecules.

$^{14}NH_3$ and $^{15}NH_3$, an absorbing cell filled with natural $^{14}NH_3$ was placed
in front of the cell with the mixture. The pressure in the ammonia filter
was selected so that the spark light at the frequency of the transition
$V' = 0$, $\widetilde{A} \leftarrow V'' = 1$, $\widetilde{X}$ (2168 Å was completely absorbed in the filter.
In this case, complete absorption occurred also on all shorter-wave
ammonia tranistions $V' = 1, 2, ..., \widetilde{A} \leftarrow V''$, $= 0$, $\widetilde{X}$.

Special attention in the experiment was focused on careful preparation of the isotopic mixture. After filling a cylinder, the mixture was
held there more than 24 hours in order that the isotopic molecules could
be completely mixed. Then, prior to irradiating, the mixture was analyzed with the mass spectrometer for subsequent determination of the
ratios.   During the experiments the following mixtures were irradiated: 10 Torr $^{14}NH_3$ + 10 Torr $^{15}NH_3$ and 5 Torr $^{14}NH_3$ + 5 Torr $^{15}NH_3$
+ a buffer gas (Ne or Xe) at a pressure of to 250 Torr.

The results of the checking experiment, when the mixture was
irradiated by UV pulse with neither ammonia filter nor IR pulse, are
given in Figure 9a. The measured increments for the lines of mass spectrum with m/e = 29 and 30 are in the ratio 2:1 to a high accuracy. This
result corresponds precisely to the theoretical values for the non-selective

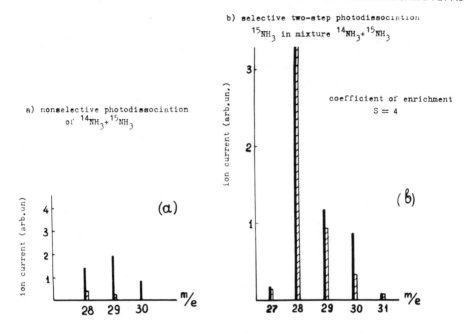

Fig. 9. Mass spectra, a) nonselective photodissociation of
($^{14}NH_3$ + $^{15}NH_3$); b) selective two-step photodissociation
of ($^{14}NH_3$ + $^{15}NH_3$).

photodissociation of the mixture of $^{14}NH_3$ and $^{15}NH_3$ with similar pressures. Figure 9b illustrates a typical form of lines with m/e = 29 and 30
on a mass spectrogram, which were obtained after the mixture had been
irradiated simultaneously by pulses of IR radiation of the CO_2 laser and
by pulses filtrated with the spark UV radiation. The line with m/e = 30
corresponding to $^{15}N_2$ was increased by much more than the line with
m/e = 29 ($^{14}N^{15}N$). This is sufficient proof of the selective two-step
photodissociation of the $^{15}NH_3$ molecules.

In the case when final products differ from initial ones, we can
introduce a separation degree S with a chosen isotope. The degree of
separation, for example, with the ^{15}N isotope will be determined as the
ratio between the total number of the ^{15}N and ^{14}N atoms in the molecular
nitrogen, with their initial numbers in the ammonia. Thus, we have:

$$= \frac{2\,[\,^{15}N_2] + [^{14}N^{15}N]}{2\,[\,^{14}N_2] + [^{14}N^{15}N]} = \frac{2I_{30} + I_{29}}{2I_{28} + I_{29}} \quad ; \quad (5)$$

Here, the brackets denote concentration of the enclosed molecules; I is the increase in ionic current of the mass spectrometer, proportional to the molecular concentration. In this case, due to collisional statistics, the ratios between line intensities in the mass spectrum always obey the relation:

$$\frac{I_{30}}{I_{29}} = \frac{1}{4} \frac{I_{29}}{I_{28}} \; ; \tag{6}$$

Thus, the degree of separation is related to the experimentally measured ratio between intensity values for the lines I_{29} and I_{30} by:

$$S = \left[2 \; \frac{I_{30}}{I_{29}} + 1 \right] \left[\frac{1}{2} \; \frac{I_{29}}{I_{30}} + 1 \right]^{-1} \tag{7}$$

In the experiment, the results of which are given in Figure 9, $I_{29}/I_{30} = 0.50 \pm 0.03$. So, the coefficient of separation here $S = 4 \pm 0.2$. This corresponds to a content of 80% of ^{15}N and 20% of ^{14}N atoms in the molecular nitrogen after the mixture is irradiated, with their concentrations being equal prior to irradiating.

In various experiments the separation degree ranges from 2.5 to 6 even with the main characteristics being under control.

ON SOME POSSIBILITIES

The experiments performed on ammonia have corroborated the effectiveness of the method of two-step photodissociation for the selective breaking of bonds of molecules with preset isotopic composition. The selectivity of the initial molecular excitation and that of secondary photochemical processes in a mixture after molecular dissociation are imperative conditions for the selectivity of the method. It should be mentioned, that there are some more possibilities available to obtain the selectivity when the above said conditions are not complied with.

The selectivity of molecular excitation is impossible when there are no non-overlapping spectral lines belonging to molecules with different isotopes in the absorption spectrum on vibrational and electronic transitions. This may be also true for polyatomic molecules with a complex spectral structure, if the isotope shift is less than the absorption band width and the mean distance between adjacent rotation-vibration lines in the band is less than the Doppler width. In this case the resonant method of particle excitation in a low pressure gas in standing wave field may prove to be effective, that was given consideration to in reference[3]. As stated in reference[23], the density of excited particles in a standing wave field has its resonance minimum during a fine tuning to the center

of the Doppler contour. With the case of a strong saturation the depth of
the minimum may reach 30%. In a mixture of the molecules A and B with
overlapping lines the denisty of excited molecules of each sort, as the
standing wave frequency function ν, has the form illustrated in Figure 10 a.

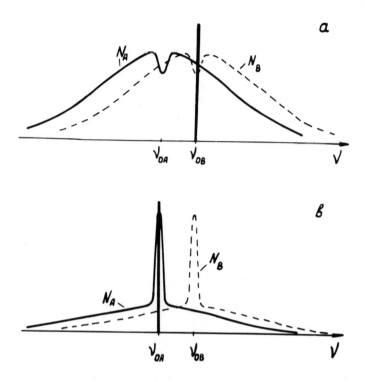

Fig. 10. Frequency dependence of one-quantum (a) and two-quantum
          (b) non-linear absorptions at the Doppler-broadened tran-
          sition in a standing wave field.

By tuning the laser frequency to the transition line center of a molecule
which one does not wish to excite, their density can be decreased by 30%.
The selectivity of this method is rather limited relative to excitation of
molecules of the wanted sort. The maximum relation between the density
of excited particles of the chosen type A and the density of the unwanted
particles B $N_A/N_B < 1.3$. In this respect, the method of two-photon

particle excitation in a standing light wave[24] is more effective. Its application in a selective excitation of particles with overlapping lines has
been considered in paper[25]. In this method the excited particle density
at the center of the Doppler line increases by a considerable value, about
$\frac{ku}{r}$. Tuning the standing wave frequency to the line center of chosen

particles we can attain a high excess of their density in excited state compared with the density of excited particles of the other sort (Figure 10b). Their density ratio may reach the values:

$$\frac{N_A}{N_B} \simeq \frac{ku}{r} \gg 1 \qquad (8)$$

When there is no possibility to blend chemically the products of molecular dissociation in a gas mixture without disturbing the selectivity then the chemical cycle may be dropped. In this case photodissociation products can be separated through a molecular directed gas jet. The dissociation scatters the molecular products with their velocity distribution being isotropic. Under a low gas pressure in the region of photodissociation molecular fragments escape from the jet and may be collected by standard methods. The method under study allows one to carry out the selective photodissociation for molecules having a low vapor pressure and calling for high temperatures.

REFERENCES

1. H. Hartley, A.O. Pender, E.J. Bowen, and T.R. Merton, Phyl. Mag. 43, 430 (1922).

2. W.B. Tiffany, H.W. Moos, and A.L. Schawlow, Science 157, 40 (1967).

3. R.V. Ambartzumian, and V.S. Letokhov, IEEE Journ. Quant. Electr. 7, 305, (1971); Appl. Optics. 11, 354 (1972).

4. R.V. Ambartzumian, V.S. Letokhov, G.N. Makarov, and A.A. Puretzki, JETP Lett. 15, 709 (1972).

5. R.V. Ambartzumian, V.S. Letokhov, G.N. Makarov, and A.A. Puretzki, JETP Lett. 17, 91 (1973).

6. V.S. Letokhov, Chem. Phys. Lett. 15, 221 (1972).

7. E.S. Yeung, and C.B. Moore, Appl. Phys. Lett. 21, 109 (1972).

8. V.S. Letokhov, Science 180, 451 (1973).

9. C. Bradley Moore, Accounts of Chemical Research 6, 323 (1973).

10. H.M. Mould, W.C. Price, and G.R. Wilkinson, Spectrochimica Acta, N 5, 313 (1959).

11. F.O. Shimizu and T. Shimizu, J. Mol. Spectr. 36, 94 (1970).

12. A.D. Walsh and P.A. Warsop, Trans. of Farad. Soc. 57, 345 (1961).

13. J.R. McNesby and H. Okabe, in "Advances in Photochemistry," Vol. 3; New York (1964).

14. W.R. Harshbarger, J. Chem. Phys. 53 (1970).

15. T.L. Cottrell and J.C. McCoubrey, Molecular Energy Transfer in Gases, Butterworths, London, 1961 p. 98.

16. R.V. Ambartzumian, V.S. Letokhov, G.N. Makarov, and A.A. Puretzki. Chem. Phys. Lett. 16, 252 (1972).

17. T. Shimizu, F.O. Shimizu, R. Turner, and T. Oka, J. Chem. Phys., 55, 2824 (1971).

18. V.S. Letokhov, A.G. Platova, and O.A. Tumanov, Optics and Spectroscopy (to be published).

19. A.L. Schawlow and C.H. Townes, Microwave Spectroscopy, McGaw-Hill, N.Y. (1955).

20. V.S. Letokhov and A.A. Makarov, JETP. Lett. 63, 2064 (1972).

21. H.W. Bass and T.G. Winter, J. Chem. Phys., 56, 3619 (1972).

22. C. Bradley Moore, in "Advances in Chemical Physics", Vol. 23, p. 41 J. Wiley and Sons, N.Y. (1973).

23. V.S. Letokhov and B.D. Pavlik, JETP. Lett. 64, 804 (1973).

24. L.S. Vasilenko, V.P. Chebotaev, and A.A. Shishaev, JETP. Lett. 12, 161 (1970).

25. V.P. Chebotaev, A.L. Golgher, and V.S. Letokhov, J. Chem. Phys. (to be published).

ASTROPHYSICS AND GEOPHYSICS

QUANTUM ELECTRONIC TECHNIQUES IN ASTROPHYSICS: PRESENT

DEVELOPMENTS AND FUTURE PROSPECTS

D. M. Rank

Lick Observatory, Board of Studies
in Astronomy and Astrophysics
University of California
Santa Cruz

The rapid development of quantum electronic devices in the
past decade has extended many conventional radio frequency techniques through the infrared region of the electromagnetic spectrum
into the visible spectral region. Subsequent development has led
to many new devices and techniques which have no radio frequency
counterparts. Quantum optics has also stimulated interest in the
development of even better conventional optical parts and materials
for use in quantum electronic devices. Much of the astronomical
data which we have today has been obtained with optical telescopes;
therefore, the new advances in optical techniques and quantum
electronics will result in new astronomical data of higher quality
than has been possible with the conventional astronomical equipment of the previous decades. Some of this improvement in astronomical observations will come from laser servocontrol mechanisms,
improved diffraction gratings, optical coatings, interference
filters, and data processing. However, quantum electronics offers
the potential for applying new optical and infrared detectors to
astronomical problems. Optical heterodyne receivers, photon
upconverters, and spatial interferometers are a few of the exciting
new types of instrumentation which quantum electronics has made
available to astronomers.

Photon detectors can be divided into two general classes:
1) linear amplifiers (phase preserving detectors), 2) photon
counters. An example of a linear amplifier is the maser which
amplifies the number of input photons, n, and preserves the phase
ϕ of these photons. Such devices are limited in the limit of
large gain by the uncertainty principal $(\Delta n \Delta \phi \gtrless \frac{1}{2})$ to an input
noise photon per Hz of the amplifiers bandwidth. This noise can
be expressed as a temperature $T_{noise} = h\nu/k$ where, h is Planck's

633

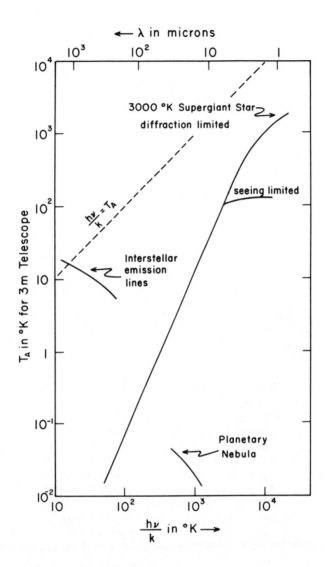

Figure 1. Antenna temperature T_A (3M telescope) for selected astronomical sources as a function of wavelength. The dashed line $h\nu/k = T_A$ is the theoretical limit for the equivalent noise temperature of phase preserving photon amplifiers.

constant, ν the photon frequency, and k Boltzmann's constant.
Therefore, all phase preserving detector systems; masers, lasers,
and heterodyne detectors have an output noise approximately equal
to one photon per Hz of bandwidth at the input even when there are
no input photons present. Photon counters, on the other hand, can
in principle have an equivalent input noise which is zero when
there are no input photons. Photomultipliers and photon upconversion systems are examples of photon counter detectors. Figure 1.
shows the strength of some astronomical signals on a 3M telescope
in units of antenna temperature T_A compared to the theoretical
noise temperature limit for linear amplifiers $h\nu/k$. Inspection
of the figure shows that all of the sources are considerably less
intense than the input noise of such amplifiers. The general
equation which determines the minimum detectable signal ΔT in
temperature units for a typical linear amplifier is

$$\Delta T \simeq \frac{T_{noise}}{\sqrt{B\tau}}$$

where T noise is the input noise of the amplifier, B is the bandwidth of the amplifier, and τ is the observation time. For many
astronomical applications such as spectroscopy B can be equal to

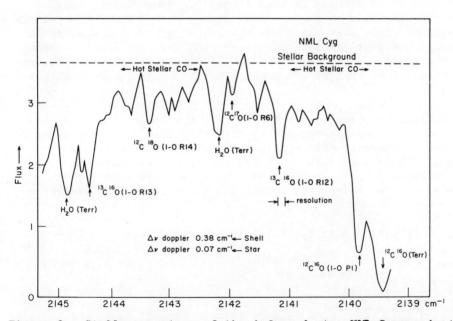

Figure 2. Stellar spectrum of the infrared star NML Cygnus in the
5 micron region. Doppler shifted CO absorption lines in a circumstellar shell are indicated with arrows. Regions of hot band CO
absorption in the stellar atmosphere are also indicated by the
horizontal arrows.

10^8 to 10^9 Hz which means that all of the signal sources in Figure 1 could be detected (at their theoretical limit) in a few seconds of observation time. Therefore, such linear amplifiers offer the possibility of obtaining very high resolution spectrscopic observations of a number of astronomical objects in the infrared and visible regions of the spectrum.

An example of conventional high resolution infrared spectroscopy in astronomy is shown in Figure 2. This spectrum was taken with a Fabry Perot interferometer at a resolution of $0.10 \mathrm{cm}^{-1}$. The carbon monoxide absorption lines in this star originate in the hot stellar atmosphere and in a cooler circumstellar shell. The line shapes are not resolved and clearly the number of lines present indicate that higher resolution would be desirable. In fact, the line widths of the absorption lines in Figure 2 are about 10^9 Hz which is a rather convenient bandwidth for heterodyne techniques of quantum electronics. A number of individual spectral lines in such an object could be studied at very high resolution and provide a wealth of information about kinetics in the stars' atmospheres and radiative transport and kinetics in the circumstellar shell. Table 1 is a general list of typical astronomical sources of line radiation and their line widths $\Delta \nu / \nu$. The actual linewidth scaled to a wavelength of 10 microns is also listed for comparison. Most of these objects could be studied with high spectral resolution within the optical bandwidths which are available for quantum electronic devices. The most pressing problems which need further development work are those of building more tunable infrared lasers and heterodyne maser-laser systems which are close to their theoretical noise limits.

Optical heterodyne experiments in the visible region of the spectrum have been carried out by Nieuwenhuijzen[1]. He has detected a number of stars at 6328 A with a bandwidth of approximately 200 MHz. These same techniques are also being extended into the infrared by Gay and Journet[2], and Johnson[3]. Both groups are applying optical heterodyne techniques to the Michelson stellar

Table 1. Linewidths of Typical Astronomical Objects

Source	$\Delta \nu / \nu$	at 10μ in Hz
Maser (interstellar)	10^{-7}	3×10^6
Star Atmosphere	10^{-5}	3×10^8
Interstellar Lines	$10^{-5} - 10^{-4}$	10^9
Nebulae Emission	10^{-4}	3×10^9

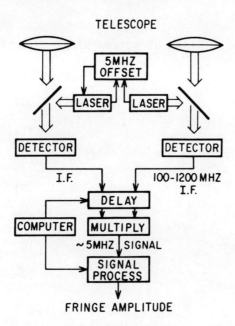

Figure 3.   Block diagram of a 10 micron heterodyne stellar interferometer using two phaselocked CO_2 lasers as local oscillators.

interferometer. A basic block diagram of such a system is shown
in Figure 3. Instead of measuring the spatial coherence of light
through two apertures by combining two beams on a single detector
as Michelson did, the infrared technique uses phaselocked lasers
to provide a phase reference and a single heterodyne detector at
each aperture. Interference fringes from the two apertures are
obtained from beating the infrared frequencies of the two heterodyne detectors together in the subsequent electronics. Such
systems are capable of using aperture separations of hundreds or
even thousands of kilometers since they require electronic stability rather than mechanical stability for interference of the
output signals. Normal astronomical observations are limited to

Table II.   Sizes of Astronomical Objects

Source	Size in Seconds of Arc
Supergiant Stars	.03
QSO's	?(.001)
Galactic Nuclei	1
Radio Masers	.01
Dust Shells	0.5
Planetary Nebulae	10
Cool Gas Clouds	10^2

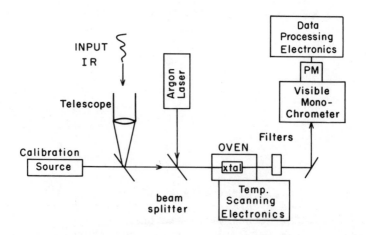

Figure 4. Block diagram of an infrared upconversion spectrometer using an argon laser at 5000 A as a pump. Wavelength scanning is achieved by varying phase matching in the nonlinear crystal with a variable temperature oven.

a resolution of approximately one second of arc or 5 x 10 radians.
A 10 micron stellar interferometer with an aperture separation of
100 meters will produce an angular resolution of 10^{-7} radians.
Table II gives some typical sizes for many common astronomical objects which could be studied with spatial interferometers. It is
clear that even modest sized spatial interferometers will give
astronomers detailed information about a wide variety of interesting objects which were beyond the limits of resolution of present
large telescopes.

    While coherent detectors such as lasers and heterodyne systems
offer very high spatial resolution or spectral resolution, they do
not have the ultimate sensitivity of photon counting systems. Photon counters need not have any intrinsic noise source other than the
fluctuations in the statistics of the input photons. In the absence
of an input signal, their theoretical noise limit is zero. A diagram of an infrared photon counting system, Smith and Townes[4], is
shown in Figure 4. The system uses a nonlinear phase matched crystal pumped by a 5000 A argon laser to convert 3 micron infrared
photons into the blue region of the visible spectrum. The visible
photons are then detected by a photomultiplier tube. The phase
matching of the infrared and optical photons is accomplished by

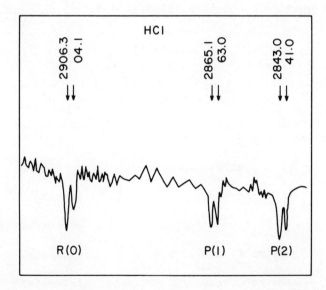

Figure 5. HCl absorption cell spectrum obtained from the 3.5 micron band of HCl with an upconversion spectrometer. The resolution
of the spectrometer $\simeq \frac{1}{2}$ cm^{-1} clearly shows the two most abundant
isotopic species of chlorine. Arrows indicate the positions of
the absorption lines in cm^{-1}.

varying the temperature of the nonlinear crystal. Phase matching conditions are satisfied for infrared photons with a bandwidth $\Delta \nu = c/L$ where L is the length of the crystal in cm. and c is the velocity of light. Since only a narrow spectral range of infrared frequencies is matched at a given crystal temperature, the upconversion detector is a narrow band tunable filter. With crystal lengths of one to a few cm., the device has a resolution comparable to some of the best present-day infrared astronomical spectrometers. The infrared quantum efficiency of an upconverter is proportional to the laser power density and the nonlinear coefficient of the crystal. Present laser technology allows an upconversion efficiency of approximately 10^{-4}. However, orders of magnitude of improvement in efficiency should be possible in the coming years. If this can be achieved, infrared photon counters will near the theoretical limit of sensitivity as many visible and ultraviolet photomultipliers are today. Figure 5 shows a spectrum of the 3 micron band of HCl taken with the Smith, Townes upconverting detector. Note that the resolution of the instrument easily separates the two common chlorine isotopes of this molecule. Such resolution is quite adequate for many interesting astronomical problems.

Quantum electronics will play an important role in the development of astronomy principally by improving detectors and optical systems in the visible and infrared spectral regions. The areas of astronomy which should benefit most directly are those involving high spatial resolution and high spectral resolution.

Table III. Astronomical Lines of Interest for Quantum Optical Systems in the Infrared

Microns	Species	Source
10.5	CO_2 hot band	planets
10.5	S IV fine structure	planetary nebulae
17	H_2 rotational $J = 1 \to 3$	Interstellar gas
28	H_2 rotational $J = 0 \to 2$	Interstellar gas
112	HD rotational $J = 0 \to 1$	Interstellar gas
230	H_2O $6_{34} - 5_{41}$	radio masers
790	H_2O $4_{14} - 3_{21}$	radio masers

For the latter, Table III lists a number of astronomical spectral lines which span the infrared spectrum and should be particularly suitable for detection by quantum optical techniques. Sensitive optical and infrared systems which will surely be built on our quantum electronic technology should advance astronomical observations and greatly improve our understanding of the universe.

References

1. H. Nieuwenhuijzen, Mon. Not. R. astr. Soc. <u>150</u>, 325 (1970)

2. J. Gay and A. Journet, Nature, <u>241</u> 280 (1973)

3. M. A. Johnson, Bull. Opt. Soc. Am. 23   (1972)

4. H. A. Smith and C. H. Townes, Polarisation, Matrère, et Rayonnement Volume Jubibaire en l'honneur d'Alfred Kastler, Presses Universitaires de France, 1969, p. 467

EXPERIMENTS OF ASTROPHYSICAL SIGNIFICANCE

USING A LASER STRAINMETER[†]

Judah Levine[*]

Quantum Electronics Division, National Bureau

of Standards, Boulder, Colorado 80302

INTRODUCTION

We are currently operating a 30 m laser strainmeter in an unworked gold mine near Boulder, Colorado. We are using the instrument to study earth tides, earth normal modes, and other problems of geophysical interest. In addition we are conducting a series of experiments of astrophysical significance. It is this latter group of experiments that we wish to report on here.

Before describing the experiments, we shall briefly review the operating principles of the strainmeter.

DESCRIPTION OF THE APPARATUS

Figure 1 shows a block diagram of the strainmeter and its associated electronics.

A 30-meter evacuated Fabry-Perot interferometer is mounted on piers and is oriented along a horizontal axis approximately 7° west of north. The interferometer is approximately 60 meters below ground level.

---

[*]Member, Joint Institute for Laboratory Astrophysics and Professor Adjoint, Department of Physics and Astrophysics, University of Colorado, Boulder, Colorado 80302.
[†]Work supported in part by the National Science Foundation under grant number GA-36365.

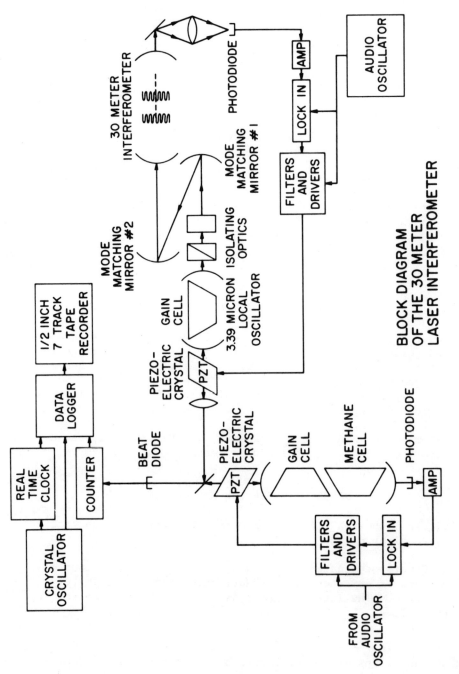

BLOCK DIAGRAM
OF THE 30 METER
LASER INTERFEROMETER

Fig. 1

The interferometer is illuminated by a 3.39 μm single mode helium-neon laser. A servo loop piezoelectrically tunes the laser to keep its wavelength coincident with one of the transmission maxima of the long interferometer. The frequency of the laser is therefore related to the length of the interferometer by

$$f = \frac{nc}{2L} \tag{1}$$

where n is an integer, c is the velocity of light and L is the length of the interferometer. Then

$$\frac{\Delta f}{f} = - \frac{\Delta L}{L} \quad . \tag{2}$$

[Note: Equation (2) is still valid if Eq. (1) is replaced by the more exact expression which includes excitation of off-axis modes of the long interferometer provided that the amplitudes of the off-axis modes do not change in time. See reference 1.]

A second 3.39 μm laser is stabilized using saturated absorption in methane.[1,2] The beat frequency between these two lasers is extracted for further processing.

Thus the beat frequency is given by

$$f_{beat} = f_{CH_4} - f$$

where f_{CH_4} is the frequency of the methane stabilized laser. Therefore

$$\frac{\Delta f_{beat}}{f} = \frac{\Delta L}{L} \quad .$$

A more detailed description of the apparatus may be found in reference 1.

SENSITIVITY OF THE SYSTEM

The sensitivity of the system is characterized at any frequency by the noise equivalent strain power; that is, by a fiducial generator in the earth which would produce the observed system noise.

For times less than or on the order of one second, the sensitivity of the system is limited by various white noise sources

within the system such as Johnson noise in the photodetectors,
shot noise in the laser beam, etc. We therefore expect the noise
in the system to be inversely proportional to the square root of
the integrating time. This is in fact the case. Using an integration time of 16 msec, for example, the system noise is on the
order of $8 \times 10^{-28} (\Delta L/L)^2 /Hz$. For integration times much longer
than a few seconds the sensitivity may be limited by "flicker"
noise.[3-5]

FREQUENCY RESPONSE

The frequency response of the system is governed by the servo
loop response. The servo loops have a unity gain point near 1 kHz
and the gain increase at 9 dB/octave below 1 kHz. The servo responds to a unit step in approximately 1.5 msec giving a system
response from dc to over 100 Hz.

GRAVITATIONAL WAVE EXPERIMENTS

The unambiguous detection of gravitational radiation from
either terrestrial or extraterrestrial sources is a difficult but
very important problem. In order for the experiment to be unambiguous it is highly desirable that the source be identified.
Weber[6] has shown that the construction of a terrestrial generator
of a detectable flux of gravitational waves is beyond the current
state of the art. We are therefore limited to extraterrestrial
sources. If we insist that the source be identifiable, we probably have to further restrict ourselves to objects which are visible in the optical or radio regions of the spectrum. This requirement discriminates rather strongly against impulsive sources.
The probability of detecting the arrival of gravitational and
optical signals from a cosmic explosion is small (if only the
Chinese had known to look for the gravitational signal from the
Crab pulsar in its youth).

The number of reasonably intense sources that meet the requirements outlined above is rather small. It includes the faster
pulsars, the fastest rotating binaries, and perhaps one or two
other types of oscillating systems. In all of the cases known
today, the frequencies of oscillation are too low to permit the
construction of a low-noise resonant antenna.

Two of the possible types of non-resonant antennas are long
bars and the earth itself. We discuss below the possibility of
using these detectors to detect gravitational waves from pulsars
or rotating binary stars.

Pulsars

The possibility that pulsars may emit gravitational radiation prompted us to search for the strain fluctuations that this radiation would induce when striking the earth. The gravitational radiation emitted by a rotating object is given by[7]

$$ E = \frac{G}{45c^5} \dddot{D}_{\alpha\beta}^{\,2} $$

where D is the quadrupole tensor, G the gravitational constant and c the speed of light. Thus if a pulsar has a varying component of its quadrupole moment tensor perpendicular to its axis of rotation, it will emit gravitational radiation at twice its rotational frequency. The radiated power will be proportional to the sixth power of the rotation frequency, suggesting that, other things being equal, the fastest pulsars stand the best chance of being detected.

The interaction of this radiation with the earth has been treated by Dyson,[8] Weber[6] and Misner et al.[9] Using Dyson's estimate, we can relate the displacement in the earth's surface, Δy, to the incident gravitation flux, W. We obtain

$$ (\Delta y)^2 = \frac{16\pi G S^2}{\omega^4 c^3} W \tag{3} $$

where S is the seismic velocity, which is on the order of 5 km/sec.

Thus, although the radiated power from a pulsar goes up very rapidly with rotation frequency, much of this advantage is lost because the absorption cross section of the earth drops so rapidly. In addition, the system noise increases for increasing frequencies. We conclude that for pulsar-type sources the detectability is proportional to $\omega^{3/2}$, so that fast pulsars are still favored, but not nearly as strongly as one might expect at first.

Although a pulsar is not the optimum source of gravitational radiation, we nevertheless searched for a pulsar coherent strain amplitude generated by the radiation from the Crab pulsar. Our conclusion[10] is that if a pulsar coherent strain signal is present, its amplitude is less than $8 \times 10^{-17} (\Delta L/L)$.

The dominant noise source in the Crab pulsar experiment is the servo loop noise both in the long path lock and in the methane stabilizer. The contributions of both of these sources could be reduced if a gravitational radiator at a lower frequency could be used. We therefore consider other sources.

Rotating Binary Stars

Unlike the Crab pulsar which is nearly spherically symmetric, rotating binary systems possess macroscopic quadrupole moments. A class of rapidly rotating binary stars has been discovered with periods on the order of hours.[11-14] Although these objects rotate slowly compared to the Crab pulsar, their macroscopic quadrupole moments more than make up for their slow rotational frequency. There are several estimates in the literature of the gravitational flux reaching the earth from these rotating binary systems.

Kraft et al.[14] estimate that the flux reaching the earth from wz Sagittae (period ~81 minutes) may be as high as 10^{-9} ergs/sec-cm^2. Estimates[9,14,15] in which the lower mass companion is assumed to be degenerate predict lower fluxes--perhaps as small as 10^{-13} ergs/sec-cm^2.

We may calculate an upper bound to the strain signal this flux would produce in the earth using Eq. (3)

$$\frac{\Delta L}{L} \sim 1.5 \times 10^{-16}$$

Weber[6] shows that the expected strain in a bar antenna is given by

$$\frac{\Delta L}{L} \sim \frac{c^2 R_{oioi} Q}{\omega^2}$$

where c is the velocity of light, ω the incident frequency, R the Riemann tensor and Q the Q of the antenna. Using Q = 1 we find that the signal is several orders of magnitude less than that obtained above.

Although somewhat slower in period, i Bootes (period ~6 hours) presents a more favorable case. If we use the estimated[9,15] flux of 10^{-10} ergs/sec-cm^2, we would expect a strain

$$\frac{\Delta L}{L} \sim 5 \times 10^{-14}$$

induced in the earth.

Although both wz Sagittae and i Bootes generate strain fluctuations that are larger than the threshold for detectability established in the Crab pulsar work,[10] neither is detectable in a reasonable time using the current apparatus. This is due to the fact that the limitation in these experiment would not come from the instrumental noise but from the noise in the earth itself.

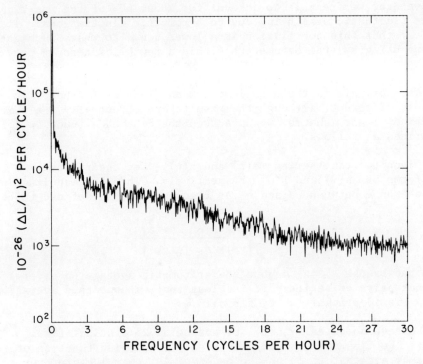

FOURIER TRANSFORM   (1-30 cph BANDPASS)

Fig. 2

In Fig. 2 we show a Fourier transform of the earth noise in the
frequency region 1-30 cycles/hour obtained using the laser strainmeter. The power estimate in each frequency band has the stability equivalent to a χ^2 distribution with 32 degrees of freedom.
At 1.5 cycle/hour, the earth noise is about $10^{-22}(\Delta L/L)^2$/cycle/
hour suggesting that an integration time on the order of 10^5
hours would be required to achieve a signal-to-noise ratio of order
unity on the detection of radiation from i Bootes.

   Although Fig. 2 might lead one to believe that the earth
noise near 1 cycle/hour is generated by random fluctuations in the
earth strain field, this is not the case. The earth noise at frequencies below several cycles/hour is neither stochastic nor stationary. In fact, a significant fraction of the earth noise in
the frequency region from several cycles/hour down to several
cycles per week is due to fluctuations in the local barometric
pressure.

   We have computed the effect of barometric pressure fluctuations on strain using conventional least-squares techniques. On

the average, we are able to account for about 60% of the non-tidal strain fluctuations as being due to barometric fluctuations. It is clear that this correlation is a lower bound to what one might achieve using several barometers within a few kilometers of the mine.

The analysis in the normal mode band (1-30 cycles/hour) is somewhat different, although the results are essentially the same. Instead of least squares, we have used the coherence function. It is defined as follows.

Given two time series a(t) and b(t). Let their Fourier transforms be A(ω) and B(ω) respectively. Then the magnitude of the coherence between a and b at frequency ω is defined to be

$$| \gamma_{ab}(\omega) | = \left\{ \frac{|<A(\omega) B^*(\omega)>|^2}{<|A|^2><|B|^2>} \right\}^{1/2}$$

where the brackets <> denote the arithmetic average of several distinct pairs of series. It is clear that without this averaging $\gamma_{ab}(\omega)$ is identically unity for all ω.

We have computed $\gamma_{ab}(\omega)$ by averaging 25 distinct time series. We find that $\gamma_{ab}(\omega)$ is a slowly changing function of frequency and that it is on the order of 40% near 1 cycle/hour. Figure 3 shows a typical determination of $\gamma_{ab}(\omega)$. It is computed using 25 pairs of time series, each 200 points long.

Again, this coherence is a lower bound to what might be achieved using an array of barometers. In any case, coherences of this magnitude, though they will likely have great impact on certain areas of geophysics, do not fundamentally change the integration time requirement outlined above.

One might speculate on other techniques which might reduce the random strain fluctuations sufficiently to make the integration time more reasonable. One very interesting possibility is to construct a second interferometer of similiar design perpendicular to the existing one. It is difficult to conceive of any source of low frequency noise in the earth which would affect the two arms in totally different ways. One might reasonably expect that by subtracting the signals in the two arms, one would cancel some of the noise and multiply the gravitational signal by 2 (assuming a wave incident along the normal to the plane containing the two arms). This technique would reduce the need for barometric pressure arrays, since one might expect the two arms to be affected in roughly the same way by fluctuations in barometric pressure. If the coherence between the two arms was 0.90 or better, it would reduce the integration time to a reasonable interval (a month or

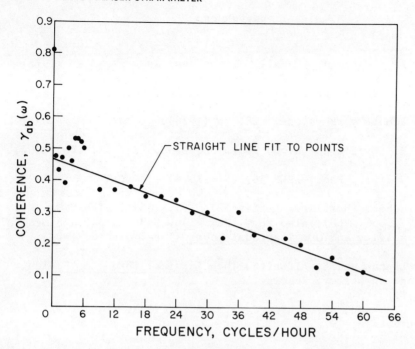

Fig. 3. Average coherence between strain fluctuations and fluctuations in the local barometric pressure as a function of frequency.
The slope and intercept of the straight line shown have been determined using least-squares techniques.

less). Although this level of coherence is not at all unreason-
able in two interferometers spaced perhaps 50 m apart, it has never
been observed to the author's knowledge.

CONCLUSIONS

The detection of gravitational waves from an identifiable
source would be a discovery of great significance. We have investigated various possibilities and we have concluded that rotation binary stars probably provide the most favorable experimental
situation. The low frequencies of these sources make it reasonable to consider using the earth as a detector, but some means
must be found to decrease the background noise. We have found at
least one major source of this noise, namely fluctuations in barometric pressure, and we are currently exploring other possible
sources of noise in the earth.

The most promising possibility for the detection of gravitational waves at the moment seems to be the coherent averaging of
several instruments located in the same general area so as to substantially reduced the earth noise.

REFERENCES

1. J. Levine and J. L. Hall, J. Geophys. Res. 77, 2595 (1972).

2. R. L. Barger and J. L. Hall, Phys. Rev. Letters 22, 4 (1969).

3. W. Schottky, Phys. Rev. 28, 75 (1926).

4. A. H. Hoffait and R. D. Thornton, Proc. IEEE 52, 179 (1964).

5. D. Halford, Proc. IEEE 56, 251 (1968).

6. J. Weber, Phys. Rev. Letters 21, 395 (1968); also Phys. Rev. 117, 306 (1960); also Phys. Rev. 146, 935 (1966), and General Relativity and Gravitational Waves (Interscience, 1961).

7. L. Landau and L. Lifshitz, The Classical Theory of Fields (Addison-Wesley, Reading, Mass., 1962), p. 366.

8. F. J. Dyson, Astrophys. J. 156, 529 (1969); also private communication, 1972.

9. C. W. Misner, K. S. Thorne and J. A. Wheeler, Gravitation (Freeman, San Francisco, in press).

10. J. Levine and R. T. Stebbins, Phys. Rev. D 6, 1465 (1972).

11. W. Krzeminski and R. P. Kraft, Astrophys. J. 140, 921 (1964).

12. S. C. Vila, Nature Physical Science 230, 39 (1971).

13. J. Faulkner, Astrophys. J. Letters 170, L99 (1971).

14. R. P. Kraft, J. Mathews and J. L. Greenstein, Astrophys. J. 136, 312 (1962).

15. V. Braginskii, Sov. Phys. Usp. 8, 513 (1966).

16. J. Levine and R. T. Stebbins, Proc. Roy. Soc. (in press).

PARTICIPANTS

Laser Spectroscopy Conference
Vail, Colorado
June 25-29, 1973

Abella, I., JILA, University of Colorado, Boulder, CO.
Abrams, R. L., Hughes Research Laboratory, Malibu, CA.
Aggarwal, R., Francis Bitter National Magnet Laboratory, Cambridge,
    MA.
Barger, R. L., NBS, Boulder, CO.
Behringer, R., ONR, Pasadena, CA.
Bender, P. L., JILA, University of Colorado, Boulder, CO.
Bergman, A., Dept. of Chemistry, Tel-Aviv University, Ramat Aviv,
    ISRAEL
Berman, P., Physics Department, New York University, Bronx, NY.
Bernhardt, A. F., Lawrence Livermore Lab., University of California,
    Livermore, CA.
Bjorklund, G., Stanford University, Stanford, CA.
Blum, F. A., Texas Instruments, Dallas, TX.
Bordé, C., JILA, University of Colorado, Boulder, CO.
Bourdet, G. L., OPM, Paris, FRANCE
Boyer, K., Los Alamos Scientific Laboratory, UC, Los Alamos, NM.
Bradley, D. J., Dept. of Physics, Imperial College, London, ENGLAND
Brewer, R. G., IBM Research Laboratory, San Jose, CA.
Bridges, T. J., Bell Telephone Laboratories, Holmdel, NJ.
Brillet, A., Laboratoire de'l Horloge du C.N.R.S., Orsay, FRANCE
Broida, H., Dept. of Physics, Univ. of Calif., Santa Barbara, CA.
Brossel, J., Laboratoire de Physique, de l'Ecole Normale Supérieure,
    Paris, FRANCE
Brueck, S., MIT, Lincoln Laboratory, Lexington, MA.
Bunker, P. R., Physics Division, NRC, Ottawa, Ontario, CANADA
Busch, G., Chemistry Dept., University of Colorado, Boulder, CO.
Byer, R., Microwave Laboratory, Stanford University, Stanford, CA.
Carlsten, J. L., Harvard College Observatory, Cambridge, MA.
Cohen-Tannoudji, C., l'École Normale Supérieure, Paris, FRANCE
Curran, A. H., Columbia University, NY.
Dana, L., Spectra-Physics, Mountain View, CA.
Deslattes, R., NBS, Molecular Spectroscopy Section, Washington, D.C.

653

Dows, D., Dept. of Chemistry, Univ. of Southern Calif., Los Angeles, CA.
Drullinger, R. E., NBS, Boulder, CO.
Eisenthal, K. B., IBM Research Laboratory, San Jose, CA.
El-Sayed, M. A., Dept. of Chemistry, U.C.L.A., Los Angeles, CA.
Emmett, J., Lawrence Livermore Lab., Univ. of Calif., Livermore, CA.
Eng, R. S., MIT, Lincoln Laboratory, Lexington, MA.
Evenson, K. M., NBS, Boulder, CO.
Ezekiel, S., MIT, Cambridge, MA.
Feld, M. S., Department of Physics, MIT, Cambridge, MA.
Fetterman, H. R., MIT, Lincoln Laboratory, Lexington, MA.
Flygare, W., Dept. of Chemistry, Univ. of Illinois, Urbana, IL.
Foster, K. L., IBM Research Laboratory, San Jose, CA.
Freed, C., MIT, Lincoln Laboratory, Lexington, MA.
Galeener, F. L., Xerox Corp., Palo Alto Research Center, Palo Alto, CA.
Gibbs, H., Bell Telephone Laboratories, Murray Hill, NJ.
Giordmaine, J. A., Bell Telephone Laboratories, Murray Hill, NJ.
Grischkowsky, D., IBM, T. J. Watson Research Ctr., Yorktown Hgts., NY.
Hahn, E. L., Dept. of Physics, Univ. of Calif., Berkeley, CA.
Hall, J. L., JILA, University of Colorado, Boulder, CO.
Hänsch, T. W., Dept. of Physics, Stanford University, Stanford, CA.
Haroche, S., Dept. of Physics, Stanford University, Stanford, CA.
Harrington, M. C., General Physics Div., AFOSR, Arlington, VA.
Harris, C. B., Dept. of Chemistry, Univ. of Calif., Berkeley, CA.
Harris, S. E., Microwave Laboratory, Stanford University, Stanford, CA.
Herbst, R., Microwave Laboratory, Stanford University, Stanford, CA.
Herzberg, G., Div. of Physics, NRC, Ottawa, CANADA
Hessel, M., Quantum Electronics Div., NBS, Boulder, CO.
Hill, R., Physics Div., Stanford Research Institute, Menlo Park, CA.
Hinkley, E. D., MIT Lincoln Laboratory, Lexington, MA.
Hordvik, A., Air Force Cambridge Research Laboratories, Bedford, MA.
Hsu, D., NBS, Boulder, CO.
Ippen, E. P., Bell Telephone Laboratories, Holmdel, NJ.
Jacobs, R., Lawrence Livermore Lab., Univ. of Calif., Livermore, CA.
Jarrett, S., Coherent Radiation, Inc., Palo Alto, CA.
Javan, A., Dept. of Physics, MIT, Cambridge, MA.
Johns, J. W., Div. of Physics, NRC, Ottawa, CANADA
Kelley, P. L., MIT, Lincoln Laboratory, Lexington, MA.
Kidder, R., Lawrence Livermore Lab., Univ. of Calif., Livermore, CA.
Kleppner, D., Dept. of Physics, MIT, Cambridge, MA.
Kuizenga, D., Microwave Laboratory, Stanford University, Stanford, CA.
Kung, A., Microwave Laboratory, Stanford University, Stanford, CA.
Kurnit, N., Dept. of Physics, MIT, Cambridge, MA.
Lamb, W. Jr., Dept. of Physics, Yale University, New Haven, CT.
Lambropoulos, M., JILA, Boulder, CO.
Lang, R., JILA, University of Colorado, Boulder, CO.

ax, B., Dept. of Physics, MIT, Cambridge, MA.
ee, P. H., University of California, Santa Barbara, CA.
eonard, D., AVCO, Everett, MA.
eone, S., Dept. of Chemistry, Univ. of Calif., Berkeley, CA.
erner, D., Cornell University, Ithaca, NY.
evenson, M., Div. of Engr. and Appl. Phys., Harvard Univ.,
    Cambridge, MA.
evin, L., AVCO, Everett, MA.
evine, J., NBS, Boulder, CO.
evy, J., Dept. of Chemistry, MIT, Cambridge, MA.
iberman, S., CNRS, Orsay, FRANCE
ineberger, C., JILA, University of Colorado, Boulder, CO.
itvak, M., Smithsonian Inst., Astrophysical Observatory, Cambridge,
    MA.
oy, M., IBM, T. J. Watson Research Ctr., Yorktown Heights, NY.
untz, A. C., IBM Research Laboratory, San Jose, CA.
agde, D., Washington State University, Pullman, WA.
ahr, H., Dept. of Physics, Cornell University, Ithaca, NY.
arburger, J., Univ. of Southern Calif., Los Angeles, CA.
ay, D., Dept. of Physics, Univ. of Toronto, Toronto, Ontario,
    CANADA
cWhorter, A. L., MIT, Lincoln Laboratory, Lexington, MA.
elngailis, I., MIT, Lincoln Laboratory, Lexington, MA.
enzies, R. T., Jet Propulsion Laboratory, Cal Tech, Pasadena, CA.
ontgomery, P., General Motors Research Laboratories, Warren, MI.
ooradian, A., MIT, Lincoln Laboratory, Lexington, MA.
orawitz, H., IBM Research Laboratory, San Jose, CA.
ill, K. W., MIT, Lincoln Laboratory, Lexington, MA.
ka, T., Div. of Physics, NRC, Ottawa, CANADA
uhayoun, M., Universite Paris - Nord, Saint Denis, FRANCE
arks, J. H., Univ. of Southern Calif., Los Angeles, CA.
atel, C. K. N., Bell Telephone Laboratories, Holmdel, NJ.
eterson, F. R., NBS, Boulder, CO.
eterson, O., Lawrence Livermore Lab., Univ. of Calif., Livermore,
    CA.
ickett, H. M., Dept. of Chemistry, Univ. of Calif., Berkeley, CA.
illoff, H., Naval Research Laboratories, Washington, D.C.
inard, J., CNRS, Orsay, FRANCE
orto, S. P. S., Dept. of Physics, Univ. of Southern Calif., Los
    Angeles, CA.
adere, F., Institut d'Optique, Faculte des Sciences, Orsay, FRANCE
atesi, R., Laboratoria di Elettronica Quantistica, Florence,
    ITALY
adford, H., Smithsonian Inst., Astrophysical Observatory, Cambridge,
    MA.
amsey, N. F., Dept. of Physics, Harvard University, Cambridge, MA.
ank, D., Dept. of Astronomy, Univ. of Calif., Santa Cruz, CA.
ausch, H., Laser Focus, Newton, MA.
odes, C. Lawrence Livermore Lab., University of California,
    Livermore, CA.

Richards, P., Dept. of Physics, Univ. of Calif., Berkeley, CA.
Riley, S. J., Dept. of Chemistry, Univ. of Calif. San Diego,
    La Jolla, CA.
Robinson, C. P., Los Alamos Scientific Laboratory, UC, Los Alamos,
    NM.
Schlossberg, H., Air Force Cambridge Research Labs., Bedford, MA.
Schmeltekopf, A., NBS, Boulder, CO.
Schmidt, J., IBM Research Laboratory, San Jose, CA.
Scully, M., Optical Sciences Ctr., Univ. of Arizona, Tucson, AZ.
Shimizu, F., Dept. of Appl. Phys., Univ. of Tokyo, Tokyo, JAPAN
Shimizu, T., Dept. of Physics, Univ. of Tokyo, Tokyo, JAPAN
Shimoda, K., Dept. of Physics, Univ. of Tokyo, Tokyo, JAPAN
Shoemaker, R. L., Optical Sciences Ctr., Univ. of Arizona, Tucson,
    AZ.
Smith, P. W., Bell Telephone Laboratories, Holmdel, NJ.
Smith, S. D., Dept. of Physics, Heriot-Watt Univ., Edinburgh,
    Scotland, U.K.
Sorem, M. S., NBS, Boulder, CO.
Stoicheff, B. P., Dept. of Physics, Univ. of Toronto, Toronto,
    Ontario, CANADA
Strauss, H., Dept. of Chemistry, Univ. of Calif., Berkeley, CA.
Svanberg, S., Dept. of Physics, Columbia University, NY.
Svelto, O., Laboratorio di Fisica del Plasma ed Elettronica Quant.,
    Milan, ITALY
Swerdlow, M., Air Force Office of Scientific Research, Arlington, VA
Szabo, A., Physics Division, NRC, Ottawa, Ontario, CANADA
Szöke, A., Dept. of Chemistry, Tel-Aviv Univ., Ramat Aviv, ISRAEL
Takami, M., Microwave Physics Laboratory, Wako, Saitama, JAPAN
Takeuchi, N., Department of Physics, Columbia University, NY.
Teller, E., Lawrence Livermore Lab., Univ. of Calif., Livermore, CA.
Vidal, C. R., Max-Planck Institut, Munich, GERMANY
Wallace, R., Chromatix, Inc., Mountain View, CA.
Walther, H., Erstes Physikalische Inst., Univ. zu Köln, Köln,
    GERMANY
Wang, S. C., Stanford University, Stanford, CA.
Weingarten, R. A., Columbia University, NY.
Wilson, K., Dept. of Chemistry, Univ. of Calif., San Diego, La Jolla
    CA.
Wing, W. H., Dept. of Physics, Yale University, New Haven, CT.
Wolff, P., Dept. of Physics, MIT, Cambridge, MA.
Wood, L. L., Lawrence Livermore Lab., Univ. of Calif., Livermore, CA
Wynne, J. J., IBM, T. J. Watson Research Ctr., Yorktown Heights, NY.
Yardley, J. T., Dept. of Chemistry, Univ. of Illinois, Urbana, IL.
Young, J. E., Microwave Laboratory, Stanford University, Stanford,
    CA.
Zimmerman, P., MIT, Cambridge, MA.

INDEX